Physics Research and Technology

Interferometry Principles and Applications

PHYSICS RESEARCH AND TECHNOLOGY

Additional books in this series can be found on Nova's website under the Series tab.

Additional E-books in this series can be found on Nova's website under the E-books tab.

Physics Research and Technology

Interferometry Principles and Applications

Mark E. Russo
Editor

Nova Science Publishers, Inc.
New York

Copyright © 2012 by Nova Science Publishers, Inc.

All rights reserved. No part of this book may be reproduced, stored in a retrieval system or transmitted in any form or by any means: electronic, electrostatic, magnetic, tape, mechanical photocopying, recording or otherwise without the written permission of the Publisher.

For permission to use material from this book please contact us:
Telephone 631-231-7269; Fax 631-231-8175
Web Site: http://www.novapublishers.com

NOTICE TO THE READER

The Publisher has taken reasonable care in the preparation of this book, but makes no expressed or implied warranty of any kind and assumes no responsibility for any errors or omissions. No liability is assumed for incidental or consequential damages in connection with or arising out of information contained in this book. The Publisher shall not be liable for any special, consequential, or exemplary damages resulting, in whole or in part, from the readers' use of, or reliance upon, this material. Any parts of this book based on government reports are so indicated and copyright is claimed for those parts to the extent applicable to compilations of such works.

Independent verification should be sought for any data, advice or recommendations contained in this book. In addition, no responsibility is assumed by the publisher for any injury and/or damage to persons or property arising from any methods, products, instructions, ideas or otherwise contained in this publication.

This publication is designed to provide accurate and authoritative information with regard to the subject matter covered herein. It is sold with the clear understanding that the Publisher is not engaged in rendering legal or any other professional services. If legal or any other expert assistance is required, the services of a competent person should be sought. FROM A DECLARATION OF PARTICIPANTS JOINTLY ADOPTED BY A COMMITTEE OF THE AMERICAN BAR ASSOCIATION AND A COMMITTEE OF PUBLISHERS.

Additional color graphics may be available in the e-book version of this book.

Library of Congress Cataloging-in-Publication Data

Interferometry principles and applications / [edited by] Mark E. Russo.
    p. cm.
 Includes index.
 ISBN 978-1-61209-347-5 (hardcover)
1. Interferometry. I. Russo, Mark E.
QC411.I575 2010
535'.470287--dc22
                   2010051595

Published by Nova Science Publishers, Inc. † New York

Contents

Preface		vii
Chapter 1	Speckle Methods for Material Analysis *Félix Salazar Bloise*	1
Chapter 2	Interferometer Based Methods for Research of Piezoelectric Materials *Miroslav Sulc*	55
Chapter 3	The Production and Accurate Measurement of a 1 Millimeter Step Standard Using a Commercial and a Laboratory White Light Interferometer *Børge Holme, Arnt Inge Vistnes, Joachim Seland Graff and Juhi Bhatnagar*	95
Chapter 4	Cyclic Path Interferometric Configuration: Some Applications *Sanjib Chatterjee*	123
Chapter 5	Single-Shot Phase-Grating Phase-Shifting Interferometry *Gustavo Rodriguez-Zurita and Noel-Ivan Toto-Arellano*	155
Chapter 6	Sar Interferometry Fundamentals and Historic Evolution in Terrain Movements Applications *Paz Fernández-Oliveras*	193
Chapter 7	High Contrast Schlieren Diffraction Interferometry *Raj Kumar*	219
Chapter 8	Diffracted Beam Interferometry *Elena López Lago, Héctor González Núñez and Raúl de la Fuente*	243
Chapter 9	Binary Grating Interferometry with Two Windows *Gustavo Rodriguez-Zurita and Cruz Meneses-Fabian*	269
Chapter 10	Electronic Speckle Pattern Interferometry: Principles and Applications *Jiong-Shiun Hsu, Chi-Hung Hwang and Wei-Chung Wang*	295

Chapter 11	Periodic Error Measurement for Heterodyne Interferometry *Tony L. Schmitz and Hyo Soo Kim*	**319**
Chapter 12	Maximum Likelihood Estimation of Optical Signal Parameters *V. S. Sobolev*	**337**
Chapter 13	Optical Interferometers: Principles and Applications in Transport Phenomena *Sunil Verma, Yogesh M. Joshi and K. Muralidhar*	**353**
Chapter 14	Phase-Stepping Algorithms: Overview and Simulations *Jan A. N. Buytaert and Joris J. J. Dirckx*	**415**
Chapter 15	Generalized Carré Multi-Step Phase-Shifting Algorithms *Jiří Novák, Pavel Novák and Antonín Mikš*	**447**
Chapter 16	Interferometric Methods Applied to Polymeric Analysis *Gustavo F. Arenas, Nélida A. Russo and Ricardo Duchowicz*	**471**
Chapter 17	Application of Optical Interferometry for Measurement of (Thermo) Diffusion Coefficients *A. Mialdun and V. Shevtsova*	**491**
Chapter 18	Modern Artwork Documentation Qualitative Evaluation of Secondary Interference Fringes: A Standalone Structural Diagnostic Tool in Artwork Documentation *Vivi Tornari*	**513**
Chapter 19	Principle and Application of Optical Interferometrs for Investigating the Refractive Index Homogeneity, Birefringence, Optical Indicatrix, and Surface Features of Crystals *Sunil Verma, S. Kar and K. S. Bartwal*	**537**
Index		**561**

PREFACE

Interferometry makes use of the principle of superposition to combine separate waves together in a way that will cause the result of their combination to have some meaningful property that is diagnostic of the original state of the waves. This new book presents current research in the principles and applications of interferometry. Topics discussed include speckle methods for material analysis; using White Light Interferometry for accurate topographic measurements of surfaces; cyclic path interferometric configuration applications; phase-stepping algorithms; periodic error measurement for heterodyne interferometry and high contrast Schlieren diffraction interferometry.

Chapter 1 – Techniques based on the speckle phenomenon have played an increasingly important role in applied optics because of their potential to be used in different scientific and technical fields. This article presents some speckle techniques for elastic and surface characterization of materials. In particular, the elastic deformation caused by different external agents, such as temperature, external forces and the magnetic field, is studied. Moreover, from the standpoint of surface analysis, the influence of corrosion on the surface roughness of a metallic sample is determined. In order to accelerate the natural process of corrosion, the sample is treated with chemical compounds, and after that, by using angular speckle correlation, its roughness is measured.

Chapter 2 - The applications of laser interferometry in research of piezoelectric materials are presented. Piezoelectric, electrostrictive, electro- optical and thermal expansion coefficients were measured by these methods.

The usage of the Michelson single interferometer to measure piezoelectric induced strain in bulk samples is described. The more sophisticated Mach – Zehnder double beam interferometer is used for the investigation of the piezoelectric thin film deposed on the substrate to compensate substrate vibration. Both homodyne interferometers are measuring amplitude of piezoelectrically induced sample vibrations. Use of the lock-in amplifier technique enables very small amplitudes of the order of 10^{-12} m to be measured. The experimental problems and limits of this method used for piezoelectric research were analyzed in detail.

The new system was developed for the measurement of coefficients in a wide temperature range. The fundamental interferometer measurement problems, such as temperature instability and mechanical vibration of the temperature chamber are solved. All parts of the Michelson interferometer are placed inside the temperature chamber, so the relative motion of interferometer parts is almost suppressed in this way. The new variation of the double beam interferometer with a minimum of elements was constructed and located inside the temperature chamber. Study of the phase transitions at relaxor crystals illustrates the benefit of this method.

The linear Pockels effect was investigated too, and electro-optic coefficients of $LiNbO_3$ crystal were measured in the setup of a Mach-Zehnder interferometer in the temperature range from 150 K to 330 K. The principal problem of this type of measurement is that electric fields induce not only refractive index change but also a change of the sample length along the path of the laser beam due the piezoelectric effect. This effect can induce the errors of electro-optical coefficients' measurement on the order of tens of percents. The new method was proposed and tested for direct compensation of this piezoelectric effect on both the Michelson and Mach -Zehnder interferometers' arrangement.

The interferometry method was also adapted to the thermal expansion coefficients measurement. Relatively small samples, with dimensions of about some millimeters, were investigated. The very small sample expansion was separated from holder movement by suitable interferometer arrangement. This method was successfully tested on well-known metallic materials at first and consequently applied to the PZT piezoelectric ceramics research.

Chapter 3 - White Light Interferometry (WLI) is widely used for rapid and accurate topographic measurements of surfaces. By creating three dimensional images with lateral sizes from several millimeters to some tens of micrometers, both large and small samples can be studied.

The introduction of a separate laser interferometer in the scanning unit was meant to increase the accuracy of the depth scale from 1 % to about one hundredth of this value. This would mean that WLI could be used for accurate measurements of the vertical dimensions of millimeter sized components, and not only to quantify their surface roughness.

In order to investigate the capabilities of the Wyko NT9800 white light interferometer and to check its accuracy for large step measurements, we built our own step standard and measured the step height independently by a white light interferometer with a reference laser interferometer set up in the lab.

A thorough investigation of possible sources of error for the lab measurements was carried out. We further investigated the stability of the step standard over time and checked the reproducibility of the NT9800 measurements. Our results indicate that the accuracy of the NT9800 is about 0.05 %.

Chapter 4 - Some useful applications of cyclic path interferometric configuration have been discussed. A cyclic path optical configuration (CPOC) can be formed with a beam splitter (BS) and two plane mirrors, M1, M2, which are inclined at 45° to each other. BS splits up an incident collimated beam into transmitted (T) and reflected (R) components, which reflect off M1 and M2 and traverse triangular paths (TP) in opposite directions. With a polarizing beam splitter (PBS), the T and R components become plane polarized in orthogonal directions, namely, p and s polarized. When M1 and M2 are symmetrically placed, at equal distances, with respect to the beam splitting plane, the T and R components traverse the same TP in opposite directions, for a particular angle of incidence of the incident beam and emerge along the same path. A lateral shear between the beam components can be introduced by shifting either M1 or M2, along the direction of the normal to the mirror, from the symmetrical position. As the T and R components traverse identical TP, the optical path difference (OPD) between the emergent beams remains zero. The counter propagating p and s polarized components, in a CPOC, can be utilized for the external measurement of right angle of optical components A CPOC, adjusted for a lateral shear between the T and R components, can be coupled to a Fizeau interferometer (FI) for the measurement of wedge angle (δ) and index of refraction (μ) of the material of the optical glass window. A CPOC, adjusted for a lateral shear, produces two spatially separated (real) point images, outside the cavity, for a convergent input beam. The coherent point images can form Young's fringes. A CPOC with a PBS and linearly polarized divergent laser beam input from a microscope objective spatial filter combination is used to produce spatially separated coherent virtual point sources of light with linear orthogonal polarization at the back focal plane of a Fizeau interferometer objective for polarization phase shifting interferometry. A converging lens, placed in the hypotenuse arm of the TP of a CPOC adjusted for zero lateral shear, focuses the counter propagating collimated beams at two different points having longitudinal separation, which depends on the position of the lens in the TP cavity. Light diverging from the focal points interfere to produce Haidinger type fringes of equal inclination. Since the focal points are coherent, high contrast interference fringes can be obtained with quasi-monochromatic or broad band (white) light sources. An important application of the setup is for the measurement of centering error of lenses. Using polarized light and PBS, the same CPOC set up, which produces two point sources with longitudinal separation, corresponding to two focal points of the emergent beams having orthogonal linear polarizations, can be used to measure focal length of lenses using Newton's formula.

Chapter 5 - Phase-shifting interferometry requires of several interferograms of the same optical field with similar characteristics but shifted by certain phase values to retrieve the optical phase. This task has been usually performed by stages with great success and requires of a series of sequential shots. However, time-varying phase distributions are excluded from this schema. Several efforts for single-shot phase-shifting interferometry have been tested successfully, but some of them require of non-standard components and they need to be modified in some important respects in order to get more than four interferograms. Two-windows grating interferometry, on the other hand, has been proved to be an attractive technique because of its mechanical stability as a common-path interferometer. Moreover, gratings can be used as convenient phase modulators because they introduce phase shifts through lateral displacements. In this regard, phase gratings offer more multiplexing capabilities than absorption gratings (more useful diffraction orders because higher diffraction efficiencies can be achieved). Furthermore, with two phase gratings with their vector gratings at 90° (grids) there appear even more useful diffraction orders. Modulation of polarization can be independently applied to each diffraction order to introduce a desired phase-shift in each interference pattern instead of using lateral translations. These properties combine to enable phase-shifting interferometric systems that require of only a single-shot, thus enabling phase inspection of moving subjects. Also, more than four interferograms can be acquired that way. A simple interferogram processing enables the use of interference fringes with different fringe modulations and intensities. In this chapter, the basic properties of two-windows interferometry with phase gratings and modulation of polarization is reviewed on the basis of the far-field diffraction properties of phase gratings and grids. Phase shifts in the diffraction orders can be used as an advantage because they simplify the needed polarization filter distributions. Examples of experimental set-ups (such as basic configurations, lateral-shear and radial-shear) are shown and discussed. It is finally remarked, that these interferometers are compatible with interference fringes exhibiting spatial frequencies of relative low values and, therefore, no great loss of resolution is related with several interferograms when simultaneously using the same image field of the camera.

Chapter 6 - This chapter includes an introduction about fundaments of the InSAR technique and its historical evolution focus on terrain movement applications. The fundamentals include the SAR image parameters and characteristics that are important to consider in the use of InSAR techniques in the study of different types of terrain movements (subsidence, landslides, volcanic activity, earthquakes...). The Chapter starts with a short introduction about SAR images wavelength and main advantages of active sensors against passive sensor in the images formation (solar radiation independence, cloud penetration). After this introduction, it includes an explanation about the SAR images geometric distortions, due to the sensor *line of sight* (LOS), and their influence in the terrain movement detection. The next section corresponds to the phase and amplitude components and their use, the definitions of Interferometry and Differential Interferometry and the DInSAR fundamental equations necessary to apply this technique to the terrain movement investigations. After that the chapter includes a review of the Advanced InSAR techniques (called A- InSAR or A-DInSAR) and Multi-interferogram techniques existing actually, that constituted a great advance in the use of Interferometric Image in the quantitative assessment of terrain movements.

Finally, the chapter conclude with a review of the historic evolution of the InSAR technique related to the use of SAR images to detect quantify and study the evolution of the different types of terrain movements, and in deep review analisis of subsidence and landslide applications.

Chapter 7 - Schlieren techniques are among the simplest and oldest known optical methods for visualizing refractive index gradients in transparent media. Conventional schlieren methods are generally used to obtain first hand qualitative information about the test field. To obtain quantitative information these techniques are transformed into interferometers.

In schlieren diffraction interferometry, the position of schlieren diaphragm/diffracting element is adjusted in such a way that it diffracts a part of the incident unperturbed geometrical light. This diaphragm diffracted light serves as reference beam while other part of geometrical light modulated with test media serves as object beam.

Interference of these beams generates the schlieren interferogram which could be used to get required information about the test media. In this chapter, methods for enhancing contrast and sensitivity of schlieren diffraction interferometer are described. A comparative study of various schlieren diffracting elements is presented.

Since schlieren, shadowgraphy and interferometry provides information related to different aspects of the test media, thus, a combined system, using holographic optical elements, is described which could simultaneously provide information related to systems involving a wide range of index or density gradients. This combined system is particularly useful for studying the highly transient phenomena such as plasma. Applications of these methods in conventional and new emerging fields are discussed.

Chapter 8 - Diffracted beam interferometry (DBI) is a self referenced characterization technique which was originally thought to reconstruct the phase of a beam starting from the interference data between the beam and its diffracted copy. The phase is recovered indirectly by means of an iterative algorithm relating the irradiances of the interfering beams and the phase difference. The first experimental demonstration of DBI was implemented on a Mach-

Zehnder interferometer which incorporated an afocal imaging system in each arm, in order to form an image of a common object in different planes at the output of the interferometer. The irradiance and phase difference data were picked up from one of the image planes and entered into the iterative algorithm. Later modifications of the iterative algorithm made DBI able to characterize both the phase and the amplitude simultaneously. This new algorithm allows faster data acquisition which makes the method less influenced by environmental disturbances.

Chapter 9 - Along this review, by grating interferometry the authors will understand an interferometer which includes at least one grating (*i.e.*, a periodic amplitude transmittance) as an essential component of the system. Limiting the scope to the case of monochromatic or quasi-monochromatic cases, the task performed by such gratings includes at least one of the following: beam division, beam combination, beam replication (multiplexer) and phase shifter.

Chapter 10 - Since Electronic speckle pattern interferometry (ESPI) has the attractive merits such as non-contact, full-field, highly sensitive, etc., it has been a powerful tool for the measurement in practice, especially for the object with diffuse surface. In this chapter, the optical theories of ESPI including static and vibration measurements will first be introduced. Then the different optical arrangements in ESPI will be described and their advantages and disadvantages will be compared. Finally, some examples of ESPI respectively applied in-plane, out-of-plane and vibration measurements will be given.

Chapter 11 - Displacement measuring interferometry offers high accuracy, range, and resolution for non-contact displacement measurement applications. One fundamental accuracy limitation for the commonly selected heterodyne (or two frequency) Michelson-type interferometer is periodic error, which is caused by frequency mixing/leakage between the reference (fixed) and measurement (moving) paths. The periodic error level for a given setup can be measured using the discrete Fourier transform of time-based position data. Alternately, it can be determined using "velocity scanning", where the optical interference signal is observed during constant velocity target motion using a spectrum analyzer and the spectral content is used to calculate the periodic error magnitudes. In this chapter, these techniques are described and demonstrated on experimental data. Using this information, the optical setup can either be adjusted to reduce the periodic error magnitudes or compensation can be applied.

Chapter 12 - Development of algorithms for the optimal estimates of the optical signals parameters in the present time is the task number 1. This is due to the rapid development of optoelectronics and, in particular, the such its directions, as fiber communications, optical disk memory, optical location and interferometry. In the last case it is very imported, as it is not possible to use corrected codes. The subtlety of the problem is that, unlike radio and radar signals, where the noise and the signals are statistically independent, receiving optical signals is accompanied inevitably by so far shot noise, which variance is strictly proportional to the intensity of the signal itself. With this in mind, the entire rich arsenal of existing optimal algorithms for estimating the parameters of signals in noise for receiving optical signals can not be directly applied. This paper is results of the long-term research aimed the solving the problem of obtaining optimal estimates of optical signals parameters with the above especially the accompanying noise. On an example of the Gaussian optical video and radio pulses are deduced and solved the likelihood equation and get the expressions for the boundaries of the Cramer-Rao determining the quality of fetched ratings. The problem is solved for the three main photodetection methods: counting the number of photoelectrons emitted at specified time intervals, fixing the time of emission of each photoelectron and analog detection . The reliability of the algorithms and expressions for the boundaries of the Cramer-Rao confirmed by computer simulation. Its results are given in the article.

Chapter 13 - Optical techniques are extensively used for high precision diagnostics and process monitoring in physical, biological, and engineering sciences. Interferometry falls in one such class of diagnostics. It relies on changes in the refractive index in the medium arising from variations in the material density. The physical region in which imaging is being carried out is required to be transparent. The light source best suited for an interferometer is a laser. Owing to its features such as greater accuracy, resolution, instantaneous response and non-intrusive nature, interferometry proves to be advantageous and extensively utilized in a broad spectrum of applications. The present chapter deals with the description of laser interferometers in visualization and monitoring of processes involving fluid flow, heat transfer, and mass transfer.

The chapter is divided into two sections. In the first section, we discuss the basic principles of interference and fringe formation. It includes the principles and operations of various interferometer configurations such as Michelson, Mach-Zehnder, holography, phase-shifting, speckle, schlieren and dual-wavelength interferometry. Interferometers can provide vivid images of temperature and solutal concentration fields. Their real utility is in the quantitative determination of transport properties in addition to heat and mass fluxes. The second section describes the applications of interferometry in studying transient heat conduction, buoyancy-driven convection in a rectangular cavity and superposed fluid layers, and crystal growth from an aqueous solution. These illustrate the utility of interferometry in engineering and research.

Chapter 14 – The authors present a (non-exhaustive) overview of phase-shifting algorithms often used in interferometry. When performing phase-stepping (or -shifting), phase differences in a periodic intensity profile are changed stepwise (or continuously), and the resulting irradiance distributions are recorded at each step (or bucket). The wanted phase can be obtained from the arctangent of the ratio between two combinations of the observed irradiances, according to the phase-shifting algorithm (PSA) used. There are many such combinations and thus different PSAs, each with specific performance and properties.

The authors briefly discuss some error sources which might influence the performance and quality of interferometry measurements. The robustness against these error sources is strongly dependent on the PSA used.

They intently gathered as many popular PSAs and some of their properties from literature as we could find. Many discrepancies, however, were found between authors' statements, not to mention typos in the published articles and chapters. The authors meticulously sorted the typos out, listed up the algorithms and ran several computer simulations on all of them to confirm which algorithms perform best in the presence of some straightforward error sources.

Chapter 15- Phase shifting is a well-known technique which is used in many areas of science and engineering. Phase-shifting algorithms are used extensively in optical interferometry. This chapter describes a group of multi-step phase-shifting algorithms for phase evaluation of interferometric measurements. Phase shifting algorithms are introduced and analyzed, with a constant but arbitrary phase shift between captured frames of the irradiance of the interference field. The phase-shifting algorithms are similarly derived as so called Carré algorithm, which was firstly described in 1966. The phase evaluation process is not dependent on linear phase shift errors using these algorithms. An advantage of the described algorithms over common phase-shifting algorithms is their ability to determine the phase shift value at each point of the detection plane. Moreover, a complex error analysis of proposed algorithms is performed and the algorithms are compared to several common error compensating phase stepping algorithms.

Chapter 16 - Several epoxies and photo- or thermal- cured polymers found their way in almost every field of structures manufacturing. These materials can compete with metallic ones and even substitute them in several applications. However, it is well known that their mechanical properties are highly dependent upon the curing process of the matrix. Curing evolution is connected directly to the contraction process occurring in the material. Relevant information include the evolution and final conversion (degree of cure) related to the amount of the chemical cross-linking occurring during cure, the gel point where a phase change from liquid-like to solid-like occurs, the glass transition temperature and the induced residual strain produced during the curing process causing structure distortion and intrinsic strain accumulation. Several cure monitoring techniques have been proposed and applied in the past. Among these techniques, optical approaches seem to be the best candidate in polymer based materials manufacturing monitoring. On this way, different methods based on interferometric techniques by using fiber optics technologies have been developed. In this work, we discuss the feasibility of different approaches. First, we analyze optical sensors based on a Fizeau fiber optic interferometer to measure polymer contraction that occurs during cure with a measure resolution better than 100 nm; second, we discuss the simultaneous application of a cantilever and the Fizeau interferometer, and, finally, the use of a pair of fiber Bragg grating based sensors to uncouple strain and temperature (assuming a thermal related cure process). Properties and relevant information extracted from of the different techniques are discussed.

Chapter 17 – The authors report on the successful application of the digital interferometry for measuring diffusion and Soret coefficients in transparent organic fluids. The unique feature of this method is that it traces the transient path of the system in the entire diffusion cell. In this way it is applicable not only for studying thermodiffusive and diffusive transport mechanisms, but also for exploring convective motion. Presently, this method is not widely used for above purpose and, in their view, not because of fundamental limitations but rather due to a lack of properly developed experimental procedures and raw data post-processing. Thus, in this paper our attention is focused on the successive analysis of different steps: the fringe analysis, the choice of reference images, the thermal design of the cell and multi-parameter fitting procedure. Using the interferometry we have measured the diffusion and the Soret coefficients for three binary mixtures composed of dodecane ($C_{12}H_{26}$), isobutylbenzene (IBB), and 1,2,3,4 tetrahydronaphtalene (THN) at a mean temperature of 25°C and 50 wt% in each component. These measurements were compared with their benchmark values and show an agreement within less than 3%.

Chapter 18 - The direct result of visible interference processes is seen as the macroscopic effect of formation of interference fringes or interference fringe patterns. These are generated due to coherence phenomena concerned in wave physics in general and in coherent physics in particular. In regards to the nowadays most common sources to generate coherence phenomena one should reasonably think of a laser light source. This chapter considers the fringe patterns which are generated after interference of coherent light beams and their visual effect of alternate dark and bright fringes as a direct basis for qualitative analysis in Cultural Heritage documentation.

Interferometry is a well known technique for quantitative measurement of shape deformation due to field wise observation of object point's displacement. In optical interferometry a single light beam is divided in two beams travelling separate paths in space and recombining by an optical element to create the phenomenon of interference captured in a screen or detector. In holographic interferometry the process is repeated twice with an initial single beam primarily coherent light divided to an object and a reference beam paths from which the object carries the information of illuminated surface recombined to reference beam in detector plane without use of an optical element. Overlapping of the interference field produces the visual effect of secondary fringe patterns. The holographic interference is possible under strict experimental settings and principles. However demanding the process it is has become possible to use holographic interference in a number of different industrial and medical applications and more recently to be included as a competitive technique in the structural documentation and diagnosis of art objects.

An artwork consist a unique piece of human kind and preservation to the next generations has always been a demand. Since antiquity history-witnessing objects were subjected to the effort of preservation for the next generations to come. In this context structural inspection and diagnosis of Cultural Heritage items requires highly sensitive and accurate techniques which can retrieve inborn and upcoming deterioration well before it becomes visible to the eye. Phase information is the coded quantity in holography interference fringe patterns and offers a unique sensitive detector of structural displacements due to any externally induced factor. Phase changes provide high information content allowing tracing invisible defects under the surface with unbeatable quality and clarity compared to any other known method including x-rays and tomographic techniques. In fact the result of phase encoding in the holographic interference process through the fringe pattern secondary distribution of intensity turns to visual evidence the underneath surface activity of defects. Each defect produce own set of localized secondary interference fringes among the general fringe pattern such that each one uniquely indicate its subsurface effect on the surface. Exact location, size, shape and value of hidden defect and the profile of its deformation can be extracted directly by naked eye within a reasonable error appraisal. The generated fringes have to be separated by the general interference field, isolated, identified and allocated too an internal cause. The basis required understanding the fringes of holographic interferometry and assigning them to underneath effects and shape deformation are given in this chapter. Automatic processes are also considered and difficulties for their implementation specifically in the field of art conservation are described.

Holographic recording can be performed in a variety of optical geometries depending on the investigation aim without interacting with the precious surface or requiring any intervention prior to illumination. It is consider a fully non destructive and not interventive method as ethics of treatment in Cultural Heritage field presuppose. The protocols of investigation involved in this application are presented and explained.

Examples of results are given in characteristic case studies in sufficient range of art objects variety.

The objective of this chapter is to familiarize the reader with the complexity of secondary interference fringes and produce the evidence of the unique source of information that is unfolded in their formation prior to or without the electronic post-processing routines that are usually implemented in the analysis of interferograms.

It should also be highlighted the fact that the artwork application due to the uniqueness of artworks, the strict requirements for their preservation, and last but not least the complexity of the results which cannot feed the known automated routines, is considered through a different approach than other known applications of the same techniques aiming in other fields.

Chapter 19 - High quality crystals are required for research and development in the field of optical frequency conversion, optoelectronics, electro-optics and acousto-optics. The crystals are grown by various techniques such as growth from solution, melt, flux and vapor. The as-grown crystal is cut and polished along specific directions to obtain an element for use in the desired device. However, before deploying a crystal element for a particular application it is necessary to assess its optical quality and measure its important optical parameters. In this respect optical interferometric techniques assume great significance. The present chapter deals with the application of the conoscopic interferometry for investigating the optical quality and the optical indicatrix of the crystal, the Mach-Zehnder interferometer for non-destructive assessment of the optical homogeneity of the crystal, the birefringence interferometer for measurement of the birefringence of the crystal along any desired direction. Additionally the potential of Michelson interferometry is presented for surface characterization. The first half of the chapter presents the principle and optical instrumentation of the interferometers. The second half presents the application of these interferometric techniques for investigating the above mentioned properties of various classes of crystals such as inorganic, semi-organic and organic.

In: Interferometry Principles and Applications
Editor: Mark E. Russo

ISBN 978-1-61209-347-5
© 2012 Nova Science Publishers, Inc.

Chapter 1

SPECKLE METHODS FOR MATERIAL ANALYSIS

Félix Salazar Bloise[*]

Departamento de Física Aplicada. ETSI Minas. Universidad Politécnica de Madrid.
Madrid, Spain

ABSTRACT

Techniques based on the speckle phenomenon have played an increasingly important role in applied optics because of their potential to be used in different scientific and technical fields. This article presents some speckle techniques for elastic and surface characterization of materials. In particular, the elastic deformation caused by different external agents, such as temperature, external forces and the magnetic field, is studied. Moreover, from the standpoint of surface analysis, the influence of corrosion on the surface roughness of a metallic sample is determined. In order to accelerate the natural process of corrosion, the sample is treated with chemical compounds, and after that, by using angular speckle correlation, its roughness is measured.

1. INTRODUCTION

In many fields of technology, such as in the aerospace industry, civil engineering, and architecture, the analysis of deformations of the elements of the structures is of vital importance in order to guarantee proper operation. There are many methods for the analysis of deformation. The use of a specific experimental technique for determining the state of strain or stress of a body depends on different factors, such as the geometrical characteristics of the sample, the kind of material, and the precision with which the measurement needs to be done. All these characteristics give us several possibilities with which to establish a classification of the methods. However, a general and simple classification of these techniques could be to divide them into "contacting" and "non-contacting" procedures. A long time ago, different contacting methods for measuring elastic properties were developed

---

[*] Tel: 00 34 91 3364179, fax: 00 34 91 3366952, E-mail: felixjose.salazar@upm.es.

and are still used today. Testing machines, extensometers, and ultrasound techniques are widely employed for the mechanical characterization of materials. In the case of thermal analysis, conventional and differential dilatometry are precise procedures used to determine the thermal expansion coefficient of a solid. In these methods, the sample is located in a holder of glassy quartz (because of its low thermal expansion coefficient $\approx 10^{-7} \, K^{-1}$) or another material with similar characteristics. When the temperature increases, the sample deforms and contacts a transducer which transmits an electrical signal to be converted. In a similar way, the study of the strain produced in magnetic materials due to the presence of magnetic field has been of great importance in the last three decades because of its applications. In this context, the measurement of the magnetostriction of materials is an important property which is applicable to sensor technology, to actuators, and in micro- and nanoelectronics. The aforementioned elastic properties refer to the volume of the material. However, surface characteristics such as polish, roughness, flatness, straightness, and surface finish are topics of great interest too. In the applied sciences and in many industrial sectors, materials must satisfy a series of requirements in order to enable the workpieces of the structures to work according to the designer's aim. As a result, different techniques for surface characterization were developed in the past. In this chapter we are also interested in the study of roughness. The measurement of the roughness of a surface can be done by several techniques. One of the most used methods is the contact profilometer. This procedure uses a diamond stylus which displaces across the sample surface and gives the profile along straight lines. This information is processed and displayed, and roughness parameters as R_a (roughness average), R_q (root mean square) are obtained. The advantages of this method are the precision and versatility with reflective surfaces, where other methods do not work. However, because of its horizontal resolution, profilometers need more time than other methods to perform the analysis of a surface. On the other hand, the measurements are influenced by the relative position between the stylus and specimen.

These experimental techniques have been revealed to be appropriate for analyzing elastic properties of solids. They do not cause any damage on the material object of study, however, contact between the specimen and the measurement system is needed, which is a disadvantage when the sample to be analyzed is difficult to access. Parallel to the development of these methods for volume and surface measurement, optical procedures have become a fundamental tool of analysis in the context of non-destructive techniques. The use of light as basis of operation gives them certain advantages with respect to the conventional contacting techniques. Thus, their speed of execution of the experiments, accuracy, versatility under different conditions, and reasonable economic cost make them very useful for solving many scientific and technical problems. These features mentioned above explain why they are increasingly used, as they clearly offer better solutions to technical problems.

2. MEASUREMENT OF ELASTIC PROPERTIES

2.1. Basic Ideas of Elasticity

Elastic analysis plays a significant role in both science and technology. The deflection of a beam, the strain on the wings of an airplane, the strains due to dynamic stresses in pipes, and the structures in civil engineering are typical cases that demand a detailed study of the behavior of materials when subject to external loads.

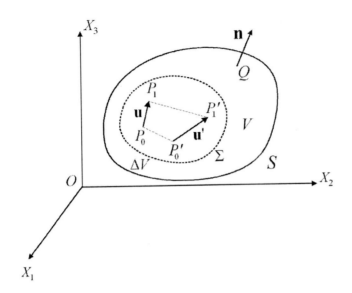

Figure 1. Elastic body in 3D. **u** and **u'** are the displacement functions for the points P_0 and P_0', respectively.

Let us suppose an elastic solid of arbitrary form occupying a volume V delimited for a surface S. Let P_0 be any point inside V of coordinates (x_1, x_2, x_3) in the reference system $OX_1X_2X_3$, and P_0' be another point located at a distance infinitely near to P_0 but of coordinates (x_1', x_2', x_3').

If external agents act on the solid (such as applied forces, temperature changes, the presence of magnetic or electric fields, phase transformations, etc.), all its parts suffer a displacement, and then P_0 and P_0' reach the positions P_1 and P_1' (Figure 1). This means that, in general (except in the case of pure translation or pure rotation), the relative distance between P_0 and P_0' varies. Let us denote by u the displacement function of the point P_0, which depends on the coordinates (x, y, z), and by u_i the component i (i=1,2,3) of this displacement. In the same way, let u' be the displacement associated to the with point P_0', u_i' being its i-component. For small deformations of the solid, expanding the displacement by Taylor's series about P_0', we have

$$u_i' = u_{0i} + \sum_{j=1}^{3} \frac{\partial u_i}{\partial x_j}\Delta x_j + \frac{1}{2}\sum_{j=1}^{3}\frac{\partial^2 u_i}{\partial x_j^2}\Delta x_j^2 + \ldots \approx u_{0i} + \sum_{j=1}^{3}\frac{\partial u_i}{\partial x_j}\Delta x_j + \mathcal{9}((\Delta x_j)^2), \quad i,j=1,2,3 \qquad (1)$$

where higher order terms have been neglected.

This expression may be rewritten by decomposing the series into two terms, obtaining

$$u'_i = u_{0i} + \frac{1}{2}\sum_{j=1}^{3}\left(\frac{\partial u_i}{\partial x_j} - \frac{\partial u_j}{\partial x_i}\right)\Delta x_j + \frac{1}{2}\sum_{j=1}^{3}\left(\frac{\partial u_i}{\partial x_j} + \frac{\partial u_j}{\partial x_i}\right)\Delta x_j. \qquad (2)$$

The first addend u_{0i} represents the component i of a translation, which is common for all P'_0 belonging to the small volume ΔV. The second term

$$\Omega_{ij} = \frac{1}{2}\sum_{j=1}^{3}\left(\frac{\partial u_i}{\partial x_j} - \frac{\partial u_j}{\partial x_i}\right) \quad i,j=1,2,3 \qquad (3)$$

is the component i, j of a skew-symmetric tensor that represents the rotation of a small element around an axis crossing P_0, and the third term

$$\varepsilon_{ij} \equiv \sum_{j}\frac{1}{2}\left(\frac{\partial u_i}{\partial x_j} + \frac{\partial u_j}{\partial x_i}\right), \quad i,j=1,2,3 \qquad (4)$$

is a symmetric tensor corresponding to the strain. Therefore if two points P_0 and P'_0 are selected within an infinitesimal region ΔV bounded by a surface Σ, their displacements are related by the equation

$$u(P') = u(P) + \Omega \times \Delta r + \varepsilon \, \Delta r. \qquad (5)$$

In general, the tensors Ω and ε have different values at each point of the elastic body, then being functions on the coordinates. However, if displacements $u(x, y, z)$ are linear functions of the coordinates, then Ω and ε may be considered constant in the small region ΔV. Therefore the Eq. (4) means that, by choosing P_0 as kinematic reduction center of the movement, neighboring points of P_0 in ΔV, such as P'_0, undergo the same translation as P_0, plus a rotation Ω around an axis passing through P_0, and a strain ε. From this point of view, it is logical that P_0 suffers only a translation, since it has been taken as reduction center.

Once the elastic tensor is defined by the Eq. (4), the stress tensor may be deduced by using thermodynamic relations. Starting from the general equation of the free energy for linear homogeneous elastic solids [1], i.e.,

$$F = \frac{1}{2}C_{ijkl}\varepsilon_{ij}\varepsilon_{kl}, \qquad (6)$$

C_{ijkl} being the elastic tensor of fourth order, the stress tensor can be defined through the gradient of F as follows,

$$\sigma_{ij} = \left(\frac{\partial F}{\partial \varepsilon_{ij}} \right).$$ (7)

At the same time, at each point Q of the body surface in equilibrium, the external forces on this surface are compensated to the internal stress. This fact may be expressed mathematically by the relation [1-3]

$$f_i = \sigma_{ij} n_j, \, i, j = 1, 2, 3,$$ (8)

where n_i are the components of a unity normal vector **n** located at any point on the surface S of the solid (Figure 1), and f_i are the projections of the force **f** per unit of area exerted at Q on the body.

2.2. Measurement of Young's Modulus of Plates by Phase Shifting Speckle Interferometry

There are numerous methods for obtaining the Young modulus of a material based on either static or dynamic response of a component to external perturbation. Among these techniques, optical methods are particularly important due to the advantages that they possess, some of which have already been discussed in the introduction. In our case, we will apply the phase shifting speckle technique.

2.2.1. The Phase Shifting Method

Phase-shifting speckle interferometry (PSSI) is a common measurement technique for surface shape analysis and surface deformation. It has been applied for solving many different problems such as the determination of mechanical and thermal properties of thin films [4,5], the measurement of Young's modulus of teeth dentin [6], deformation and displacement analysis [7,8], or shape measurement [9-18]. One advantage of this technique is that the fringe pattern is shown on a video monitor without the use of any photographic process, allowing real-time measurements. Moreover measurements can be performed without contact with the sample to be analyzed, and information about the entire surface examined may be extracted (whole field inspection) and archived electronically for posterior manipulation. However, special attention requires the stability of the entire system, which may be perturbed by different causes, such as variations of temperature, vibrations, and local pressure changes. On the other hand, phase-shift miscalibration and nonlinearities are systematic sources of errors inherent to this technique and should be taken into account to ensure reliability of the measurements.

As it is well known, the basic idea involved in this technique is to reconstruct the phase map of a waveform by changing the phase difference between the radiation beams corresponding to the sample to be analyzed and the reference [19]. For modifying the phase experimentally different procedures may be used. One possibility is to employ a moving mirror as reference controlled by a piezoelectric transducer (PZT), which changes the position with high accuracy and then the phase [11,20]. However, there are other ways to change this phase such as, for example, by stretching a glass fiber through which the light is sent [21], by changing the polarization of light [11], by modification of the wavelength [22,23], or with the help of a plane plate [11]. There are different algorithms for calculating the phase of the modulating interference pattern registered on the detector [24-31]. Basically, they differ in the number of images required to reconstruct the phase map.

In order to explain the general characteristics of the technique, let us consider an optical system composed by a reference and the sample object of study, like a Michelson interferometer, for example. If both specimen and reference are at the same time illuminated by a homogeneous monochromatic beam, the following relation holds:

$$I(x,y) = A(x,y) + \gamma(x,y)\cos(\varphi(x,y)) = A(x,y) + \gamma(x,y)\cos(\delta(x,y) + \varphi_0). \quad (9)$$

In this identity $A(x,y)$, $\gamma(x,y)$, and $\delta(x,y)$ are unknown; at least three equations are needed to calculate the value of the phase. To solve this problem, the phase φ_0 is moved in well defined steps, obtaining an equation system that allows $\delta(x,y)$ to be calculated. For this purpose mathematical algorithms must be used. Within the possibilities we may choose, there are two kinds of procedures that are commonly employed, namely three-step methods and four-step techniques.[1]

The three-step algorithm needs three exposures for different phases φ_0, and it is the algorithm that requires the minimum number of frames for obtaining the phase function $\delta(x,y)$. Hariharan-Schwieder's algorithm employs four patterns by changing the phase in $\pi/4$ [33]. The so-called Carré's algorithm also employs four frames in which the phase shifts are unknown but equal to each other [34]. The advantage of this last procedure is its non-dependence on miscalibration, however, both algorithms are sensitive to phase shifter nonlinearities. In order to avoid these problems some algorithms for compensating errors have been developed [35-37], and others for non-linearities [38,39].

At the same time, another problem for most of the algorithms is related to the time required for the wrapping when real time processing is needed. Due to the phase algorithms' work with the arctangent function, they need a relatively long time to perform the calculus. In order to inprove the speed of processing, algorithms have recently been developed to reduce this time [40,41].

With the aim of applying the phase shifting technique for obtaining elastic properties of materials, we have used a four-step phase-shifting algorithm, in which the resulting function for the phase is given by the following formula:

$$\delta(x,y) = \arctan\left(\frac{I_4 - I_2}{I_1 - I_3}\right), \quad (10)$$

where I_1, I_2, I_3, and I_4 are the intensities registered on the CCD camera for four values of φ_0. The result obtained for the phase is ambiguous because of the periodic characteristics of the

---

[1] There are more possibilities [32].

function arctangent. With the aim of solving this problem, a compensation of the values of $\delta(x,y)$ is required, which is known as phase unwrapping. In order to eliminate 2π ambiguities, the intensity between adjacent pixels may be compared. If the phase difference is greater than π, 2π is added or subtracted until the difference between neighbor pixels is less than π. However, for that process to work correctly, it is necessary that the wavefront between adjacent pixels not change by more than π.

Once the aforementioned problems have been taken into account, the actual shape of the the surface under study may be reconstructed by the relation

$$h(x,y) = \frac{\lambda}{4\pi} \delta(x,y), \qquad (11)$$

$h(x,y)$ being the surface height.

2.2.2. The Young Modulus of Thin Plates

There are many known classical and modern techniques to measure elastic properties of materials. However, the specific characteristics of each problem limit the applicability of an experimental procedure. Therefore, if size, shape, and/or physical magnitude are to be investigated, these are some aspects which require a detailed analysis before an experimental method can be chosen. In our case, we are interested in the elastic properties of a thin rectangular plate because of its importance in industrial applications. The knowledge of the elastic properties of plates is of considerable importance in different branches of civil engineering and machinery, among others.

Taking into account the geometric characteristics of the sample, some of the most common optical devices cannot always be used, since the specific disposition of the specimen to be analyzed is not compatible with the possibilities offered by the setup. Other times it may be that the optical system allows the measurements to be carried out, but the layout of the sample is very complicated, or needs additional tools, making its use impractical. In particular, when studying plates or shells, it is not possible to use the common press for compression. It would produce a flexion in the sample, making the measurement of its elastic properties unfeasible. It would also be possible to use a tensile force instrument like a press, capable of producing stretching in the specimen by holding the material by its ends. However, there is another possibility that is simpler and cheaper. It consists of placing the material plate like a cantilever beam and putting an appropriate load at its end. This load is easy to perform by means of a mass. The weight of the mass produces a bending, making able the partial analysis of the elastic properties. This is the method we will present in this section.

2.2.2.1. Thin Plates. The Cantilever Beam

The problem of a thin isotropic plate with one champed end and the others free, may be solved by using the general biharmonic differential equation [1-3], [42-45]

$$\nabla^4 W(x,y) = \frac{q(x,y)}{D}, \qquad (12)$$

with the corresponding boundary conditions, where $q(x, y)$ is the load per unit area and D is the local flexural stiffness constant defined as

$$D = \frac{Eh^3}{12(1-\nu^2)}, \qquad (13)$$

h being the thickness of the plate, and E and ν the elastic modulus and Poisson's ratio of the material, respectively (we identify $u_3=w$ - see section 2.1). On the other hand, the constitutive equations for the bidimensional problem are

$$\varepsilon_x = \frac{1}{E}\left(\sigma_x - \nu\sigma_y\right)$$
$$\varepsilon_y = \frac{1}{E}\left(\sigma_y - \nu\sigma_x\right)$$
$$\varepsilon_{xy} = \frac{2}{E}(1+\nu)\sigma_{xy} \qquad (14)$$

and for bending moments per unit length:

$$M_x = \int_{-\frac{h}{2}}^{\frac{h}{2}} z\sigma_x dz \qquad M_y = \int_{-\frac{h}{2}}^{\frac{h}{2}} z\sigma_y dz \qquad M_x = \int_{-\frac{h}{2}}^{\frac{h}{2}} z\sigma_{xy} dz.$$
(15)

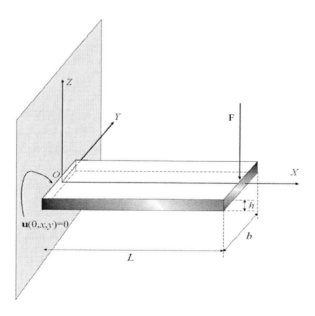

Figure 2. Thin plate clamped in one of its edges. L, length; b, wide; h, thickness. F, force per unit length. The force is applied at the middle point of its end.

An analytical solution of this equation system is in most cases not possible, so numerical or semi-analytical methods are needed to find a solution. However, if the length of the thin plate is much greater than its width, i.e., $L>>b$, the response of this system to bending can be approximated in the case of a slender of the same length as a rectangular cross-section $b.h$, but with flexural rigidity $D = \dfrac{Eh^3}{12}$ [1].

Under the preceding conditions, let us suppose a cantilever of length L is built-in at one of its ends ($L>b$). If at the free-end edge, at the middle point, a force per unit of length F is applied (Figure 2), a deformation of all parts occurs (except where the bar is clamped). Taking as reference the origin O, the static deflection w of the neutral axis of the beam is, approximately,

$$w(x) = -\dfrac{F}{EI}\left(L\dfrac{x^2}{2} - \dfrac{x^3}{6}\right), \qquad (16)$$

where F is the applied load, L is the length of the cantilever beam, E is the Young modulus of the material, I is the moment of inertia of the beam cross section, and x is the distance from the loaded point to O. For the dimensions of the cantilever beam shown in Figure 2 the moment of inertia is $I = \dfrac{1}{12}bh^3$. The value w static deflection S_F at the end of the cantilever is

$$w(L) = S_F = -\dfrac{FL^3}{3EI}, \qquad (17)$$

in which the negative sign arises because the force is applied in the negative direction of the OZ- axis. The expression (17) offers a way to obtain approximately the Young modulus of a material by measuring the displacement w.

2.2.2.2. Measurement of the Elastic Modulus E

We propose a general procedure to measure the Young modulus of thin plates ($L<b$) by using the phase shifting method. With this aim, the experimental setup shown in Figure 3 was used. A laser beam is directed onto the mirror M which reflects the light in the direction where there is a pinhole PH. The emergent beam is expanded with the help of a lens L, obtaining a collimated beam which passes through a beam splitter. At this point, the light is divided into two branches and follows different optical paths. One of them strikes the reference R, and the other one goes to the mirror M(PZT) controlled by a piezoelectric transducer, which enables the change of the phase. Once this beam is reflected by M(PZT), it reaches the sample S by means of the tilted mirror MS. The resulting reflected light from the reference and sample are recorded by a CCD camera and stored in a computer. By means of a program, the resulting phase map is filtered with a mask, thus improving the fringe contrast of the pattern and eliminating the image noise as well. Observe that due to the form of piece, and its position (parallel to the plane OXY-see also Figure 2), the inclusion of the mirror MS in the arrangement is needed (Figures 3, 4). It permits the easy location of the sample S, then the

measurement of the deformation by the proposed technique. In order to minimize the errors, certain experimental aspects should be taken into consideration. The sample must be well clamped in order to guarantee that all points of the specimen located on the plane OYZ do not suffer any displacement (Figure 2), but at the same time, the grips G (Figure 3) in contact with the specimen must not produce local tensions in the plate, which would invalidate the measures. Both conditions are essential for applying the equation (17). On the other hand, a homogeneous illumination of the sample and reference is required.

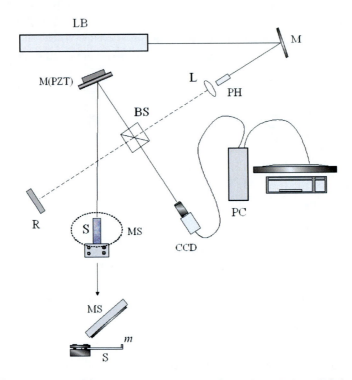

Figure 3. Experimental layout. LB, laser beam; M, mirror; PH, pinhole; L, lens; BS, beam splitter; R, reference; M(PZT), piezoelectric transducer with mirror; S, sample; MS, tilted mirror; CCD, camera; PC, personal computer; m, mass.

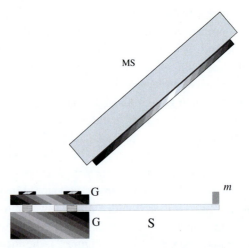

Figure 4. Partial view of the tilted mirror MS, the grips, the sample S, and the load m.

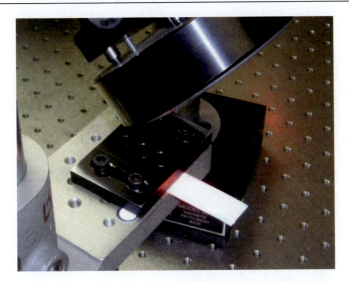

Figure 5. Actual view of the tilted mirror MS, and sample.

Once the geometrical and optical parameters of the device are adjusted, the entire system is calibrated with help of the piezoelectric transducer mirror M(PZT), and proceeds as follows.

A digital image of the sample without perturbation is taken, which will serve as a reference. After that, a calibrated mass m is placed at the end of the plate in its midpoint (Figures 2,4), producing a bending of the piece. Before making another recording, the system is left a reasonable time to deform until reaches its equilibrium state, and then a second image is archived. By using the phase shifting technique, and subtracting digitally these patterns corresponding to the free and deformed states, a phase map consisting of dark and clear fringes is obtained.

2.2.2.3. Experimental Results

To test the validity of the proposed method, two metallic plates were employed. The first material studied was a pure aluminium specimen. The dimensions of the sample are the following: length L=53.00 mm; width a=18.85 mm; and thickness h=4.05 mm. We conducted a series of experiments for different loads; the results appear in Table 1. Figure 6 shows the phase map of the deformed plate corresponding to "experiment 2" in Table 1 (see the end of the section). In order to obtain a better contrast, the phase bands are filtered, obtaining an improved image. As it may be seen, a definite set of fringes appear. With these data, by applying Eq. (11), the phase pattern is unwrapped and scaled, obtaining a graphic as shown in Figure 7. In this illustration all values of the displacement suffered by the sample are represented in color scale. The color blue corresponds to points of small displacements and the color red to parts of larger displacements. The right side of this figure shows a color scale to quantify the values of displacements in micrometers. Figure 8 represents the final shape for the sample in 3D. Thus, after obtaining the fringes, we calculate the state of deformation at each point of the sample of aluminium. With the value of the displacement of a point at the end of the specimen, and knowing the load, through Eq. (17) the value of Young's modulus is determined. As it can be checked, the values obtained for each experiment are very similar to the calibrated value done for the supplier. The mean value is E=71.2 GPa, but taking into consideration the systematic uncertainty of the method, the experimental results obtained are satisfactory.

A second set of experiments were performed with a plate of aluminium alloy. By proceeding in a similar way, the values measured are displayed in Table 2 (see the end of the section). The result in this case is E=42.2 GPa, being the calibrated 43 GPa. The corresponding 3D-graphic for this sample is shown in Figure 9. Finally, Figure 10 exhibits the one dimensional function of the elastic curve, which is calculated averaging the results of all points of the surface piece.

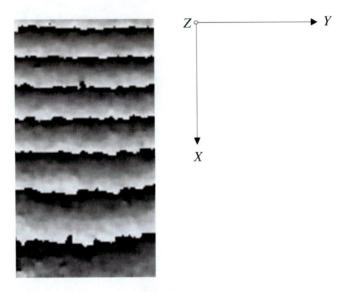

Figure 6. Phase map corresponding to a loaded thin plate.

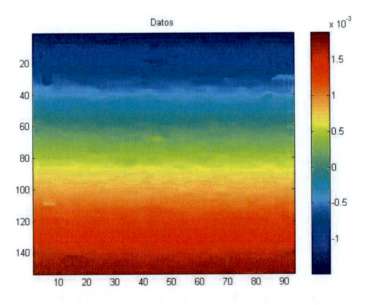

Figure 7. Displacement fields. Blue color corresponds to small displacements and red color to higher deformations.

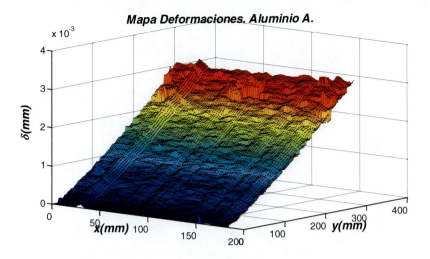

Figure 8. Shape of the aluminium sample after loading.

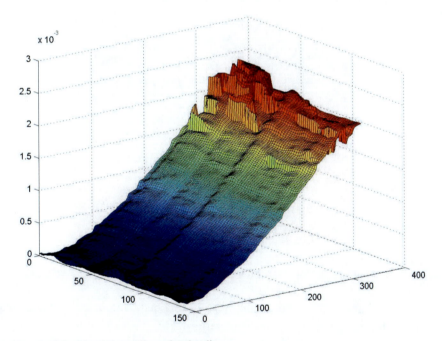

Figure 9. Shape of the aluminium alloy after loading.

If we examine Figures 7 and 8, a homogeneous color for each row of values in the image may be observed. This color homogeneity means that the deformation corresponds to a pure bending.

On the other hand, the color on the top of the pattern is blue, which shows that the points very near to the grips G (Figure 4), where the plate is clamped, did not suffer any displacement, satisfying the boundary conditions. Figure 11 represents a failed experiment. The mass m used to load the plate was located at a non-symmetric point at the end. As a result a displacement map is also obtained but formed of twisted lines.

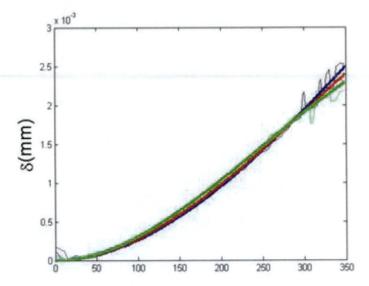

Figure 10. Deflection curve for the aluminium alloy after averaging.

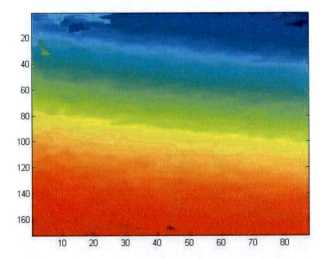

Figure 11. Plate loaded asymmetrically. A non-parallel displacement field appears.

To test the validity of the experimental results obtained, simulation of the displacement and stress fields for the aluminium plate were performed.

To this aim, we simulated the bending of a plate of the same characteristics of our sample (with our experimental Young's modulus), and we calculated the displacement fields by employing the program ANSIS. If the value of the modulus measured is correct, the displacement field at the end of the plate must be the same as we obtained in the laboratory. Indeed, if we analyze the results given by simulation (Figures 12,13), the deflection at the end of the aluminium plate is $w(L)=3.2$ μm. Comparing this value with the one in Table 1 (second experiment), we see that the value measured was 3.1 μm, which agrees within the uncertainty of the method (section 2.2.2.4.). Figures 13 and 14 represent the stress σ_x and σ_{xz}, respectively.

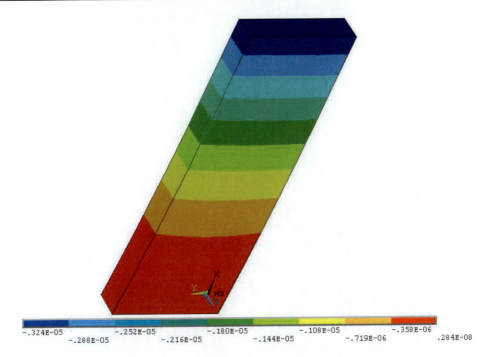

Figure 12. Displacement z-fields for the aluminium specimen. In this figure red color corresponds to small displacements, and blue color to higher displacements (contrary to Figures 7-9).

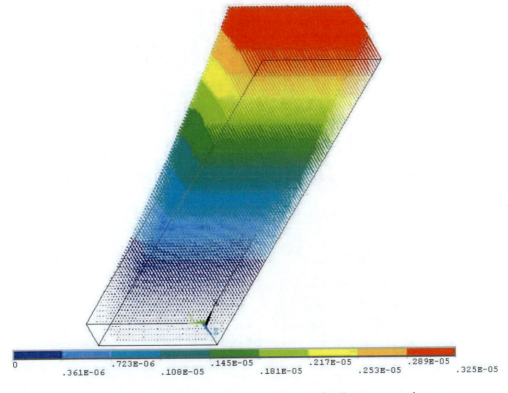

Figure 13. Displacement vector fields at each point in *z* direction for the same sample.

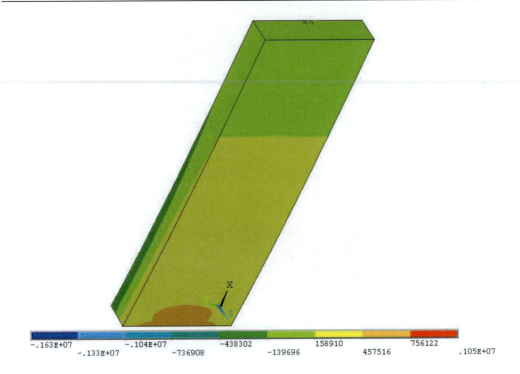

Figure 14. Stress distribution (σ_x).

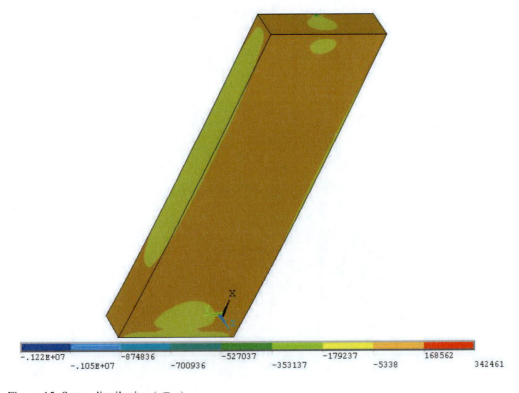

Figure 15. Stress distribution (σ_{xz}).

Table 1.

Experiment	Defection at the end (µm)	Young's Modulus (GPa)	Systematic uncertainty (GPa)	Young's Modulus(GPa) (calibrated)
1	3,1	68,8	± 3.0	
2	3,1	68,8	± 3.0	
3	1,9	73,5	± 3.2	70±1
4	1,9	73,5	± 3.2	
5	1,7	71,2	± 3.1	

Table 2.

Experiment	Defection at the end (µm)	Young's Modulus (GPa)	Systematic uncertainty (GPa)	Young's Modulus(GPa) (calibrated)
1	1,0	44,1	± 1.9	
2	0,8	39,9	± 1.8	
3	1,4	41,1	± 1.9	
4	1,4	41,1	± 1.9	43±1
5	1,8	41,2	± 1.8	
6	1,8	41,2	± 1.8	
7	2,1	44,1	± 1.9	
8	2,3	44,9	± 2.0	

2.2.2.4. Systematic Uncertainty

In this section we estimate the uncertainty of the measurement technique. To this end, the basic theory of errors must be applied [46-48]. Considering Eq.(17), the expression for obtaining the E modulus is

$$E = -\frac{FL^3}{3wI} \qquad (18)$$

Using this equation, the uncertainty for E is

$$U_s(E) = \sum_{i=1}^{N} \left|\frac{\partial E}{\partial q_i}\right| U_s(q_i), \qquad (19)$$

where E is the Young modulus, q_i represents each variable that appears in Eq.(17), and $U_s(q_i)$ is the systematic uncertainty of q_i. Observe that E is a known function of the primary quantities (F, L, w, and I). Propagating Eq. (19), and dividing the expression of $U_s(E)$ by E, the relative systematic error of the Young modulus has the form

$$\frac{U_s(E)}{E} = \left|\frac{U_s(F)}{F}\right| + \left|\frac{3U_s(L)}{L}\right| + \left|\frac{U_s(w)}{w}\right| + \left|\frac{U_s(I)}{I}\right|. \qquad (20)$$

In this equality $U_s(I)$ is not known because it depends on the variables b and h. For this reason, the uncertainty of the moment of inertia must be independently calculated, i.e.,

$$U_s(I) = \sum_{i=1}^{N} \left|\frac{\partial I}{\partial q_i}\right| U_s(q_i) = \frac{1}{12}\left(\left|h^3\right| U_s(b) + \left|3bh^2\right| U_s(h)\right). \tag{21}$$

In order to give a numerical estimation of the uncertainty of E, we introduce in Eqs.(20) and (21), the values of experiment 2 of Table 1, with the following uncertainties: $U_s(F)=9.8 \times 10^{-6}$ N, $U_s(L)=0.05$ mm, $U_s(w)=0.1 \times 10^{-6}$ m, $U_s(b)= U_s(h)=0.01$ mm,

$$\frac{U_s(E)}{E} = 1.38 \times 10^{-5} + 2.83 \times 10^{-3} + 0.032 + 9.17 \times 10^{-3} = 0.044 = 4.4\%$$

If this result is examined, it can be seen that the term which contributes more to the uncertainty is the third, which corresponds to the measure of the deflection at the end. To diminish the errors of the measurement method, an improvement of the phase shifting would be necessary.

2.3. Application of the Speckle Photography for Measuring Elastic Deformation

2.3.1. Thermal Expansion

Thermal expansion of materials is a common property of all elastic bodies. This is important in many technical and scientific fields such as, for example, in machinery, where the structures need an exact adjustment to guarantee its operation. Another important example refers to aerospace industry. Aircrafts are subjected to high temperature gradients which produce thermal stress. As a result, materials and junctions suffer failures, thus making the device unusable. In the cases of satellites and space stations, the components which form part of the structure have different thermal expansion properties (such as the junctions between lens and metal or composite materials). This fact may lead to fractures of the components due to stress and fatigue. However, the importance is not limited to this type of industry. In building construction, ornamental stones (granite, marble, sandstone, slate, etc.) are normally employed. In this case, anisotropy plays an important role, since the behavior of the stone tiles on the walls of buildings depends on the direction in which they were cut. The temperature changes deform and contract the tiles, causing them to crack and eventually break, leading to an economic loss, as well as a possible risk to the safety of people. These examples, among others, show the importance of the thermal characterization of solids.

There are different ways to measure thermal deformation of solids: mechanical methods (push-rod dilatometers), electrical, X ray diffraction, optical, etc. Most of them require contact with the sample, which makes it not always efficient. Moreover, the measurement systems are normally complex and the applications are limited. Among the non-interacting procedures, optical methods based on speckle phenomenon have played an important role in the last three decades because of their advantages compared with the traditional techniques. Generally, speckle methods have a wide range of applicability and do not interact with the specimen to be analyzed. In our case, for the experimental analysis of the thermal expansion tensor, we will employ the speckle photography method.

Speckle photographic procedure is a non-destructive and non-interacting technique, precise, and its range of measurement is easy to adjust. Furthermore, focused speckle photography, with a simple monochromatic light beam, allows the measurement of the in-plane displacement in a general deformation experiment, being insensitive to out-of-plane movements. This characteristic of the technique is useful for application in experiments in which deformation or/and out-of-plane displacements are present (as we will see in the next paragraphs).

Speckle photography has been applied for strain analysis [49-62], thermal deformation [63-67], particle image velocimetry [68-70] stellar problems [71], determination of time-varying in-plane displacement fields [72], measurement of small tilts [73], information processing [74], analysis of crack [75], gas temperature measurement [76], analysis of thermal flow [77], study of surface micrereliev variation and contact pressure measurement [78], measurement of the magnetostriction coefficient [79], and roughness measurement [80-88], among others. In this section is a review of the speckle photography technique used to measure the thermal expansion tensor of an anisotropic body is presented.

A possibility for measuring the tensor expansion tensor of an anisotropic material is to cut six slender bars in six different directions and calculate the deformation of each one for a known temperature increment [89].

By defining the unitary strain, per unit temperature increase ΔT for each sample as α_k, and supposing a linear response of the material, a system of six equations is obtained where the unknowns are the thermal expansion coefficients of the anisotropic body. To apply this procedure a conventional dilatometer can be employed, however, contact between the sample and the sensor is required. On the other hand, as six samples are needed to measure the coefficients, a long time is spent.

As an alternative, optical methods may be used. A possibility is the double-exposure holographic interferometry [90-92], or the speckle photography technique [64-66]. These procedures have demonstrated to be adequate to measure thermal expansion of solids. They are precise and do not need much time, however a precise cut of the bars from the block material is needed.

In order to simplify the preparation of the samples, without loss of precision in the measurements, a simple procedure may be carried out. It consists of cutting three perpendicular plates from the material object of study and measuring the in-plane displacements for different points on each of them. As we shall demonstrate, for an adequate election of points, it is possible to obtain all components of the thermal expansion tensor of any material.

2.3.2. Equations of the Model

Let us suppose a homogeneous anisotropic body has a fixed point O to which the coordinate origin is assigned. Let r be the position vector of a point P having components (x_1, x_2, x_3). Let u(r) be the displacement of P due to the temperature increase ΔT.

With the help of Eqs. (1-3), we define the coefficients of thermal expansion tensor as follows:

$$\alpha_{ij} = \frac{\varepsilon_{ij}}{\Delta T}, \qquad (22)$$

then Eq.(4) becomes

$$\alpha_{ij} = \frac{1}{2\Delta T}\sum_j \left(\frac{\partial u_i}{\partial x_j} + \frac{\partial u_j}{\partial x_i}\right), \quad i,j=1,2,3. \qquad (23)$$

From Eq. (1), the displacement suffered by a point P of the solid is

$$\Delta \mathbf{r} = \widetilde{\alpha}\, \mathbf{r}\, \Delta T, \qquad (24)$$

or in tensorial notation,

$$\Delta x_i = \alpha_{ij} x_j \Delta T. \qquad (25)$$

As we are interested in measuring the symmetric tensor $\widetilde{\alpha} = \alpha_{ij}$ by employing plates, it is necessary to know the displacement of points on the sample surface. It implies the projection of the total displacement of the examined point on its corresponding plane (Figure 16).

The equation for determining the displacement of a point P in a specific direction $\mathbf{e}_n$ in the space is

$$(\Delta \mathbf{r}, \mathbf{e}_n) = (\widetilde{\alpha}\, \mathbf{r}\Delta T, \mathbf{e}_n) = (\widetilde{\alpha}\, \mathbf{n}_p, \mathbf{e}_n)\, r\Delta T = (\mathbf{n}_p, \widetilde{\alpha}\, \mathbf{e}_n)\, r\Delta T, \qquad (26)$$

where $\mathbf{n}_p$ is a unitary vector in the direction of r. By using Eq.(26) we shall demonstrate that, with the analysis of the displacements of three perpendicular plates of the anisotropic material, there is sufficient information to calculate all components of the thermal expansion tensor.

In effect, let us suppose geometry as depicted in Figure 16. It shows three mutually perpendicular rectangular slabs, S_1, S_2 and S_3, cut from the material bulk of volume V. The slabs may be considered separately for the analysis of deformation, and then we can study their behavior as three independent plane problems.

In effect, let us suppose the slab S_3 depicted in the Figure 17.a. It represents a rectangular piece of side L and small thickness. Let us consider two separate points P_0 and P_0' on the sample (for simplicity two vertices have been chosen).

If an increase of temperature occurs, the sample changes its shape (transformation $T(x_1,x_2)$), but, additionally, it can suffer a rigid movement. Using the Figure 17.b, the significance of the coefficients α_{ij} may be understood.

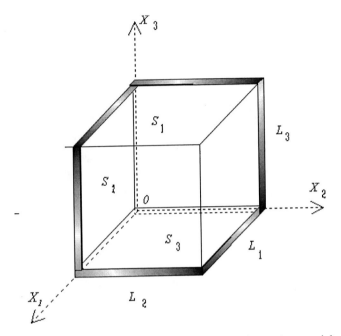

Figure 16. Three perpendicular plates are cut from a block of an anisotropic material.

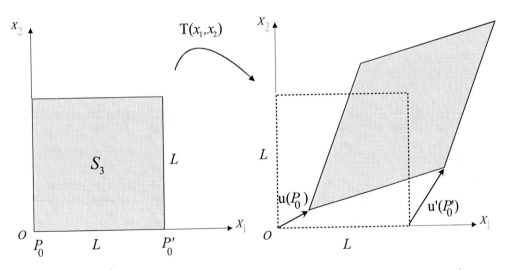

Figure 17.a. Slab S_3 before deformation. (b) Plate S_3 after deformation when the temperature is increased.

Let us consider, for example, the coefficient α_{11}.

Employing the Eq. (23), it follows

$$\alpha_{11} = \frac{1}{\Delta T} \frac{u_0'(P_0')_{x_1} - u_0(P_0)_{x_1}}{L}, \qquad (27)$$

where $u_0'(P_0')_{x_1}$ and $u_0(P_0)_{x_1}$ are the projections of the displacements of points P_0' and P_0 over the axis x_1, respectively. Similar expressions may be found for the other α_{ij} coefficients.

Due to the experimental constraints we can control in the experiments, any one of the three slabs will be deformed by the effect of the temperature, but each of them may also translate in its plane and rotate around an axis perpendicular to its own plane. If the sample is very thin, out-of-plane deformation due to thermal expansion can be ignored, however.[2]

Taking into account the Eq. (5) and using Eqs. (25-26), for the present case of a plane slab, the movement consists of the superposition of a plane translation, a rotation around an axis normal to the plane passing through the origin, and a deformation.

Let P be a point of the slab whose position vector with respect to the coordinate origin O is $\mathbf{r}$. If $\alpha \mathbf{r} \Delta T$ is the displacement due to homogeneous thermal deformation, $\mathbf{t}$ is the common translation for all points of the slab and $-\mathbf{r} \times \Delta\Omega$ is the displacement of point P due to a very small rotation Ω around the origin, the total displacement of P may be written as

$$\Delta \mathbf{r} = \mathbf{t} + \Delta\Omega \times \mathbf{r} + \alpha \mathbf{r} \Delta T. \qquad (28)$$

The Eq.(28) represents the fundamental relation of the model proposed.

2.3.3. The Speckle Photography Method

The speckle photography method consists in taking two photographs of the sample on the same photographic plate, before and after the perturbation is applied (in our case the increment of temperature), using a monochromatic light source (Figure 18).

Once the plate is developed, we can determine the displacements on the surface of the material by means of the diffraction pattern obtained when a point on the specklegram is illuminated with a narrow laser beam (Pointwise filtering technique), as shown in Figure 19. The mentioned pattern is characterized for maxima and minima of intensity similar to those obtained in the Young's experiment of double slit, but modulated by the so-called diffraction halo, which depends on the diameter D of the lens used to take the photograph and on the distance z_0 from this lens to the photographic plate. The extension of the diffraction halo may be calculated, approximately, by the expression $\sin \theta_h \approx \dfrac{D}{z_0}$ (Figure 19).

The characteristics of the lens are important because they are directly related to the sensibility of the method. So, the minimum displacement we can measure by this technique is

---

[2] For example, for a plate of 4 mm thickness the displacement auto-of-plane is of the order of 0.5 μm, which is much smaller than the longitudinal speckle diameter ($\sigma_l \approx 8\lambda(z_0/D)^2$ - see Fig.(18)), causing no decorrelation in speckle pattern.

of the order of the average speckle size, i.e., $\sigma_s \approx 1.2\lambda \frac{z_0}{D} = 1.2\lambda F(1+M)$, where F is the f-number of the lens (f/#) and M the magnification factor, which relates the lineal dimensions of the sample to its image recorded on the photographic emulsion (PP).

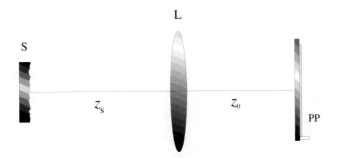

Figure 18. Double-exposure speckle photography. Z_s and Z_0 are the distances from sample to lens and lens to photographic plate, respectively.

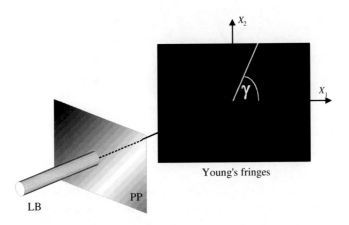

Figure 19. Pointwise filtering technique. A laser beam (LB) is diffracted by a developed photographic plate (PP), in which a double-exposure has been recorded. The plane where the Young fringes are observed is parallel to the developed plate (OX_1X_2 system). In general the direction of displacement forms an angle γ with the axis OX_1.

Focused speckle photography technique allows us to measure the in-plane displacements, while being insensitive to out-of-plane displacements. The expression which allows the measurement of in-plane displacements $|\Delta\mathbf{r}|$ by the intensity maxima of the diffraction pattern is [93]

$$|\Delta\mathbf{r}|\sin\theta_n \cos\xi = n\lambda m \qquad (29)$$

where $|\Delta\mathbf{r}|$ represents the displacement module, θ_n is the angle formed by the OZ axis and the direction of observation of the nth order maximum, m is the demagnification factor, and ξ

is the angle between the in-plane displacement direction OA (perpendicular to the fringes) and the direction defined by the centre of the diffraction halo O and the observation point B (Figure 20). When the observation point B is taken on the straight line through O and perpendicular to the nth intensity fringe, the equation (29) is transformed into the well known $|\Delta \mathbf{r}|\sin\theta_n = n\lambda m$.

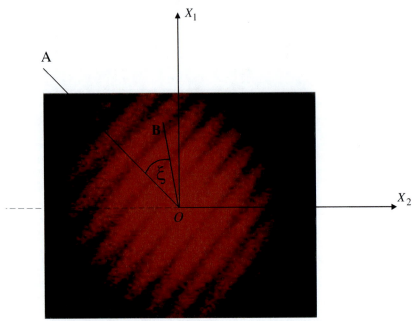

Figure 20. The Young fringes and diffraction halo.

2.3.4. Experimental Determination of the Coefficients α_{ij}

We now demonstrate mathematically that the analysis of the plane displacement of three plane samples provides enough information to measure the thermal expansion tensor.

As we explained in the previous paragraph, we will consider three mutually orthogonal slabs extracted from a block of an anisotropic material (Figure 16).

With the aim of obtaining the equations for measuring the coefficients α_{ij} we choose first the sample S_3 containing the OX_1X_2 coordinate plane. Let us take the point O corresponding to the coordinate origin as the reduction center of the rigid plane movement. In this model we suppose that tilt is not present. If there were a small tilt of the piece, the Young fringes could deform, or might ever disappear within the diffraction halo, making nonsense of the measurements, besides which, the model we are proposing would no longer apply.

Under these conditions the displacement $\Delta\mathbf{r}$ of a point $P(x_1,x_2)$, in general, will have two components.

Projecting the equation (28) onto the OX_1 and OX_2 coordinate axes, and taking into account the equation (29) that gives experimentally the modulus of the in-displacement, the following relation for the sample S_3 is found

$$\frac{nm\lambda \cos\gamma}{\sin\theta_n} = (\alpha_{11} x_2 + \alpha_{12} x_2)\Delta T + t_{13} - x_2 \Omega_3$$

$$\frac{nm\lambda \sin\gamma}{\sin\theta_n} = (\alpha_{12} x_1 + \alpha_{12} x_2)\Delta T + t_{23} + x_1 \Omega_3 \quad , \tag{30}$$

where t_{13} is the projection of the translation vector of the sample S_3 over the OX_1 axis, t_{23} is the projection of **t** over the OX_2 axis, Ω_3 is the rotation of the of this sample around an axis perpendicular to its plane passing through the origin O, and γ is the angle between the direction of the displacement and the OX_1 axis (Figure 19). Proceeding in the same way as above but for the plate S_2 (OX_1X_3), another system of equations is obtained:

$$\frac{nm\lambda \cos\gamma}{\sin\theta_n} = (\alpha_{11} x_1 + \alpha_{13} x_3)\Delta T + t_{12} - x_3 \Omega_2$$

$$\frac{nm\lambda \sin\gamma}{\sin\theta_n} = (\alpha_{13} x_1 + \alpha_3 x_3)\Delta T + t_{32} + x_1 \Omega_2 , \tag{31}$$

and for S_1 (OX_2X_3)

$$\frac{nm\lambda \cos\gamma}{\sin\theta_n} = (\alpha_{22} x_2 + \alpha_{23} x_3)\Delta T + t_{21} - x_3 \Omega_1$$

$$\frac{nm\lambda \cos\gamma}{\sin\theta_n} = (\alpha_{23} x_2 + \alpha_{33} x_3)\Delta T + t_{31} + x_2 \Omega_1 \tag{32}$$

Each system of equations contains, as unknowns, three components α_{ij} of the thermal expansion tensor, two components of the translation vector over two coordinate axes, and one corresponding to the rotation Ω_i. It is therefore a system of two equations with six unknowns. Summarizing, we have six equations with six thermal unknowns, six translations, and three rotation unknowns, i.e., a total of 15 unknowns. In order to make this system of equations determinate, we use the theory developed herein to three points P, P' and P'' for each sample. This procedure gives three systems of equations similar to (30-32) for each one of the slabs, resulting in a system of 18 equations with 15 unknowns. It would be possible to examine two points in only one of the samples because there are more equations than unknowns, i.e., we have redundant equations. In this case we would have a system of 16 equations with the 15 unknowns, being still possible to ignore one of the equalities (compatible). However, although Cramer's solution is correct, it is not the more appropriate procedure when working with experimental data.

Once the measurement of the displacements is made, there is another similar mathematical procedure to calculate the thermal expansion coefficients. Instead of using equations (30-32), the idea is to calculate the differences of the projections of the displacement for two points corresponding to the same slab, such that the components of each translation vector vanished. This does not affect the final result or simplify the calculus. The new equations have the following form for the sample S_3:

$$\frac{m}{\Delta T}\left(\frac{n\cos\gamma_i}{\operatorname{sen}\theta_i} - \frac{k\cos\gamma_j}{\operatorname{sen}\theta_j}\right) = (x_{i1} - x_{j1})\alpha_{11} + (x_{i2} - x_{j2})\alpha_{12} - \frac{x_{i2} - x_{j2}}{\Delta T}\Omega_3, \qquad (33)$$

where the subscripts i and j refer to different points for this sample, and n and k are the diffraction orders for these points when the Pointwise filtering technique is employed.

2.3.5. Statistical Analysis of Data

The experimental study of the movement of 8 points leads to 16 equations with 15 unknowns. Theoretically, one of them could be eliminated. However, the data of the equations are obtained in the laboratory, and hence, the reliability of the results is not ensured if Cramer's solution is employed. To avoid problems deriving from the inaccuracy of the measurements, multiple linear regression may be applied [94, 95].

With the aim of explaining this procedure, let us suppose a system of Q equations with R unknowns ($Q>R$). In our case, this set of equations is obtained empirically. From a mathematical point of view, the problem reduces to calculate the hyperplane of best fit for the data measured. Suppose that the use of all measured data is desired, i.e., $Q>R$. In this case, a system of equations of type

$$\delta = \widetilde{F}\,\alpha \qquad (34)$$

is obtained, where δ represents a vector in q-dimensional space. The components of this vector correspond to the projections of the infinitesimal displacements of each point analyzed in each sample plane along the coordinate axes ($nm\lambda\cos\gamma/\sin\theta_n, nm\lambda\sin\gamma/\sin\theta_n$). $\widetilde{F}$ is the matrix obtained by arranging the coordinates x_i and the products $x_i\Delta T$, and α is the vector in r dimensional space containing all of the unknowns as components (i.e., the terms of tensor α_{ij}, the translations t_{ij}, and the rotations Ω_i). The best value of α which solves the system (30-32) must verify

$$\alpha = \left(\widetilde{F}^t.\widetilde{F}\right)^{-1}\widetilde{F}^t\,\delta, \qquad (35)$$

where $\widetilde{F}^t$ represents the transposed matrix of $\widetilde{F}$. If the equations (33) are employed, the solution (35) can be also used. However, the components of the vectors δ and α, and the F_{ij} terms of (34) must be changed accordingly to Eq. (33).

2.3.6. Experimental Setup

To verify the proposed model, three slabs of 0.4 cm thick were cut from a rectangular parallelepiped with sides of approximately 5 cm, as shown in Figure 16. These three plates constitute samples S_1, S_2 and S_3. Samples found to not be very reflective and diffusive are scratched and spread with graphite on the side to be studied. This is better than using white paint, as we have experimentally demonstrated. The samples are located at the end of the supports (a, b and c) in such a way that their legs pass through boreholes drilled in the double bottom of the sample holding furnace (Figure 21). These legs are bars of glassy quartz since this material has a small thermal expansion coefficient, and, hence, it does not contribute to the out-of plane movements of the samples. The bath consisted of a copper container with a double bottom in which water circulated in closed circuit, passing through a thermostat which can regulate the temperature with a precision of 0.1 °C. The container top is made of transparent glass. The table with the samples is placed in the container in such a way that its legs pass through boreholes drilled in the double bottom and rest on a vibration-free table. The monochromatic light of a 15 mW He-Ne laser (L), filtered through a beam expander and pinhole (SF) and reflected off a mirror (B), illuminates the samples (Figure 21). The light diffused by the samples is reflected in the mirror, passes through a photographic objective and is recorded on a photographic plate.

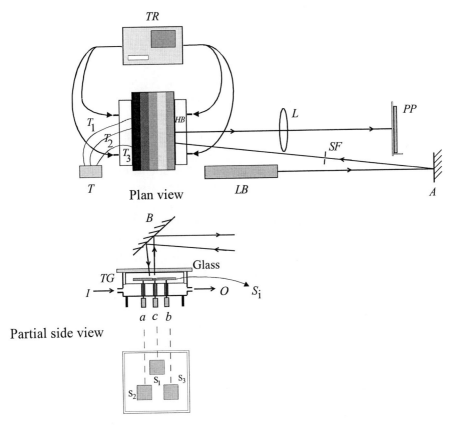

Figure 21. Experimental set-up. (a) Plan view. LB Laser (He-Ne); A and B, mirrors; SF spatial filter; HB sample holding bath; T_1, T_2, T_3, thermocouples; T, electronic thermocouple device; TR, thermostatic regulator, L, lens, PP, photographic plate. (b) Partial side view. TG, transparent glass, I, water inlet; O water outlet; a, b and c, supports of samples and legs.

The first exposure was performed with the thermostatic control set to supply water at an appropriate temperature. Temperature T_1 of the slabs was measured by *J*-type thermocouples attached to the slabs in the stationary state. The second exposure is performed with the samples at temperature T_2. Once the photographic plate is developed, the displacements at different points on the photographed samples are carried out.

The temperature increase ΔT can be calculated as the difference T_2-T_1, or better still, by placing three standard OFHC-type copper sheets, of known thermal expansion coefficient, on all the samples. Since the copper shells are in contact with the sample, their temperatures will coincide to a very high degree of precision (Figure 22). By applying the same methodology as explained in the preceding sections, the temperature may be indirectly determined. In this case, the equations to be employed for measuring the temperature of the sample S_3 are the following:

$$\frac{nm\lambda \cos\phi}{\sin\theta_n} = \alpha x_1 \Delta T + \tau_{13} - x_2 \omega_3$$

$$\frac{nm\lambda \sin\phi}{\sin\theta_n} = \alpha x_2 \Delta T + \tau_{23} + x_1 \omega_3 \quad ,$$

(36)

in which ϕ, τ_{13}, τ_{13}, and ω_3 are the angle between the modulus of the displacement over the coordinate axes, the components of the translation, and the rotation, respectively (observe that, in general, these magnitudes will be different that t_{13}, t_{23}, and Ω_3). Contrary to what happens in the case previously seen for the coefficients of the thermal expansion tensor, in this system of equations (36) only the temperature increment, and not α, is the physical quantity to be determined. We need at least the study of two points in order to have enough equation to solve the problem.

For all experiments conducted in the laboratory, the double exposure on the photographic plate is carried out with demagnification unity. This data could be checked or subsequently corrected by measuring the size of a sample and its image on the plate.

The specklegram is illuminated at the points in question with a read-out laser beam. As stated previously, the read-out laser beam diameter has to be small enough [96,97]. In order to reduce the measurement errors, the specklegram may be mounted on a high-precision *XY* translation table so that the plane of the image is perpendicular to the explorer beam. The mechanism allows the coordinate differences between any two points to be measured with a precision of 1 μm.

A specklegram rotation system allows the angles, formed by the direction of displacement of a point with one of the coordinate axes, to be measured. This rotation system can also be used to measure the angles θ_n at which the diffraction maxima occur.

Figure 22. Sample of marble with cooper calibrated sheet. By applying the described method, the temperature may be directly measured at the same time that the thermal expansion coefficients (photo courtesy of F. Ortega).

2.3.7. Experimental Results

With the aim of demonstrating the validity of the exposed technique, the thermal expansion tensor of an anisotropic sample such as natural quartz was studied.

The study of the physical properties of this mineral is important because of its practical applications in electronics as a frequency control, microphones, wave filters in telephone communications systems, in pressure gauges, and as parts of precision instruments (balances, galvanometers, gravimeters, etc), among others. On the other hand quartz is the only natural silica mineral used in significant quantities in the industry. For example, the sand that is an essential ingredient of concrete and mortar is largely quartz, as are the sandstone and quartzite used as building stones. Crushed sandstone and quartzite are used for road and railway construction, roofing granulates, and riprap. Its physical, mechanical and thermal properties lead to its use as sandpaper abrasive, sandblasting, polishing, and cutting glass too. From a thermal point of view, pure quartz is commonly used in refractory products, such as insulation and firebricks, electrical insulators, etc., because of the combination of its high melting temperature, low coefficients of thermal expansion, and low cost.

In order to measure the thermal coefficients, three mutually perpendicular plane samples were cut (Figure 23). The cuts were made so that one of the resulting slabs was perpendicular to the optical axis of the crystal, and the other two contained this axis, which is an axis of symmetry. We performed various experiments with a temperature increase to the order of 30 degrees, but without obtaining any result. Two reasons may be adduced for this. First, since quartz is very transparent to the radiation used, the reflection is very weak. Second, speckle is collected not only from the anterior surface, but also from the posterior surface and from the internal discontinuities, so that the information is overlapped on the photographic plate, and therefore, the proposed theory can not applied.

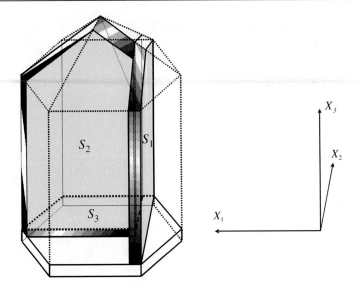

Figure 23. The three slabs S_1, S_2, and S_3 indicated by broken lines were cut from a quartz crystal. Only one double exposure of the three slabs together is needed to calculate a thermal expansion tensor.

By using the system of equations (33) and a regression model to fit the data, the values for the components of the thermal expansion coefficients of quartz are obtained were:

$\alpha 11$	11.6
$\alpha 22$	9.7
$\alpha 33$	5.0
$\alpha 12$	-0.2
$\alpha 13$	2.0
$\alpha 23$	-0.2
$\Omega 1$	72.4
$\Omega 2$	1.1
$\Omega 3$	49.9
Eigenvalues	
αI	4.5
αII	9.7
αIII	12.2

($\alpha \times 10^6 K^{-1}$; $\Omega \times 10^6 rad$).

For this reason, we spread with graphite the surfaces of three slabs in such a way that the speckle generated resulted exclusively from the light scattering from their anterior surfaces.

These results give evidence of the anisotropy of this mineral. The maximum eigenvalue of the thermal expansion tensor represents the maximum value of the thermal expansion coefficient α among all those corresponding to any spatial direction in the material. On the other hand, the trace of the thermal expansion tensor is related with the volume of the material as follows:

$$tr(\alpha_{ij}) = \alpha_{11} + \alpha_{22} + \alpha_{33} = \alpha_{\perp} + \alpha_{\perp} + \alpha_{\parallel} \approx \frac{V - V_0}{V_0 \Delta T}. \tag{37}$$

This relation says that the addition of the diagonal elements of the tensor is, approximately, the relative increase in volume per unit temperature. In our case, taking the values of Table 2, we obtain for the volume increment

$$\frac{V - V_0}{V_0 \Delta T} = \frac{\Delta V}{V} \approx 768 \cdot 10^{-6}.$$

In relation with the eigenvalues found in the literature, they differ slightly from one author to another [94], although they are around

$\alpha_I = 8 \cdot 10^{-6}$ K^{-1}, $\alpha_{II} = \alpha_{III} = 13 \cdot 10^{-6}$ K^{-1}.

The result measured by us for α_I, α_{II} and α_{III} differ from these, which could lead to the conclusion that the model should be rejected for uniaxial crystals. However, measurements made by holographic interferometry on the same samples [97] yielded values very close to ours.[3] For this reason, we are led to assume that the origin of the discrepancy is in the quartz we used for the measurements.

In fact, when examining the three samples employed, they show the presence of macroscopic inhomogeneities as well as dislocations, cracks and breakages running throughout the crystal, something probably caused by the cutting necessary to prepare the specimens. For this reason, we believe that we have measured the true thermal expansion coefficients of the quartz crystal. It can be extracted. The important conclusion is that, in general, the deformation undergone by a elastic body due to the variation of the temperature is influenced in some manner by the number of defects it possesses. On the other hand, if Table 2 is examined, another result may found. In effect, by considering the values of α_{ij} ($i \neq j$), we can see that the slabs corresponding to the S_3 and S_1 planes have been cut so that their corresponding OX_1X_2 and OX_2X_3 axes practically coincide with the principal directions, since their α_{12} and α_{23} coefficients are practically zero. Since the eigenvectors of a symmetric tensor are perpendicular, this implies that the α_{13} component of the thermal expansion tensor must also be approximately zero; however, this is not the case. This means that sample OX_1X_3 is not perpendicular to the other two, or equivalently, the cut to prepare this plate corresponding to this plane was not performed correctly. This, in turn, affected the calculated eigenvalues, since the equations obtained for the model presented are only valid if the samples are perpendicular.

Both of the issues mentioned may contribute to yield results that do not coincide exactly with those of other authors.

---

[3] Information about holographic interferometry and applications can be found in P. Hariharan, Optical Holography. Cambridge University Press (1984), G.Wernicke, W.Osten, Holographische Interferometrie, Physik-Verlag (1982). Y.I.Ortrovsky, V.P.Shchepinov, V.V.Yakovlev, Holographic Interferometry in Experimental Mechanics, Springer-Verlag (1991).

The methodology shown in this section is applicable to any anisotropic material. To see other examples, the references may be consulted [64,66].

2.4. Measurement of Magnetostriction

2.4.1 The Importance of Magnetostriction

In this section we deal with magnetic properties of material; concretely, we are interested in the phenomenon of magnetostriction.

Magnetostriction could be defined as the variation in the dimensions of a material produced by a change in the value and in the direction of its magnetization. This phenomenon plays an important role because of its scientific and technical applications. For instance, the discovery of the giant magnetostriction has opened the door to new applications in science and technology. The application of these materials for industrial use has been important in many fields such as automobiles, robotics, and scientific instrumentation. Another important group of magnetostrictive materials are the metallic glasses. These amorphous solids are very sensitive to mechanical stress, and this property can be directly applied in technology for sensors and actuators, which are extensively used in machine-tool sliding tables, semiconductor fabrication, aircrafts, and others. Magnetostrictive transducers make use of a type of magnetostrictive material in which an applied magnetic field H perturbs the positions of the atoms of the body together; if this external field varies with time it may create a periodic distortion in the length of the magnetic material and, thus, produce an entire elastic vibration. Unlike what happens with the piezoelectric transducers, the actuators based on the magnetostrictive effect are used primarily in the lower frequency range.[4] In this sense, the magnetostrictive characterization of materials is important for new applications as described above.

One of the most important parameters of a magnetic material is the saturation magnetostriction coefficient λ_s. For isotropic materials the saturation magnetostriction at an angle θ to the direction of the magnetic field is defined as follows [98,99]:

$$\lambda_s(\theta) = \frac{3}{2}\lambda_s\left(\cos^2\theta - \frac{1}{3}\right). \tag{38}$$

This equation shows that a magnetic field produces not only a longitudinal deformation $\lambda_{s||}$ in the direction of the magnetic field, but also a transversal strain $\lambda_\perp$. Introducing the value of $\theta = 0$ and $\theta = 90°$ into Eq. (38), we obtain

$$\lambda_s = \frac{2}{3}\left(\lambda_{s||} - \lambda_\perp\right). \tag{39}$$

---

[4] Piezoelectric transducers are used over the entire range of frequencies and at all output levels.

This results shows that the saturation magnetostriction λ_s may be obtained from the difference between the saturation magnetostrictions measured parallel λ_s and perpendicular $\lambda_\perp$ to the magnetic field applied. The value of λ_s given by Eq. (39) remains independent of the demagnetizated state [100], while determining λ_s from only the measurement of $\lambda_{s\parallel}$, according to Eq. (38), is sensitive to the presence of anisotropies in the demagnetized state.

Direct and indirect tecniques have been employed to measure λ_s such as, for example, capacitance methods, strain gauge, cantilever capacitance procedure (thin films), tunnelling microscopy, and acoustical and optical tecnniques [101-104]. Among them, techniques based on optical interferometry have been revealed to be suitable for measuring the magnetostriction of thin films [105-107], steel sheets [108], and ribbon samples [109].

The purpose of this section is to obtain the saturation magnetostriction coefficient λ_s in only one experiment by measuring simultaneously the coefficients $\lambda_{s\parallel}$ and $\lambda_{s\perp}$. For this purpose we use the speckle photography technique (2.3.3). Our analysis deals with the measurement of strain when the external perturbation is a magnetic field. Rigid movements of the sample during the experiment are undesirable because they lead to incorrect values of the physical unknowns. By means of the speckle photography technique, as we saw earlier, from the total displacement measured, it is possible to separate the rigid movements of the in-plane strains, which gives the actual value of the coefficients.

2.4.2. Model for Measuring λ_s

Let us suppose there is a prismatic bar whose dimensions are L_1, L_2 and L_3 along the coordinate system $OXYZ$, respectively (Figure 24). Let us consider a point $P_i(x_i, y_i, 0)$ on the surface S_1 defined by L_1 and L_2, whose position vector with respect to the origin O is $\mathbf{r}_i$ (Figure 25). If we apply a magnetic field along the direction OX of the sample, it produces a displacement of the point P_i to another position $P_i'(x_i', y_i', z_i')$. Due to the conditions under which the experiments were made, the sample suffers, in general, not only a strain but

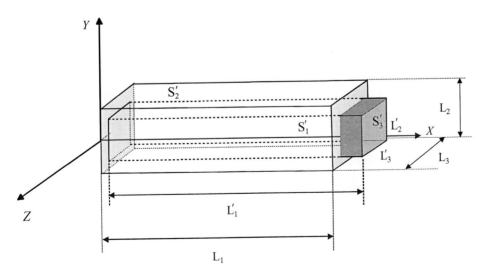

Figure 24. Prismatic bar of dimensions L₁, L₂, and L₃, before deformation. Observe the change in shape as a consequence of the magnetic field.

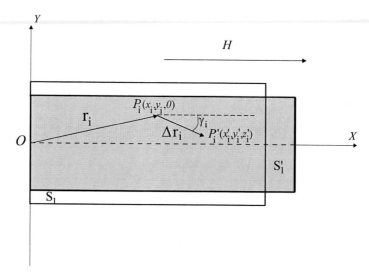

Figure 25. When the material is deformed, a point P_i of coordinates $(x_i, y_i, 0)$ displaces to position (x_i', y_i', z_i').

also a translation and a rotation. In our case, rigid motions perpendicular to the plane S_1 are impeded by the experimental constraints and by the shape of the specimen. This means in the presented model no tilt is allowed.

If we choose the origin O as the reduction center and we suppose small continuous homogeneous deformations, when the applied field H is large enough to saturate the specimen, the projection of the displacement on the surface S_1 suffered by P_i may be expressed as functions of the coefficients $\lambda_{s||}$ and $\lambda_{s\perp}$ as follows:

$$\Delta x_i = \lambda_{s||} x_i + t_x - \Omega_z y_i$$

$$\Delta y_i = -y_{s\perp} x_i + t_x + \Omega_z x_i ,$$

(40)

where t_x and t_y are the components of translation, Ω_z is a small rotation around the OZ axis perpendicular to the sample surface, and $\lambda_{s||}$ and $\lambda_{s\perp}$ are the aforesaid saturation magnetostriction coefficients along the direction parallel and perpendicular to the applied field H, respectively. By introducing Eq. (29) in Eq. (40) (for $\xi = 0$ - Figure 20), the following system of equations is obtained:

$$\frac{n\lambda m}{\sin \theta_n} \cos \gamma_i = \lambda_{s||} x_i + t_x - \Omega_z y_i ,$$

(41)

$$\frac{n\lambda m}{\sin\theta_n}\sin\gamma_i = -y_{s\perp}x_i + t_x + \Omega_z x_i,$$

where γ_i is the angle between the direction of the in-plane displacement on S_1 and the axis OX (Figure 25). These equations have five unknowns, i.e. $\lambda_{s\parallel}$, $y_{s\perp}$, t_x, t_y and Ω_z, which means we need to measure at least the displacements of three points of the sample to obtain such values.

However, as we work with experimental data, it is possible that the results for the unknowns differ from the actual values if the measurement of the displacement of these three points is not very accurate.

Moreover, it is necessary to separate these points as much as possible and the disposition must not be in a straight line. The reason is that, in this case, the systematic uncertainty of the calculation would be quite large. A possibility to avoid that problem is that which was explained in section 2.3.4., i.e., to use a linear regression model in order to find the best values for all these values.

By using the identity (35), the general expression for the matrix $\widetilde{F}$ for calculating the magnetostriction coefficients, and the other unknowns, has the following form:

$$\begin{pmatrix} \frac{n_1\lambda m}{\sin\theta_{1n}}\cos\gamma_1 \\ \frac{n_1\lambda m}{\sin\theta_{1n}}\sin\gamma_1 \\ \frac{n_2\lambda m}{\sin\theta_{2n}}\sin\gamma_2 \\ \frac{n_2\lambda m}{\sin\theta_{2n}}\cos\gamma_2 \\ \vdots \\ \frac{n_j\lambda m}{\sin\theta_{jn}}\cos\gamma_j \\ \frac{n_j\lambda m}{\sin\theta_{jn}}\sin\gamma_j \end{pmatrix} = \begin{pmatrix} 1 & 0 & -y_1 & x_1 & 0 \\ 0 & 1 & x_1 & 0 & y_1 \\ 1 & 0 & -y_2 & x_2 & 0 \\ 0 & 1 & x_2 & 0 & y_1 \\ \cdot & \cdot & \cdot & \cdot & \cdot \\ \cdot & \cdot & \cdot & \cdot & \cdot \\ \cdot & \cdot & \cdot & \cdot & \cdot \\ 1 & 0 & -y_j & x_j & 0 \\ 0 & 1 & x_j & 0 & y_j \end{pmatrix} \begin{pmatrix} t_x \\ t_y \\ \Omega_z \\ \lambda_{s\parallel} \\ \lambda_{s\perp} \end{pmatrix} \quad (42)$$

in which the subscript j refers to a generic point P_j.

2.4.3. Experimental Layout

The experimental setup employed is described in Figure 26. A laser beam (LB) (He-Ne, 37 mW) is directed through a series of mirrors (M1 and M2) to a pinhole (PH) which expands the beam (LB). The expanded laser is then directed to the sample plane (S). The rough specimen scatters the beam (producing speckles) and the resulting scattered light passes through the lens (L) reaching the photographic plate (PP). This experiment corresponds to the first exposure and the sample is demagnetized. Afterwards, a magnetic field H is applied along the longitudinal direction of the sample, and a second photograph is taken for this new state of the object on the same photographic plate. This means that two photographs are superimposed on the same photographic emulsion. By developing the plate, we are able to measure the displacements suffered by any point of the sample plane through the diffraction pattern (Young's fringes) described in section 2.3.3.

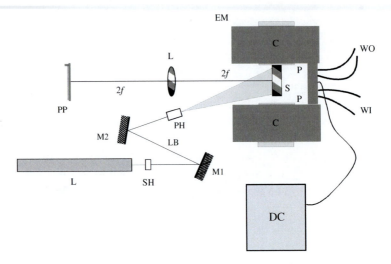

Figure 26. L, laser; SH, shutter; M1, mirror; LB, laser beam; M2, mirror; PH, pinhole; S, sample; L, lens; PP, photographic plate; EM, electromagnet; C, coil; P, pole; WI, water inlet; WO, water outlet; DC, current generator.

Figure 27. Frontal view of the electromagnet with sample (Photo courtesy of F. Ortega).

In order to avoid great rigid displacements and tilt of the sample between exposures, a face of the specimen was glued to an iron pole between the poles of the electromagnet (Figures 26 and 27).

To generate the magnetic field we used an electromagnet refrigerated by water (Newport Instruments) which may produce a magnetic field up to 1 T. The magnetic field of this electromagnet was previously calibrated by measuring the strength field with a magnetometer (Walker Scientific Mg-4D) and with a transversal sound type HP.14S (see Figure 27).

The applied optical system to register the displacement of the sample is corrected of aberrations. The lens has an F-number of 2.8 and the demagnification factor m for all experiments of unity. With this device, the minimum displacement we can detect is about 4.2 μm (the average diameter of speckle for this layout).

To check the validity of the proposed technique, samples of nickel (99.75%) in the form of a prismatic bar 125 mm in length and square section of 40 mm height were investigated. The mechanical treatment that followed was cold drawn and straight length, and for the heat treatment the sample was annealed at 900°C in a continuous furnace. The magnetic field was applied in the longitudinal direction of the bar (direction of axis OX, Figure 25). We have seen that only three points are needed to obtain the five unknowns. However, it is recommendable to employ a high number of points and use Eq.(42) referring to the multiple lineal regression method. With this aim, we have measured the displacement of ten points which resulted in 20 equations and 5 unknowns. Applying Eq.(42), the values of $\lambda_{s||}$ and $\lambda_{s\perp}$ are simultaneously determined and then the value of the saturation magnetostriction coefficient λ_s (Eq.39). The solutions we have obtained from the experimental results were

$$\lambda_{s||} = -54.7 \times 10^{-6}, \quad \lambda_{s\perp} = 25.6 \times 10^{-6}, \quad t_x = 10.7 \text{ μm}, \quad t_y = 5.3 \text{ μm}, \text{ and } \Omega_z = 192.9 \text{ μrad}.$$

With these values, Eq.(39) yields

$$\lambda_s = \frac{2}{3}(\lambda_{s||} - \lambda_{s\perp}) = -53.5 \times 10^{-6}.$$

This experimental result for λ_s is similar to that obtained by other authors [110]. The value of the magnetostriction λ_s for nickel depends on the chemical composition and mechanical and thermal treatment. On the other hand, because the sample used is finite, the form effect should be taken into consideration [111-116]. The value estimated for our paralepipedic sample results is of the order of 0.3×10^{-6}, which may be neglected as compared with the magnetostriction coefficient measured by the speckle photography technique.

2.4.4. Measurement Errors

To calculate the errors of the methodology employed, the expressions (39) and (41) must be analyzed. In the equations (41), the appearance of translations and rotations correspond to rigid plane movements of the sample. However, as a first approximation, for estimating the uncertainty of λ_s, rotations and translations may be not considered. This procedure considerably simplifies the calculation without loss of generality and gives the order of magnitude.

For determining the errors of λ_s, we expand the uncertainty from Eq. (39), i.e.,

$$U_s(\lambda_s) = \left|\frac{\partial \lambda_s}{\lambda_{s||}}\right| U_s(\lambda_{s||}) + \left|\frac{\partial \lambda_s}{\lambda_{s\perp}}\right| U_s(\lambda_{s\perp}) = \frac{2}{3}\left(\left|U_s(\lambda_{s||})\right| + \left|U_s(\lambda_{s\perp})\right|\right) \tag{43}$$

This expression shows the dependence of the $U_s(\lambda_s)$ on the uncertainties of $\lambda_{s||}$ and $\lambda_{s\perp}$. Therefore, for determining the systematic error of λ_s, the values of $U_s(\lambda_{s||})$ and $U_s(\lambda_{s\perp})$ must be calculated.

To this goal, let us consider Eq.(41) in which only the term corresponding to strain remains. For this case we have

$$\lambda_{s||} = \frac{n\lambda m}{x\sin\theta_n}\cos\gamma, \tag{44}$$

and its uncertainty may be expressed as follows:

$$U_s(F) = \sum_{i=1}^{N}\left|\frac{\partial \lambda_{s||}}{\partial q_i}\right| U_s(q_i), \tag{45}$$

where, in the habitual sense, the parameters q_i correspond to the variables of $\lambda_{s||}$.

$$U_s(\lambda_{s||}) = \sum_{i=1}^{N}\left|\frac{\partial \lambda_{s||}}{\partial q_i}\right| U_s(q_i) = \left|\frac{nm}{x\sin\theta_n}\cos\gamma\right| U_s(\lambda) + \left|\frac{n\lambda m}{x^2\sin\theta_n}\cos\gamma\right| U_s(x) + \left|\frac{n\lambda m}{x\sin\theta_n}\sin\gamma\right| U_s(\gamma)$$
$$+ \left|\frac{n\lambda m}{x\sin\theta_n}\cos\gamma\frac{1}{\tan\theta_n}\right| U_s(\theta_n).$$

Introducing Eq.(44) in this expression, it is obtained:

$$\frac{U_s(\lambda_{s||})}{\lambda_{s||}} = \left|\frac{U_s(\lambda)}{\lambda}\right| + \left|\frac{U_s(x)}{x}\right| + \left|\tan\gamma U_s(\gamma)\right| + \left|\frac{U_s(\theta_n)}{\tan\theta_n}\right|. \tag{46}$$

The angle θ_n refers to the intensity maxima of the diffraction pattern. To measure θ_n the highest order of fringes n may always be chosen (see Figure 20). It means we can measure the fringe on the border of halo with the aim to reduce the uncertainty. Taking into consideration that the limit of diffraction halo verifies

$$\sin\theta_n = \frac{D}{z_0} = \frac{D}{2f},$$

the following addition can be written:

$$\frac{U_s(\lambda_{s\|})}{\lambda_{s\|}} = \left|\frac{U_s(\lambda)}{\lambda}\right| + \left|\frac{U_s(x)}{x}\right| + |\tan\gamma U_s(\gamma)| + \left|\frac{2f}{D}\right| U_s(\theta_n) \qquad (47)$$

If instead of the angle θ_h being formed by the direction of the intensity maximum on the border of the halo and the central maximum the angle θ_h is measured between the maxima of order n and $-n$ on the diffraction pattern, then the last term of equation (47) is divided by 2, obtaining for the relative uncertainty

$$\frac{U_s(\lambda_{s\|})}{\lambda_{s\|}} = \left|\frac{U_s(\lambda)}{\lambda}\right| + \left|\frac{U_s(x)}{x}\right| + |\tan\gamma U_s(\gamma)| + \left|\frac{f}{D}\right|U_s(\theta_n) = \left|\frac{U_s(\lambda)}{\lambda}\right| + \left|\frac{U_s(x)}{x}\right| + |\tan\gamma U_s(\gamma)| + |F|U_s(\theta_n), \qquad (48)$$

where F is the f-number of the lens.

Assuming, $\lambda = 632$ nm, $x = 0.10$ m, $\gamma \approx 45$, $f/2.8$, $m=1$, $U_s(\lambda)=0.1\times10^{-9}$ m, $U_s(x)=0.01$ mm, and $U_s(\gamma) = U_s(x) = 0.017$ rad, the equation (47) gives the result

$$\frac{U_s(\lambda_{s\|})}{\lambda_{s\|}} = 0.00001 + 0.0001 + 0.017 + 0.047 = 0.064,$$

that is, about 6.4% of relative uncertainty.

Proceeding in the same way with the parameter $\lambda_{s\perp}$

$$\lambda_\perp = \frac{\Delta y}{y} = \frac{n\lambda m}{y\sin\theta_n}\sin\gamma \qquad (49)$$

it yields to

$$\frac{U_s(\lambda_\perp)}{\lambda_\perp} = \left|\frac{U_s(\lambda)}{\lambda}\right| + \left|\frac{U_s(x)}{y}\right| + \left|\frac{U_s(\gamma)}{\tan\gamma}\right| + |F|U_s(\theta_n) \qquad (50)$$

Introducing the same values of the variables used for $\lambda_{s||}$, Eq.(50) gives

$$\frac{U_s(\lambda_\perp)}{\lambda_\perp} = 0.00001 + 0.005 + 0.017 + 0.047 = 0.065,$$

that is, 6.5%. The main contribution to the uncertainty of $\lambda_{s||}$ and $\lambda_{s\perp}$ is the term corresponding to the diffraction angle θ_h. This means that the systematic error of λ_s could be minimized by choosing a lens with a smaller F-number.

Once the uncertainties of $\lambda_{s||}$ and $\lambda_\perp$ are known, the value of λ_s is obtained by means of the equation (43), resulting in

$$U_s(\lambda_s) = \frac{2}{3}\left(\left|U_s(\lambda_{s||})\right| + \left|U_s(\lambda_{s\perp})\right|\right) = 3.4 \times 10^{-6},$$

and its relative uncertainty is

$$\frac{U_s(\lambda_s)}{\lambda_s} = 0.064 = 6.4\%.$$

3. SURFACE ANALYSIS BY SPECKLE CORRELATION

3.1. Surface Roughness

In different fields of science and industry the properties of the surface of materials are of great importance. Processing changes surface roughness, which affects the final shape of the products. For example, in machinery pieces such as gears and bearings, or the milling cutters, the characteristics of the surface play an important role affecting the operation of the system. In architecture the surface finish (polish) of ornamental stones (such as granite, marble, etc.) is of crucial importance, not only because of esthetical reasons, but also for its daily use. In civil engineering surface wear, fluid flow in pipes (for oil conduction, for example) and functioning of vacuum seals are examples in which roughness has a large influence. On the other hand, in the context of material technology, the state of the surface is crucial since properties such as corrosion resistance and cleaning ability of the metal surfaces are dependent on surface roughness [117]. But this importance is not limited to these industrial sectors. Thus, in medicine it has been demonstrated that rougher surfaces improve an implant's ability to attach to bone tissue [118]. In the case of the pharmaceutical industry the processing of tablets by compression depends on particle size and surface roughness [119]. Industrial production in chain requires that all the specimens have the same characteristics within the tolerance allowed. The products must satisfy some standard requirements to be used, and at the same time, they must preserve the criteria of quality and safety. When there are defects in the production, the roughness is one of the most variable parameters. For this reason, many different experimental techniques have been developed for its characterization and analysis.

There are different methods to measure roughness. The most conventional technique is the profilometer, which is based on the registration (usually electronic) of the microscopic surface heights by means of a fine tip. Taking a large amount of data, and through its statistical processing, it is possible to measure parameters such as Ra, Rq, skewness, and kurtosis, among others. This method has proved to be very useful, but the displacement of the stylus along many parallel lines on the surface under study is time consuming. In addition, as a method of contact, it is very sensitive to the relative position between the stylus and the sample and may yield results that deviate significantly from the actual values. Other disadvantages of this procedure are the inability to use it on soft surfaces where the contact pressure may cause damage or in machinery where the pieces should be examined in-situ. In this context, optical methods play an important role because they are non-contact, thus, there are no risks, and, moreover, they are quick and precise. Among them, optical techniques based on the phenomenon of speckle have been used in the past twenty years because of its advantages.

It is well known that there are different procedures based on speckle interferometry for measuring roughness [120-131]. Basically, there are two general procedures in the branch of the speckle techniques: devices operating with a wavelength but varying the incidence angle of the radiation and those that use two different wavelengths without modification of the geometrical incidence of the laser beam. This section deals with the angular speckle correlation technique (ASC). This method belongs to the first category and it has the advantages of other optical techniques, but also some disadvantages. For instance, it requires an accurate alignment of the optical system, and it is very sensitive to environmental vibrations.

The aim of this work is the investigation of the influence of corrosion on the surface roughness of a metallic material.

3.2. The Problem of Corrosion

Corrosion is a great problem in many fields of industry. The appearance of the corrosion causes great damage to materials, which decreases the lifetime of the structures. Pipes, engines, bridges, buildings, aircrafts, parts of underwater steel structures (ships hulls, piles) [132], buried and concrete structures, welding, platforms, and dam walls are some examples in which corrosion is a big problem. For this reason, many techniques have been developed for the protection of structures against corrosion, which, on the other hand, involves economic savings.

Corrosion is defined as "the deterioration of a material as a result of electrochemical attack by their environment." Many materials are unstable when exposed to corrosive atmospheres (oxygen, salt, carbon dioxide, etc.). This process can be understood as the general trend of the materials to adopt the most stable state that corresponds to the minimal amount of energy.

The process of corrosion can occur in several complex forms. Thus, the way in which a material is corroded depends on its specific physical and chemical characteristics, its shape, and the substances present in the environment. Some of the most common types of corrosion that we can distinguish are, uniform or generalized, galvanic, pitting, crevice, graphitic, concentration cell, and microbiologically influenced corrosion, respectively. Each type has its particularities, the causes of its appearance are very different, and the explanation of all phenomena involved is difficult. As we have commented, the objective of this section is the study of the relation between corrosion and roughness. For this reason it is necessary to first establish a method to quantify the degree of corrosion of a material and then to investigate its relation with the roughness. There are different procedures for analyzing corrosion. A possibility is by means of the impedance method which uses piezoelectric crystals [133]. Due to the electrical impedance of the piezoelectric sensor being coupled with the mechanical impedance of the structure to be analyzed, any change in the mechanical impedance in the sample to be analyzed leads to variations in the measured electrical impedance of the piezoelectric crystal. Therefore, by measuring the electrical impedance of the piezoelectric material, damage may be observed by studying the impedance changes between different states of a specimen. Recently, a new technique that uses electrodes has been developed. This method is based on the measurement of noise caused by fluctuations in the electrochemical reactions that occurs in the corrosion process [134]. This procedure is able to differentiate between various kinds of localized corrosion, such as crevice, pitting, and stress corrosion. However, optical methods based on interferometry have been also applied to study and quantify the degree of corrosion. For example, double-exposure holographic interferometry has been employed for the measurement of the degree of corrosion in metallic samples [135,136]. Speckle interferometric tecniques were employed to determine the stress corrosion cracking of stainless steel in hydrochloric acid solution [137]. Furthermore, a procedure based on the mass loss of the sample to be analyzed has been investigated. This technique is based on the holographic interferometry, and it is able to analyze the degree of corrosion by means of the interference fringes that appear in the experiment [138]. Holographic interferometry was also applied to measure other parameters related to corrosion. For instance, this procedure was used to study the variation of the corrosion current density of a metal in aqueous solutions [139]. Speckle method techniques were also applied to determine the etch depth in metallic samples [140]. The changes in the microtopography of a metal surface during a corrosion process are also measured by decorrelation of the scattered speckle fields under coherent illumination [141]. Detection of very small cracks and cavities in materials has been determined using laser diffraction patterns [142]. These small cracks behave as diffraction apertures. By analyzing the intensity patterns obtained when a laser beam is diffracted by these small apertures, the characteristic of these defects are determined. The subjective speckle method has been employed to analyze the processes of stress corrosion in microregions by measuring the local surface roughness of the specimens [143]. Employing this procedure, the initiation stage of stress corrosion of an aluminium alloy specimen was investigated. Other methods based on the speckle phenomenon, but by digital processing, have been also employed. So, electronic speckle pattern interferometry (ESPI) and digital speckle correlation (DSPI), are used to detect the existence of crevice corrosion, which is usually undetectable from outside by visual inspection [144]. The ESPI setup is configured to measure out-of-plane displacement and not only the location of defects, but also their approximate sizes. Digital speckle correlation technique takes advantage of laser speckle properties: the speckle decorrelation due to out-of-plane displacement in the region with crevice corrosion is greater than without defects. The use of DSPI is also suitable for monitoring oxide layers growing in metallic structures [145]. Corrosion resistance of various industrial components depends on the performance of paints, and ESPI is a useful technique in detection of decohesion of some paints that are used in the industry [146] Within the speckle methods, speckle photography has been also employed to determine the influence of corrosion on the elastic properties of metallic bars [147].

3.3. Theory

Speckle correlation techniques consist of calculating the correlation between two speckle patterns obtained, either for two different illuminating angles (angular speckle correlation-ASC) or for two wavelengths (spectral speckle correlation-SSC). By ASC a coherent beam is directed on the rough sample, under an angle α, and is referred to as the first exposition; the second pattern is registred in the same way but the incidence angle of the beam is angularly displaced by a quantity $\delta\alpha$. Assuming a Gaussian distribution for the surface heights, the correlation of the speckle fields for two angles in the Fourier plane is [122,123]

$$C(\delta\alpha, f, k, \sigma_h \Delta\xi, \Delta\eta) = \exp\left[-\left(\left(\sin\alpha_i\delta\alpha + \sin\alpha_o \frac{\Delta\xi}{f}\right)k\sigma_h\right)^2\right]\exp\left[-\left(\frac{Lk}{2f}\right)^2\left(\left(\Delta\xi - \frac{\cos\alpha_i}{\cos\alpha_o}f\delta\alpha\right)^2 + (\Delta\eta)^2\right)\right], \quad (51)$$

where σ_h is the variance of the surface, L is the half-width of the beam, f the focal distance of the lens, k the wave number of the laser beam, , and α_i and α_o the angles of incidence and observation, respectively. The coordinate frame is referred to by ξ and η on the Fourier plane (CCD array- Figure 29). To find a simple relation between roughness and correlation, the following conditions must be fulfilled:

$$\Delta\xi - \frac{\cos\alpha_i}{\cos\alpha_o}f\delta\alpha = 0 \quad (52)$$

and

$$\Delta\eta = 0. \quad (53)$$

In our case we set, $\alpha_i = \alpha_o$ (see next section), then Eq. (52) becomes

$$\Delta\xi = f\delta\alpha, \quad (54)$$

and the correlation of the two patterns

$$C = \exp\left(-(2k\sin\alpha_i\delta\alpha\sigma_h)^2\right), \quad (55)$$

which depends directly on the variance σ_h, and k and the incidence angles that are controlled in the experiment. Eq. (52) and Eq. (53) show that the speckle pattern shifts by an amount

$f\delta\alpha$ as a consequence of change the incidence angle for the laser beam. This means that we must move the position of the camera in order to find the maximum correlation value of the two speckle fields. This is presented schematically in Figure 28. The first exposure is made and registered by the camera under the incidence angle α. It corresponds to the pattern A1 (Figure 28). The second exposure is performed, but by changing the illuminating angle by an amount $\delta\alpha$ (B1 in Figure 28).

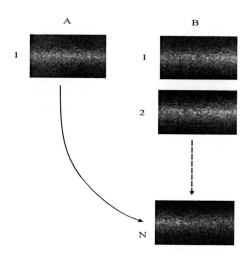

Figure 28. Speckle patterns for the first (A) and second illuminating angles (B), respectively. Observe that it is necessary to register a series of N patterns for different location of the CCD camera.

The correlation of these two speckle fields is very poor, and it does not fulfil the conditions imposed by the Eq.(52) and Eq.(53). In order to carry out the experiments correctly, the camera must be displaced (Eqs. 2,3). This modus operandi is explained in Figure 28. A series of exposures must be registered (B2, B3,…,BN) and correlated with the pattern A1. By displacing the system a small amount (as small as possible) each time, the maximum value of the correlation may be found and then the roughness of a sample determined.

3.4. Experimental Layout and Results

Figures 29 and 30 show the optical setup used for the measurements. A sample of iron was tested. A laser LS was directed onto a beam splitter which divided the radiation in two parts. The first one (1) passed through a variable reflective mirror and struck on the sample. While performing this experiment, the shutter (SH) closed the second laser beam (2). A part of the light scattered by the surface was collected by the lens and registered on the CCD array. For the second exposure the procedure was the same, but the shutter (SH) hampered the first beam; then the second beam (2) was reflected in the mirror M and reached the specimen S. In order to regulate the intensity of each beam, a gradual mirror GM was tuned in both expositions (Figure 29).

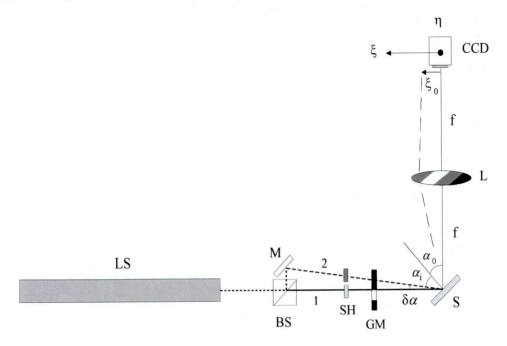

Figure 29. LS, laser beam; BS, beam splitter; M, Mirror; SH, shutter; GM, gradual mirror; S, sample; L, lens; CCD, camera; $\delta\alpha$, difference angle; f, focal length; ξ and η, CCD reference frame; ξ_0, displacement of the CCD camera.

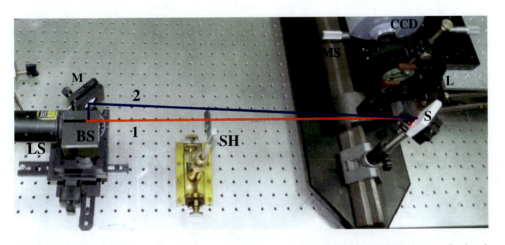

Figure 30. Experimental device. LS, laser beam; BS, beam splitter; M, Mirror; SH, shutter; S, simple; L, lens; CCD, camera; $\delta\alpha$ (Photo courtesy of F.Ortega).

With the aim of studying the influence of corrosion on the surface roughness, a specimen of iron was tested. However, due to corrosion being a phenomenon that occurs very slowly over time, we produced artificial corrosion in the laboratory through a chemical reaction with nitric acid. By this procedure, the aforementioned process was accelerated, then allowing for the investigation. This type of corrosion is known as hydrogen type. The specimen of iron was attacked with nitric acid solutions of several concentrations. The corrosion level was determined by the measurement of the mass loss of the sample. The roughness of the iron piece was measured first, without chemical treatment. Then, it was corroded by the procedure explained above and its roughness measured. This process was repeated seven times, obtaining the experimental values shown in the Figure 31. The upper curve depicts the values measured of R_q versus the degree of corrosion for a specimen of iron (blue graphic). The first value (denoted as 0 in the abscissa axis) corresponds to the sample without corrosion and it is for us the initial state. After attacking the piece with diluted acid, the values of the roughness changed. The curve shows a non-linear behavior, i.e., it grows when increasing corrosion to a value of mass loss of 2.37%, but from this point roughness varies slowly and is around a constant value. This means that the roughness increment happens in the first stage of appearance of corrosion, when the material surface has not yet been fully attacked. When the action of acid increases, the surface becomes rougher, but when all material has been strongly attacked the surface reaches a maximum of roughness. From this moment the chemical solution continues to erode the material surface, but the roughness does not vary significantly. To test these results, two complementary experiments were made. In the first one, the roughness of the sample was measured with a confocal microscope. By this technique the specimen is scanned vertically by a light beam in many steps, so that the light reflected from a small part of the surface passes through a pinhole to the focus plane of a CCD camera. The surface height at each pixel location on the detector is found by collecting the peak of the narrow axial response.

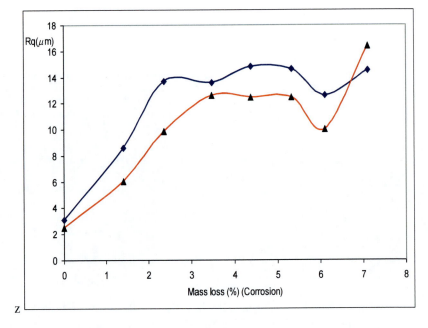

Figure 31. Values of roughness versus degree of corrosion for a sample of iron. The blue graphic (squares) represents the values obtained by the ASC method, and the red curve (triangles) corresponds to the results be means of the confocal microscope.

The final confocal image is formed by recording the information of different acquisition planes during the raster process. The advantage of this method of measurement is that it gives the highest lateral resolution that may be achieved by an optical profiler, at least within the FOV of the objective used. In order to have high resolution, an appropriate selection of both the magnification objective and the number of planes is needed. With this system a resolution with highs better than 4 nm rms is possible. The results obtained by this technique for R_q are shown in the lower curve of Figure 31. Figure 32 corresponds to the same analysis but for the parameter R_a. As observed in Figure 31, when comparing the results determined by both methods, there are some differences between them, however the trend of both curves is very similar. The first reason refers to the systematic uncertainty of the speckle method. This is of the order of 14.2%, for the conditions under which the experiments were conducted. If we take this fact into consideration, the curve measured by confocal microscopy is within the range of uncertainty of the ASC technique, approximately. Thus, both results are coherent. The second additional reason that may explain the differences has to do with the area of the sample studied. In fact, comparing both procedures shows that the surface examined by the speckle method was about 3 mm in length. However, the region scanned by the microscope was 1.2 mm. The second experiment to test the veracity of the results was carried out with an electron microscope (500x magnification). A microscopic analysis of the sample for each state of corrosion was made. Figure 33 shows the results obtained for the first phase of corrosion process. In Figure 33.a appears the original specimen before corrosion occurs, sample 0. In 33.b the surface is attacked with diluted acid (1.41% mass loss), and it shows the appearance of the first signs of corrosion. Figure 33.c presents the next state, when attacking the material for more time. Comparing these three photos corresponding to the evolution of the material, they depict the growing of cavities on the surface. These stages correspond to the principle of the graphic (32) in which the curve is increasing. Figure 34.a exhibits the surface topography when the mass loss is 3.47, 4.38, and 5.33%, respectively. Comparing the photos, (a), (b), and (c) of Figure 34, it is noted that the grooves are similar in shape and are also distributed homogeneously. The only difference between them is the location of the holes. Figure 35 corresponds to the least two points represented in the graphic (32).

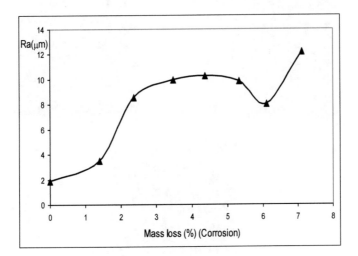

Figure 32. Results for R_a measured by the confocal microscope.

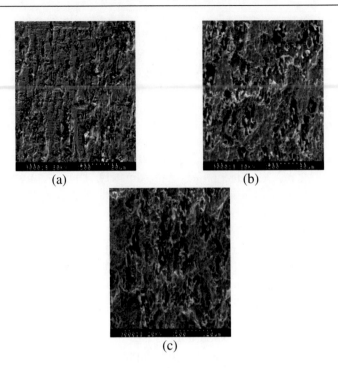

Figure 33. Sample of iron alloy photographed with SEM: (a) Initial state (sample 0). Not attacked with nitric acid (first value of the graphic 31- squares). (b) Corroded with acid diluted (mass loss of 1.41%). (c) Attacked (2.37%).

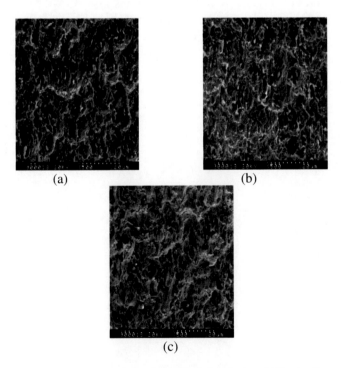

Figure 34. Photographs corresponding to the experimental data 3, 4 and 5 in the figure 32. (a) Mass loss 3.47%. (b) Corroded. Mass loss of the sample, 4.38%. (c) Idem, 5.33%. Observe that no variation in the appearance of the surface is appreciated. The grooves have the same shape.

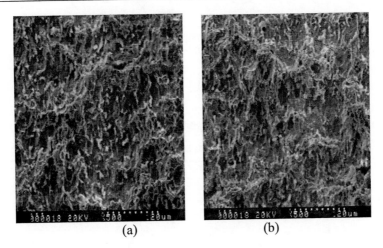

(a)   (b)

Figure 35. Surface of the specimen for the last two points in Figure 32. (a) Mass loss, 6.11%. (c) Mass loss, 7.11%.

Analyzing the photos of this figure, no variation on the surface may be observed either. This confirms the experimental results obtained. Therefore, from the state of corrosion corresponding to the inflection point (Figure 32 and Figure 33) (3.37%), the only effect of the external attack on the surface is an increase in the corrosion/erosion, but roughness remains the same, approximately.

3.5. Precision of the Method

In the same way as used in other sections, we calculate the experimental systematic errors of the method employed for measuring roughness. With this goal, the theory of uncertainties [46-48] is applied to the expression of σ_h, adopting the following form:

$$U_s(\sigma_h) = \sum_{j=1}^{N} \left| \frac{\partial \sigma_h}{\partial x_j} \right| U_s(x_j) \tag{56}$$

where x_j are the different variables of the function σ_h. Taking into consideration Eq. (55), the value of the variance of the surface is

$$\sigma_h = \frac{\lambda \sqrt{-\ln C}}{4\pi \delta \alpha \sin \alpha} \tag{57}$$

Applying Eq.(56) to this equation, we have

$$U_s(\sigma_h) = \left| \frac{\sqrt{-\ln C}}{4\pi\delta\alpha \sin \alpha} \right| U_s(\lambda) + \left| \frac{\cos\alpha \sqrt{-\ln C}}{4\pi\delta\alpha(\sin\alpha)^2} \right| U_s(\alpha) + \left| \frac{\lambda\sqrt{-\ln C}}{4\pi(\delta\alpha)^2 \sin\alpha} \right| U_s(\delta\alpha) \tag{58}$$

and its relative uncertainty:

$$\frac{U_s(\sigma_h)}{\sigma_h} = \left|\frac{U_s(\lambda)}{\lambda}\right| + \left|\frac{U_s(\alpha)}{tg\alpha}\right| + \left|\frac{U_s(\delta\alpha)}{\delta\alpha}\right|. \tag{59}$$

For the conditions in which the experiments were conducted we have $\delta\alpha = 9\times10^{-3}$ rad, $U_s(\delta\alpha) \approx 1.2\times10^{-3}$ rad, $U_s(\lambda)=0.1$ nm and $U_s(\alpha)= 1.7\times10^{-2}$ rad. Using these values Eq. (59) gives

$$\frac{I_s(\sigma_h)}{\sigma_h} = 1.6\cdot10^{-4} + 1.7\cdot10^{-2} + 124\cdot10^{-2} \approx 14.2\cdot10^{-2} = 14.2\%.$$

It can be noted that in the least addition the third term contributes more to the uncertainty than the rest. It corresponds to the angular difference between incident laser beams. To diminish this error and improve the method, it would be necessary to have more precision in the measurement of the angular difference $\delta\alpha$.

ACKNOWLEDGMENTS

I want to dedicate this work to the memory of my parents, Carmen and Félix, and to the memory of my uncle Ramón.

REFERENCES

[1] Landau L.D., Lifshitz E.M., *Theory of elasticity*; Course of Theoretical Physics; Pergamon: Oxford, UK, (1975); Vol 7, Chap.1,2.
[2] Love A.E.H., *A treatise on the mathematical theory of elasticity*, Dover: New York, USA, (1944), Chap. 1-7,15,22,24.
[3] Timoshenko S. P., Goodier J. N., *Theory of elasticity*; International Student Edition; McGraw-Hill Kogakusha: Tokyo, Japan, (1970), Chap. 1-3,7,9.
[4] Chuen-Lin T., Cheng-Chung L., Kie-Pin C., Cheng-Chung J., (2000) *J. Mod. Opt.* 47, 1681-1691.
[5] Chuen-Lin T., Cheng-Chung L., Yu-Lung T., Wen-Shing S., (2001) *Opt. Commun.* 198, 325-331.
[6] Zaslansky P., Currey J. D., Friesem A. A., Weiner S., (2005) *J. Biomed. Opt.* 10 024020.
[7] Morimoto Y., Nomura T., Fujigaki M., Yoneyama S., Takahashi I., (2005) *Experimental Mechanics* 45, 65-70.
[8] Voronyak T. I., Kmet A. B., Muravs'kyi1 L. I., (2009) *Materials Science* 45, 372-377.
[9] Chang M., Hu C. P., Lam P., Wyant J. C., (1985) *Appl. Opt.* 24, 3780-3783.
[10] Creath C., Cheng Y. Y., Wyant J. C., (1985) *Optica Acta.* 32, 1455-1464.

[11] Creath C., in *Progress in Optics*; Wolf. E.; Ed.; XXVI; Elsevier Science Publishers B.V.: Amsterdan, The Netherlands (1988) 349-393.
[12] Hack E., Frei B., Kästle R., Sennhauser U., (1998) *Appl. Opt.* 37, 2591-2597.
[13] Creath C., (1987) *Appl. Opt.* 26, 2810-2816.
[14] Wang L.S., Krishnaswamy S., (1996) *Meas. Sci. Technol.* 7, 1748–1754.
[15] Berger E, von der Linden W, Dose V, Jakobi M, Koch AW, (1999) *Appl. Opt.* 38 4997-5003.
[16] Voronyak T. I., Kmet A. B., Muravs'kyil L. I., (2009) *Materials Science* 45, 372-377.
[17] Yamaguchi I., Jun-ichi K., Matsuzaki H., (2003) *Opt. Eng.* 42, 1267-1271.
[18] Yamaguchi I., Ida T., Yokota M., (2008) 44 *Strain*, 349-356.
[19] Owner-Petersen, M., Damgaard Jensen, P., (1988) *NDT int* 21, 422-426.
[20] Möltner T., *PhD Thesis* Technische Universität München, Germany (2000).
[21] Brozeit A., Hinsch K.D., (1996) *Proceedings of SPIE* Vol 2860, 144–149.
[22] Ishii Y., Onodera R., (1991) *Optics Letters* 16, 1523-1525.
[23] Wiese P., *PhD Thesis* RWTH Aachen, Germany (1996).
[24] Angel J.R.P., Wizinowich P., (1988) *ESO Proceedings* Vol 30, 561–567.
[25] Wizinowich P.L., *Applied Optics* (1990) 29, 3271-3279.
[26] Tang S., (1996) *Proceedings of Laser Interferometry VIII: Techniques and Analysis* SPIE, Vol 2860, 33-44.
[27] Guzhov V.I., Nechaev V.G., Mishina E.M., (1997) *Optoelectronics, Instrumentation and Data Processing* 4.
[28] Hibino K., Oreb B.F., Farrant D.I., Larkin K.G., (1997) *J. Op. Soc. Am.* A 14, 918–930.
[29] Zhu Y., Gemma T., (2001) *Applied Optics* 40, 4540–4546.
[30] Kao C., Yeh G., Lee S., Lee C., Yang C., Wu. K., (2002) *Appl. Opt.* 41, 46–54.
[31] Huang P.S., Hu Q.J., Chiang F., (2002) *Appl. Opt.* 41, 4503–4509.
[32] Novak J., (2003) *Optik* 114, 63-68.
[33] Schwider J., Falkenstorfer O., Schreiber H., Zoller A., Streibl N., (1993) *Optical Engineering* 32, 1883-1885.
[34] Carré P., (1966) *Metrologia* 2, 13–23.
[35] Hariharan P., Oreb B.F., Eiju T., (1987) *Appl. Opt.* 26, 2504-2505.
[36] Surrel Y., (1993) *Appl. Opt* 32, 3598-600.
[37] Schmit J., Creath K., (1995) *Appl. Opt.* 34, 3610–3619.
[38] de Groot P., *Appl. Opt.* (1995) 34, 4723-4730.
[39] Hibino K., Oreb B. F., Farrant D. I., Larkin K. G., (1997) *J. Op. Soc. Am.* A 14, 918-930.
[40] Zhang S., Yau S., (2007) *Optical Engineering* 46, 113603-1-6.
[41] Huang P.S., Zhang S., (2006) *Applied Optics* 45, 5086–5091.
[42] Timoshenko S., Woinowsky-Krieger S., *Theory of plates and shells* McGraw-Hill: New York USA.
[43] Mansfield E. H., *The bending and stretching of plates* (1989) Cambridge University Press, Cambridge USA (1959), Chap. 1,2,4-6.
[44] Kang F. and Zhong-Ci S., *Mathematical Theory of Elastic Structures*, Springer Verlag, Berlin Germany (1991), Chap. 2, 4-6.
[45] Sader J. E. and White L., (1993) *J. Appl. Phys.* 74, 1-9 (See references therein).
[46] Taylor, J.R., *An introduction to error analysis*, University Science Books, Mill Valley CA (1982), Chap. 1-3.

[47] Squires G. L., *Practical physics* Cambridge University Press, Cambridge UK (1989), Chap. 2-5.
[48] Schrüfer E., *Elektrische Messtechnik* Hanser München Germany (2009), Chap. 1.
[49] Mallick S., (1973) *Nov. Rev. Optique* 4, 267-272.
[50] Cloud G., (1975) *Appl. Opt.* 14, 878-884.
[51] Chiang Fu-Pen., Juang Ren-Ming., (1976) *Appl. Opt.* 15, 2199-2204.
[52] Khetan R. P., Chiang F. P., (1976) *Appl. Opt.* 15, 2205-2215.
[53] Hung, Y. Y. In *Speckle Metrology*; Erf R. K.; Ed, Academic Press, New York USA (1978), Chap 4.
[54] Brdicko, J., Olson, M.D., Hazell, C. R., (1979) *Experimental Mechanics* 19, 160-165.
[55] Chiang, F. P., Adachi, J., Anastasi, R., Beaty, J., (1982) *Opt. Eng.* 21, 379-390.
[56] Holoubek J., Sedláček B., (1984), *J. Macromol* CI.-PHYS., B23(1), 143-152.
[57] Chiang, F. P. and Li, D. W. 1985, *Opt. Eng.* 24, 937-944.
[58] Li, D. W., Chen, J. B., and Chiang, F. P. 1985, *J. Opt. Soc. Am.* A, 2, 657-666.
[59] Schwieger, H., und Streubel, R. 1983, *Materialprüf.*, 25,105-112.
[60] Chiang, F. P., and Li, D. W., (1985) *Opt. Eng.* 24, 936-943.
[61] Tippur, H.V., Chiang F-P., (1991) *Appl. Opt.* 30, 2748-2756.
[62] Gascón, F., and Salazar, F. 1996, *Opt. Commun*, 123, 734-742.
[63] Chiang F. P., Anastasi R., Beaty J., Adachi J., (1980) *Appl. Opt.*, 19, 2701-2704.
[64] Salazar F., Gascón F., (1993) *Opt. Commun.* 103, 235..
[65] Gascón F., and Salazar, F., (1993) *Rev. Sci. Ins.* 64, 2241-2244.
[66] Salazar F., and Gascón, F., (1995) *Opt. Commun.*, 119, 361-372.
[67] Vikram C. S., (1997) *Meas. Sci.Technol.*, 8, 1156-1159.
[68] Pickering, C.J.D., Halliwell, N.A. (1984) *Appl. Opt.*, 23, 2961-2969.
[69] Hinsch, K.D., In *Speckle Metrology*, Siorhi R.S.; Ed.;, Marcel Dekker: New York USA (1993), Chap. 6.
[70] Hinsch, K., (1989) *Appl. Opt.* 28, 5298-5204.
[71] Dainty, J. C., In *Laser speckle and related phenomena,* Dainty J.C.; Ed.; Vol 9 *of Topics in Applied Physics of Topic in Applied Physics*; Springer Verlag: Berlin Germany (1984), 255-320 (see references therein).
[72] Huntley, J. M., Field J. E., (1986) *Appl. Opt.* 25, 1665-1669.
[73] Tiziani H. J., (1972), *Opt. Commun.* 5, 271276.
[74] Françon M., In *Laser speckle and related phenomena,* Dainty J.C.; Ed.; Vol 9 *of Topics in Applied Physics of Topic in Applied Physics*; Springer Verlag: Berlin Germany (1984), 172-202.
[75] Erf R. K., In *Speckle Metrology*, Erf R. K.; Ed.; Academic Press: New York USA (1987), Chap. 9.
[76] Farrel P. V., Hofeldt D. L., (1984) *Appl. Opt.* 23, 1055-1059.
[77] Kihm K. D., Cheeti S. K. R., (1994) *Exp. Fluids* 17, 246-252.
[78] Otrovsky Y. I., Shchepinov V. P., In *Speckle Metrology*, Siorhi R.S.; Ed.; Marcel Dekker, New York USA (1993), Chap. 11.
[79] Salazar F., Bayón A., Chicharro JM., (2009) *Opt. Commun.*, 282, 635-639.
[80] Crane R. B., (1970) *J. Opt. Soc. Am.* A 60, 1658-1663.
[81] Tribillon G., (1974) *Opt. Commun.* 11, 172-174.
[82] Léger D., Mathieu E., Perrin J. C., (1975) *Appl. Opt.,* 14, 872-877.
[83] Pedersen H. M., (1976) *J. Opt. Soc. Am.* A, 66 1204-1210.

[84] Léger, D., Perrin J. C., (1976) *J. Opt. Soc. Am.* A 66, 1210-1217.
[85] Kadono H., Asakura T., Takai N., (1987) *Appl. Phys.* B, 44, 167-173.
[86] Asakura T., In *Speckle Metrology*, Erf R. K.; Ed.; Academic Press: New York USA (1987), Chap. 3.
[87] Tay C. J., Toh S. L., Shang H. M., Zhang J., (1995) *Appl. Opt.* 34, 2324-2335.
[88] Cheng C., Liu C., Zhang N., Jia T., Li R., Xu Z., (2002) *Appl. Opt.* 41, 4148-4156.
[89] Gascón F., Balbás M., (1986) *Boletín Geológico y Minero* XCVII-VI, 793-802.
[90] Balbás M., Fraile D., Gascón F., Varadé A., Vilarroig P., (1989) *Appl. Opt.* 28, 5065-5068.
[91] Balbás M., Fraile D., Gascón F., Varadé A., Vilarroig P., (1992) *Appl. Opt.* 31, 876-880.
[92] Fraile D., Gascón F., Varadé A., (1992) *Appl. Opt.* 31, 7371-7374.
[93] Salazar F., and Gascón, F., (2003), *Opt. Commun.* 221, 279-288.
[94] Nye J. F., *Physical properties of crystals*, Clarendon Press: Oxford UK (1990), Chap. 9.
[95] Myers, R. H., *Classical and modern regression with applications*, PWS-KENT: Boston USA (1990), Chap. 2, 3.
[96] Gascón, F Salazar F., (2003), *Opt. Commun.*, 172, 77-84.
[97] Fraile D., *PhD Thesis* Politechnic University of Madrid, Spain (1988).
[98] Cullity B. D., *Introduction to magnetic materials*, Addison-Wesley: Massachusetts USA (1972), Chap. 8.
[99] Chikazumi S., *Physics of Magnetism*, Krieger: Malabar USA (1964), Chap. 8.
[100] E. Trémolet, *Magnetostriction*: Theory and applications of magnetoelasticity, CRC Press: Florida USA (1993), Chap. 3.
[101] Squire P. T., (1994) *Mess. Sci. Technol.* 5, 67-81.
[102] Vlasák G., (2000) *J. Mag. Mag. Mat.* 215-216, 479-481.
[103] Xuesong J., Kim C.O., Lee Y.P., Zhou Y., Huibin Xu, (2001) *Appl. Phys Lett.* 79, 650-652.
[104] Rotter M., Müller H., Gratz E.,Doerr M., Loewenhaupt M., (1998) *Rev. Sci. Instrum.* 69, 2742-2746.
[105] Tam A. C., Schroeder H., (1989) *IEEE trans. Mag.* 25, 2629-2638.
[106] Bellesis G. H., Harlle P. S., Renema III, A., Lamberth D. N., (1993) *IEEE Trans. Magn.* 29, 2989-2991.
[107] Harlle III P. S., Bellesis G. H., Lamberth D. N., (1994) *J. Appl. Phys.* 75, 6884-6886.
[108] Nakata T., Takahashi N., Nakano M., Muramatsu K., Miyake M., (1994) *IEEE Trans. Magn.* 30, 4563-4565.
[109] Gates L., Stroink G., *Rev. Sci. Instrum.* 63 (1992) 2017-2021.
[110] Chen Y., Kriegermeier-Sutton B. K., Snyder J. E., Dennis K. W., McCallum R. W., Jiles D. C., (2001) *J. Magn. Magn. Matter.* 236, 131-138.
[111] Osborn J.A., (1945) *Phys. Rev.* 67 (11-12), 351-357.
[112] Stoner E.C., (1945) *Phil. Mag.* 36 (7) 803-821.
[113] E. W. Lee, (1955) Magnetostriction and magnetomechanical effects, *Repts. Progr. in Phys.* 18, 184-229.
[114] Joseph R. I., Schlömann E., (1965) *J. Appl. Phys.* 36, 1579-1584.
[115] Zheng G., Pardavi-Horath M., Huang X., Keszei B., Vandlik J., (1996) *J. Appl. Phys.* 79, 5742-5744.

[116] Du-Xing C., Pardo E., Sánchez A., (2002) *IEEE Transactions on Magnetics* 38, 1742-1752.
[117] Cochrane, D.J., (2000) Symposium Stainless steel in Architectur, Berlin, Germany (www.euroinox. org/pdf/paper/Cochrane_EN.pdf).
[118] Suzuki, K., Aoki, K., Ohya, K., (1997) *Bone* 21(6), 507-514.
[119] Eiliazadeh B., Briscoe B.J., ShengY., Pitt K., (2003) *Particul.Sci.Technol.* 21, 303-316.
[120] Goodman J. W., (1975) *Opt. Commun.* 14, 324-327.
[121] Glio M., Musazzi S., Perini U., (1979) *Opt. Commun.* 28, 166-170.
[122] Ruffing B., (1986) *J. Opt. Soc. Am.* A 3, 1297-1304.
[123] Ruffing B., *PhD Thesis* University of Karlsruhe, Germany (1987).
[124] Pearson U. (1993) *Wear* 160, 221-225.
[125] Spagnolo G. S., Paoletti D., (1996) *Opt. Commun.* 132, 24-28.
[126] Pérez Quintián F., Rebollo M. A., Nogert E. N., Landau M. R., Gaggioli N. G., (1996) *Opt. Eng.* 35, 1175-1178.
[127] Lehmann P., Patzelt S., Schöne A., (1997) *Appl. Opt.* 36, 2188-2197.
[128] Tay C. J., Toh S. L., Shang H. M., and Zhang J.: "Whole-field determination of surface roughness by speckle correlation", *Appl. Opt.* vol. 34, 1995, pp. 2324-2335.
[129] Dalmases F, Cebrián R, Buendía M, Romero C, Salvador R, Montilla J., *Phys. Med. Biol.*, vol. 33, 1988, 913-922.
[130] Yamaguchi I., Kobayashi K., and Yaroslavsky L., (2004) *Opt. Eng.* 43, 2753-2761.
[131] Persson U., *J. Mater. Process. Tech.* 80, 2006, 233-238.
[132] Guibert A., Chadebec O., Coulomb J.-L., Rannou C., (2009) *IEEE Trans. Mag.* 45, 1828-1831.
[133] Park, G., Sohn, H., Farrar, C., Inman, D., (2003) *Shock and Vibration Digest*, 35, 451-463.
[134] Cottis B., (1999) *Materials World* 7, 482-483.
[135] Petrov K.N., Presnyakov, Yu.P., (1978) *Opt. Spectrosc.* (USSR) 44, 309-311.
[136] Vikram C.S, Vedam K., (1980) *Optik* 55, 407-414.
[137] Huang Y., (2002) *Bulletin of Materials Science*, 25 , 47-51.
[138] Sajan M.R., Radha T.S. Ramprasad B.S., Gopal E.S.R., (1991) *Optics and Lasers in Engineering* 15, 183-188.
[139] Habib K., (1990) *Applied Optics* 29 , 1867-1868.
[140] Radha T.S., Ramprasad B.S. (1985), *J.Electrochem. Soc. India* 34, 149-150.
[141] Fricke-Begemann T., Gülker G., Hinsch K.D., Wolff K., (1999) *Appl. Opt.* 38, 5948-5955.
[142] Li X., Soh A.K., Huang C., Yang C.H., Shi H., (2002) *Optical Engineeering* 41, 1295-1307.
[143] Shi H., Ruan C., Li X., (2006) *Materials Science and Engineering* A 419 , 218-224.
[144] Jin, F., Chiang, F.P., (1995). *American Society of Mechanical Engineers*, Aerospace Division (Publication) AD 47, Structural Integrity in Aging Aircraft, 221.
[145] Mayorga-Cruz D., Padilla-Sosa P., Cerecedo-Núñez H.H. (2006) *Proceedings of SPIE-The International Society for Optical Engineering* 6341, Speckle06, 634122.
[146] Singh Raman R.K., Bayles R., (2006) *Engineering Failure Analysis* 13, 1051.
[147] Medina R., Salazar F., Pérez O., In *Speckle Photography and Speckle Interferometry to Mechanic Solid Problems*; Salazar F.; Ed.; Kerala, India (2008) Research Signpost, 65-76.

In: Interferometry Principles and Applications
Editor: Mark E. Russo

ISBN 978-1-61209-347-5
©2012 Nova Science Publishers, Inc.

Chapter 2

INTERFEROMETER BASED METHODS FOR RESEARCH OF PIEZOELECTRIC MATERIALS

Miroslav Sulc[*]

*Department of Physics, International Center for Piezoelectric Research,
Technical University of Liberec, Liberec, Czech Republic*

ABSTRACT

The applications of laser interferometry in research of piezoelectric materials are presented. Piezoelectric, electrostrictive, electro- optical and thermal expansion coefficients were measured by these methods.

The usage of the Michelson single interferometer to measure piezoelectric induced strain in bulk samples is described. The more sophisticated Mach – Zehnder double beam interferometer is used for the investigation of the piezoelectric thin film deposed on the substrate to compensate substrate vibration. Both homodyne interferometers are measuring amplitude of piezoelectrically induced sample vibrations. Use of the lock-in amplifier technique enables very small amplitudes of the order of 10^{-12} m to be measured. The experimental problems and limits of this method used for piezoelectric research were analyzed in detail.

The new system was developed for the measurement of coefficients in a wide temperature range. The fundamental interferometer measurement problems, such as temperature instability and mechanical vibration of the temperature chamber are solved. All parts of the Michelson interferometer are placed inside the temperature chamber, so the relative motion of interferometer parts is almost suppressed in this way. The new variation of the double beam interferometer with a minimum of elements was constructed and located inside the temperature chamber. Study of the phase transitions at relaxor crystals illustrates the benefit of this method.

The linear Pockels effect was investigated too, and electro-optic coefficients of $LiNbO_3$ crystal were measured in the setup of a Mach-Zehnder interferometer in the temperature range from 150 K to 330 K. The principal problem of this type of measurement is that electric fields induce not only refractive index change but also a change of the sample length along the path of the laser beam due the piezoelectric effect.

---

[*] miroslav.sulc@tul.cz.

This effect can induce the errors of electro-optical coefficients' measurement on the order of tens of percents. The new method was proposed and tested for direct compensation of this piezoelectric effect on both the Michelson and Mach-Zehnder interferometers' arrangement.

The interferometry method was also adapted to the thermal expansion coefficients measurement. Relatively small samples, with dimensions of about some millimeters, were investigated. The very small sample expansion was separated from holder movement by suitable interferometer arrangement. This method was successfully tested on well-known metallic materials at first and consequently applied to the PZT piezoelectric ceramics research.

1. INTRODUCTION

The importance of piezoelectric materials has increased in recent years. These materials are basic elements of many devices. We can observe a lot of various kinds of new applications of intelligent smart elements that are a realized coupling of electrical and mechanical or optical quantities. They are used as actuators, sensors, transducers, transformers, switches and so on [1-6]. Also the electro-optical materials have become widely used in most of the applications; in optoelectronic devices such as optical phase modulators, optical intensity modulator or optical switches; lenses of controllable focal length; adaptive optics; etc. [7,8]. The most important applications of electro-optical materials are in telecommunication [9].

An electric field applied on the sample can invoke displacement of ions in materials. It manifests itself as converse piezoelectric effect - observable macroscopic displacement of material surfaces, described as a strain. The electric field can also invoke an electro-optical effect. It changes the refractive index of electro-optical materials. It is important to know all material characteristics, such as piezoelectric and electro-optic coefficients, including the temperature characteristics of these materials in high-tech applications. Both the effect of piezoelectric induced displacement and the refractive index change are very small, and a very sensitive method must be used to investigate them.

The research of piezoelectric and electro-optical effects has been done by various methods. The resonant method, piezoresponse forces microscopy, elipsometry and set of next ones are very useful for progress in this field. The laser interferometry methods are also very useful. The effect of piezoelectric induced displacement is very small, obviously in range below the wavelength of light, and very sensitive methods must be used. They can be adapted for measurement of different kinds of materials of different compositions, shapes and dimensions, from very small thin layers with thickness below 1 μm to bulk 1 cm materials.

The first applications of interferometry for piezoelectric elements studies appeared eighty years ago, and some measurements were performed during the following years [10-13]. However, the real progress was connected with the use of an He-Ne laser and lock-in amplifier for this type of measurement. This Michelson interferometer was built in 1988 [14]. It was a very successful instrument for bulk materials research [15], but some disadvantages came at thin films measurement. The new double-beam Mach-Zehnder interferometer was arranged for compensation of vibrations of thin film substrate [16,17]. They were also improved [18] so they were able to detect the amplitude of piezoelectrically induced vibrations as small as 1 picometer. The above described SE interferometers were homodyne and were made from discrete optical elements. These types are presently the most used in piezoelectric oriented research. The heterodyne interferometers also had been used with similar sensitivity [19-24]. Variations of previous interferometers, using optical fibers as a light guide, started to be applied in later years [25-27].

We built an interferometer set-up at our laboratory at 1997. We adapted known interferometric method for our research, but some special experimental problems had to be solved. A detailed analysis of possible sources of errors was made. Afterward we tried to improve the accuracy of our interferometer. With a special arrangement of beam alignment, we have the possibility of measuring displacements of on the order of 10^{-12} m, with accuracy better than 10 %. This is very useful for the scanning of profile deformation of small electrodes deposed on piezoelectric thin films.

Some technical applications of piezoelectric materials are used in various temperature environments, from cryogenic to very high temperatures, so it is important for technicians to know how the properties of these materials can change in range of useful temperatures. This type of research is also important from a scientific point of view. It enables one to investigate motion of piezoelectric domains and to estimate this (external) influence on final piezoelectric response. Very interesting research of phase transitions can be done on studied materials, too. Investigation of piezoelectric coefficients in wide temperature ranges by interferometer was never done before. We built a miniaturized Michelson interferometer, and placed it inside a cryostat or temperature chamber. Therefore, this interferometer is very stable for the measurement of temperature dependency of the piezoelectric coefficient. It is very successful for research of phase transition in relaxor crystals. A new, special type of double beam interferometer was arranged for thin films measurement. It was also placed in the temperature chamber. It can compensate for substrate vibration and it works in a wide temperature range.

Interferometry method is evidently also useful for electro-optical effects measurements. The temperature dependency of the linear electro-optical coefficients of $LiNbO_3$ crystal was not well known, so we tried to measure it in wide temperature range. One of the main problems was separation of piezoelectric and electro-optical effects in electro-optical crystals. It was executed at that time only by taking piezoelectric correction into account, but it is not convenient for temperature dependency measurement. We found a method for full compensation of piezoelectric effect and this solution is presented.

The interferometric method can be used also for research of the thermal expansion of piezoelectric materials. It is not so trivial a measurement, but we fully compensated for dilatation of the sample holder. The method was at first successfully tested for known metallic materials. Afterward, some kinds of pizoceramics PZT were investigated, and the results of this research are presented.

2. PIEZOELECTRIC AND ELECTROSTRICTIVE COEFFICIENTS MEASUREMENT

2.1. Piezoelectric and Electrostrictive Coefficients

The piezoelectric and electrostrictive coefficients are important characteristics of piezoelectric materials. It is possible to determine them from piezoelectrically induced displacement measurement, performed by the interferometry method.

Due to anisotropy of some materials (such as crystals), this displacement, induced by converse piezoelectric effect is not generally parallel to an applied electric field. This is why the tensor equation is used for describing this phenomenon. The mechanical strain S_{ij} can be induced not only by electric field E_k, but also by applying of mechanical stress T_{kl} or by temperature change ΔT. We can describe it by the equation for linear response materials:

$$S_{ij} = s^E_{ijkl} T_{kl} + d_{kij} E_k + \alpha^E_i \Delta T , \qquad (1)$$

where s^E_{ijkl} is the mechanical compliance of the material measured at zero electric field ($E = 0$), d_{kij} is the piezoelectric coefficient, α^E_i is the thermal expansion coefficient under zero electric field, and ΔT is the change of temperature.

The strain tensor is the second order symmetrical tensor with six independent components indicated by indexes ij. They can be replaced by one index p (ij~p: 11~1, 22~2, 33~3, 23~4, 13~5, 12~6). The tensor of piezoelectric coefficients is the third order tensor with 6x3 coefficients, but the number of independent ones is many times smaller due to material symmetry. The piezoelectric coefficients are calculated from strain measurement on an unclamped sample, without mechanical stress, at a constant temperature. They are given by the first derivation of the strain tensor at almost zero electric field

$$d_{kp} = \frac{\partial S_p}{\partial E_k} . \qquad (2)$$

Geometrical illustration of these equations is presented in Figure 1. When the applied electric field is parallel to the direction of material polarization P, longitudinal expansion Δl (characterized by coefficient d_{33}) is accompanied by transversal contraction Δt (characterized by coefficient d_{31}). If the electric field is perpendicular to the direction of polarization, the shear deformation (characterized by coefficient d_{15}) is observed.

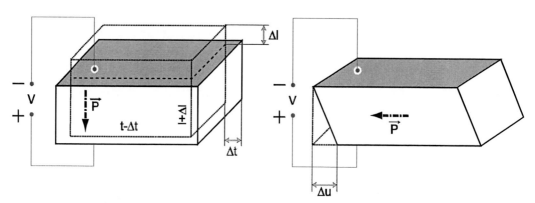

Figure 1. Piezoelectrically induced longitudinal Δl and transversal Δt displacement for electric field parallel to the direction of polarization P and shear deformation Δx induced by electric field perpendicular to the polarization.

For material research, a consideration of longitudinal and transversal displacements is often sufficient. For the converse piezoelectric effect, Eq. (2) can be simplified for transversal coefficient,

$$d_{31} = \frac{\Delta t}{t} \frac{l}{V}, \tag{3}$$

and for longitudinal coefficient where electric field is applied along sample,

$$d_{33} = \frac{\Delta l}{l} \frac{l}{V} = \frac{\Delta l}{V}, \tag{4}$$

where V is voltage applied to sample, t and l thickness and length of sample, and $\Delta t/t$ and $\Delta l/l$ relative transversal and longitudinal deformation – strain.

For large electric field, not only linear converse piezoelectric effect is observed, but the quadratic electrostrictive effect appears too. S_{ij} can be written as a power series in the electric field E_k. If we take into account only quadratic terms, this equation is valid:

$$S_{ij} = d_{kij}E_k + M_{klij}E_kE_l, \tag{5}$$

where the first term in the equation represents the contribution of the converse piezoelectric effect and the second term with M_{ijkl} electrostrictive coefficient describes electrostriction. The nonlinearity of strain and applied electric field is observable only for a large electric field. This is why the effect is observable only for thin piezoelectric layers in practice and almost immeasurable for bulk materials.

2.2. Michelson Interferometer

This is a basic interferometer for piezoelectric research. It is relatively simple, but with lock-in amplifier detection, it is very precise. The coefficients can be determined by measurement of piezoelectric displacement ΔL by laser interferometry. This displacement changes the path length of the beam in the interferometer arm about $2\Delta L$. Detected light intensity I in interferometer (both in Michelson and Mach-Zehnder type) is a harmonic function of the phase difference and, therefore, also of the displacement ΔL:

$$I = I_s + I_r + 2\sqrt{I_sI_r}\cos(4\pi\Delta L/\lambda) \tag{6}$$

where I_s, I_r are intensities in sample and reference interferometer branches, and λ is the wavelength of used light.

Standard interferometry methods work with sensitivity of about 1 nm. The more precise lock-in technique with harmonic vibrating samples is necessary for the measurement of still smaller displacements.

The optical set-up of the interferometer for piezoelectrically induced displacement is classical [14]; see Figure 2. A stabilized He-Ne laser with a wavelength of 632.8 nm, is used as a light source. The laser beam is going through the diaphragm and Faraday rotator. Both elements are allowed to stop backwards reflected beams. The beam is afterward divided by a beamsplitter to two interferometer arms. The sample with the mirror is placed at the first arm. The sample can vibrate with the frequency of the applied electric field from AC source. An active element, holding a constant phase difference $\pi/2$ between waves from both arms, is situated in the second reference arm. Beams are joined after reflection, and the interference pattern is projected to the photodiode by the lens. The lock-in amplifier and, simultaneously, the scope are used for signal detection. A computer is used for experiment control and data acquisition.

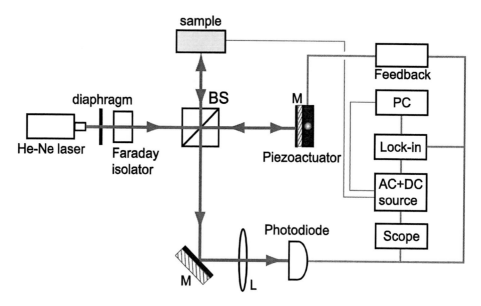

Figure 2. Schematic drawing of Michelson interferometer for piezoelectric coefficient measurement.

The above mentioned active element must be used for this type of measurement with a lock-in amplifier. Piezoelectric actuators are often used for this purpose, but electro-optical modulators or servo-transducers can be used too. This element with a feedback loop keeps the length of the arm at the so-called working or $\pi/2$ point. There is maximal change of light intensity by phase shift (or by displacement) at working point. This change is also linear. If the intensity of light, detected by photodiode, is decreasing or increasing due to the thermal or mechanical optical path fluctuation, the feedback loop compares detected intensity with adjusted value. Afterward it changes the voltage on the piezoelectric actuator to compensate for fluctuation of arm length. This can compensate for all slow fluctuations. Eq. (6) can be overwritten as

$$I = 1/2(I_{max} + I_{min}) + 1/2(I_{max} - I_{min}) \cdot \sin(4\pi \Delta L/\lambda), \qquad (7)$$

where I_{max}, I_{min} are maximal and minimal intensities of interfering light. A displacement periodically changes by applying harmonic voltage to the piezoelectric sample:

$$\Delta L(t) = d_0 \cos \omega t, \qquad (8)$$

where d_0 is the amplitude of sample vibrations and ω is the frequency of the applied harmonic electric field. We can replace sine function in (7) by argument of one for small displacement. Detected harmonic voltage at the photodiode is proportional to the intensity so we can measure voltage on the photodiodiode (with the interferometer at working point) as

$$U(t) = U_0 + U_{p-p} \cdot 2\pi \cdot d_0 / \lambda \cdot \cos \omega t , \qquad (9)$$

where $U_{(t)}$ is detected voltage, U_0 is voltage corresponding to intensity of light at working point, and U_{p-p} is peak-to-peak voltage corresponding to the intensities difference $I_{max} - I_{min}$.

This harmonic part of the signal from Eq. (9) can be detected and emphasized by lock-in amplifier technique. The lock-in amplifier compares the detected voltage with the signal supplied to reference the input of the lock-in amplifier from the voltage generator. This reference signal has a frequency of the electric field applied on the sample. The output voltage from the lock-in amplifier is proportional to the product of measured and sinusoidal reference voltages, so only a signal of reference frequency is detected. This comparison of signals can be made at a desired integration time, much longer than is the period of applied AC voltage. This is why this device is able to detect a very small signal of reference frequency from a high-noise signal, where the noise level is about three orders of magnitude higher. This amplifier measures effective voltage so the amplitude of harmonic vibration of the sample can finally be calculated from the equation

$$d_0 = \frac{U_{out}}{U_{p-p}} \cdot \frac{\lambda}{\sqrt{2} \cdot \pi} , \qquad (10)$$

where U_{out} is the voltage detected by the lock-in amplifier.

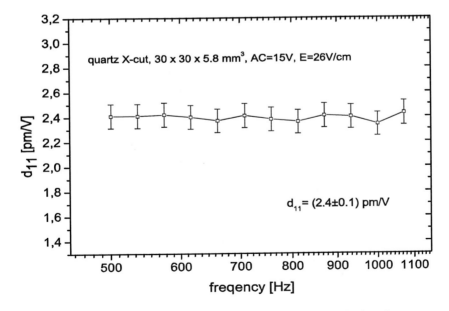

Figure 3. The piezoelectric coefficient d_{11} measurement used for checking the interferometer accuracy.

There is a big advantage to using the dynamic lock-in amplifier method, which allows us to measure amplitude of sample vibrations in the range of 10^{-12} m. The sensitivity was checked, by measurement of a 2 pm amplitude of vibrations of the bulk quartz sample, with adequate accuracy; see Figure 3. The error of measurement was about 5 % and the determined d_{11} coefficient corresponded with the published one in the range of experimental error. The false noise signal of the same frequency as a reference (without field on the sample) was observed by the lock-in amplifier and was adequate for apparent amplitude of vibration smaller than $1 \cdot 10^{-13}$ m.

Interferometer is very stable for long time measurement, as is presented in Figure 4. The amplitude of the sample vibration is almost unchanged for some hours.

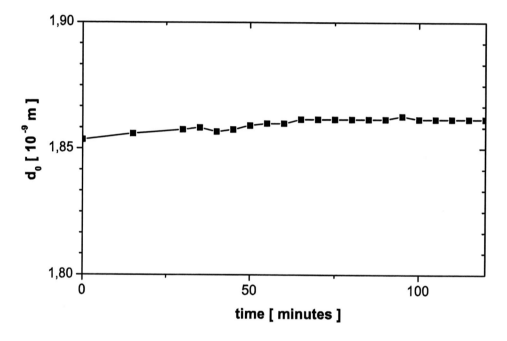

Figure 4. Interferometer temporal stability, illustrated by long time measurement of d_{31} coefficient for piezoelectric ceramic PZT APC 888

2.3. Mach-Zehnder Interferometer

The main problem of piezoelectric-induced displacement measurement by the Michelson interferometer is that samples exhibit not only this displacement but also forced oscillation of the whole sample. This effect is observable with bulk shape samples too, but is very significant in the case of piezoelectric thin films on substrates. The amplitude of forced vibrations of substrate (thickness about 300 μm) can be almost ten times higher than the piezoelectric ones of thin films (thickness about 1 μm). The vibration of the sample can be partially damped by appropriate sample clamping, especially by gluing the sample to a massive holder. However, it is not generally applicable in all cases, and we must use the Mach-Zehnder interferometer arrangement for exact subtraction of both types of vibrations. The laser beam is reflected from the front side of sample and passes around it to fall to the rear side of sample. The longer path at the front side, due a substrate tilt, is compensated for by a shorter way at opposite side [18,28]. For sample mounting details, see Figure 5. The influence of forced substrate vibrations on amplitude Δx is compensated.

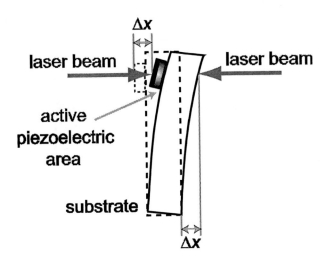

Figure 5. Compensation of vibration of substrate with island of piezoelectric thin film. Path of the laser beam is shorter, longer about Δx for reflection on the front, rear side due a substrate vibration.

Complete arrangement of the Mach-Zehnder interferometer is described in Figure 6. This type of interferometer is more complicated that the Michelson one. The electronic part is the same as with the previous case. The piezoelectric actuator with feedback holds the interferometer at the working point, but the polarizing optic elements must be used to aim the beams in the right directions [18,28]. The beamspliters BS-1 to BS-3 are polarizing, BS-4 unpolarizing. The half-wave plate is used to balance intensity distribution to the reference and sample arms by the first beamsplitter. Quarter-wave plates change the linear polarization to left-handed (for example) circular polarization. The reflection of the beam changes the wave polarization to right-handed polarization. The second passing of beam through the quarter-wave plate changes the polarization to linear, orthogonal to the previous one. The beam is passing through a polarizing beamsplitter toward right direction now. The last beamsplitter BS-4 is unpolarizing, to enable the connection of beams and creation of the interference pattern. Lenses focus the laser beam to selected parts of sample. The alternative arrangement can be built where beamsplitter BS-2 and the quarter-wave plate are substituted by a tilted mirror. The reflected beam passes along mirror. The polarizer is used to observe the interference pattern of orthogonally polarized beams.

The key problem of this measurement is sample mounting. Bulk type samples were clamped in their center by the pressing of a tip against a small cylinder; see Figure 6. The samples were pooled in a direction perpendicular to the sample surfaces, and the polarization P is symbolized by dash-and-dot arrow. The applied field is between the gray colored electrodes. Laser beams are falling to the glued mirror on the sample from both the front and rear sides. This holder, (c), at Figure 7 enables untypical measurement of shear coefficient d_{15}. The other method for shear coefficient measurement, used commercially, is the Zygo interferometer, and the result was published [29].

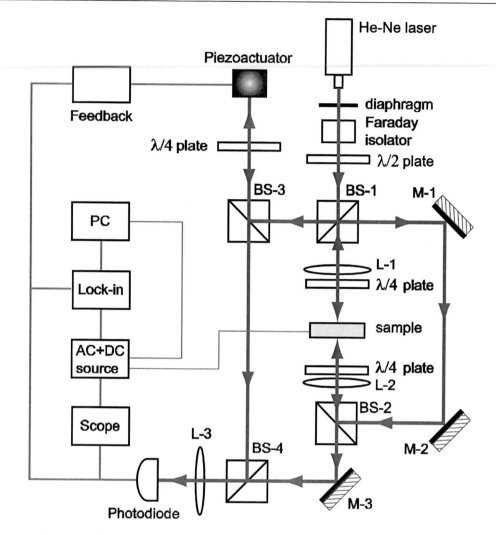

Figure 6. Mach-Zehnder interferometer arrangement.

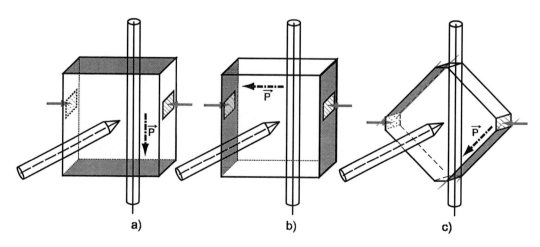

Figure 7. Samples mounted for measurement of a) d_{31}, b) d_{33}, c) d_{15} coefficients.

Plenty of measurements were made for PZT ceramics [30,31] and relaxor crystals PMN-PT [32,33] and PZN-PT [34]. Our results are similar to another measurement for both ceramics and PMN-PT [35] and PZN-PT [36] crystals.

Thin film measurement is still more complicated. The typical piezoelectric films are deposited continuously on the substrate with the bottom electrode. The small upper electrodes, with diameters ranging from 0.05 mm to1 mm, were sputtered through a mask. They can reflect light very well. The gold or platinum reflecting layers were sputtered on the rear side of the substrates to reflect the laser beam.

Samples are held to a special holder with five degrees of freedom by two springs in our experiments. The electric signal is applied to the electrode by an electric probe with three degrees of freedom and a small 10 µm tip. The arrangement of this small tip and focusing of the laser beam on the electrode are monitored by an optical camera system with sufficient magnification. The approximate alignment of beams is sufficient for bulk samples measurement, but very precise alignment is necessary for perfect compensation of the sample vibration with amplitudes about tens pm at the thin films' research. It is shown in Figure 4 that the beams must be almost parallel. Transversal shift of the beams must be as small as possible to compensate substrate vibration. Details are described later.

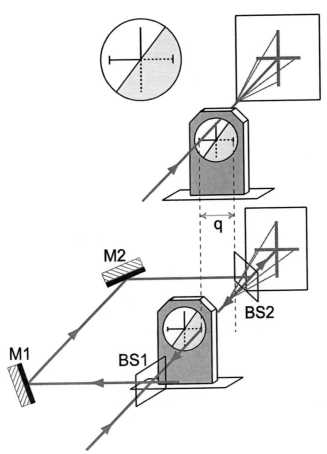

Figure 8. Diffractive element and adjusting procedure of alignment of two beams in Mach-Zehnder interferometer.

The special diffractive element was proposed and used for exact beam alignment. This element is a circular plate, with the same thickness as the sample substrate, 0.3 mm. There are two cross marks, engraved with very well known distance on the rear surface. This element is subdivided into two halves. The first left one is transparent; the second right one has a deposited mirror layer (see Figure 8). This element is placed at the same plane as is standard sample position. The direction of the first beam from beamsplitter BS-1 is aligned. The element holder moves to receiving position, where the best diffraction pattern appears at first. The cross mark lies at the focus of the first lens at that time. Finally, the element is moved about known cross-marks-distance, so the second cross is now exactly at the previous position of the first one. The beam is reflected from the front surface now and can go around the sample to the rear side of the sample. The direction of the beam is aligned to fall exactly on the position of the second cross mark, and diffraction can be observed too. We are able to adjust focused beams with position accuracy of 30 μm, so the substrate vibration is almost suppressed by this way.

2.4. Thin Piezoelectric Films Measurements

This type of research is complicated because films have a thickness below 1 μm and smaller piezoelectric response than bulk sample from the same material [37-39], generally. So the typical amplitudes of oscillations are from picometers [40,41] to hundreds of picometers. Perfect compensation of substrate vibration must be done. We are able to measure only the longitudinal transversal piezoelectric coefficient d_{33} and electrostrictive coefficient M_{33}.

The measured layer is clamped to the substrate so we measure the small effective d_{33} coefficient, done by

$$d_{33}^{eff} = d_{33} - \frac{2 \cdot d_{31} \cdot s_{13}^E}{s_{11}^E + s_{12}^E}, \qquad (11)$$

where s_{11}^E, s_{12}^E are the mechanical compliance at zero electric field. This correction is due to deformation also in transversal direction, which induces elastic response of the surrounding material. The effect of substrate is substantial [42,43]. The Eq. (11) is a simple model, and measurement is more complicated. There is also influence from the materials of the electrodes, so the effective coefficient depends also on the size of the top electrode [44].

When the voltage of 1 V is applied to thin film of thickness 1 μm, the electric field inside the film is relatively strong and electrostriction effects can appear. The electrostrictive coefficient describes piezoelectric response as dependent on the square of electric field and also to the square of applied voltage V. From equations (4), (5), (8) concludes

$$\Delta L = d_{33} V \cos \omega t + M_{33} \cdot \cos^2 \omega t \cdot V^2 / y \qquad (12)$$

because this identity is valid

$$\cos^2 \omega t = 1/2 + 1/2 \cos 2\omega t \qquad (13)$$

and

$$\Delta L = d_{33} V \cos \omega t + M_{33} \cdot (1 + \cos 2\omega t) \cdot V^2 / 2y \qquad (14)$$

voltage of second harmonic frequency 2ω appears on the diode. We can measure the effective voltage of the second harmonic frequency (we filter out the first harmonics) by lock-in amplifier. We can determine the electroctrictive coefficient from

$$M_{33} = d_0 = \frac{U_{out}}{U_{p-p}} \cdot \frac{\lambda}{\sqrt{2} \cdot \pi} \cdot \frac{2y}{V^2}. \qquad (15)$$

This sample response is linear for low electric field (measured at the same frequency, as applied to the sample). The electrostrictive response seems to be quadratic (measured at second harmonics frequency), see Figure 9, when we made a measurement with strong electric field, similarly as in [45].

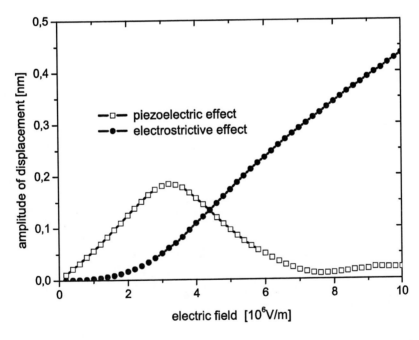

Figure 9. Amplitude of displacement induced by piezoelectric and electrostrictive effects on thin PZT layer.

The PZT [46-48] and ZnO [49,50] thin films were investigated. Deformation was also observed for different waveform of applied electric signal, mostly low frequency (10 Hz); a triangle signal increased to 1000 times the small amplitude higher frequency signal (1000 Hz). The typical shapes of hysteresis loops were observed; see Figure 10. Because electrostrictive effect depends on the electric field square, the displacement is only in one direction – only the extension is observed. Similar results were obtained by other laboratories and are presented [51-53].

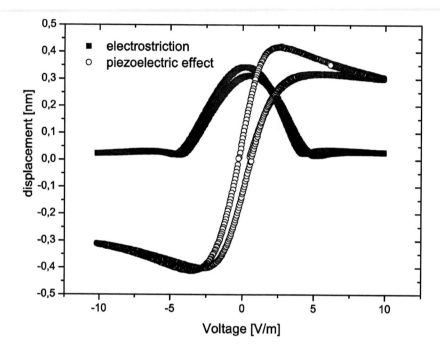

Figure 10. The typical shape of hysteresis loops for piezoelectric effect. The same for electrostrictive effect, sample of thin PZT ceramics layer.

The laser Doppler interferometry (LDI) is frequently used in thin film research [54-56]. The published coefficients are sometimes higher than the above described double-beams interferometer [38,39]. It can be done by sample clamping, too. The LDI method is frequently used for sample profile scanning [57,45]. It has been demonstrated that the scanning of the vibration of the top electrode area can give better information than a measurement made at a single point [58]. Another theoretical analysis and experimental measurements indicate [59] that by bonding thin film piezoelectric samples to a substantial holder, the substrate bending can be minimized to a negligible level by gluing. Single-beam LDI scanning vibrometry can give results very similar results to the double-beam interferometer.

Profiles of thin film deformation near to small upper electrodes were measured by our set-up too. Our results were checked by a single-beam Doppler vibrometer at another laboratory [60], and they are in good agreement.

2.5. The Wide Temperature Range Measurements

Using of an interferometer in a wide temperature range is complicated. The change of temperature can change not only properties of investigated materials [61], but also clamping, optical path I elements in interferometer, etc. The fundamental measurement problem is temperature instability. It causes drift of wave phases in interferometer branches. Another mechanical vibration problem appears when the closed loop cryostat is used for cooling the sample. This is why this temperature measurement was only exceptionally performed at piezoelectric research oriented laboratories.

We performed this type of measurement at first at our laboratory. Two temperature stabilized chambers were used here. The first one was a cryogenic cooling system (Oxford Instruments). It is based on a closed circuit of helium gas and was used for interferometric measurement in range from - 150 °C to + 60 °C. The measurement was not performed at temperatures lower than – 150 °C due problems with interferometer stabilization. An unhomogeneous dilatation of parts of the interferometer (there are elements of brass, quartz, glue, samples) caused deviation of the beams in the interferometer and cancelled the interference pattern. The second one, a heated evacuated chamber for the measurements above room temperature (from 20 °C to 250 °C), was developed and constructed at the university.

The problems with an interferometer, temperature instability and mechanical vibration of cryostat, were solved by placing all of the Michelson interferometer inside the temperature chambers. All parts of the interferometer were miniaturized and placed on a holder plate of 40 mm diameter. The spherical shape holders of both samples and mirror were used. The sample was glued on spherical shape holder. This ball with cylindrical hole in the middle was mounted on a narrower screw. It enabled small vertical tilts and full horizontal too. The ball was fastened on an annular hole by the nut. The position of the holder was very accurately fixed and very stabile; see Figure 11.

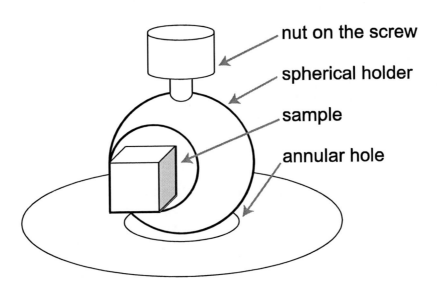

Figure 11. Spherical shape holder with glued sample.

The Michelson interferometer arrangement is presented in Figure 12. Contrary to Figure 2, the polarizing beamsplitter is used, and waves have orthogonal polarizations in reference and sample arms. Electro-optical modulator changes phase shift between waves to keep $\pi/2$. It stabilizes the working point of interferometer. The modulator is located in the outside chamber, so it is independent of temperature change. The position of the modulator can be both at front of and behind the interferometer. It is driven by the same high voltage source as the piezoactuator at standard measurement, described in the previous chapter, to keep the working point of interferometer. The analyzer, turned to about 45°, must be used for interference pattern detection. The interferometer is very compact, and the relative motion of parts of the interferometer is almost suppressed in this arrangement. The resulting interference pattern is very stable.

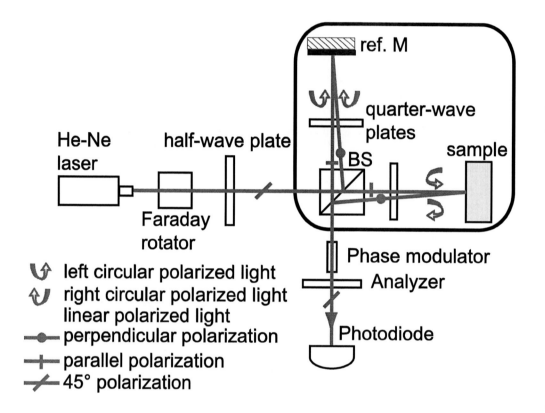

Figure 12. Schematic drawing of Michelson interferometer for measurement inside the cryostat.

The reference and monitoring beams' misalignment (caused by deformation of mirrors, etc.) can appear at temperatures far from room temperature. The interference pattern can be suppressed and visibility is very small. There is a possibility to partiallycorrect this problem. The electro-optical modulator can be replaced by the classical Mach –Zender interferometer with piezoactuator in one branch. The light from the temperature chamber goes to the input polarizing beamsplitter. So, we have the reference wave in one branch and the measured wave in the second one. The actuator can hold phase distance $\pi/2$ between both waves. The interference can be observed through the analyzer, rotated about 45° with respect to polarization of both waves.

This method was used for phase transition research of relaxor crystals with phase transition in the temperature range from 100 °C to 200 °C. This method gives much better and more exact results than dielectric constant measurement or Raman spectroscopy. The illustration of received results is presented in Figure 13 for crystal PZN-PT 0.92Pb $(Zn_{1/3}Nb_{2/3})O_3$-$0.08PbTiO_3$. Cryostat was used for cooling and a heated chamber for heating of t this sample. Phase transition temperatures are clearly detectable [32-34].

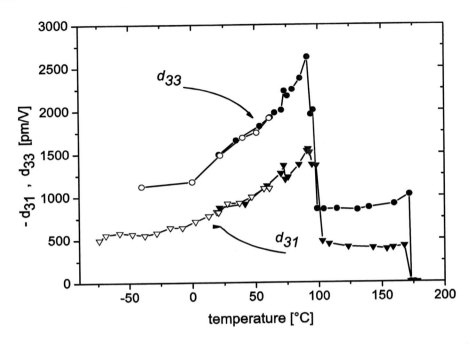

Figure 13. Piezoelectric coefficients d_{33} and d_{31} of PZN-PT crystal measured at cryostat (open symbol) and heated temperature chamber (solid symbols).

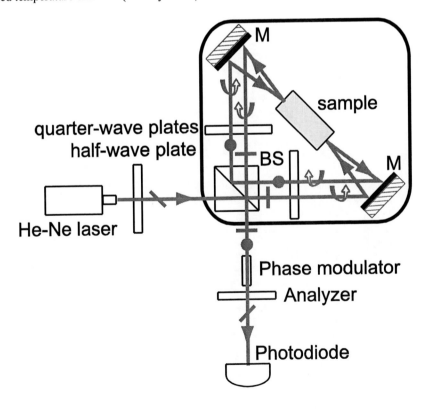

Figure 14. Schematic drawing of double beam interferometer for measurement inside the cryostat.

The Mach-Zehnder double beam interferometer is essential for measurement of thin piezoelectric films. However, placing it into a small chamber or cryostat is very problematic. The classical interferometer has a lot of optical elements. Some modification of the interferometer can reduce this number, but it is still not sufficient. This is why a new type of interferometer was proposed [6]. The schematic drawing is in Figure 14.

It has only one polarizing beamsplitter. Linearly polarized (45° angle) light from the laser is divided into two orthogonally linearly polarized beams by the beamsplitter. The first reference beam goes directly towards the detector after reflection on beamsplitter. The second monitoring beam goes first to the front surface of the sample and is reflected there. It changes polarization two times by quarter-wave plate passing. This is why it is reflected by the beamsplitter afterward and it goes to the sample from the rear side. It changes polarization after reflection there by double passing through the quarter-wave plate and is directed to the detector. The electro-optical phase modulator is used as a feedback for working point control and the analyzer is required for interference visibility. Substrate vibration is subtracted in this way.

The method was used for PZT thin films measurement [62,63] in the range of 240 °C-340 °C. The typical result with about 5 % error is presented in Figure 15.

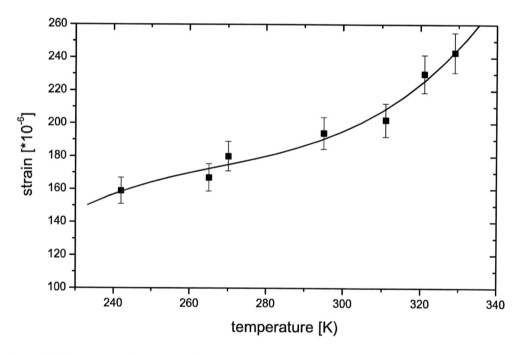

Figure 15. Temperature dependency of strain, measured on thin PZT film.

2.6. Experimental Problems and Limits of Laser Interferometers for Piezoelectric Measurement.

The laser interferometry is concededly a very useful method for measurement of piezoelectrically induced strain. However, there were great expectations about the sensitivity of this method at the start of using of interferometers in piezoelectric materials research. The lock-in amplifier detecting technique was very successful, so authors had presented sensitivity of method to be in range between 10^{-12} m and 10^{-15} m. They had taken into account only electronics' noise which can be suppressed to a level, corresponding to 10^{-15} m amplitude of piezoelectric material vibrations. It was the reason that we published a critical general review of optical sources of noise and instabilities in interferometers used for piezoelectric research [64].

There are two optical and mechanical noise sources. The first one has its origin in the used optical and mechanical elements, and the second one in the used method. They both can raise the noise and make parasite signals with the same frequency as the correct detected signal. It follows that the measurement (by the lock-in amplifier technique) of very weak signals with the frequency of the driving voltage on the sample does not mean that this signal is caused only by the piezoelectric sample displacement towards the reference mirror.

Experimental problems with laser interferometric measurements of piezoelectric response are caused by used elements. The key element of an interferometer is the laser. The He-Ne laser, 632.8 nm, is frequently used for this type of measurement. An output of unstabilized He-Ne laser usually consists of three or four frequencies. The thermal instability of a laser cavity length causes a shift of frequencies, and unstabilized lasers exhibit power fluctuations of up to 10 %. In a stabilized laser the allowed frequencies are kept centered on the gain curve by the feedback control system. Only the single frequency appears on the laser output, while other frequencies are suppressed. The intensity variation is smaller than 0.1 %; the stability of the frequency is better than ± 1 MHz.

The smallest detectable amplitude of displacement d_{min} is from Eq. (10),

$$d_{min} = \frac{\lambda \cdot U_{min}}{\sqrt{2}\pi \cdot U_{p-p}}, \tag{16}$$

where U_{min} is the smallest detectable voltage by lock-in amplifier, and U_{p-p} is peak-to-peak voltage difference measured by photodiode in the interferometer. It corresponds to the difference of maximal and minimal interference intensities $I_{max} - I_{min}$. Visibility V of the interference pattern is proportional to this difference, too.

$$V = \frac{I_{max} - I_{min}}{I_{max} + I_{min}} \tag{17}$$

Visibility is also proportional to a complex degree of coherence, and, so, it increases with increasing laser coherency. The stabilized laser has large degree of coherence. Fluctuations of visibility and of U_{p-p} are many times higher in the case of using an unstabilized laser, so there are reasons why the interferometers based on stabilized lasers have much higher sensitivity and stability than interferometers based on unstabilized ones. Unstabilized lasers were used for the first simple measurements in the 1990s. For accurate or time consuming measurement (such as a measurements of frequency or temperature dependencies), it is necessary to utilize stabilized laser.

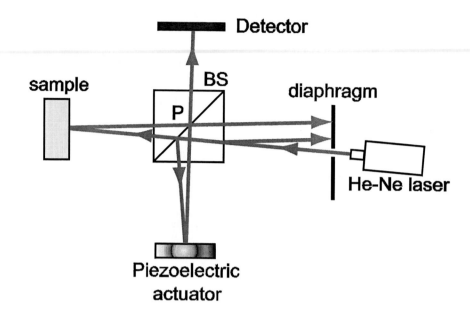

Figure 16. Michelson interferometer arrangement. Both beams must go through one point P in the same direction.

Laser gain can be destabilized by light, which is reflected from the interferometer back to laser cavity. This reflected light is the source of instabilities of both intensity and frequency of the output laser beam. The visibility and peak-to-peak voltage can fluctuate in the range of ten percent, so the measurement is impossible in this case. This is why the back-reflected light must be suppressed to zero. The deflection of reflected light outside the diaphragm hole placed before laser output is the simplest way to eliminate this problem and is used very often. It can be achieved by a small tilt of the beamsplitter in the Michelson interferometer (or the first beamsplitter in the Mach-Zehnder interferometer), so the incident ray from the laser is not perpendicular to the front side of beamsplitter. The reference and monitoring beams from the beamsplitter must go through one point P in the same direction to receive a homogenous interference pattern in a plane of detector; see Figure 16. The results (from these requirements for a little tilted beamsplitter surface) are that the lengths of the interferometer arms must be practically identical.

Another possibility of how to block the reflected beam from returning to the laser cavity is application of a Faraday rotator. This rotator turns a plane of light polarization about 45° for the output beam and about 45° further (in the same direction, for example left-handed) for the reflected beam. So reflected light has orthogonal polarization with respect to output one after double passing thru the Faraday rotator. This reflected beam is afterward deflected by a small polarizing beamsplitter. The reflected light intensity is decreased to be smaller than one-thousandth of the previous reflected one. It can protect laser stability very effectively.

The laser beams interfere in some part of space, and the interference maximum and minimum are not localized. The observed interference pattern is the plane section of the system of the coaxial rotatating hyperboloids and depends on the mutual position of real (or virtual – after reflection on mirrors in interferometer arms) light sources. The arrangement, where the sources are on one line, is advisable for the interferometric measurement, as you can see in Figure 17, detector position Dc. However, in this case, there will be an almost homogenous distribution of the light intensity of the interference pattern on the detector.

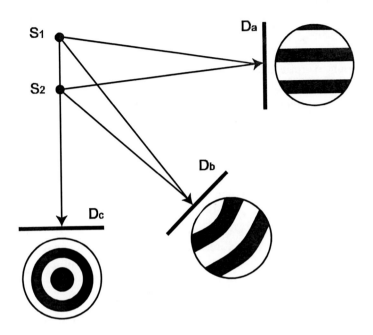

Figure 17. The interference pattern in the planes of detectors D_a, D_b, D_c and its dependence on the relative position of the sources S_1 and S_2.

This arrangement of the interferometer is not so critical. When the position oscillations of source S_1 with respect to S_2 are not parallel with line S_1 and S_2, the measured amplitude of oscillations remains the same, as for in-line oscillations. However, only one fringe is tolerable on the photodetector. The smaller part of the fringe at the detector is the better solution, of course.

All optical elements in the interferometer reflect part of a light. These reflections create new sources S_i with random positions with respect to S_1, corresponding to the position of the mirror sample. The set of the fringes, with different space frequencies, orientation and shapes, can appear also on the detector, because it is impossible to deflect all unwanted reflected beams from another direction towards the detector. When a piezoelectric sample starts to oscillate, all fringes are move in the plane of detector. The direction of movement of unwanted fringes can be arbitrary with respect to the movement of fringes from the reference mirror. The waves are not in the same phase too. If the phase difference between the sample and reflected waves is $-\pi/2$, the inverse change of interference light intensity with sample displacement appears. This is the reason smaller measured values of amplitude of piezoelectric induced oscillations.

The interference pattern depends on arrangement of all parts of interferometer. The intensity of interference light is not a harmonic function of sample phase shift, but it is superposition of harmonic functions with different amplitudes and phase shifts due reflections from other elements. It can dramatically change the slope of the intensity curve - derivation of intensity – in the working point. The error of piezoelectrically induced amplitude measurement cans reach tens of percent due this effect. It is very suitable to check the shape of this curve before measurement. It can be done by applying triangle voltage to the piezoelectric actuator to receive displacement of more than one wavelength. The deviation from ideal harmonic curve and curve smoothness in the working point can be observed on the scope and checked also on computer. If this curve is not ideal, rearrangement of the interferometer must be made.

The total amount of reflected light is high, and substantially contributes to interference intensity at the detector. For the crown glass with refractive index of $n = 1.52$, one surface reflects 4.2 % of incident light; for silica with $n = 1.46$ this reflectivity is 3.4 %. Single - layer antireflection coating from MgF_2 reduces the one-surface reflectivity to 1.4 %. For special multilayer coatings the reflectivity can be smaller than 0.1% on one surface.

The transmittance (with reflections taken into account) of four parts of the double beam interferometer, i.e., the part before the first beamsplitter, the part behind the last beamsplitter and the parts in both sample and reference, arms are summarized in Table I.

Very good polarization optics can eliminate some reflections. Only the reflected light with the right polarization can go through the quarter-wave plates and polarizing beamsplitters system backwards. Therefore the placing of the quarter-wave plates between the lens and the sample is better than between the lens and the beamsplitter. The use of optical elements with antireflective coatings can eliminate unwanted reflections. We used optical elements with single - layer antireflection coatings, and less than 6 % of the detected intensity had an indefinite phase in an interferometer detector. Multilayer coating is still more effective but also more expensive.

Although unwanted reflection can be partially eliminated by using polarizing optical elements, the interference always appears between the sample and the front side of quarter-wave plates in both the reference and the sample arms. This parasite signal has the same frequency as the driving voltage, and it is not possible to eliminate it by the lock-in amplifier. Because the surfaces of these elements are not in $\pi/2$ position to the sample, incorrect response decreases the measured displacement. The real detected signal from the photodetector has a smaller error than the sum of the errors described above, because all reflecting surfaces are in random phase shift, not only in $-\pi/2$ shift. The influence of different reflections can be partially spontaneously compensated by this way. It can reduce this error by about one order of magnitude.

Another contribution to interference pattern change is given by changing the shape of the wave, reflected by the sample after application of an electric field of specific frequency. It can vary the visibility of interference pattern with the same frequency. This is why the detected signal doesn't correspond only to simple piezoelectric sample oscillations.

Table 1. The transmittance of Mach-Zehnder interferometer parts. Optical elements are considered without, with single layer, with multilayer coating

Elements	sample arm	reference arm	before BS	behind BS
without coating	0.40	0.70	0.90	0.81
single layer coating	0.71	0.87	0.96	0.93
multilayer coating	0.976	0.990	0.997	0.995

For better understanding of the interferometric problems, the Gaussian shape of the laser beam must be taken into account [11]. A He-Ne laser has the symmetrical output of the intensity. The intensity distribution of the Gaussian beam in the distance z from the plane, where the wavefront is flat, is given by

$$I(z,\rho) = \frac{2P}{\pi w^2} \cdot \exp\{-2\rho^2 / w^2(z)\}, \tag{18}$$

where P is the power of a beam, ρ the distance from the beam axis and $w(z)$ - the waist radius - is the radius of $1/e^2$ intensity contour in the distance z. In order to achieve visibility $V = 1$, two beams with the same intensity, the wavefront radii of curvature $R_1=R_2$ and waist radii $w_1=w_2$ must interfere. If this is not valid, the visibility is smaller according to factor K:

$$K(R_i, w_i) = \frac{1}{w_1 w_2} \frac{\sqrt{\left(\frac{1}{w_1^2}+\frac{1}{w_2^2}\right)^2 + \left(\frac{\pi}{\lambda}\left(\frac{1}{R_1}-\frac{1}{R_2}\right)\right)^2} + \frac{1}{w_1^2}+\frac{1}{w_2^2}}{\left(\frac{1}{w_1^2}+\frac{1}{w_2^2}\right)^2 + \left(\frac{\pi}{\lambda}\left(\frac{1}{R_1}-\frac{1}{R_2}\right)\right)^2} \tag{19}$$

For the small transversal relative shift x of the axes of the beams this factor is

$$K(x) = \exp\{-x^2 / 2w^2\} \tag{20}$$

For the small tilt α of the beams (for $R_1=R_2$, $w_1=w_2$) we can find

$$K(\alpha) = \exp\left\{-\frac{1}{2}\left(\frac{\pi}{\lambda}\right)^2 \alpha^2 w^2\right\}. \tag{21}$$

The flat mirror surface of the sample can be curved by an applied voltage to the radius of about $R = 2.5 \cdot 10^4$ m (for $2 \cdot 10^{-9}$ m displacement in sample center, 2.6 cm sample length [10]). This changes the wavefront radius of the reflected light from R_1 to R_2:

$$1/R_2 = 1/R_1 + 2/R. \tag{22}$$

From (19) we can see that this small difference between R_2 and R_1 can vary the peak-to-peak voltage by about $1.2 \cdot 10^{-5}$ % (in case of $R_1=1$ m, $w=1.4$ mm) with driving voltage frequency. The deviation of the axes of angle α gives similar contribution to the error. The influence of the transversal shift of beams can be neglected. All changes of the shape of a reflecting mirror surface on the sample can modulate wavefronts and it can vary the detected signal. This effect is not possible to eliminate and represents the optical origin of interferometer limits.

The theoretical analysis of the sources of systematic errors and inaccuracies pointed to that the limits for amplitude of piezoelectrically induced vibration measurement is of the order picometers, both for single and double beam interferometers. It is also confirmed by the set of our experimental experiences. It appears that some of the published results of electric-field induced displacements of order of hundredths of picometers are problematic.

3. ELECTRO-OPTICAL COEFFICIENTS MEASUREMENT

3.1. Electro-Optical Coefficients

Applied electric field E can change the impermittivity tensor η and index ellipsoid of investigated materials. The relation between impermittivity and refractive index is described for isotropic materials by the equation

$$\eta = \varepsilon_0/\varepsilon = 1/n^2 . \tag{23}$$

If the intensity of the electric field is small, it is possible to express these components of tensor η by the first members of Taylor expansion:

$$\eta_{ij}(E) = \eta_{ij} + \sum_k r_{ijk} E_k + \sum_{kl} s_{ijkl} E_k E_l . \tag{24}$$

Linear Pockels coefficients r_{ijk} are the first derivation of the impermittivity tensor for zero electric field:

$$r_{ijk} = \partial \eta_{ij}/\partial E_k . \tag{25}$$

The impermittivity tensor is a second order tensor with six independent components according to the dielectric-constant tensor. So, the Pockels coefficients tensor has generally 18 components, but the number of independent ones is many times smaller. They are mostly zero due to crystal symmetry. The indexes ij can be replaced by one index p, as it is in the case of piezoelectric coefficients.

Quadratic Kerr coefficients s_{ijkl} are the second derivation of the impermittivity tensor for zero electric field. The coefficients are described by a fourth order tensor with 6x6 coefficients.

$$s_{ijkl} = 1/2\, \partial^2 \eta_{ij}/\partial E_k\, \partial E_l \tag{26}$$

The change of refractive index due to the applied electric field E on electro-optical crystal is described by

$$\Delta n = n(E) - n = \partial n/\partial \eta \cdot \Delta \eta(E), \tag{27}$$

if we return from tensor to scale description due to simplicity. From previous equations, it can be concluded that

$$n(E) = n - 1/2 \cdot n^3 \cdot r \cdot E - 1/2 \cdot n^3 s \cdot E^2. \tag{28}$$

The Pockels effect can be observed only for materials without a center of symmetry. The most common crystals used as Pockels cells too are $LiNbO_3$, $LiTaO_3$, ADP ($NH_4H_2PO_4$), and KDP (KH_2PO_4). The typical value of Pockel coefficients is $1 - 100 \cdot 10^{-12}$ m/V. This means, for $E = 10^6$ V/m, the typical refractive index change is very small and it is on the order of 10^{-6} to 10^{-4}. For materials with a center of symmetry, such as gases, liquids, and certain crystals, $n(E)$ must be an even function and Pockels coefficients must be equal to zero. All materials display the Kerr effect, with varying magnitudes, but it is generally much weaker than the Pockels effect, so only for centrosymmetric materials is quadratic effect not negligible. Typical values of Kerr coefficients are $10^{-18} - 10^{-14}$ m^2/V^2 in crystals and $10^{-22} - 10^{-19}$ m^2/V^2 in liquids. So for $E = 10^6$ V/m, the index change is very small – about $10^{-6} - 10^{-2}$ in crystals and $\sim 10^{-10} - 10^{-7}$ in liquids.

3.2. Pockels Coefficients Measurement

The linear Pockels coefficients r_{ijk} characterize electro-optical properties of crystals (if we can neglect the weak Kerr effect). We will describe measurement of them for one of the most widely used crystals in optoelectronics - $LiNbO_3$ crystal of congruent composition (48.5% Li, 51.5% Nb).

Crystal $LiNbO_3$ is a uniaxial rhombohedral crystal with point-group symmetry *3m*. It has only r_{13}, r_{33}, r_{22}, r_{51} non-zero electro-optical coefficients. Although $LiNbO_3$ crystals have been extensively studied, the data are still not single-valued. The temperature dependence of these coefficients is not so well known either, because there is an influence of additional effects, such as thermo-optical effect, strain effect, etc. [65]. This is why we used this crystal for testing of interferometry method.

The coefficients r_{13}, r_{33} are mostly interesting ones for practical purposes. The applied electric field $E = (0,0,E)$ was along the optical axis z, perpendicular to the laser beam in case of their measurement. The values of the principal refractive indices with field E are

$$n_o(E) = n_o - \tfrac{1}{2}n_o^3 r_{13} E \tag{29}$$

$$n_e(E) = n_e - \tfrac{1}{2}n_e^3 r_{33} E, \tag{30}$$

where n_o, n_e, $n_o(E)$, $n_e(E)$ are refractive indexes of an ordinary, extraordinary beam without and with applied electric field.

A very detailed description of experimental techniques for Pockels effect measurement of $LiNbO_3$ is given in the paper done by Aillerie et al. [66]. The six elipsometric and one interferometric methods are compared there. It results, from this paper, that the elipsometric methods are more precise that interferometric ones. Only one advantage of interferometry is mentioned - the elipsometric methods can only measure phase difference of two orthogonally polarized beams, so the only difference of coefficients from equations (8) and (9) can be calculated from these results:

$$r = r_{33} - (n_o^3/n_e^3) \cdot r_{13}. \tag{31}$$

However, only one paper describing interferometric measurement was taken into account in comparison with elipsometric ones and this experiment was, however, not so exactly performed. The published values of Pockels coefficients in this paper differ by about 13% from generally accepted values, presented in a critical review paper [67]. All performed experiments concerned room temperature conditions and the use of a 632.8 nm He-Ne laser. The measurement can be performed by the Michelson interferometer too [68], but we used the Mach-Zehnder interferometer set-up for the measurement of the electric field induced shift between waves at the reference and sample arms; see Figure 18. This method is similar to the one described for piezoelectric coefficients research. Analogical set-ups are used for investigation of polymer films [69-71], thin layers [72-75], or crystals [76-78].

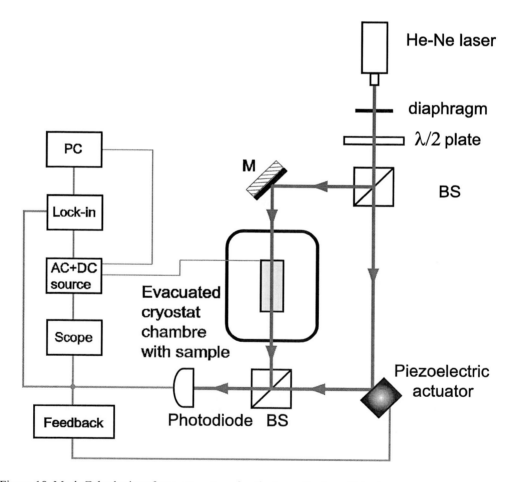

Figure 18. Mach-Zehnder interferometer set-up for electro-optical coefficients measurement.

The applied voltage on electro-optical crystal changes the refractive index and so causes electric field induced phase shift $\Delta\varphi(E)$ of laser wave in a sample:

$$\Delta\varphi(E) = 2\pi \cdot \frac{L}{\lambda}(n(E)-n) = \frac{\pi \cdot Ln^3 rE}{\lambda} \tag{32}$$

A sample is placed in one branch of the Mach-Zehnder interferometer. The half-wave plate can rotate polarization of light to enable measurement of r_{33} or r_{31} coefficients. Intensity of interference light is described by

$$I = I_s + I_r + 2\sqrt{I_s I_r}\cos(\Delta\varphi(E) + \varphi_0), \tag{33}$$

where I_s, I_r are intensities in sample, reference arms. The piezoelectric actuator holds initial phase difference $\varphi_0 = \pi/2$ in the working point. We can rewrite

$$\cos(\Delta\varphi(E) + \frac{\pi}{2}) = \sin\Delta\varphi(E), \tag{34}$$

rewrite sine for the small values of $\Delta\varphi$ as

$$\sin\left(\frac{\pi Ln^3 rE}{\lambda}\right) \cong \frac{\pi Ln^3 rE}{\lambda}, \tag{35}$$

and Eq. (27) can be rewritten in terms of maximal and minimal intensities I_{max}, I_{min}:

$$I = \tfrac{1}{2}(I_{max} + I_{min}) + \tfrac{1}{2}(I_{max} - I_{min})\frac{\pi \cdot Ln^3 r_{p3} E}{\lambda}. \tag{36}$$

For applied harmonic voltage $U_A \cdot \cos\omega t$ on a sample of thickness y,

$$E = U_A/y \cdot \cos\omega t \tag{37}$$

and detected intensity gives rise to voltage $U(t)$ on the photodiode

$$U(t) = U_0 + \frac{1}{2}U_{p-p}\frac{\pi \cdot Ln^3 r_{p3}}{\lambda \cdot y}U_A \cos\omega t. \tag{38}$$

We measure effective voltage U_{out} (with the same frequency, applied to the sample) on the output of the lock-in amplifier:

$$U_{out} = \frac{1}{2}U_{p-p}\frac{\pi \cdot Ln^3 r_{p3}}{\lambda \cdot y}U_A/\sqrt{2} \tag{39}$$

Electro-optic coefficients can be determined from the equation

$$r_{p3}^{eff} = \frac{1}{\pi \cdot U_A} U_{out} \frac{2\sqrt{2}}{U_{p-p}} \frac{\lambda}{n^3} \frac{y}{L} \tag{40}$$

It is very important to differentiate whether the sample is clamped or unclamped. The measurement of an unclamped sample is unfortunately complicated by inverse piezoelectric effect. The application of electric field in optical axis direction also induces a change of sample length ΔL along the path of the laser beam due the piezoelectric effect. It is proportional to piezoelectric coefficient d_{31}:

$$\Delta L = L \cdot d_{31} \cdot E . \tag{41}$$

The refractive index of air ($\cong 1$) is substituted by n_i ($n_i=n_o$ for r_{13} and $n_i=n_e$ for r_{33}) of the crystal on piezoelectric induced length ΔL. This causes phase shift $2\pi/\lambda \cdot (n_i \cdot \Delta L - 1 \cdot \Delta L)$. The total phase shift $\Delta\varphi_i(E)$ of linear polarized beam is

$$\Delta\varphi_i(E) = \frac{\pi L n^3 r_{p3} E}{\lambda} + 2\pi \cdot \frac{L}{\lambda} \cdot (n_i - 1) \cdot d_{31} E . \tag{42}$$

This is why only effective coefficients are measured in the above described arrangement. It is impossible to separate both piezoelectric and electro-optical contribution to the resulting values of Pockels coefficients because both effects change the phase shift with the same frequency. One must take into account this effect for the calculation of real coefficients and to make correction of measured effective coefficients from Eq. (40)

$$r_{p3}^0 = r_{p3}^{eff} + \frac{2 \cdot (n_i - 1) \cdot d_{31}}{n_i^3} , \tag{43}$$

where r_{p3}^0, r_{p3}^{eff} are real and effective (measured) Pockels coefficients.

So final equation is

$$r_{p3}^0 = \frac{1}{\pi U_A} U_{out} \frac{2\sqrt{2}}{U_{p-p}} \frac{\lambda}{n^3} \frac{y}{L} + \frac{2 \cdot (n_i - 1) \cdot d_{31}}{n_i^3} . \tag{44}$$

Our laser interferometer experiments were performed with a nominally undoped LiNbO$_3$ crystal of congruent composition, balk shape, and measuring 36x3x2 mm^3. This crystal was investigated in transversal configuration- the electric field along optical axis z, parallel to 2 mm edge, and perpendicular to the laser beam. The crystal was researched both at room temperature and at a wider temperature range of 150 K – 330 K.

The sample in cryostat was cooled (warmed) to engage temperature and was stabilized at this temperature for 10 minutes. The cryostat system was switched off to eliminate sample vibration. Afterward, coefficients r_{13} and r_{33} were measured. The beam polarization choice was made by a rotation of the half-wave plate. The measurement was performed in frequency range of applied electric field 200 Hz – 10 kHz.

The most important problem of this experiment was temperature field homogeneity and stability. The small variation of the sample temperature causes the change of refractive index Δn and the change of optical path $\Delta n \cdot L$. It changes the working point position too. When this phase shift is higher than a $\pi/2$, it is not possible to use a feedback loop to correct working point position. Therefore, it is necessary to have good temperature stabilization and a short time of measurement. For temperatures lower than 220 K, thermal instabilities are very significant and the feedback loop was not able to stabilize working position automatically in most cases.

The other problem of this temperature dependency measurement is related to the calculation of r_{13}, r_{33} from (38). There are temperature dependent quantities: n_i, y, L. The influence of the temperature dependence of ratio y/L is small and can be neglected. The temperature dependence of refractive index was not measured by us. Thermo-optic coefficient dn/dT is $4.4 \cdot 10^{-6}$ K^{-1} for n_o, and $37.91 \cdot 10^{-6}$ K^{-1} for n_e. It was estimated that the optical path changes, $\Delta n \cdot L$, given by 1 K temperature difference, is about 160 nm/K for r_{13} and 1500 nm/K for r_{33} in our experiment. That is why the r_{33} is more sensitive to the temperature instabilities, as was observed too.

The values of electro-optical coefficients at room temperature were determined from Eq. (38). The principal refractive indices are $n_o = 2.2847$ and $n_e = 2.2006$. Correction for piezoelectric effect ($d_{31} = -0.85 \cdot 10^{-12}$ m/V) was taken into account. It changed the effective coefficients about −0.2 pm/V. Resulting values are $r_{13} = (9.8 \pm 0.2)$ pm/V and $r_{33} = (30.5 \pm 0.3)$ pm/V for the room temperature. The values seem to be only a little smaller than the values published in the electro-optic parameters overview [67] for unclamped congruent LiNbO$_3$ sample ($r_{13} = (10.4 \pm 0.4)$ pm/V and $r_{33} = (31.4 \pm 1.0)$ pm/V).

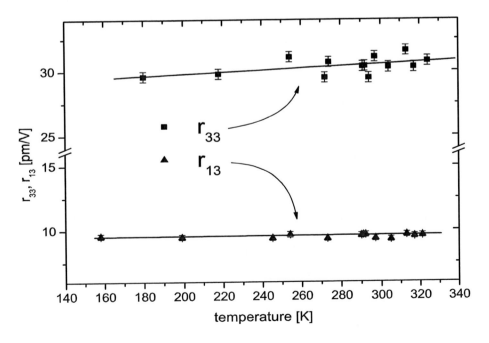

Figure 19. Temperature dependency of electro-optical coefficients r_{13} and r_{33}.

The coefficients r_{13} and r_{33} don't exhibit significant temperature dependency in the measured temperature range of 150K – 330 K; see Figure 19. It seems that electro-optical coefficient r_{33} became lower with respect to the decrease of the temperature; coefficient r_{13} seems to be constant. The increase of about 20 % is presented in the paper with the r_{33} measurement result and interferometry results [79]. However, the error of low temperature measurements in our experiment is higher than 10% and the observed tendency of r_{33} to decrease in lower temperatures cannot be confirmed by us.

3.3. Kerr Coefficients Measurement

The Kerr effect is proportional to the square of the electric field. Phase shift, induced by this effect is

$$\Delta\varphi(E) = 2\pi \cdot \frac{L}{\lambda}\left(n(E) - n\right) = \frac{\pi \cdot L n^3 s E^2}{\lambda}. \qquad (45)$$

We will here use scalar description due to simplicity. When the electric field of frequency ω is applied to the sample, the resulting voltage $U(t)$ on the photodiode coresponding to the detected light intensity from (38) is

$$U(t) = U_0 + \frac{1}{2} U_{p-p} \frac{\pi \cdot L n^3 s}{\lambda \cdot y} U_A^2 \cos^2 \omega t. \qquad (46)$$

Similarly to piezoelectric coefficient measurement, the voltage of the second harmonic frequency 2ω appears on the diode. We can measure the effective voltage of the second harmonic frequency, applied to the sample, by lock-in amplifier:

$$U_{out} = \frac{1}{4} U_{p-p} \frac{\pi \cdot L n^3 s}{\lambda \cdot y} U_A^2 / \sqrt{2}. \qquad (47)$$

Kerr coefficients can be determined from previous equation.

$$s = \frac{1}{\pi U_A^2} U_{out} \frac{4\sqrt{2}}{U_{p-p}} \frac{\lambda}{n^3} \frac{y}{L} \qquad (48)$$

So, the measurement of some important Kerr coefficients can be done by the same method as described for Pockels coefficients. It can be done by switching to a second harmonics measurement on the lock-in amplifier control panel only.

3.4. Compensation of Piezoelectric Induced Displacement

It is not so easy to make correct measurements of electro-optical coefficients at room temperature with high influence of the piezoelectric effect. The errors can be on the order of tens of percents [80]. The difficulties appear at the measurement of crystals with piezoelectric coefficient d_{31} higher than this one for LiNbO$_3$. The second term at Eq. (37) can be much higher than the real value of the electro-optical coefficient.

Two papers, describing investigation of promising electro-optical crystals, were published, and Eq. (37) was used for the correction of the Pockels coefficient measurement. The first one was a single crystal 0.88Pb(Zn$_{1/3}$Nb$_{2/3}$)O$_3$-0.12PbTiO$_3$, PZN-PT. The effective coefficients $r_{33}^{eff} = (174\pm4)$ pm/V and $r_{13}^{eff} = (44\pm1)$ pm/V are presented. However, the piezoelectric effect is relatively high: $d_{31} = -(210\pm10)$ pm/V. The real calculated values are $r_{33}^0 = (134\pm5)$ pm/V and $r_{13}^0 = (7\pm2)$ pm/V after this substantial correction. The same method was used for single crystal xBiScO$_3$-yBiGaO$_3$-(1-x-y)PbTiO$_3$ investigation [81]. The effective values from the experiment were $r_{33}^{eff} = 47$ pm/V and $r_{13}^{eff} = 19$ pm/V, piezoelectric coefficient $d_{31} = -55$ pm/V. The values after correction were $r_{33}^0 = 36$ pm/V and $r_{13}^0 = 4$ pm/V. This paper also deals with temperature dependency of Pockels coefficients at the temperature range from 20°C to 80°C but with assumption of constant piezoelectric coefficient d_{31}, most probably. Both experimental studies were performed carefully but even use of the subsequent correction is not able to eliminate the large error of coefficients determination, especially in the case of r_{13} coefficient.

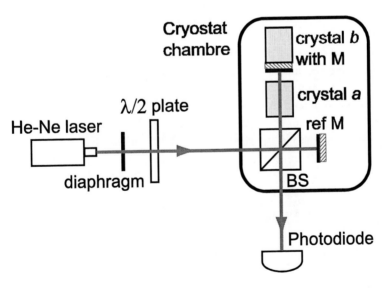

Figure 20. Michelson interferometer for electro-optical coefficients measurement with piezoelectric effect compensation.

This problem generally starts to be more complicated when we want to measure the exact temperature dependencies of electro-optical coefficients. The reason is that there are many temperature dependent parameters in Eq. (44). The independent measurement of temperature dependencies of r_{33}, r_{13}, d_{31}, n_o, n_e is very inaccurate by itself, and subsequent correction cannot give exact values.

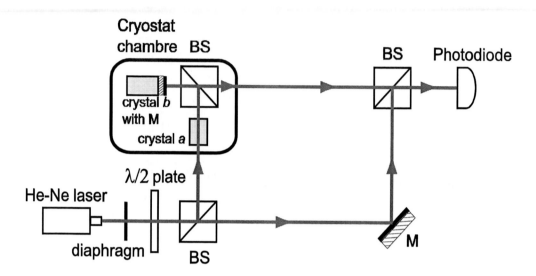

Figure 21. Mach-Zehnder interferometer for electro-optical coefficients measurement with piezoelectric effect compensation.

This is why we proposed the method for direct compensation of this piezoelectric effect influence. We used the second crystal, made from the same material as sample crystal, but with another length. The reflecting mirror is placed on the top of this crystal. If the same electric field is applied on an investigated sample (light is passing through it) and on the compensating crystal (light is reflected from this one), we can fully compensate for piezoelectric effect. This compensation can be made both in the Michelson (Figure 20) and the Mach-Zehnder interferometer arrangement (Figure 21).

Both the measured and compensating crystals are placed in the temperature chamber so they have the same temperature. The transversal electric field (with the same amplitude, frequency and phase) is applied to the both. It changes the length of a measured crystal about ΔL_a and refractive index about Δn. The length of the compensating crystal arm changes about ΔL_b. The total phase shift is, in the case of the Michelson interferometer (double pass thru measured crystals)

$$\Delta \varphi = 4\pi/\lambda \cdot \left[(n+\Delta n)\cdot(L_a+\Delta L_a)-1\cdot\Delta L_a-n\cdot L_a-1\cdot\Delta L_b\right]. \tag{49}$$

The following, similar equation is valid for Mach-Zehnder interferometer (only one pass thru measured crystal):

$$\Delta \varphi = 2\pi/\lambda \cdot \left[(n+\Delta n)\cdot(L_a+\Delta L_a)-1\cdot\Delta L_a-n\cdot L_a-2\cdot\Delta L_b\right]. \tag{50}$$

We can neglect term $\Delta n\cdot\Delta L_a$ and modify the equations for the Michelson and the Mach-Zehnder interferometers:

$$\Delta\varphi = 4\pi/\lambda \cdot [\Delta n \cdot L_a + (n-1)\cdot \Delta L_a - 1\cdot \Delta L_b] \qquad (51)$$

$$\Delta\varphi = 2\pi/\lambda \cdot [\Delta n \cdot L_a + (n-1)\cdot \Delta L_a - 2\cdot \Delta L_b]. \qquad (52)$$

The phase shift, measured by interferometer, is proportional to the refractive index change only when third terms in equations (46) or (47) compensate the second ones. Because the piezoelectrically induced change of the crystal length is proportional to crystal length, we can write

$$L_b/L_a = \Delta L_b/\Delta L_a . \qquad (53)$$

It finaly concludes that the ratio of compensating and measured crystal length must be, for the Michelson interferometer,

$$L_b/L_a = n-1 \qquad (54)$$

or, for the Mach-Zehnder interferometer,

$$L_b/L_a = (n-1)/2 . \qquad (55)$$

The refractive index of a crystal depends on the light polarization at equations (54) and (55). So the length of the compensating crystal is different for r_{13} and r_{33} measurements in general. We need to estimate the error of measurement for the case when we use only one crystal length ratio for piezoelectric effect compensation for both Pockels coefficient measurements. If we chose ΔL_b, exactly satisfying equation (49) or (50) for an ordinary ray, the relative error θ_r (the ratio of uncompensated piezoelectric induced phase shift to electro-optical induced shift) is given for an extraordinary ray at the Michelson interferometer by the equation

$$\theta_r = \frac{(n_e - n_o)\cdot \Delta L_a}{\Delta n_e \cdot L_a}. \qquad (56)$$

Because equations (25) and (35) are valid, it results that error is equal to

$$\theta_r = \frac{2d_{31}\cdot (n_e - n_o)}{r_{33}\cdot n_e^3}. \qquad (57)$$

This error is roughly 0.05 % for crystal LiNbO$_3$ and can be neglected. This error is large for PZN-PT crystal with piezoelectric coefficient $d_{31} = -(210\pm10)$ pm/V and other parameters: $r_{33} = 134$ pm/V, $n_e = 2.46$ a $n_o = 2.57$. This error is due to imperfect compensation of about 2.3 %. For very precise measurement we can use two sets of compensating crystals, but for standard measurement it is possible to perform measurement with only one compensating crystal.

The set-up described above with the compensating crystal also allows for compensating of thermal dilatation of measured crystal. There is a possibility to measure temperature dependencies of coefficients r_{13}, r_{33}, d_{31} (signal applied only to compensating crystal) and thermo-optic coefficient dn/dT in one sample arrangement.

4. THERMAL EXPANSION COEFFICIENTS MEASUREMENT

Although measurement of thermal expansion by laser interferometry is a very well known and frequently used method, adaptation of this technique to piezoelectric materials study is not trivial [82]. There are two sources of problems. The first one is that samples are often of small dimension (typically some millimeters), and the thermal dilatation of interferometer parts must be very well compensated. The second one is specific to piezoelectric materials which can exhibit a pyroelectric effect too [83]. This effect is the source of non-zero electric field in samples without a connection to an external power supply. This electric field can contribute to the change of the length of a sample by inverse piezoelectric effect too. The short-circuited samples must be used to measure thermal expansion coefficients with zero mechanical stress and zero electric field, as concluded from Eq. (1).

Two coefficients characterize pooled ceramics. The first one is the longitudinal coefficient α^E_{33}. It describes the change of sample length, characterized by strain, in the direction parallel to the sample polarization. The second is the transversal one, α^E_{11}. The change of the length is measured perpendicularly to the sample polarization. Upper index E means zero electric field:

$$S_{ii} = \alpha^E_{ii} \Delta T. \tag{58}$$

When the sample is not short-circuited, contribution due pyroelectric effect appears and coefficients are (with zero electric displacement $D=0$)

$$\alpha^D_{11} = \alpha^E_{11} - p_3 \cdot \frac{d_{31}}{\varepsilon_{33}} \tag{59}$$

$$\alpha^D_{33} = \alpha^E_{33} - p_3 \cdot \frac{d_{33}}{\varepsilon_{33}} \tag{60}$$

where ε_{33} is dielectric constant and p_3 is pyroelectric coefficient.

Some methods were tested at our laboratory, but only one was successfully adapted to this measurement. The set-up of this measurement is schematically drawn at Figure 22. The sample has a small thin mirror glued on the top. The similar reference mirror is glued alongside the sample on the holder of the sample. This part of interferometer is placed inside the evacuated temperature chamber. A laser beam with an appropriate diameter of some millimeters (a small beam expander can be used, if necessary) is directed by tilted mirror to both sample and reference mirrors. The reflected beam is separated from incident one by the beamsplitter and forwarded to CCD. The set of parallel fringes appears on CCD due a small angle tilt between the sample and reference mirrors. The thermal expansion of the sample causes movement of fringes. The dilatation of the sample holder and other parts of the interferometer is fully compensated by this arrangement.

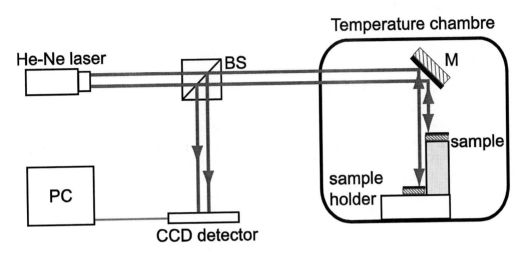

Figure 22. Schematic drawing of interferometer set-up for thermal expansion measurement.

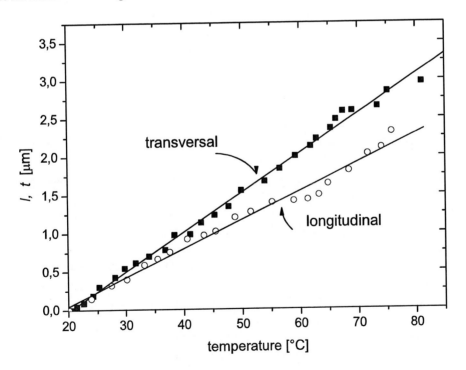

Figure 23. The change of the length of sample APC 880 due thermal expansion. Longitudinal, transversal change of the length means change in the direction parallel, perpendicular to the sample polarization.

Application of CCD to fringes detection improves the accuracy of this method. The used software can statistically interpret fringes' movement due thermal expansion and can reliably register displacement equivalent to $\lambda/20$ (about 30 nm). The distance between the detected fringes depends on the relative tilt of both the sample mirror and mirror on a holder. This random tilt is caused by the gluing of mirrors. However, the fringes' separation was acceptable in the vast majority of our measurements. It was also constant during warming, as was confirmed by our experiments.

The accuracy of the method was tested by measurement of the thermal expansion coefficients of two small prisms 2x5x5 mm^3, made from pure copper and aluminum. The obtained results have relative errors, about 4% for both samples, and were in good agreement with published values. Our research was concerned with PZT piezoceramic measurement. The set of samples from APC International, Ltd., USA, was tested. Results for APC 880 ceramic illustrate this method in Figure 23.

The linearity of this temperature dependency confirms that thermal expansion coefficients are constant at this temperature range. They are $(7.4\pm0.3)\cdot10^{-6}$ K^{-1} for longitudinal and $(10.3\pm0.3)\cdot10^{-6}$ K^{-1} for transversal expansion. Situations of real measurement are still more complicated due to the the fact that domain walls in ceramics are stabilized by this warming during the first measurement. So, the coefficients obtained from the first measurement of the sample differ from the next measurement results. However, the difference between results received by the second and, for example, fourth measurements is small and is in the range of experimental error. The measurements of samples without short-circuited electrodes were very complicated, with unacceptable great error. It was due to instabilities caused by random discharges of pyroelectrically induced charge on sample surfaces.

CONCLUSION

Laser interferometry, using a lock-in amplifier, is a very useful technique in the research of piezoelectric and electrostrictive effects. It is able to measure amplitude of piezoelectrically induced sample oscillations on the order of 10^{-12} m with accuracy between 5 - 10 %. It is limited by laser stability, homogeneity of the interference pattern in photodetector plane and quality of the sample surface. Forced oscillation of samples substrates are suppressed by using the Mach-Zehnder interferometer. Some random and systematic errors are caused by distortion of samples surfaces and cannot be quite eliminated. The miniaturized Michelson interferometer, placed inside cryostat or temperature chamber, can be used for wide temperature range measurement. This method is very successful for phase transition identification. The thin films measurement can be provided with a new, special type of double beam interferometer, also in wide temperature range. The method for measurement of temperature dependence of electro-optical effect was also tested. The problem of separation of both piezoelectric and electro-optical effects was solved by full compensation of piezoelectric effect. The interferometric method was also adapted to investigation of thermal expansion of piezoelectric materials. The solution, which compensates dilatation of sample holder, was found.

REFERENCES

[1] G. Haertling, *Journal of the American Ceramic Society* 82, 797-818 (1999).
[2] S. Tadigadapa and K. Mateti, *Measurement Science & Technology* 20, (2009).
[3] M. Maeder, D. Damjanovic, and N. Setter, *Journal of Electroceramics* 13, 385-392 (2004).
[4] N. Setter, D. Damjanovic, L. Eng, G. Fox, S. Gevorgian, S. Hong, A. Kingon, H. Kohlstedt, N. Park, G. Stephenson, I. Stolitchnov, A. Tagantsev, D. Taylor, T. Yamada, and S. Streiffer, *Journal of Applied Physics* 100, (2006).
[5] P. Muralt, *Journal of Micromechanics and Microengineering* 10, 136-146 (2000).
[6] U. Ozgur, Y. Alivov, C. Liu, A. Teke, M. Reshchikov, S. Dogan, V. Avrutin, S. Cho, and H. Morkoc, *Journal of Applied Physics* 98, (2005).
[7] K. Rao and K. Yoon, *Journal of Materials Science* 38, 391-400 (2003).
[8] C. CHEN and G. LIU, *Annual Review of Materials Science* 16, 203-243 (1986).
[9] L. Eldada, *Review of Scientific Instruments* 75, 575-593 (2004).
[10] S. TOLANSKY and W. BARDSLEY, *Proceedings of the Physical Society of London Section B* 64, 224-& (1951).
[11] M. ZONKHIEV and MYASNIKO.LL, *Soviet Physics Acoustics-Ussr* 13, 534-& (1968).
[12] D. BRUINS, C. GARLAND, and T. GREYTAK, *Review of Scientific Instruments* 46, 1167-1170 (1975).
[13] T. YAMAGUCHI and K. HAMANO, *Japanese Journal of Applied Physics* 18, 927-932 (1979).
[14] Q. ZHANG, W. PAN, and L. CROSS, *Journal of Applied Physics* 63, 2492-2496 (1988).
[15] M. GINDRE, W. URBACH, R. COURSANT, and M. FINK, *Journal of the Acoustical Society of America* 84, 11-19 (1988).
[16] W. PAN and L. CROSS, *Review of Scientific Instruments* 60, 2701-2705 (1989).
[17] W. PAN, H. WANG, and L. CROSS, *Japanese Journal of Applied Physics Part 1-Regular Papers Short Notes &* 29, 1570-1573 (1990).
[18] A. Kholkin, C. Wutchrich, D. Taylor, and N. Setter, *Review of Scientific Instruments* 67, 1935-1941 (1996).
[19] D. ROYER and V. KMETIK, *Journal De Physique Iv* 2, 785-788 (1992).
[20] D. ROYER and V. KMETIK, *Electronics Letters* 28, 1828-1830 (1992).
[21] L. Lian and N. Sottos, *Journal of Applied Physics* 87, 3941-3949 (2000).
[22] I. Guy and Z. Zheng, *Ferroelectrics* 264, 1691-1696 (2001).
[23] N. Takahashi, S. Kakuma, and R. Ohba, *Optical Engineering* 35, 802-807 (1996).
[24] V. Kulesh, L. Moskalik, and A. Sharov, *Measurement Techniques* 52, 1319-1327 (2009).
[25] H. OKAMURA and J. MINOWA, *Electronics Letters* 25, 395-397 (1989).
[26] J. Fernandes, F. de Sa, J. Santos, and E. Joanni, *Review of Scientific Instruments* 73, 2073-2078 (2002).
[27] F. Chen and U. Mohideen, *Review of Scientific Instruments* 72, 3100-3102 (2001).
[28] W. PAN, H. WANG, L. CROSS, and B. LI, *Ferroelectrics* 120, 231-239 (1991).

[29] S. Liu, W. Ren, B. Mukherjee, S. Zhang, T. Shrout, P. Rehrig, and W. Hackenberger, *Applied Physics Letters* 83, 2886-2888 (2003).
[30] L. Burianova, M. Sulc, and M. Prokopova, *Journal of the European Ceramic Society* 21, 1387-1390 (2001).
[31] L. Burianova, C. Bowen, M. Prokopova, and M. Sulc, *Ferroelectrics* 320, 629-637 (2005).
[32] M. Sulc and M. Pokorny, *Journal De Physique Iv* 126, 77-80 (2005).
[33] M. Sulc, J. Tryzna, and M. Pokorny, *Journal of Electroceramics* 19, 443-446 (2007).
[34] M. Sulc, J. Erhart, and J. Nosek, *Ferroelectrics* 293, 283-290 (2003).
[35] K. Cheng, H. Chan, C. Choy, Q. Yin, H. Luo, and Z. Yin, *Proceedings of the 2001 12th Ieee International Symposium* On 533-536 (2001).
[36] B. Mukherjee, W. Ren, and S. Liu, *Ferroelectrics* 326, 11-18 (2005).
[37] A. Kholkin, E. Colla, K. Brooks, P. Muralt, M. Kohli, T. Maeder, D. Taylor, and N. Setter, *Microelectronic Engineering* 29, 261-264 (1995).
[38] A. Kholkin, M. Calzada, P. Ramos, J. Mendiola, and N. Setter, *Applied Physics Letters* 69, 3602-3604 (1996).
[39] A. Kholkin, E. Akdogan, A. Safari, P. Chauvy, and N. Setter, *Journal of Applied Physics* 89, 8066-8073 (2001).
[40] I. Guy, S. Muensit, and E. Goldys, *Applied Physics Letters* 75, 4133-4135 (1999).
[41] C. Chao, Z. Wang, and W. Zhu, *Review of Scientific Instruments* 75, 4641-4645 (2004).
[42] H. Maiwa, J. Christman, S. Kim, D. Kim, J. Maria, B. Chen, S. Streiffer, and A. Kingon, *Japanese Journal of Applied Physics Part 1-Regular Papers Short Notes* & 38, 5402-5405 (1999).
[43] K. Prume, P. Gerber, C. Kugeler, A. Roelofs, U. Bottger, and R. Waser, *2004 14th IEEE International Symposium on Applications Of* 7-10 (2004).
[44] P. Gerber, A. Roelofs, C. Kugeler, U. Bottger, R. Waser, and K. Prume, *Journal of Applied Physics* 96, 2800-2804 (2004).
[45] S. Muensit and I. Guy, Applied Physics Letters 72, 1896-1898 (1998).
[46] J. Nosek, L. Burianova, M. Sulc, C. Soyer, E. Cattan, and D. Remiens, *Ferroelectrics* 292, 103-109 (2003).
[47] J. Nosek, M. Sulc, L. Burianova, C. Soyer, E. Cattan, and D. Remiens, *Journal of the European Ceramic Society* 25, 2257-2261 (2005).
[48] J. Nosek, M. Pokorny, M. Sulc, L. Burianova, C. Soyer, and D. Remiens, *Ferroelectrics* 351, 112-121 (2007).
[49] J. Nosek, M. Sulc, Z. Zheng, T. Radobersky, and L. Burianova, *Ferroelectrics* 370, 94-103 (2008).
[50] J. Nosek, M. Sulc, T. Radobersky, Z. Zheng, and L. Burianova, *Ferroelectrics* 389, 63-74 (2009).
[51] P. Gerber, A. Roelofs, O. Lohse, C. Kugeler, S. Tiedke, U. Bottger, and R. *Waser, Review of Scientific Instruments* 74, 2613-2615 (2003).
[52] P. Gerber, C. Kugeler, U. Bottger, and R. Waser, *Journal of Applied Physics* 95, 4976-4980 (2004).
[53] Z. Zheng, I. Guy, and T. Tansley, *Journal of Intelligent Material Systems and Structures* 9, 69-73 (1998).

[54] T. Hagimoto, S. Nam, M. Lebedev, H. Kakemoto, S. Wada, J. Akedo, and T. *Tsurumi, Transactions of the Materials Research Society of Japan,* Vol 31, No 1 31, 113-116 (2006).
[55] R. Herdier, G. Leclerc, G. Poullain, R. Bouregba, E. Remiens, and E. Dogheche, *Ferroelectrics* 362, 145-151 (2008).
[56] Z. Huang and R. Whatmore, *Review of Scientific Instruments* 76, (2005).
[57] C. Leung, H. Chan, C. Surya, W. Fong, C. Choy, P. Chow, and M. Rosamund, *Journal of Non-Crystalline Solids* 254, 123-127 (1999).
[58] K. Yao, S. Shannigrahi, and F. Tay, *Sensors and Actuators a-Physical* 112, 127-133 (2004).
[59] G. Leighton and Z. Huang, *Smart Materials & Structures* 19, (2010).
[60] R. Herdier, D. Jenkins, E. Dogheche, D. Remiens, and M. Sulc, *Review of Scientific Instruments* 77, (2006).
[61] W. Lehmann, P. Gattinger, M. Keck, F. Kremer, P. Stein, T. Eckert, and H. Finkelmann, Ferroelectrics 208, 373-383 (1998).
[62] L. Burianova, M. Sulc, M. Prokopova, and J. Nosek, *Ferroelectrics* 292, 111-117 (2003).
[63] M. Sulc, L. Burianova, and J. Nosek, *Annales De Chimie-Science Des Materiaux* 26, 43-48 (2001).
[64] M. Sulc and D. Barosova, *Ferroelectrics* 224, 557-564 (1999).
[65] P. Ney, A. Maillard, M. Fontana, and K. Polgar, *Journal of the Optical Society of America B-Optical Physics* 17, 1158-1165 (2000).
[66] M. Aillerie, N. Theofanous, and M. Fontana, *Applied Physics B-Lasers and Optics* 70, 317-334 (2000).
[67] M. Jazbinsek and M. Zgonik, *Applied Physics B-Lasers and Optics* 74, 407-414 (2002).
[68] H. ZHANG, X. HE, Y. SHIH, and S. TANG, *Optics Communications* 86, 509-512 (1991).
[69] X. Yin, Q. Pan, W. Shi, and C. Fang, *Applied Optics* 41, 5929-5932 (2002).
[70] M. Shin, H. Cho, S. Han, and J. Wu, *Journal of Applied Physics* 83, 1848-1853 (1998).
[71] F. Ghebremichael and H. Lackritz, *Applied Optics* 36, 4081-4089 (1997).
[72] V. Spirin, C. Lee, and K. No, *Journal of the Optical Society of America B-Optical Physics* 15, 1940-1946 (1998).
[73] V. Spirin, F. Mendieta, and K. No, *Ferroelectrics* 271, 1911-1916 (2002).
[74] L. Zhang, F. Zhang, Y. Wang, and R. Claus, *Journal of Chemical Physics* 116, 6297-6304 (2002).
[75] M. Tsukiji, H. Kowa, K. Muraki, N. Umeda, and Y. Tajitsu, *Japanese Journal of Applied Physics Part 1-Regular Papers Brief* 44, 7115-7118 (2005).
[76] S. Han and J. Wu, *Journal of the Optical Society of America B-Optical Physics* 17, 1205-1210 (2000).
[77] S. Han and J. Wu, *Journal of the Optical Society of America B-Optical Physics* 17, 1205-1210 (2000).
[78] H. Tian, Z. Zhou, M. Zhang, D. Liu, and L. Li, *Optics Communications* 281, 5420-5422 (2008).
[79] C. Herzog, G. Poberaj, and P. Gtinter, *Optics Communications* 281, 793-796 (2008).
[80] S. Yin, J. Wu, C. Zhan, and C. Luo, *Photorefractive Fiber and Crystal Devices: Materials, Optical* 5206, 280-289 (2003).

[81] S. Zhang, D. Jeong, Q. Zhang, and T. Shrout, *Journal of Crystal Growth* 247, 131-136 (2003).
[82] M. Okaji, N. Yamada, K. Nara, And H. Kato, *Cryogenics* 35, 887-891 (1995).
[83] W. Cook, F. Scholz, and D. Berlincourt, *Journal of Applied Physics* 34, 1392-& (1963).

In: Interferometry Principles and Applications
Editor: Mark E. Russo

ISBN 978-1-61209-347-5
© 2012 Nova Science Publishers, Inc.

Chapter 3

THE PRODUCTION AND ACCURATE MEASUREMENT OF A 1 MILLIMETER STEP STANDARD USING A COMMERCIAL AND A LABORATORY WHITE LIGHT INTERFEROMETER

Børge Holme, Arnt Inge Vistnes, Joachim Seland Graff and Juhi Bhatnagar

ABSTRACT

White Light Interferometry (WLI) is widely used for rapid and accurate topographic measurements of surfaces. By creating three dimensional images with lateral sizes from several millimeters to some tens of micrometers, both large and small samples can be studied.

The introduction of a separate laser interferometer in the scanning unit was meant to increase the accuracy of the depth scale from 1 % to about one hundredth of this value. This would mean that WLI could be used for accurate measurements of the vertical dimensions of millimeter sized components, and not only to quantify their surface roughness.

In order to investigate the capabilities of the Wyko NT9800 white light interferometer and to check its accuracy for large step measurements, we built our own step standard and measured the step height independently by a white light interferometer with a reference laser interferometer set up in the lab.

A thorough investigation of possible sources of error for the lab measurements was carried out. We further investigated the stability of the step standard over time and checked the reproducibility of the NT9800 measurements. Our results indicate that the accuracy of the NT9800 is about 0.05 %.

INTRODUCTION

Traditional Metrology and Alternative Measurement Techniques

Surface metrology is the science of measuring the spatial dimensions of surfaces. The tools available for surface metrology have developed rapidly in the past 30 years. Traditionally, surface roughness or step heights were measured by a mechanical stylus. Such an apparatus placed the tip of a sharp needle onto the surface with a small, constant force, pulled the tip sideways along the surface, and recorded the two dimensional (2D) profile $z(x)$ along a certain length, typically some millimeters. To obtain three dimensional (3D) data, several scans have to be carried out next to each other. Modern stylus instruments do this automatically [1].

Optical non-contact techniques, e.g. for measuring corrosion pit depths, have been available in most metallographic microscopes: The focusing knob has a micrometer scale which allows a manual measurement of the distance from the top surface to the bottom of the pit by taking the difference between the focusing depths when the two features are in best focus. Such 1D measurements are, however, tedious and operator dependent.

For large scale features – several centimeters or longer – one can either use a mechanical Coordinate Measuring Machine (CMM) [2], or so-called "structured light" where several stripe patterns are projected onto the component under study, while a camera captures the image from a certain angle and then calculates the shape. A CMM can be very accurate, but is slow since each point is measured in sequence by moving a probe until it touches the surface. Structured light, on the other hand, is fast since it measures each image (x,y) in parallel. The resolution is about 1/10 000 of the image size, thus for a 1 m large component, features down to 0.1 mm can be resolved [3]. Structured light requires surfaces that give diffuse reflection of light, thus mirror-like components can not be measured. Since this technique is based on projectors and cameras, there is a limit to how small objects that can be studied.

With the advent of fast, inexpensive computers with control over mechanical scanning units, a whole range of metrology instruments were invented. The Scanning Tunneling Microscope (STM) [4] and Atomic Force Microscope (AFM) are miniature versions of the stylus, with a much sharper tip and with the ability to scan in a raster to produce a three dimensional image $z(x,y)$ with atomic resolution [5]. A whole range of extensions to these instruments have been made, allowing measurements not only of the topography, but also local electrical, magnetic and chemical properties, surface softness, friction, etc. Common limitations for all the scanning probe microscopes are the slow operation – typically several minutes per image, the maximum image size which is limited by the scanning units to some hundred micrometers, and non-linearities in the scanning which produces image distortions unless the system is frequently calibrated and fine tuned.

Optical Profilers

The so-called *optical profilers* constitute another class of computerized surface metrology tools. They all use light to probe the sample surface, but do it in many different ways:

Confocal laser scanning microscopy [6] can see details inside transparent media, but is slow due to scanning along the three spatial directions.

In chromatic confocal microscopy [7], the need for a z-scan is eliminated by using the chromatic aberration of the objective to focus light of different wavelengths to different positions along the optical axis. By analyzing the color of the reflected light, the local height of the surface can be calculated [8] with an accuracy down to 20 nm and a resolution of 5 nm.

Reflected polarized light colorimetric analysis [9] can measure very small height differences on thin films with accuracy better than 1 nm, but requires transparent and thin samples, so for instance bulk metals can not be studied. The lateral resolution is about 0.5 µm at best, just like for any other technique based on optical microscopes.

Focus variation microscopy [10] is the automated 3D equivalent of the pit depth measurement technique mentioned above. It is a fast technique since all points in the image are captured in parallel. The main advantage is that it can image slopes as steep as 85°. However, smooth, polished or featureless surfaces (like glass) can not be measured. The depth resolution depends on the magnification and ranges from 2 µm to 10 nm.

White light interferometry [11] in the Vertical Scanning Interferometry (VSI) mode has few of the drawbacks of the previously mentioned techniques by being fast, accurate and able to measure any surface with a reflectivity of more than 2 %. It is not well suited for measuring steep surface facets since the maximum slope is 27°, and it can not measure transparent films thinner than a few micrometer, but overall WLI is a very versatile method. The rest of this chapter will focus on the WLI technique.

White Light Interferometry (WLI)

A WLI microscope is similar to an ordinary optical microscope, but has an interferometer built into the objective lens (Figure 1).

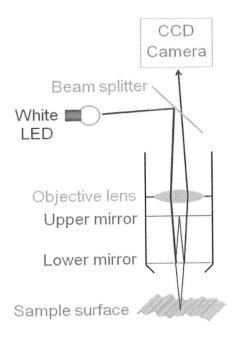

Figure 1. Schematic drawing of a white light interferometry microscope.

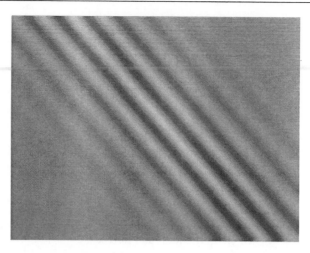

Figure 2. Image from the WLI microscope camera of a tilted glass plate, showing the white light interference fringes with the brightest fringe corresponding to the line of exact focus. The height difference between two neighboring bright lines is one half the average wavelength, or about 300 nm.

The beam from the light source is split; one part to be reflected from the sample, and the other part bouncing between the interferometer mirrors. When these two rays recombine and continue towards the grayscale video camera, they will interfere. If the optical path traveled by the two rays happens to be exactly the same, the two beams will interfere constructively and give a bright spot in the image. If the path difference is half the (average) wavelength, the light waves will interfere destructively and a dark spot appears in that part of the image. For other path differences, some shade of gray will occur, depending on the exact distance.

A monochromatic light source with infinite coherence length, represented by a pure sine wave, would give a bright spot also for a path difference of one wavelength – or any multiple of one wavelength. The reason for using white light and not a laser is that for white light, having a short coherence length, the constructive interference at ± one average wavelength is not as perfect as for a laser. Thus, the intensity from a spot on the sample which gives a path difference of one average wavelength is lower than from a spot giving zero path difference. The further away from zero path difference, the less intense are the constructive and destructive interferences. The image of a flat but tilted surface in a WLI microscope becomes a series of fringes with one brightest and then alternately dark and bright fringes, with progressively lower intensity until the interferences fade into an evenly gray surface (Figure 2).

WLI Measurements

For a well adjusted WLI microscope, the plane of best focus coincides with the plane where the path difference is zero. This means that all points on a rough surface that are in exact focus, will also have maximum brightness due to the interference effect. By scanning the lens mechanically in the z direction such that the plane of best focus passes from the highest peak to the deepest valley, all points of the imaged surface will experience a change in brightness from the normal gray, through alternating dark and bright shades with increasing intensity culminating in the brightest shade at best focus, and finally a series of dark and bright shades dimming out to the normal gray level (Figure 3). During a WLI measurement, the video camera records these intensity fluctuations in the course of the 10–50 s it typically takes to scan the lens across the entire height range of the sample. Each frame in the "video" is associated with the distance the lens has travelled since the beginning of the scan. A computer then finds the lens position corresponding to the center of the brightest fringe for each pixel in the image. The key to WLI's usefulness is that this position can be determined with an accuracy and a precision which is far smaller than the wavelength of light – just as the peak of a sine wave can be located much more accurately than half the wavelength.

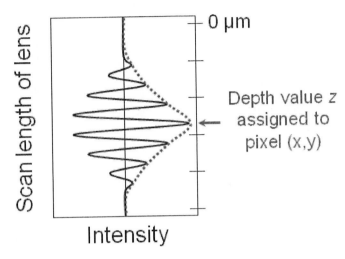

Figure 3. Finding the depth, z, of pixel (x,y) from the intensity curve obtained at that pixel during a scan of the lens. The position of the maximum interference can be determined with accuracy and precision much smaller than the wavelength of the light used.

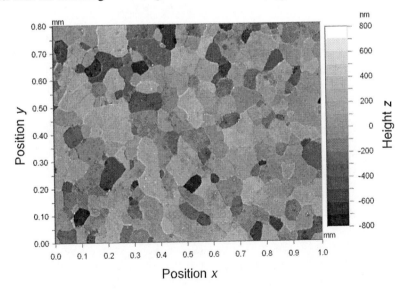

Figure 4. Typical WLI image displayed as an (x,y) matrix where the height $z(x,y)$ is denoted by a gradual scale from black to white. Here the convention was followed of using the average height as the zero level for the z-scale. The sample is made of aluminum which was polished and then etched in a special way to obtain a stepped surface with the grains at different heights.

Once the lens position (z) at maximum intensity – corresponding to the position of best focus – is known for every pixel (x, y) in the imaged area, we have a full 3D topographic map $z(x,y)$ of the surface. This map is usually represented by the computer as a rectangular image matrix (x,y) where a gradual color scale shows the height (z) of each pixel (Figure 4). However, also shaded, side-view 3D representations are possible.

White Light Interferometry with Internal Reference for Self-Calibration

From the above description we see that one of the factors limiting the accuracy of the WLI microscope is how correctly the computer assigns a depth value (z) corresponding to the position of the lens at maximum constructive interference.

Since the lens is mechanically moved towards the surface, or equivalently the sample is moved towards the lens in some systems, the microscope requires good mechanical stability, reproducible motors movements, and an accurate gauging system to tell the computer how far the lens has travelled at any given time. Such mechanical systems tend to loose accuracy when the room temperature changes, and need frequent re-calibration to stay accurate. Experience from the Wyko NT2000 WLI microscope showed that the accuracy could be off by up to 2 % a few weeks after the last calibration [12].

However, during the course of a day with a stable room temperature, the calibration could stay constant to within 10 nm when checking against a glass step standard of 10 µm height. The precision (here one sample standard deviation over 10 measurements) was also typically 10 nm.

The Wyko NT9800

In 2006 the Wyko NT9800 was introduced by Veeco (now Bruker Corporation). Here the accuracy and reproducibility was significantly improved by the introduction of a laser interferometer which gauged the movement of the objective lens. With the lens position (z) continuously measured by interferometric means, the system is much less sensitive to temperature changes or mechanical instabilities.

Such a self-calibrating instrument delivers step height measurements with a precision (average standard deviation for repeated measurements) of 7 nm on a 10 µm step standard, based on our own control measurements over more than a year.

The accuracy could not be properly checked using this step since the given uncertainty (1 σ) was 0.9 %, or 90 nm, thus much larger than the expected accuracy of the instrument itself. The maximum scan length of the NT9800 is 10 mm, which allows quite large steps to be measured. The step height is limited by the working distance of the lens, which in the case of the most used 10x lens is 7 mm.

During the development of this instrument, a WLI interferometer which was built on an optical table with a geometry similar to the NT9800 laser scan-length measurement unit showed differences of about 200 nm for repeated measurements on a 1 mm step gauge [13]. The motivation behind the present study was to determine the accuracy and precision of our NT9800 for a 1 mm step. In order to achieve this, we needed a step standard with very parallel surfaces and a precisely known height.

Step Standards

Surface steps for calibration purposes are commercially available. They are typically made of glass or silicon and produced by lithography and etching. Such step standards are available for heights between 10 nm and 250 µm, and typically cost 100 USD. For certified standards traceable to official standards organizations, the price may be 10 – 30 times as high. The step height accuracy is typically 1 nm for step heights below 25 nm, and 1 % of the height for larger steps. So-called *ultra thick step height standards* with 150 – 250 µm steps are available (VLSI Standards, Inc.), with step height accuracies of 0.12 µm (0.05 %). Such a step could have been useful, but we wanted a higher step of 1 mm, measured with an even better accuracy to match the expected accuracy of the NT9800. For larger step heights, one commonly uses so-called gauge blocks; accurately machined metal bars of different thickness. These can be put next to each other to produce a number of different step heights. By wringing two gauge blocks together, one can get a solid step held together without adhesive. However, it is difficult to obtain perfectly parallel step surfaces in this way [2].

EXPERIMENTAL

Making a Step Standard

The lack of commercially available 1 mm step standards prompted us to attempt to make our own step, and then measure it accurately. Several obvious approaches were tried before an unconventional but workable solution was found. Gluing two pieces of glass on top of each other would seem like the easiest way to produce a step standard. However, this approach has two main problems: Even microscope slides are not perfectly flat. On a 1 mm thick slide, we measured a root-mean-square local roughness (R_q) of 23 nm (standard deviation 5 nm). Secondly, a 1 mm glass slide is quite flexible, so it is difficult to get an evenly thick glue line. This results in slightly non-parallel surfaces for the two halves of the step. Then the step height is no longer a well defined value since it varies with position along the step. Letting the pieces of glass melt together in a furnace to avoid the problem with glue, caused the originally plane glass surfaces to buckle and round off near the corners. So this was not a useful way to join the two pieces.

The standard approach of lithographic etching works well for small steps, but for steps of 1 mm or larger, the etching will take a very long time and may not give a flat and parallel enough lower surface due to small local differences in the etch rate. The unconventional approach, which was chosen in this study, was to use a polished and cut block of stainless steel as the basis for the step, gluing the two pieces together under the WLI microscope to obtain the required parallelism, and orienting the cut such that any expansion of the glue line would not significantly affect the step height. Stainless steel is a common material for gauge blocks used as standards for precision measurements [14]. Mechanical grinding and polishing can give quite even and flat surfaces on sufficiently hard materials. For the stainless steel block, we obtained a local root-mean-square roughness (R_q) of 10 nm (standard deviation 2 nm) after manual metallographic grinding and polishing, using 1 µm diamond particles in the final polishing step. Since the polishing will always round off the edges of the sample, we could not use any polished edge as the basis for the step. Instead we cut the block at 85° angle to the polished surface. Exactly 90° would be preferable for the stability towards glue line swelling, but the few degrees from an orthogonal cut allowed a 100 µm overshoot of the upper surface. This geometry hides the glue line and some of the excess glue when the imaging is done from directly above the polished surface (Figure 5). The 5° deviation was taken as a compromise between these two effects.

The cutting was done on a water cooled saw with a rotating cutting disc. We then ground the freshly cut surfaces on coarse silicon carbide paper (FEPA P320) to remove regions of the polished surface that had been deformed by the cutting. Care was taken to avoid faceting or rounding off the cut surfaces.

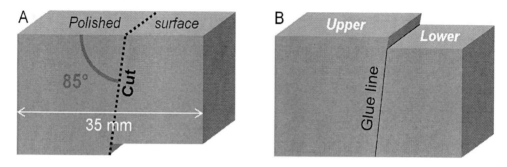

Figure 5. Schematic illustration of how the step standard was made. From a block of stainless steel a 1 mm deep region was machined away on the under side (A). The opposite side was ground and polished flat before cutting at an 85° angle. The smaller piece was shifted downwards before gluing, creating a 1 mm step in the polished surface (B).

Figure 6. The special holder designed for gluing the pieces of the steel step together under the WLI instrument to make the lower and upper mirror surfaces parallel. The polished surfaces of the step are marked with "L" and "U", respectively. The glue line begins just above the arrow.

To ensure two parallel surfaces after gluing, we mounted the two steel pieces under the WLI microscope on special holders which allowed small and controlled adjustments of their relative orientations (Figure 6). One piece was put on a surface which could be tilted around a horizontal x-axis. This piece could also be moved in the x-direction by a micrometer screw. We used the screw to move the pieces towards each other for the gluing. The other piece was fastened to a mirror holder, normally used on optical tables, where two fine-threaded screws allowed rotations around the horizontal y-axis and the vertical z-axis.

The steel pieces were fastened to the holders by minute droplets of rapidly solidifying glue (Loctite® Super Glue Gel) applied by a wooden toothpick. Three to five specks of super glue was enough to keep the steel pieces firmly in place, while allowing the finished step to be detached by a light tap with a rubber hammer.

The two pieces were glued to each other with a slow curing epoxy (EpoFix from Struers, Denmark) with a nominal curing time of 12 h. We took care not to add too much glue on the surfaces to prevent epoxy from flowing out on the lower part of the step. A broad line of glue would give a wide region between the upper and lower parts of the step which would have to be masked from the WLI images in order to get correct step measurements in the WLI microscope. However, such masking would not be possible in the lab setup (described below) for the official step height measurements. Thus, any glue on the lower surface, beyond the 100 μm overshoot of the upper surface, was undesirable. We ended up with some excess glue that was not hidden by the upper surface, but it was possible to find regions along the step with virtually no visible glue within the 0.84 mm image height (in the y direction) used for the microscope imaging.

While the epoxy solidified under the WLI microscope, we kept focusing up and down between the upper and lower surfaces of the step and then adjusted the tilt to gradually make the two surfaces more and more parallel. We did not need to make WLI images at this stage since the interference fringes gave instant information about the tilt. The height difference between a bright stripe and a dark stripe in WLI is about 150 nm. Such a tilt is easily seen in the WLI microscope. Even with the mechanical tilting mechanisms available to us, it was possible to align the two surfaces so they tilted a fraction of this amount over the 0.55 mm image half-width. Due to masking of the regions closest to the step, the effective width of the imaged regions used for step height calculations were 340 μm for the upper step and 450 μm for the lower step. The asymmetry was due to more noise in the data near the edge of the upper step. Within these regions the maximum height difference was typically 15 nm for both surfaces. This mainly resulted from a slight rounding of the surfaces since an unplanned, light polishing had to be carried out after the cutting. This was to remove organic contamination from the cooling water which had burned into the polished surface due to the high temperature during cutting. If the cutting had been done in a more controlled way, even flatter surfaces could have been obtained. Using the gluing rig described here, an alignment with a parallelism better than 5 nm over the 1.1 mm image area should be possible.

With a small but non-zero tilt, it became important to always measure the step height at the same position. To ensure this, we marked the most parallel region along the step by engraving arrows on both surfaces, just beyond the edges of the 1.1 mm image frame used for the step height measurements in the WLI microscope. The position of the arrows was marked a bit further away by dots from a permanent marker pen. This made it easy to find the same location each time the step was measured.

Lab Setup for White Light Interferometer with Laser Reference

The experimental setup consisted of two light sources, an interferometer and two detectors. We refer to Figure 7 while describing the setup in detail.

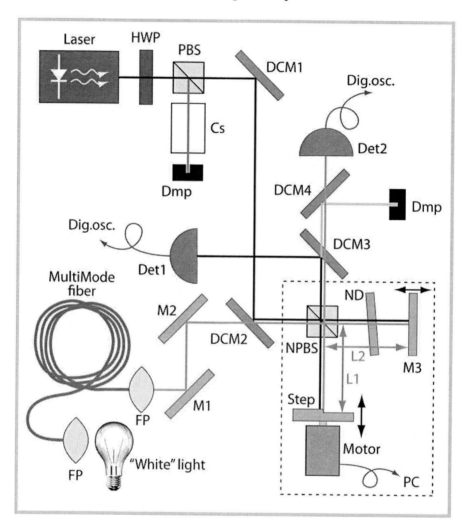

Figure 7. Schematic diagram of the lab setup built in the quantum optics lab at the University of Oslo to perform the control measurements of the steel step height.

The Laser Beam

The measuring stick in the step height measurement was the wavelength of a stable laser with sufficiently long temporal coherence. We used an infra red (IR) solid state laser, 852 nm, 140 mW (type DL100, Toptica Photonics), equipped with various control units. The wavelength could be adjusted to the D_2 absorption/emission line for cesium (Cs), which is 852.120532 nm in air [15].

The laser beam passed through a half-wave plate (HWP) and a polarizing beam splitter (PBS). The half wave plate in combination with the polarizing beam splitter made a continuously variable directional coupler. About 99 % of the incoming light intensity could be divided continuously between the forward direction and the 90° deflected path. For the forward beam the two elements acted as an attenuator.

A small sealed glass vial containing diluted cesium vapor (Cs), was placed in the deflected beam in front of an absorber (Dmp). Since light at 852 nm is not visible, we had to use an IR scope (Electro Viewer 7215, Electrophysics) in order to adjust the laser wavelength so that the cesium atoms were excited by the beam that passed through the cell. This adjustment to the excitation of the cesium D_2 line can be done with a precision corresponding to about 1 nm in the wavelength [16].

The main beam continued to a set of two dichroic mirrors (DCM1 and DCM2) that reflect 99 % of the 852 nm light when the angle of incidence is 45°. The mirror is transparent to light with other wavelengths. The latter of the two mirrors directed the IR beam into a standard Michelson interferometer. Here the beam was split 50/50 in two directions by a non-polarizing beam splitter (NPBS). The part of the beam that continued straight ahead hit a mirror (M3) mounted on an adjustable slide, and was reflected back to the beam splitter. The other half of the beam was reflected 90° by the beam splitter, hit the step and was reflected back to the beam splitter again. We tried to let the IR laser beam hit only one side of the step to get a purer interference signal.

The reflection from the reference mirror M3 was considerably stronger than the reflection from the stainless steel surfaces of the step. A weak neutral density filter (ND) was placed in the reference arm of the interferometer to compensate for the difference in the degree of reflection. The neutral density filter was placed with its axis slightly tilted away from the beam, so that reflections from the surfaces of the filter would not influence the interference.

Both of the reflected beams from the mirror+step interferometer were again divided in two when they passed the NPBS in reverse. One part was directed towards the laser and was intentionally misaligned slightly so that it did not enter the laser. The other part of the beam went towards the detectors. Roughly 99 % of this beam was reflected by a third dichroic mirror (DCM3) onto detector Det1 for IR light. The remaining 1 % of the IR beam passed through the DCM3 and headed towards the detector for white light (Det2). Since the laser signal was much more powerful than the white light beam, we had to introduce another dichroic mirror (DCM4) in order to further reduce the amount IR light that reached the white light detector.

The White Light Beam

The white light beam is much more difficult to handle than the laser beam, since no white light source has a well defined wavefront and beam. The lateral coherence is poor, i.e. the light has a "short spatial coherence length". This means that it is impossible to focus an extended white light source to a very narrow spot without loosing most of the intensity. We tried various kinds of white light sources, and ended up using a 75 W halogen lamp, normally used for illumination of samples in stereo microscopes (LQ1100, Fiberoptic-Heim). The light from this unit was sent through two flexible light cables, consisting of about 5 mm diameter optical fiber bundles. The light intensity from one of these cables was approximately 400 mW when the lamp electrical power was 72 W. The light intensity was measured by a ThorLabs PM100 power meter.

The white light source had a fan built into the unit. A fan makes vibrations, and we did not want to let these vibrations influence our measurements. The light source was therefore placed on a separate stand just next to the optical table, and the fiber bundle was not in direct contact with the fiber port.

We tried to focus the white light into a single mode fiber in order to shape the light beam, but the intensity that the fiber let through was very low indeed, and far too low for the signal detector. A multimode fiber (2 m GIF625, ThorLabs) was used instead, with a high-quality adjustable fiber port (FP), including an Olympus 20x microscope objective to focus the light beam into the fiber as efficiently as possible. The light intensity transmitted through the fiber was approximately 7 µW when 400 mW was fed into the fiber port entrance. The fiber port (FP) at the exit of the fiber was adjusted so that the beam diameter was minimum 2 mm at the interferometer. The beam intensity profile was not Gaussian, but rather rectangular.

The white light beam was given correct position and direction relative to the laser beam by means of two mirrors (M1 and M2). The beam was then fed straight through the dichroic mirror (DCM2) that reflected the IR laser beam into the interferometer. The white light beam followed the same path as the IR beam, but slightly shifted to the side so that half of the white light beam would hit the upper surface of the step and the other half the lower surface.

The two reflected beams of white light from the interferometer were directed towards the fiber port, or the detector Det2, respectively. Most of the white light went straight through the dichroic mirrors DCM3 and DCM 4 before it was absorbed by Det2.

Detectors and Miscellaneous

The detectors were "home made" analog detectors based on 10 mm * 10 mm photocells. The signals were fed into two channels on a LeCroy WavePro 7100A 1 GHz digitizing oscilloscope. The two signals were digitized simultaneously for 10 s at a frequency of 100 000 samples per second while the step moved.

The step was moved by a stepper motor (ThorLabs DRV001 stepper motor actuator) run by a control unit (ThorLabs BSC103 stepper motor controller) connected to a PC. The motor's step length was approximately 50 nm and the speed 0.125 mm/s. During a measurement, the motor moved the steel step towards the interferometer beam splitter (NPBS) over a 1.25 mm distance such that first the top surface of the steel step, then the bottom of the step would have the same path length as the reference beam. This produced two sets of white light interferences, about 8 s apart (Figure 8, right-hand side).

The temperature of the setup was monitored by a digital thermometer (Ama-digit, type "as15th") by placing the probe in thermal contact with the steel step.

Signal Processing

Two separate signals were recorded simultaneously on the digitizing oscilloscope. One signal was the interference signal for the laser beam, coming from the first detector (Det1). Since the coherence length for this laser was of the order of meters, the interference led to a sinusoidal signal with roughly the same amplitude during the full steel step movement, as illustrated to the left in Figure 8. Unfortunately, the stepper motor did not move with exactly constant velocity. The speed changed periodically with a period of approximately six interference oscillations (2.5 µm).

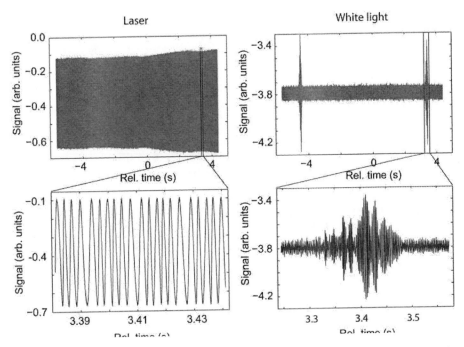

Figure 8. Signals from Det1 (left) and Det2 (right) as recorded by the oscilloscope. Plots in the lower row show details of the two signals. The IR laser interference (left) has a wavelength that changes with time since the stepper motor did not move with a constant velocity. The white light interference (right) had two strong peaks, both made up of smaller intensity fluctuations.

The signal from the white light detector (Det2) showed a relatively constant value for most of the scan with no obvious interference pattern. However, when the distance L2 between the beam splitter (NPBS) and mirror M3 was equal to the distance L1 between the beam splitter and either the top or the bottom surface of the step, a white light interference pattern appeared. The pattern stretched through many IR wavelengths, roughly 20 μm for the entire main group (right part of Figure 8). This differs from the nice and symmetric fringe pattern of Figure 2 with one main wave packet about 2.7 μm long. The reason is that the lab setup gave several, partly overlapping white light interferences. This is most likely due to multiple reflections in the two internal surfaces of the NPBS. Single reflections alone will produce four series of coherent wave trains that may all interfere and produce a complex set of interference fringes.

In Figure 8 the intensity data is plotted on the original, arbitrary scale. Before any further signal processing, the average intensity of the entire 10 s dataset was subtracted from both signals. This gave IR and white light intensities fluctuating symmetrically above and below zero.

The noisy background in the white light detector consisted of several contributions. In a preliminary test, the light source (halogen lamp) was supplied by 50 Hz AC current. The signal then had severe 100 Hz oscillations. We had to switch to a DC power supply in order to reduce this unwanted signal.

The laser light in our apparatus lead to diffusely scattered IR light with intensity oscillations in tune with the signal at the IR detector. We introduced black painted metal plates several places in the setup in order to reduce this effect. Even so, some of this laser light reached the white light detector. We could, however, remove this contribution almost completely by subtracting a tiny part of the IR signal from the original white light signal before further processing.

Measuring the Step Height with the Lab Setup and the NT9800

Measurements Using the Lab Setup

The two white light interference patterns, corresponding to the two surfaces of the steel step, were quite similar. However, they were not true copies of each other. In order to measure the step height, we had to determine the distance between the two peaks. This led to two challenges: First, to compare the two white light interferences and find their relative shift distance. Second, to translate that shift from the time scale over to a distance scale. In order to minimize the probability of making fundamental errors in the analysis, two of the authors developed independent strategies for how to carry out the measurements. The simplest, semi-automatic method will be referred to as "the minimum method" while a more complicated but fully computerized procedure will be called "the correlation method". Figure 9 shows some initial steps of the calculations that were common to both methods. The original signal (A) was Fourier filtered (B), squared (C) and then smoothed (D) by a Gaussian filter. (The Fourier filtering, which removed frequencies below about 120 Hz to get rid of noise, was only used in the minimum method.)

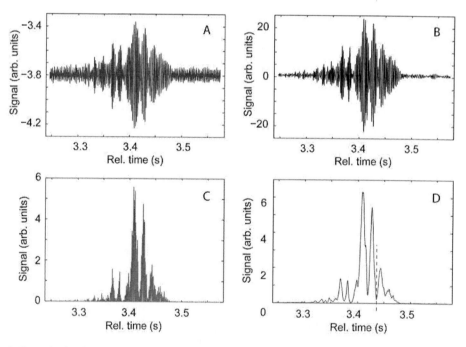

Figure 9. Steps in the data processing of the second white light interference. Original signal (A), Fourier filtered signal (B) after setting the mean intensity to zero, squared signal (C), and the same squared signal after Gaussian smoothing (D). The vertical, dashed line shows the position of the characteristic minimum used in the "minimum method" to pinpoint the position of the second interference relative to the corresponding point in the first interference.

The minimum method used a characteristic local minimum near the right-hand side of the filtered signal (dashed, vertical line in Figure 9 D) as the reference position for each interference. By a simple Matlab program and a computer, we could determine the number of wavelengths of the IR laser signal between these two reference points. One oscillation of the IR signal corresponds to the step moving a distance of half the laser wavelength in air, i.e. $\lambda/2 = 426.060$ nm, since the light travels this distance twice. Normally, the two reference points would not happen to lie at the same place in the IR cycle. In our case, it was slightly more than 2689 IR interference periods between the reference points. Exactly 2689 cycles would correspond to a distance of 1.145 676 mm.

In order to increase the accuracy of the method, we had to include the fractional IR cycle separating the reference points. This task was made difficult by the non-constant speed of the stepper motor movement. Therefore, the "local wavelength" of the IR signal had to be used. Searching from the reference point of the first interference, the nearest downward crossings of the IR signal on both sides were located. Using the indices of these three points, the fractional IR cycle before the first full cycle could be calculated. In the same way, the fractional cycle between the last full IR cycle and the reference point of the second interference was found. Adding these two endpoint fractions to the 2689 full IR cycles gave the total distance between the two white light interferences in units of $\lambda/2$. For instance, the first measurement gave 2689.814 IR cycles, which corresponds to a step height of 1.146 023 mm.

The correlation method used every half-cycle of the IR interference signal as a 213.030 nm long "yardstick" to generate a depth scale for the array of WLI data with equal spacing of about 1.5 nm. This was done in two stages by working on several arrays of numbers in the computer; one array for the common time values, one array for the IR signal, one array for the white light signal, and two empty arrays to be filled with depth data.

The first step included locating each measuring point on the time scale immediately following an upwards or a downwards crossing of the IR signal. We knew that within this time interval, the step standard had moved 213 nm towards the beam splitter, regardless of how fast the stepper motor had been moving. Making the approximation that the stepper motor velocity had been constant within this half-cycle (213 nm), we generated a linear depth scale interval of 213.030 nm length with equal spacing for all data points obtained during this time interval. Repeating this for all the half-cycles of the IR signal produced an accurate depth scale, but with stepwise unequal spacings between the points since the IR half-cycles lasted for different times.

The second step was therefore to re-map the data over to a linear, equally spaced depth scale. This was done by determining the length (in nm) of the full dataset and dividing by the number of data points to get a suitable length interval for the new depth scale. This new point spacing was typically 1.5 nm. Then the analysis program took each new depth value in turn and found the two depth values from the old scale on each side of this depth. The new white light intensity value was taken simply as the average of the two corresponding neighboring points in the white light array. Control plotting showed that this gave a sufficiently accurate reproduction of the original signal, so a more sophisticated linear interpolation scheme was not necessary here.

We now had a squared interference signal as shown in Figure 10. In the graph, the second white light interference has been shifted and is plotted on the negative scale for easier comparison. Before the analysis program compared the entire smoothed first white light interference with the whole of the second interference, the signal was smoothed by a Gaussian filter. The filter used here was a bit wider than in the minimum method in order to ensure a smooth signal with only the major intensity fluctuations present (thick lines in Figure 10).

With the two white light interferences represented as data arrays of intensity on a common, linear, equally spaced length scale, the cross-correlation of the two smoothed interferences could easily be calculated. This was done in the standard way of shifting part of the signal by a fixed amount and then summing all the products of the shifted times the non-shifted signal for all measured points along the entire length of the interferences: *Cross-correlation(shift length)* = Σ_x *WLI(x)*WLI(x+shift length)*. Repeating this for all possible shift lengths gives the Cross-correlation vs. Shift length curve of Figure 11. The global maximum of the curve lies at a shift length of 1.146 302 nm and was obtained when the two smoothed curves of Figure 10 were aligned as shown in that figure. All calculations in the analysis program were based on the very precisely defined vacuum wavelength of the IR light. The correction to the wavelength in air, which reduces the values by 0.0266 %, had not been done at this stage.

Simply taking the highest calculated value on the cross-correlation curve gives the step height with a numerical resolution corresponding to the sampling distance, 1.5 nm in this case. This was sufficient for our purposes. For data obtained in March 2009, the sampling of the oscilloscope was less dense and corresponded to 19 nm between each measured point. For those data, the precision could be slightly improved by fitting a parabola around the 21 points nearest to the peak of the cross-correlation curve and then using the position of the maximum of the parabola as the step height. With this fitting technique the numerical depth resolution is not limited by the sampling rate.

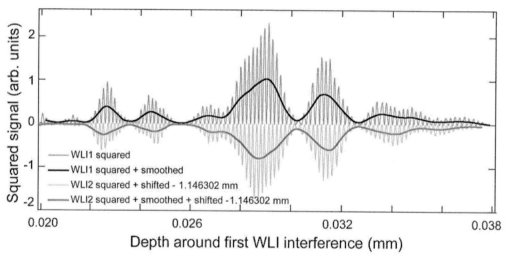

Figure 10. Stages in the data analysis of the "correlation method": The two spiky curves show the original squared white light interference signals. The second interference is plotted along the negative *y*-axis for clarity. It has been shifted a distance of -1.146 302 nm along the depth axis, since this was the shift which gave the best matching of the two interferences. The smoothed versions of the respective signals are shown in thick, dark lines.

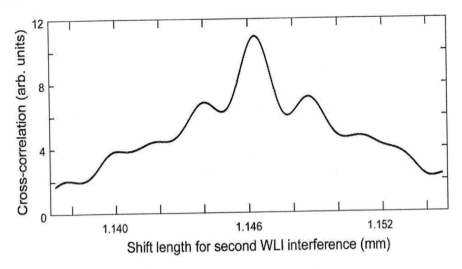

Figure 11. The cross-correlation curve obtained by shifting the two smoothed curves in Figure 10 relative to each other by the amount "Shift length" and summing the products of the two curve values at all positions along the length axis. The highest peak of this curve corresponds to the step height we wanted to measure.

Since the correlation method compares the whole length of the white light interferences, while the minimum method uses only one point, it is not surprising that the spread in the data is larger for the minimum method. The two datasets labeled "Lab 2010..." in Figure 12 show series of 17 measurements, and give a good impression of the differences among the data analysis methods. Lab data analyzed by the correlation method give quite consistent results, while the minimum method applied to the same data give more scatter and a 50 nm lower mean value (Table 1). Statistical analysis using a two-sample t-test assuming unequal variances [17] showed that the results from the two methods are significantly different (significance probability $p = 0.006$).

Measurements Using the NT9800

During the acceptance test of the NT9800 in March 2009, the step height data measured in the lab were compared to measurements made by the NT9800. The measurements appeared to be similar to within 50 nm. However, it was later found out that the lab measurements had been based on the wavelength in *vacuum* (852.347 276 nm [15]) of the cesium D_2 absorption line, to which the IR laser had been tuned. When scaling down the step height values by the 0.0266 % corresponding to the D_2 wavelength in *air*, the NT step height was 400 nm too high (temperature corrected). To get a second check of these results, we decided to repeat both the lab measurements and the NT9800 measurements 18 months after the original trial. For the new NT9800 measurements, we placed the step under the WLI microscope with the edge of the step along the y-direction, aligned the optical axis of the lens to be orthogonal to the step surfaces, and set up the software to automatically make 10 images in a row and save them on the hard drive. We used a self-made program to automatically calculate the step height from the images. The program read each new image from the disc, masked out a strip (20 % of the image width for the upper surface and 10 % for the lower surface) along the step edge to exclude glue and edge effects from the measured area, aligned the lower surface to be perfectly horizontal in case of slight non-perfect alignment of the lens, and finally determined the vertical height of the step as the distance between the mean height of the lower and upper surface. Manual step height measurements using the software for the NT9800 gave similar results but was much more time consuming.

Comparing Step Height Measurements from 2009 and 2010

The results in Figure 12 show that all the lab results are quite consistent, with a standard deviation of 60 nm, even though the data obtained in 2009 give a somewhat higher mean value. The step heights from the NT9800, on the other hand, differ by about 1 µm, lying 500 nm above the lab average in 2009 and 500 nm below the lab data in 2010. This indicates that despite the internal laser reference in the NT9800, the calibration may drift over time. There is no particular reason to assume that the measurements in March 2009 and September 2010 represent any extremes of the drift, but we can conclude that the error is at least 500 nm for a 1 mm step. This means that the accuracy of the NT9800 can be no better than 500 ppm for such step heights.

Since it is very important to know the accuracy when performing measurements based on difference images of surfaces with large height ranges [18, 19], we need to carefully consider all conceivable corrections or possible measurement errors that could explain the observed differences. The first thing to do is to compensate for differences in the temperature of the step between measurements which would lead to thermal expansion or contraction of the step.

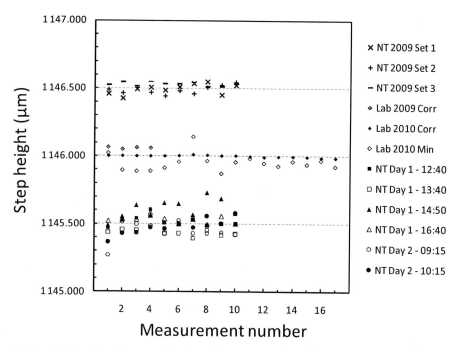

Figure 12. Step height values for several series of measurements obtained in the lab and with the NT9800 in March 2009 and September 2010. For the lab results, the abbreviations "Corr" and "Min" represent analyses based on the "Correlation method" and the "Minimum method". Data series labeled "NT Day..." were obtained on 22nd and 23rd September 2010. None of the values shown here have been corrected for the temperature differences at the time of measurement, but this is done in Table 1.

Correcting for Thermal Expansion of the Step

Table 1 summarizes the data from Figure 12 and also shows the effect of adjusting the step height values to compensate for any difference in temperature of the step between the measurements. Since the lab data from 2010 were analyzed by two independent methods, results for both are included in the table. The SD columns give one sample standard deviation of the mean values, based on the number of measurements given in the "Number of Measurements" column.

The temperature adjusted height data compensate for the known thermal expansion of steel, from the measurement temperature to 27.0 °C, which was the average temperature during the last measurement series. The column "Δh (nm)" shows how large this temperature adjustment was. We have here used the value $1.6 \cdot 10^{-5}$ K^{-1} for the coefficient of linear thermal expansion for stainless steel [20]. With this temperature correction, the step heights can be directly compared.

The temperature correction actually increases the difference between the lab measurements and the NT values for 2010 to about 560 nm. The overestimation from 2009 remains close to 500 nm. The values are compared to the lab data from 2010, analyzed by the correlation method, which is the most precise dataset.

The step temperature was recorded during the NT measurements by a digital thermometer with the sensor touching the steel stage on which the step lay, about 1.5 cm away from the step. The digital thermometer had been calibrated against ice water and boiling water.

Since the temperature also varied slightly during the two days of the NT measurements, the average of the temperature recorded at the beginning and the end of each run was used to adjust the step height for the five NT runs where the temperature deviated from the average. This correction, however, only shifted the mean step height by 1 nm as seen in Table 1.

Table 1. Summary of the step height measurements performed in the lab on 4th March 2009 and 18th September 2010, compared to the results from the Wyko NT9800 measurements on 13th March 2009 and 22nd – 23rd September 2010

Analysis run	As measured step height (nm) Mean	SD	Number of Measurements	Step temperature (°C) Mean	SD	Temperature adjusted height to 27.0 °C h (nm)	Δh (nm)	Difference from Lab 2010 Corr (nm)
NT9800 2009	1 146 504	35	30	23.4	-	1 146 569	65	503
Lab 2009 Corr	1 146 060	7	4	20.5	-	1 146 179	119	113
Lab 2010 Corr	1 145 998	6	17	23.3	0.1	1 146 066	68	0
Lab 2010 Min	1 145 949	64	17	23.3	0.1	1 146 017	68	-49
NT9800 2010	1 145 504	78	60	27.0	0.3	1 145 503	-1	-563

ERROR ESTIMATES

The general message from the data in Figure 12 and Table 1 is that the lab measurements can determine the height of the step with an accuracy of about 100 nm. The precision within one measurement series is below 10 nm when the best analysis method is used. The NT9800 has a larger spread in the measurements with a precision in the range 40 nm to 80 nm. Since the mean values tend to vary systematically over time, the accuracy can be no better than 500 nm. However, before we can draw firm conclusions regarding the accuracies of the two techniques, we need to consider all possible sources of error in the measurements.

Sources of Error in the Lab Measurements

Mode jump of the IR laser

A laser can run on several nearby frequencies. Our IR laser had an external cavity that forced the laser frequency to change in steps of 9 GHz. For a laser wavelength tuned to the cesium D_2 transition, this means that one external mode jump of the laser leads to a relative change in the wavelength of $2.56 \cdot 10^{-5}$. For a step height of 1.15 mm, this corresponds to 29 nm.

We did not regularly monitor the laser wavelength during the measurements, especially not in 2009 before proper error estimates were performed. It is therefore possible that at least some of the 110 nm difference found between the 2009 and 2010 measurements may be explained by 3 – 4 mode jumps after the initial tuning and the actual data recordings. If the laser wavelength had been better monitored during the measurements, this source of error could have been eliminated.

It is therefore important to check often during measurements that no mode jump has occurred. This can be done by viewing the Cs reference cell and making sure that the Cs vapor is excited by the laser and lights up along the laser path when viewed through the IR scope.

Uncertainties in the IR Laser Wavelength

The value used for the vacuum wavelength of the IR reference beam, corresponding to the D_2 transition in cesium atoms, is known with eleven digit precision [15]! However, due to the hyperfine structure of the corresponding energy levels, the D_2 line is actually a doublet [16]. This means that we could have tuned our IR laser to one of the two frequencies in the doublet. Since we were not in a position to determine which of the two lines we actually used, we have applied the tabulated average wavelength of the transition [15] in our calculations. The relative error introduced by this simplification corresponds to half the doublet separation of the hyperfine splitting divided by the average frequency and is about $1.3 \cdot 10^{-5}$. For our 1.15 mm step this becomes 15 nm. This error is too small to explain any of the systematic differences in our data, but it limits the absolute accuracy of our measurements. The error will be systematic for all measurements made with the same tuning of the laser, unless mode jumps also occur. It is in principle possible to choose one of the wavelengths in the doublet, but this is not necessarily attainable with our setup.

Misalignment of the Beams

If the beams in the interferometer are not perfectly aligned, and there is a difference in the alignment for the IR beam and the white light beam, repeated measurements with a new setup may give different results. It is difficult to quantify this source of error in our case, but it *might* have contributed to the difference in step height measurements in the lab between 2009 and 2010.

Change in the Refractive Index of Air due to Temperature, Pressure and Humidity

Since the calculated step height scales linearly with the wavelength used for the IR laser, any shift in the wavelength of the laser due to the atmosphere in our lab would influence the measurements. The refractive index of air will change with temperature, pressure, and with the content of CO_2 and H_2O in the air. These contributions can be calculated accurately from empirical equations given in the literature [21,22].

For a pure temperature change from 20.5 °C to 27.0 °C, the refractive index of air will be reduced by 6 ppm. This translates to a height error of 7 nm for a 1.15 mm step, which is similar to the standard deviation of our most precise measurements. So even though this is a systematic error which can be corrected, it is not required with the precision and accuracy of the measurements obtained here.

The values for the Cs D_2 wavelength in vacuum and air from [15] give a refractive index of 1.000266 for air. This corresponds to a temperature of 24.4 °C at 1 atm pressure, which is nicely centered in the temperature interval of our measurements.

The air pressure on 4th March 2009 and 18th September 2010 was measured at the Blindern meteorological station, 600 m away from the lab, as 99 110 Pa and 98 410 Pa, at the time of the respective measurements [23]. Using these air pressures and the temperatures measured in the lab on those days in the equations from [22], gives refractive indices 1.000264 and 1.000259, respectively. These values give corrections to the height of the step between 3 nm and 8 nm. Again we find effects that are small enough that they can be neglected for our data. The effect of humidity is harder to asses since the relative humidity was not recorded in the lab on the days of measurements. Using the outdoor relative humidities measured at the nearby weather station combined with the equations in [24] to find the saturated vapor pressure of water at the given pressures and temperatures, and combining again with equations in [22], gives step height corrections between 4 nm and 8 nm. In cold countries, the indoor relative humidity may range from 15 % in the winter and up to 100 % on wet days [23]. Even if using these extreme values, the corrections to the step height are between 3 nm and 9 nm for the humidity change alone. The equation used from [22] is really only valid for visible light, but our estimates indicate that the corrections due to humidity are small compared to the more serious sources of errors. Although the effect of changes to the index of refraction of air is too small to explain the differences in the step height measurements of 100 nm to 500 nm, temperature, air pressure and relative humidity should be logged continuously during future measurements in the lab.

Temporal Drift of the Optical Components

Mechanical instabilities among the optical components should only have an effect on the step height if they occur inside the interferometer, i.e. while the beams are split along the paths L1 or L2 in Figure 7. The non-linear movement of the stepper motor is taken care of in the analysis since it affects both the IR beam and the white light beam in the same way. However, any movement of the reference mirror would perturb the step height measurement. A general thermal expansion of the whole optical table can be ruled out since it would affect both paths of the interferometer equally. On the other hand, to produce a 10 nm change in the 83 mm long interferometer arms L1 or L2 (Figure 7) would only take a temperature increase of 0.008 K on the steel table. Thus, any source of temporary thermal gradients that might spread across the optical table during a measurement should be avoided. Such sources could be placing a hand on the table near the optical components during adjustments, local heating from electronic devices like motors or detectors, or having a non-stable air circulation in the lab. Although thermal gradients can not explain the 500 nm differences between the lab and the NT9800 measurements, they might have played a role in the 24 nm steady drop in "Lab 2010 Corr" dataset between measurement 7 and 17, barely visible in Figure 12. To sustain the required 12 nm drift per 8 s measurement interval would – over the 15 minutes between these datasets – only cause a 1 K temperature difference. This is only an estimate and not a thorough calculation, but it shows that even small temperature fluctuations should be taken into account when using a setup with macroscopic dimensions to measure microscopic distances. Vibrations of the optical components due to sound waves seems an unlikely source of error since we would always stop talking or walking in the lab during a measurement.

Systematic Errors in the Data Analysis Programs

Although we used two different methods for analyzing the lab data from September 2010, the results of the analyses gave mean values that were significantly different. So there seems to be a systematic error in at least one of these methods. A careful examination of the software which performed the cross-correlation analysis did not reveal any sources of bias in the step height measurements. The most likely explanation for the differences is some kind of inaccuracy in the "minimum method" which gave the lower values and the much larger spread in the results. We think that the lack of proper correction for the uneven speed of the stepper motor is the most likely problem with this method.

Stability of the Step Standard

A solid block of stainless steel can be considered quite inert and stable over time. When protected from aggressive media or strong mechanical forces, the only dimensional changes expected are due to thermal expansion. For our step, however, where the pieces are glued together by epoxy, we need to estimate the step height changes due to thermal expansion of the epoxy and of swelling due to changes in the humidity. Figure 13 shows the geometry used in the following analyses.

Thermal Expansion

By grinding the sides of the step, the glue line was revealed and its thickness could be measured by an optical microscope. Near the polished surface we found 20 µm on one side and 90 µm on the other (Figure 13). At the base of the steel blocks, the glue line was 140 µm and 120 µm thick. We may therefore consider the glue line as a 20 µm wide, parallel film lying next to a somewhat irregular "wedge" with thickness 0, 70, 100 and 120 µm in the four corners. Thermal expansion of the parallel film will give a pure translation of the lower part of the step, whereas the homogeneous expansion of the "wedge" will cause the lower surface to rotate. Since only rotation out of the plane of the mirror surfaces will affect the step height, we will only estimate the effect of this rotation.

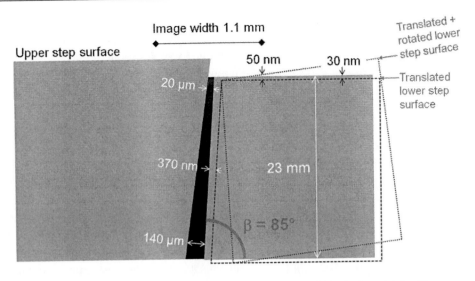

Figure 13. Schematic side view of the 1 mm step with the two steel blocks (gray) joined by a non-parallel glue line (black). The given values correspond to swelling of the glue due to water uptake when changing from a dry to a wet atmosphere. The wedge-like glue line gives both a translation and a rotation of the lower step surface relative to the upper surface. For our step, these two effects happened to cancel almost completely within the imaged area for the WLI microscope, only causing a rotation of the lower step near the glue line. The various parts of the drawing are not to scale.

Values for the coefficient of thermal expansion for epoxy vary in the literature [20,25,26]. The average value of $3.8 \cdot 10^{-5}$ K^{-1}, which is twice as much as for steel, will be used here. For a 1 K temperature increase, the 1.15 mm step will be 18 nm higher due to the thermal expansion of steel. At the same time the thickness t of the 20 μm parallel component of the glue line will increase by $\Delta t = 1$ nm. With a cutting angle $\beta = 85°$, a simple trigonometric calculation gives that the step height will change by $\Delta h = \Delta t \cos \beta = 0.1$ nm due to the expansion of the glue line alone. This contribution constitutes only 0.4 % of the thermal expansion of the steel step and can safely be ignored when correcting the step height due to different temperatures at the time of two measurements.

The wedge-like component of the glue line is at most 120 μm at the base, straight below where the "wedge" has zero thickness at the top. When this "wedge" expands, its top thickness will of course remain zero, while the bottom part widens by 5 nm for a 1 K temperature increase. For the 23 mm high steel block, this gives a rotation of $2 \cdot 10^{-7}$ radians which will lift the outer end of the lower mirror surface somewhat. Since 0.55 mm of the lower surface was imaged by the NT9800, the far side of the step will at most be lifted by 0.1 nm for a 1 K thermal expansion. Again an effect that can be safely ignored since the local roughness of the mirror surfaces is 100 times larger.

Swelling of Epoxy in Humid Atmosphere

Any organic glue will tend to swell by taking up water from the surroundings. Since the indoor relative humidity around the step was not monitored, we will use the extreme range from 15 % to 100 % in our estimates. With this humidity range, the equilibrium moisture content of an epoxy will vary between 0.5 wt% and 7 wt% [26]. This moisture content is independent of temperature. The corresponding linear swelling of the epoxy will be between 0.05 % and 2 %, relative to a perfectly dry sample. For the 20 μm parallel part of the glue line, the swelling is between 10 nm and 380 nm for dry and wet conditions, respectively. The same trigonometric expression as above can be used to calculate the influence of the step height to 1 nm – 33 nm. Thus, if the step is placed in a dry atmosphere and later in a very humid, the maximum expected *translation* of the lower step would be about 30 nm.

Similar considerations as above on the relative *rotation* of the step surfaces due to swelling, gives a maximum lift at 100 % relative humidity of about 50 nm for the outer part of the lower surface within the 1.1 mm wide area imaged by the NT9800 for the step height measurements. Since the two effects of glue swelling are opposite in this case, i.e. downwards translation and upwards rotation of the lower step surface, the effects would tend to cancel each other out almost exactly for images made by the NT9800. For the lab setup where the beam diameter was 2 mm, an increased humidity would make the step appear shorter. However, a change from 15 % to 100 % relative humidity would only shorten the step by about 50 nm on average. Swelling due to humidity requires diffusion of water in the epoxy over about 1 cm to reach the center of the glue line. Therefore the changes in the step height are expected to be slow. The *outdoor* weather data over the last month before the measurements in 2009 and 2010 show relative humidities fluctuating between 44 % and 96 % in both cases, with an average of 84 % in 2009 (standard deviation 10 %) and average 77 % in 2010 (standard deviation 14 %). So although the *indoor* humidity was not logged during storage and measurements in this case, there is no reason to expect large differences in the shape of the step due to humidity. Thus, a change in humidity can not explain the 110 nm systematic difference between the (temperature adjusted) measurements in 2009 and 2010.

Creep of Glue Line due to Shear Forces during Storage

The way that the step was constructed resulted in small height differences also on the under side of the step. When placed on a flat surface, the larger piece would not fully rest on the surface. This height difference is indicated in Figure 13 by the slightly higher bottom surface of the larger piece. During the one and a half year of storage between the two measurement series, the weight of the larger piece will have caused downward shear forces onto the glue line. Since even weak forces may lead to deformation due to creep, one would expect the step to have become somewhat shorter over the last 1.5 years that it was stored with the polished surface facing up. Lacking data for the creep properties of the Epofix resin used as glue, we have not been able to estimate the expected creep. But this effect *may* be one explanation for the roughly 100 nm lower step height measured in 2010.

Flatness of the Measured Step and Actual Areas of Measurement

The lab measurements were based on a beam of white light 2 mm in diameter, while the NT9800 data were obtained from a 1.1 mm * 0.84 mm rectangular area. Figure 14 shows the position of these regions superimposed on a WLI image of the step standard where the height difference between the upper and lower surfaces has been eliminated. In this way the flatness of the two surfaces can be assessed. The two arrowheads on the left and right had been engraved with a needle in the polished surface and act as reference points to align the imaged region in the NT9800. The position of the rectangle is therefore accurate to a tenth of a millimeter or better. The position of the white light spot during the lab analysis is harder to determine. The size and position of the circle is based on photos made in the lab where the light spot is visible.

The region selected for step height measurements was chosen because it lies in the bottom of a shallow valley at a position where there is no excess glue on the lower surface. With the orientation of this enlarged image, the non-masked surfaces in the NT9800 area are somewhat tilted due to the general topography of the shallow valley. But, as mentioned above, both surfaces are parallel to within 15 nm. The root-mean-squared roughness of the non-masked areas is 10 nm. The somewhat asymmetric position of the lab beam still includes all of the NT9800 area. Further, since it encompasses the region from the bottom of the valley and up one of the hills, it is actually quite flat. Roughness measurements from the circular region in Figure 14 gives a root-mean-square roughness between 20 nm and 25 nm, depending on whether the excess glue is excluded from the analysis or not. The largest height differences within each half are 60 nm to 70 nm. However, since the reflected light which forms the interference signals is an average over the entire semi-circle for each half, we will still get an accurate determination of the average step height within the encircled region. With the area for NT9800 measurements being included in the circle illuminated in the lab, and with both areas being part of the same, relatively flat region in both step halves, different measured areas could not be the explanation for the 500 nm height difference measured by the two methods.

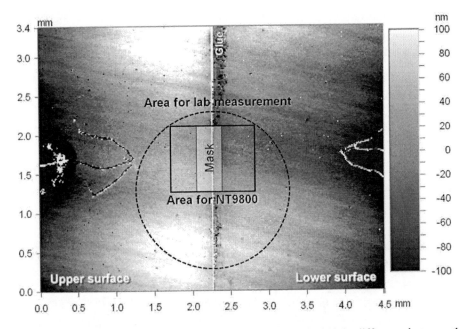

Figure 14. Low magnification image made by the NT9800 where the height difference between the two surfaces has been removed. Only in this way can we observe the detailed topography of the upper and lower surfaces on a common depth scale. The areas that were measured in the lab and with the NT9800 are marked in the image. The positions were determined from the engraved arrows and a blue dot drawn on the upper surface (black region with white specks near the left-hand edge of the image).

Summary of the Error Estimates

Despite considering a large number of factors that might affect the accuracy of the step height measurements, we have not found any external sources of error that can explain the 500 nm discrepancy between the lab measurements and the values from the NT9800. It was beyond the scope of this study to examine any sources of error in the NT9800 instrument itself. We feel, however, that the present study gives a solid foundation on which to build when looking for the causes of systematic errors in the self calibrating white light interferometer NT9800.

Conclusion

Based on the above descriptions, experiments and considerations, the following conclusions can be drawn:

- With a certain amount of effort, it is possible to make a 1 mm high step standard out of stainless steel and have it measured in the lab with an accuracy better than 100 nm (0.01 %) and a precision below 10 nm (0.001 %).

- Using this step standard to assess the performance of our Wyko NT9800 white light interferometer indicates that the accuracy is not better than 500 nm on a 1 mm step (>0.05 %) and that the precision may range from 40 nm to 80 nm (0.004 – 0.008 %).

- The most important correction to consider when comparing measurements made with different instruments at different times on the same steel step is the thermal expansion of the material.

- Other sources of systematic errors seem to lie in the 10 – 30 nm range. To obtain measurement accuracies better than this, very careful and controlled experiments must be performed.

References

[1] Morrison, E. *Nanotechnology* 1996, 7, 37-42.
[2] Kruger, O.A. *Metrologia* 2001, 38, 237-240.
[3] Skotheim, Ø.; Couweleers, F. Proceedings of 12th International Conference on Experimental Mechanics (ICEM12), Bari, Italy, McGraw-Hill (August, 2004), ISBN 88 386 6273-8.
[4] Binnig, G.; Rohrer, H.; Gerber, Ch.; Weibel, E. *Phys. Rev. Lett.* 1982, 49, 57-61.
[5] Binnig, G.; Gerber, Ch.; Stoll, E.; Albrecht, T.R.; Quate, C.F. *Surface Science* 1987, 189/190, 1-6.
[6] Cremer, C.; Cremer, T. *Microscopica Acta* 1978, 81, 31-44.
[7] Molesini, G.; Pedrini, G.; Poggi P.; Quercioli, F. *Optics Communications* 1984, 49, 229-233.
[8] http://www.stilsa.com/EN/prin.htm (Downloaded 20th September 2010.)
[9] http://www.nano-lane.com/nanotechnology-sarfus-presentation.php (Downloaded 20th September 2010.)
[10] http://www.alicona.com/home/products/InfiniteFocus/Fokus-Variation.en.php (Downloaded 20th September 2010.)
[11] Caber, P.J. *Applied Optics* 1993, 32, 3438-2441.

[12] Holme, B.; Hove, L.H.; Tveit, A.B. *Measurement* 2005, 38/2, 137-147.
[13] Der-Shen Wan, Tony L. Schmitz, and Erik Novak, Proceedings of the 20th Annual Meeting of the American Society for Precision Engineering, October 9-14, 2005, Norfolk, VA, (http://www.aspe.net/publications/Annual_2005/ POSTERS/ 3METRO/ 5MEASU/1778.PDF)
[14] Tano, K., 2002, http://www.freepatentsonline.com/6427355.html (Downloaded 20th September 2010.)
[15] Daniel A. Steck, "Cesium D Line Data," available online at http://steck.us/alkalidata (Revision 2.1.2, 12 August 2009. Downloaded 20th September 2010.).
[16] Andalkar, A; Warrington, R. B. *Physical Review A* 2002, 65, 032708, 7 pages.
[17] Ruxton, G.D. *Behavioral Ecology* 2006, 17, 688-690.
[18] Hove, L. H.; Holme, B.; Young, A.; Tveit, A. B. *Acta Odontol. Scand.* 2007, 65, 259-264.
[19] Hove, L. H.; Holme, B.; Young, A.; Tveit, A. B. *Caries Research* 2008, 42, 68-72.
[20] Tennent, R.M. Science Data Book; Oliver and Boyd: Edinburgh, UK, 1983; p 60.
[21] Birch, K. P.; Downs, *M. J. Metrologia*, 1994, 31, 315-316.
[22] Tables of Physical and Chemical Constants (16th edition 1995). Kaye and Laby Online, The National Physical Laboratory, UK, 2005, Version 1.0, 2.5.7 Refractive index of gases. www.kayelaby.npl.co.uk (Downloaded 20th September 2010.)
[23] www.eklima.no (Downloaded 20th September 2010.)
[24] Buck, A.L. *Journal of Applied Meteorology* 1981, 20, 1527-1532.
[25] Adamson, M.J. *Journal of Materials Science* 1980, 15, 1736-1745.
[26] McKague, E. L.; Reynolds, J. D.; Halkias, J. E. *Journal of Applied Polymer Science* 1978, 22, 1643-1654.

In: Interferometry Principles and Applications
Editor: Mark E. Russo

ISBN 978-1-61209-347-5
© 2012 Nova Science Publishers, Inc.

Chapter 4

CYCLIC PATH INTERFEROMETRIC CONFIGURATION: SOME APPLICATIONS

Sanjib Chatterjee[*]
Raja Ramanna Centre for Advanced Technology,
Department of Atomic Energy, Indore-452013, India

ABSTRACT

Some useful applications of cyclic path interferometric configuration have been discussed. A cyclic path optical configuration (CPOC) can be formed with a beam splitter (BS) and two plane mirrors, M1, M2, which are inclined at 45° to each other. BS splits up an incident collimated beam into transmitted (T) and reflected (R) components, which reflect off M1 and M2 and traverse triangular paths (TP) in opposite directions. With a polarizing beam splitter (PBS), the T and R components become plane polarized in orthogonal directions, namely, p and s polarized. When M1 and M2 are symmetrically placed, at equal distances, with respect to the beam splitting plane, the T and R components traverse the same TP in opposite directions, for a particular angle of incidence of the incident beam and emerge along the same path. A lateral shear between the beam components can be introduced by shifting either M1 or M2, along the direction of the normal to the mirror, from the symmetrical position. As the T and R components traverse identical TP, the optical path difference (OPD) between the emergent beams remains zero. The counter propagating p and s polarized components, in a CPOC, can be utilized for the external measurement of right angle of optical components A CPOC, adjusted for a lateral shear between the T and R components, can be coupled to a Fizeau interferometer (FI) for the measurement of wedge angle (δ) and index of refraction (μ) of the material of the optical glass window. A CPOC, adjusted for a lateral shear, produces two spatially separated (real) point images, outside the cavity, for a convergent input beam. The coherent point images can form Young's fringes. A CPOC with a PBS and linearly polarized divergent laser beam input from a microscope objective spatial filter combination is used to produce spatially separated coherent virtual point sources of light with linear orthogonal polarization at the back focal plane of a Fizeau interferometer objective for polarization phase shifting interferometry. A converging lens, placed in the

---

[*] E-mail: schat@rrcat.gov.in.

hypotenuse arm of the TP of a CPOC adjusted for zero lateral shear, focuses the counter propagating collimated beams at two different points having longitudinal separation, which depends on the position of the lens in the TP cavity. Light diverging from the focal points interfere to produce Haidinger type fringes of equal inclination. Since the focal points are coherent, high contrast interference fringes can be obtained with quasi-monochromatic or broad band (white) light sources. An important application of the setup is for the measurement of centering error of lenses. Using polarized light and PBS, the same CPOC set up, which produces two point sources with longitudinal separation, corresponding to two focal points of the emergent beams having orthogonal linear polarizations, can be used to measure focal length of lenses using Newton's formula.

1. INTRODUCTION

Cyclic path optical configurations (CPOC) are widely used for optical metrology. A CPOC can be formed with a beam splitter (BS) and two plane mirrors, which are inclined at 45° to each other [1]. BS splits up an incident collimated beam into transmitted (T) and reflected (R) components which reflect off the CPOC cavity mirrors and traverse triangular paths (TP) in opposite directions. With a polarizing beam splitter (PBS), the T and R components become plane polarized in orthogonal directions, namely, p and s polarized. When the cavity mirrors are symmetrically placed, at equal distances, with respect to beam splitting plane, the T and R components traverse the same TP in opposite directions, for a particular angle of incidence of the incident beam and emerge along the same path. A lateral shear between the beam components can be introduced by a number of ways, such as by shifting one of the cavity mirrors, forming the CPOC, along the direction of normal to the mirror, from the symmetrical position. As the T and R components traverse identical TP, the optical path difference (OPD) between the emergent beams remains zero. As the component beams travel same optical circuit, therefore the setup is relatively insensitive to external mechanical vibrations. A CPOC can be easily adjusted for zero OPD between the parallel emergent beams with variable lateral shear [1]. In Ref. [2] a polarization CPOC has been used to generate a pair of parallel pencil beam with small lateral separation for a long trace profiler. A compound prism CPOC has been described, for using in a multi pass optical configuration for measuring residual wedge angle of high optical quality transparent near parallel optical glass/fused silica plate, in [3].

J A. Ferrari et al [4] presented a polarization based technique for external phase shifting of a polarization CPOC. In this technique, a pockel cell is used to modulate the phase difference between the orthogonal polarization components of an expanded collimated linearly polarized laser beam which is launched in a polarization CPOC. The incident collimated beam (plane wave) is divided into top and bottom halves. A test phase object is placed at one half of the incident beam while the other half is used as reference beam. By adjusting the lateral shear between the emergent beams from the CPOC, it is possible to make full or partial superposition of the object and the reference beams and obtain two beam Fizeau or lateral shearing type interference fringes, respectively, from which the phase information can be extracted.

Y. H. Lo et al [5] described a polarization CPOC/Sagnac lateral shearing interferometer for wave front measurement. A broad band PBS and three plane mirrors form the CPOC/Sagnac interferometer. Using an achromatic half wave plate in the CPOC, the counter propagating beams with orthogonal polarizations are made to travel exactly same path and undergo exactly same number of reflections and transmissions in the CPOC cavity for broad band (white light) operation. Two identical positive lenses are placed in the cavity in such a way so that their common focus falls on a cavity mirror which can be tilted to produce lateral shear between the emergent waves. Geometric phase shifts are introduced by a combination of an achromatic quarter wave plate and a linear achromatic polarizer.

A CPOC can be used as both lateral and radial shearing interferometers. Kothiyal et al [6] discussed polarization phase shifting radial and lateral shearing interferometers. An afocal system with non unity afocal magnification can be used for radial shearing [7].

Hariharan et al [8] described a geometric phase interferometer using a polarization CPOC. Cheng et al [9] combine a telescopic imaging system with a lateral shearing CPOC and applied phase shifting interferometry (PSI) for measuring the complex spatial coherence function of a linearly polarized optical field. Darling et al [10] described a self referencing cyclic shearing interferometer for testing beam collimation.

In Section 2 of this chapter, we discuss the basic principles of CPOC. Section 3 describes the applications of CPOC for the following cases: 3.1. External measurement of dihedral right angle of optical components [11], 3.2. Measurement of the wedge angle and refractive index of the material of the transparent optical glass/fused silica optical windows simultaneously [12], 3.3 Generation of white light Young's fringes [13], 3.4 Polarization phase shifting Fizeau interferometer [14], 3.5 Generation of white light Haidinger fringes [15], 3.6 Measurement of centering error of lenses [16], 3.7 Determination of the focal length of lenses [17]. Section 4 concludes the chapter with a brief discussion.

2. BASIC PRINCIPLE

Optical schematic of the CPOC, formed by a BS and plane mirrors (M1) and (M2), is shown in Figure 1. M1 and M2 are inclined at 45° to each other and are symmetrically placed at equal distances from the BS. An incident ray of light, represented by (AO), is split up into R and T rays by the BS. The angle of incidence of the incident ray AO on the BS is assumed to be 45°. The R and T rays denoted by (OQ) and (OP), respectively undergo reflections on M1 and M2 and follow the same triangular path (TP), OQPO/OPQO in opposite directions and again split up into R and T components by the BS. Thus pairs of beam components emerge along OA and OR (perpendicular to OA) with zero OPD and without any lateral shear. Lateral shear between the beam components can be introduced by shifting either M1 or M2 in a direction normal to the plane of the mirror. In Figure 1, M2 is shifted in a direction normal to its plane and its new position is denoted by M2'. The R and T rays originated from the incident ray AO due to the splitting by the BS now traverse identical parallel paths, in opposite directions, namely (OQQ2O2) and (OP1Q1O1), in the CPOC and split up again into R and T rays by the BS. Thus two pairs of laterally sheared parallel rays, with zero OPD between them, denoted by (O1B1, O2B2) and (O1R1, O2R2), emerge out from the CPOC.

The rays emerge through the input port of the BS traverse parallel to the incident ray but in opposite direction while that coming out of the output port of the BS travel along a direction normal the incident ray direction. It can be shown by applying simple geometry that for a shift s_0 of the cavity plane mirror, either M1 or M2, in a direction normal to the plane of the mirror, the lateral shear introduced between the emergent rays is given by, $s=\sqrt{2}s_0$. Thus for a particular angle of incidence of an incident collimated beam on the BS, when the angle between the R and T beams produced due to the splitting up of the incident collimated beam by the BS, is equal to the combined angular deviation due to M1 and M2, the emergent collimated beams with lateral shear always have zero OPD between them. As the angle of incidence changes say by Δi, an angle of 2 Δi is introduced between the R and T rays in the CPOC. The emergent beams although remains parallel to each other, would be inclined at an angle Δi with the direction of the emergent beams with zero OPD. An OPD which is directly proportional to the lateral shear and change in the angle of incidence (for small change) is introduced between the emergent beams. Thus the zero OPD setting of CPOC, for a particular angle of incidence can be easily ensured by adjusting the emergent beams along the incident beam direction. So far as the emergent beams in the input side is concerned, the CPOC acts as a plane mirror and the emergent beams suffer identical reflections and transmissions on BS while for the beams coming out through the output port, it is RR and TT. i. e., one of the emergent beams suffers two reflections while the other undergoes two transmissions on the BS. Therefore in the input side, the intensities of the emergent beams would be always equal whereas the emergent beam intensities in the out put side depend on the splitting ratio of the BS. In a polarization CPOC, the BS is generally replaced by a cube type polarization beam splitter and input beam with linear polarization at an appropriate azimuth (45°) is allowed to fall on the PBS. The R and T components as splits up by the PBS are polarized in orthogonal planes (say p and s). The p and s components counter propagate in the CPOC and comes out only through the output port of the CPOC. In this case the p and s components suffer unequal phase changes on reflections and hence there would be some residual OPD between the emergent beams in the output side. By using a half wave plate at an appropriate orientation in the CPOC it is possible to make the beam components with orthogonal linear polarizations suffer identical number of reflections and transmissions and thus undergo equal phase shifts and emerge out along the direction of the incident beam with zero OPD [5].

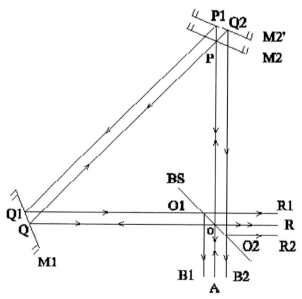

Figure 1. Ray paths in a CPOC.

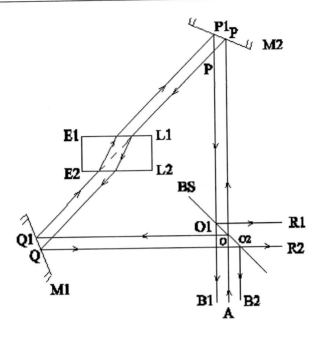

Figure 2. Lateral shear due to a tilted parallel plate.

Lateral shear between the emergent beams can also be introduced by tilting a parallel glass plate (E1L1L2E2) as shown in Figure 2. The glass plate method of lateral shearing has the disadvantage of causing astigmatism for non collimated input beams. In this section we have considered the use of CPOC in collimated beam. Uses of CPOC with non collimated beams are discussed in specific applications.

3.1. External Measurement of Dihedral Right Angle of Optical Components

Right angled optical components such as corner cubes and right angled prisms are widely used in optical systems. The system performances, in most cases, depend on the accuracy of the 90° dihedral angles. Internal measurement of 90° angle is relatively easier [18-22]. In certain situations, the external measurement is necessary [23-25]. In this section, an external measurement technique is discussed. [11] The optical setup for the external measurement of 90° angle of a right angled optical component is shown in Figure 3. Expanded collimated He-Ne (λ=632.8nm) laser beam, which is linearly polarized with its plane of polarization at 45°, from a Fizeau interferometer (FI) is allowed to fall on a polarization beam splitter (PBS) at 45° angle of incidence. The collimated beam is represented by plane wave front (ABC). The incident beam denoted by the central ray (BO1) is split up by the PBS into R and T components, shown by (O1Q) and (O1P), respectively with orthogonal linear polarizations, i. e., s and p polarization components. PBS along with two plane mirrors (M1) and (M2), which are inclined at 45° to each other forms a CPOC-1 Initially, CPOC-1 needs to be set for counter propagating beam components (R and T) with zero lateral shear. M1 and M2 are adjusted to make the p and s components traverse the same triangular path in 180° opposite direction in CPOC-1. The right angled component is placed with its right angle edge perpendicular to the plane of the Figure 3, in such a way, so as to deflect the counter propagating p and s components of CPOC-1 in near normal direction, as shown in Figure 3, by adjusting the angle of incidence at 45° on the surfaces forming the right angle. Now, with the right angled optical component placed in the hypotenuse arm of CPOC-1, in the above manner, the counter propagating s and p components represented by (QR) and (PS), respectively suffer reflections on the plane surfaces (90° angular deviation) and travel nearly parallel to each other along a near normal direction. A Cartesian co-ordinate system, whose Y axis lies along the direction of the counter propagating breams in the hypotenuse arm of CPOC-1., the Z axis along a direction, perpendicular to the plane of the Figure 3., is considered. For a perfect 90° angle of the right angled component, the emergent s and p beam components denoted by (RB1) and (SB1'), respectively would be parallel to each other and have a lateral shear between them. The lateral shear depends on the transverse position of the right angled component in CPOC-1. In the presence of an angle error (δ) in the test right angle, there would be an angle error $\varepsilon = 2\delta$, between the emergent p and s components, that lie in the plane of the Figure 3. The emergent p and s components fall on PBS1, which along with plane mirrors (M4) and M5 forms CPOC-2, which is used to remove the lateral shear between the p and s components. Thus ε can be measured from the spacing of the Fizeau fringes which would be formed due to interference of the beams (brought to the same polarization state as they travel through a quarter wave plate QWP at 45° with the direction of polarization of the p and s components and a linear polarizer) as they are recombined. CPOC-2 is not subject to the same external alignment requirement, only internal alignment and stability.

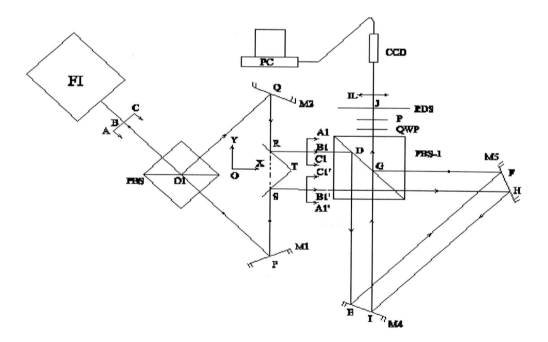

Figure 3. Schematic of a setup for the external measurement of dihedral right angle.

The counter propagating wave part of the interferometer is used to nullify lateral shear and the input angle of the laterally sheared beam components (RB1 and SB1') is irrelevant. Hence the problem of the external measurement of 90° angle is transformed to measurement of air wedge angle between the plane wave fronts corresponding to the exit beam components from CPOC-2.

The longitudinal position of the right angled component along the hypotenuse arm of the CPOC-1 would decide the relative phase delay between the emergent p and s components, i. e., which of the p and s components suffers the phase lag. Thus for a particular position of the right angled component, the direction of the air wedge depends on the sign of the error δ. Thus it is possible to relate the sign of δ with the direction of air wedge between emergent plane wave fronts. It is evident that the wedge direction would remain unchanged, as the beams suffer even number of reflections in CPOC-2, at the exit end, when the beams overlap over the full aperture as the lateral shear is removed. Hence, sign of δ can be known by determining the sign of the air wedge between the plane wave fronts at the exit end. The wedge direction is commonly determined from the direction of the fringe movement, linear or angular as a known perturbation, such as temporal phase shifting or a spatial phase biasing is introduced between the interfering plane wave fronts. The QWP transforms the state of polarization of the exit beam components (p and s) from CPOC-2 to right and left circular polarizations. The polarization components selected by the linear polarizer (P) interfere to produce two beam Fizeau fringes on a rotating diffuser screen (RDS). It is possible to introduce polarization phase shifts between the interfering components by adjusting angular orientation of P. An imaging lens (IL) transfers the fringe image on to the plane of a two dimensional CCD array, which along with a frame grabber card and a PC captures the fringe image. Thus δ can be determined accurately from the tilt of the OPD plane by applying phase shifting interferometry. [26].

So far we have assumed that the right angle edge of the optical component is normal to the plane of the figure. In practice, the right angle edge may have some residual angular tilt. It is evident that residual tilt of the right angle edge, i. e., departure from perpendicularity, in a plane parallel to YZ plane would produce a relative angular tilt between the emergent beams in an orthogonal direction. This would cause a fringe rotation. The effect can be obviated and hence δ can be determined by taking a component of the resultant wedge in the desired direction. [27]. Figure 4 shows the nature of fringes due to angle error δ and an orthogonal tilt component.

Accuracy of measurement of δ primarily depends upon the angular accuracy of the counter-propagating beams in the CPOC-1. A CPOC-1 coupled to a FI, can be easily aligned to very high order of accuracy in fraction of arc second. A highly stable CPOC-1, where the optical components are cemented on a Penta prism type base plate has been discussed. [11] CPOC-2 is relatively insensitive to vibration as the two interfering beams travel around the same closed circuit in opposite directions.

Figure 4. Nature of fringes due to presence of right angle error and an orthogonal tilt component.

3.2. Measurement of the Wedge Angle and Refractive Index of the Material of the Transparent Optical Glass/Fused Silica Optical Windows

Refractive index (n) and wedge angle (δ) are important parameters for near parallel, high optical quality, transparent optical windows. Methods for measuring both n and δ simultaneously are discussed in [28-30].

In this section, a technique for simultaneous measurement of n and δ of high optical quality transparent window is discussed. [12]

The optical schematic of the setup for the measurement of δ and refractive index n of the material of a high optical quality transparent optical glass/fused silica window/ plate, is shown in Figure 5. The expanded collimated, He-Ne ($l=632.8$nm) laser beam, represented by plane wave front (ABC), from a Fizeau interferometer (FI) (not shown in Figure 5) is allowed to fall at an angle of incidence of 45° on a beam splitter (BS). BS along with plane mirrors (M1) and (M2), which are inclined at 45° to each other, forms a CPOC. The collimated beam is represented by central ray (BO). M1 and M2 are symmetrically placed at equal distances with respect to BS. Thus the CPOC is set for counter propagating beams with zero lateral shear.

A mask, as shown in Figure 6, with two circular apertures (openings) is placed in the path of the incoming collimated beam from the FI. The centers of the apertures are denoted by A and C. The test wedge plate (WP) is placed, with its wedge direction normal to the plane of the Figure 5, in the path of the beam passing through the aperture centered at C, as shown in Figure 5.The collimated beam , passing through the aperture centered at A, falls directly on BS. Thus the direct beam and that passing through WP can be denoted as A and C beam, respectively.

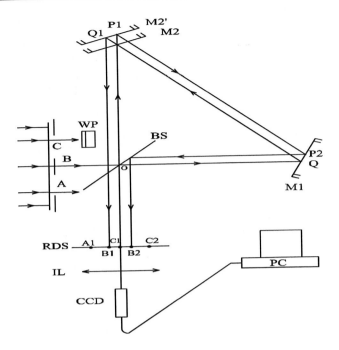

Figure 5. Schematic of the optical setup.

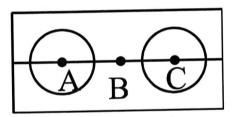

Figure 6. Nature of the beam mask used on the incident beam.

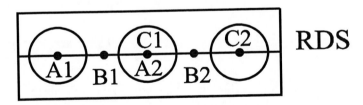

Figure 7. (A1B1C1) and (A2B2C2) are due to transmitted and reflected components.

The reflected (R) and transmitted (T) components of the incident beam, produced due to splitting up of the incident collimated beam, counter propagates same triangular path in the CPOC and emerge completely overlap on each other from the other port of the BS. Lateral shear between the R and T beam components is introduced by shifting M2 without changing its orientation. The degree of overlapping between the two pairs of emergent beams can be controlled by adjusting the shift of M2. Suppose M2' is the shifted position of M2 such that the C beam (the beam passing through WP) of transmitted component, completely overlaps the A beam (direct beam) of the reflected component. This is illustrated in Fig 7, which shows the output beam positions on a rotating diffuser screen (RDS). In Figure 7, C1 is the center of the C beam of the transmitted component and A2 is that of the A beam of the reflected component.

The collimated beam passing through WP suffers an angular deflection $(n-1)\delta$. An imaging lens (IL) transfers the two beam Fizeau fringes formed on RDS, due to interference between the direct beam and that passes through WP, on to the plane of a two dimensional CCD detector. The fringe image is digitized and captured by a Frame grabber and Personal computer (PC). The spacing of the two beam Fizeau fringes is given by

$$d_t = 1 / [(n-1)\delta] \qquad (1)$$

The spacing of the reflection Fizeau fringes, formed due to the interference of Fresnel reflected beams from the surfaces of the WP, in the FI is given by

$$d_r = 1 / 2n\delta \qquad (2)$$

From Eq. (1) and (2), n can be written as

$$n = d_t / (d_t - 2d_r) \qquad (3)$$

Substituting the value of n in Eq. (1) and from the known values of the other parameters, δ can be determined.

The main advantage of the technique is that the effect of external mechanical vibrations on the measurements is very less and expensive vibration isolation is not required.

3.3. Generation of Young's Fringes

The first practical demonstration of interference of light was given by Thomas Young. Young's fringe's established the wave theory of light. The classical optical setup for the production of Young's fringes and the principles are explained in [31]. There are other standard methods, based on division of wave front, such as Fresnel's biprism, Lloyd's mirror, Billet's split lens etc, for the production of Young's fringes. [31, 32]

In this section, a CPOC based technique for the production of Young's fringes with a low coherence broad band (white light) light source and its application for the determination of the temporal coherence of low coherence light sources are discussed. [13].

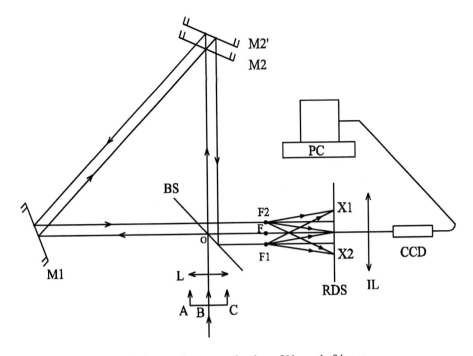

Figure 8. Schematic of the optical setup for the production of Young's fringes.

A schematic of the optical setup is shown in Figure 8. Expanded collimated light beam denoted by plane wave front (ABC) is allowed to pass through an aberration corrected telescope objective lens (L). The converging beam coming out from L is allowed to fall on a beam splitter (BS). BS along with plane mirrors (M1) and (M2), which are inclined at 45° to each other, forms a CPOC. Initially, L is removed, the angle of incidence of the collimated beam on BS is set at 45° and M1, M2 are adjusted at equal distances with respect to BS, so as to set the CPOC at zero lateral shear between the reflected and transmitted beam components produced due to splitting up of the incident beam by the BS. On introduction of L in the path of the incident collimated beam, the R and T components of the incident convergent beam produced by BS counter propagate in the CPOC and come to focus at a point (F) outside the CPOC, as shown in Figure 8, which shows the paths of the central rays. Separation between the focal spots, formed by the emergent beams, in transverse direction, can be introduced by shifting either M1 or M2 as for introducing lateral shear between collimated bean components. In Figure 8, M2 is shifted to M2' and the focal points shown by (F1) and (F2) have a separation of d= $\sqrt{2}s_0$, where s_0 is the longitudinal shift of M2. Since the beam components forming F1 and F2 travel identical optical paths, they are fully coherent sources, irrespective of the spectral composition of the input radiation. Rays of light emerging from F1 and F2 interfere and form non-localized Young fringes. An imaging lens (IL) transfers the Young fringe images formed on a rotating diffuser screen (RDS) on to the plane of a two dimensional CCD detector array. The image is grabbed by a frame grabber and Personal computer (PC). The spacing of the Young fringes corresponding to wavelength 1 is given by

f = D 1/ d   (4)

where D is the distance between RDS and the transverse plane containing F1 and F2. It is evident that for the output port of the BS, the emergent beams are RR and TT types, i.e., one beam component suffers two reflections on BS while the other undergoes two transmissions, thus there may be a drop in visibility/contrast of the fringes due to intensity mismatch of the emergent beams through the output port of the BS. Also one beam component suffers an extra phase of p and for that reason the center of the Young fringes is dark. The emergent beams coming out through the input beam port of the BS, suffer identical reflections and transmissions on BS and form high contrast Young fringes with bright center. F1 and F2 also suffer a longitudinal shift with respect to F by s_0. It is evident that F1 and F2 are actually the images of the source of light used for the setup. For gas laser such as He-Ne, they are very small and there is no drop in the visibility of the fringes due to the finite size of the source. With an incoherent source, the source size is not as small as the laser source. Each point of a source with finite extension would form its own interference fringes and hence there would be an incoherent superposition of interference fringes, which are shifted with respect to each other, from numerous point sources of the extended source and there is a degradation of the fringe contrast.

It can be shown from simple consideration [13] that this effect can be minimized by satisfying the following relation t<<D l/ d, where t is the effective source dimension. The Young's fringes thus obtained can be used for the estimation of the temporal coherence of low coherence broad band light source by concentrating the light from the source on to a small pin hole. The pin hole light source or its real /virtual images which form the sources for the Young's fringes should be small enough (t<<D l/ d). The variation in fringe visibility as the OPD, Δ is increased (as one moves away from the center of the interference field) can be obtained from the grabbed digitized fringe intensity values. The value for which visibility falls to a poor level gives a measure of the coherence length of the source. The lower limit for the spectral bandwidth for the coherence length, which can be determined, depends on the lowest compatible pin hole dimension, pixel size, spacing and effective size of the detector array.

A division of amplitude type technique, for producing Young type fringes using a CPOC based setup has been discussed. Advantages of the techniques are: 1) useful for both coherent and incoherent sources 2) relatively insensitive to vibration as the two interfering beams travel around the same closed optical circuit in opposite directions. Thus both the interfering beams are equally affected by external mechanical vibrations and the effects of vibrations cancel out 3) separation between the coherent point sources can be easily varied and thus the fringe spacing is easily adjustable 4) zero order can be maximum or minimum and is easily identifiable 6) high throughput

3.4. Polarization Phase Shifting Fizeau Interferometer

In this section, a polarization phase shifting (PPS) Fizeau interferometer (FI) [14] is discussed. FIs [33, 34] are widely used in optics laboratories for the measurements of surface forms, parallelism, homogeneity etc of optical components.

Though FIs are not exactly common path interferometers, the interfering waves from the reference and the test surfaces travel nearly same path through the interferometer system (from the reference surface).

Because of the common path of the reference and the test waves through the interferometer system, the measurement results are much less affected by aberrations of the interferometer system and the FI systems are much less susceptible to external mechanical vibrations.

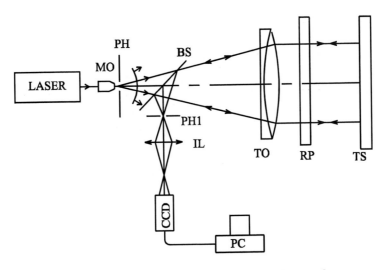

Figure 9. Optical schematic of a Fizeau interferometer for testing of plane surfaces.

Optical schematic of generic FI system for measurement of surface form of plane optical surfaces is shown in Figure 9. Light from a He-Ne laser source is allowed to pass through a spatial filtering arrangement consisting of a microscope objective (MO) and an appropriate pinhole (PH) placed at the front focal point of the MO. PH is also situated at the back focal plane of a well- corrected telescope objective (TO) and thus an expanded collimated beam is produced. FI cavity is formed between a reference plane (RP) (uncoated front surface of a high optical quality fused silica reference plate) and a plane test surface (TS). The RP and TS are mounted in high quality tip, tilt mirror mounts. The reference plate is wedged (0.5°) to eliminate the effect of unwanted Fresnel reflections from its rear surface, which is anti-reflection coated. The collimated beams reflected from the RP and the TS interfere and produce two-beam Fizeau fringes. The thin plate beam splitter (BS) directs the interfering waves towards an imaging lens (IL) and a CCD camera. An appropriate pin-hole (PH1) placed at the rear focal plane of the TO blocks unwanted light. Fizeau fringes are grabbed by means of the CCD camera, Frame grabber and Personal Computer (PC) arrangement.

Phase shifting interferometry (PSI) [35-37] has been regarded as the most useful and accurate method for quantitative evaluation of phase information. PSI technique is extensively used in FIs. Piezo-electric transducer (PZT) based mechanical phase shifters are most common. The main disadvantage of PZT is that the displacement is not always a linear function of the applied voltage. Ref. [38] discussed the effect of PZT non-linearity on PSI. In PPS[39, 40] , generally, the reference and the test waves with linear orthogonal polarizations are transformed to opposite circular polarizations and are allowed to pass through a linear polarizer. The (geometrical) phase shift is introduced by varying the angular orientation of the pass direction of the polarizer. The most important advantage of PPS is that it is possible to capture all the necessary phase shifted interferograms simultaneously [41, 42], thereby reducing the vibration susceptibility to a minimum and extending the scope of the measurement to dynamic interferometery [43-45].

In FI, as the input beam is common for both the reference and the test surfaces, it is not simple to code the waves from the reference and the test surfaces with orthogonal polarizations and some manipulations of the basic system is needed for implementation of the PPS technique. In ref. [40], a linearly polarized He-Ne (632.8 nm) laser source is used and a quarter wave plate (QWP) is inserted between the reference and the test surfaces for getting the desired linearly polarized beams with orthogonal polarizations. The residual aberration, if any, due to QWP is finally subtracted by adopting a two step measurement method. The technique is most suitable for measurements of the surface forms of plane surfaces and is limited by the aperture diameter of the QWP.

Barnes [46] discussed a heterodyne FI for testing plane surfaces in which two frequency shifted beams, which are +1 and -1 diffraction orders, generated by a rotating radial grating are used to produce two slightly tilted collimated beams in the FI cavity. The reference and the test surfaces are adjusted so that in the focal plane of the FI objective the focus of -1 order reflected from the test surface is superimposed on the focus of +1 order reflected from the reference surface. A spatial filter in the focal plane allows only the central superposed beams which are made to interfere. In the following, a PPS type FI is discussed [14]. A CPOC is used to generate two coherent (virtual) point sources at the back focal plane of a corrected telescope objective (TO) for producing a pair of tilted collimated beams with linear orthogonal polarizations in the FI cavity. Orthogonal linear polarization components reflected from the reference and the test surfaces are utilized for PPS.

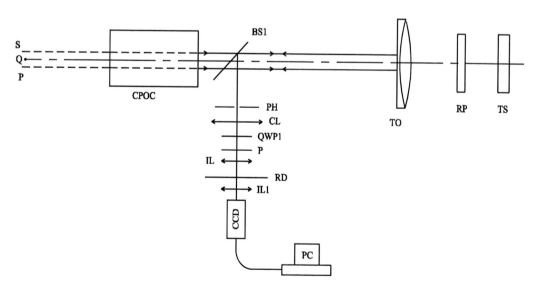

Figure 10. Schematic of polarization phase shifting Fizeau interferometer.

Optical schematic of the PPS type FI system for the measurement of surface form of plane optical surfaces is shown in Figure 10. A CPOC, shown as black box, is used to generate two coherent (virtual) point sources (P and S) at the back focal plane of a corrected telescope objective, TO, for producing a pair of tilted collimated beams with linear orthogonal polarizations in the FI cavity. The center line denotes the axis of TO. Orthogonal linear polarization components (p and s) reflected from the reference plane, (RP), and the test surface, (TS), are utilized for PPS. By adjusting the angular orientations of the reference and the test surfaces, it is possible to align the desired interfering components along nearly same direction. A thin plate beam splitter, BS1, at 45° orientation deflects a portion of the converging p and s components in the perpendicular direction and the real images are formed at the focal plane of the TO on the reflection side of the BS1. An appropriate pin-hole, PH, allows only the superimposed central spots and blocks the rest. A lens (CL) collimates the beams passing through PH. Quarter wave plate, (QWP1), transforms the state of polarizations of the beams to opposite circular polarizations and the linear components along the direction of pass axis of the polarizer (P) interfere. An imaging lens (IL) is used to focus the plane of the TS and the interferogram is formed on a rotating diffuser screen (RD). A second imaging lens (IL1) transfers the fringe-image on to the plane of a two dimensional CCD detector where it is captured by means of a Frame grabber and Personal computer (PC).

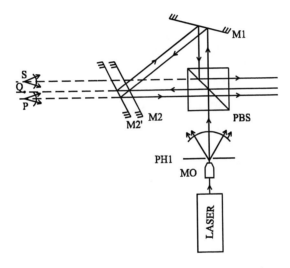

Figure 11 Schematic of a polarization CPOC for generation of coherent point sources.

Now, we consider Figure 11 for explaining the function of the CPOC. Light from a linearly polarized He-Ne (632.8 nm) laser source is allowed to pass through a spatial filtering arrangement consisting of a microscope objective, MO, and an appropriate pin-hole, PH1, placed at the front focal point of the MO. The diverging beam from PH1 is split up into transmitted (p) and reflected (s) components by the beam splitting plane of a polarization beam splitter (PBS), which together with plane mirrors (M1) and (M2) forms a CPOC. For simplicity, we have shown the path of the central ray. M1, M2 can be adjusted to make the central rays of the p and s components traverse the same triangular path (TP) but in opposite directions. Thus the CPOC is set for zero shear condition and the emergent beams from the exit side of PBS appear to diverge from a virtual image point Q. M2 is the position of the mirror at zero shear condition and M2 coincides with the position of virtual image of M1 at the beam splitting plane of the PBS. Lateral shear between the counter-propagating central rays in TP can be introduced by shifting one of the plane mirrors from zero shear position, as shown by M2' in Figure 11. On introduction of lateral shear, the emergent p and s beam components appear to diverge from the virtual image points P and S, respectively. The lateral separation, (d) between P and S is given by d= √2s₀, where s₀ is the separation between M2 and M2'. P and S also suffer a small longitudinal shift, equal to s_0, with respect to Q. P and S are adjusted to be at the back focal plane of a corrected TO and are kept equally separated with respect to the optical axis of the TO (Figure 10). Thus, as shown in Figure 12, two sets (p and s) of expanded collimated beams denoted by plane wave fronts W_p and W_s emerge from TO. Each set of collimated beams is angularly tilted by an angle (α), with respect to the optical axis of the TO, given by α= $\tan^{-1}$ (d/2f), where f is the focal length of the TO. W_p, W_s suffer Fresnel reflections on the surface of the reference plane (RP) placed normal to the optical axis of the TO. The reflected p and s wave fronts represented by W_{rp} and W_{rs} after passing through the TO would converge towards S and P (Figure 10), respectively. The relative positions of the focal spots on the focal plane of the TO (on the reflection side of BS as shown in Figure 10) is illustrated in the inset of Figure 12, where the focal points for the Fresnel reflected p and s components from the RP are shown by R_p and R_s, respectively.

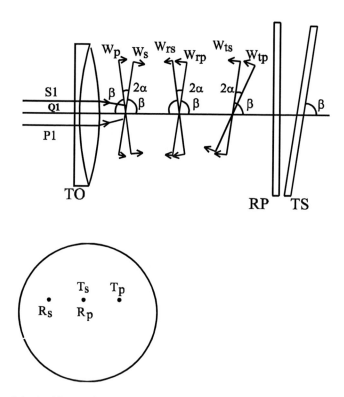

Figure 12. Nature of the incident and Fresnel reflected wave fronts. Relative positions of the focal spots are shown in the inset.

The test surface (TS) can be tilted to Fresnel reflect, nearly normally, one of the incident collimated beam components, either p or s. As shown in Figure 12, the TS is tilted at an angle β, where β = (90°- α) and W_s undergoes a normal reflection on the TS. The reflected p and s wave fronts from the TS are denoted by W_{tp} and W_{ts}, respectively. Now, W_{ts} is (nearly) parallel to W_{rp} and they come to focus (almost) at the same point in the focal plane of the TO.

In the inset in Figure 12, T_p and T_s denote the focal points due to Fresnel reflected p and s components from the TS. In actual practice, T_s is made to coincide with R_p by adjusting the tilt of the TS.

The optical path difference (OPD) between the interfering waves can be expressed as

$$\Delta_1 = [W_{ts}] - [W_{rp}] \tag{5}$$

where the square brackets are used to indicate the OPD. As the interfering waves are nearly parallel to each other, the return path of the beams through the system would be same for both and hence cancels out. We can write

$$[W_{ts}] = [W_s] + [2W_t] \tag{6}$$

$$[W_{rp}] = [W_p] + [2W_r] \tag{7}$$

where $[2W_r]$ and $[2W_t]$ are the changes in the OPDs due to the reflections on the reference and the test surfaces, respectively. Substituting Eq. (6) and (7) in Eq. (5) and simplifying, we get

$$\Delta_1 = 2\{[W_t] - [W_r]\} + \{[W_s] - [W_p]\} \tag{8}$$

It is necessary to reduce the effect of the off-axis aberrations to negligible value by reducing the separation between the (virtual) source points at P and S to a minimum. The minimum separation between the (virtual) source points is restricted by the size of the focal spot of the beam reflected from the TS. It is evident that the focal spot size depends on the wave front aberrations and the diffraction. The diameter of the PH2 should be approximately 1.8 to 2 times bigger than the spot size, so as to allow the passage of the superimposed central spots without any diffraction and block the other, and the minimum separation between the source points can be taken as equal to the diameter of the PH2. Taking f=1000mm, and d= 50 μm, we obtain $\alpha \approx 5.0$ arc seconds. Thus the angle of inclination α is very small in comparison to the angular field of view, which is of the order of few degrees, for a well corrected telescope objective. Hence the effect of the off-axis aberration is practically negligible.

Moreover, the virtual point sources at P and S are symmetrical around the optical axis of the TO and except for the errors in fabrications of the TO, which may give rise to asymmetric wave front aberrations, the nature of the incident wave fronts W_p and W_s would be same. Thus for a good quality TO, $[W_p]$ is practically same as $[W_s]$. Hence the surface form measurement is not affected by the instrumental aberration and the function of the FI is similar to that of an on-axis one for all practical purposes.

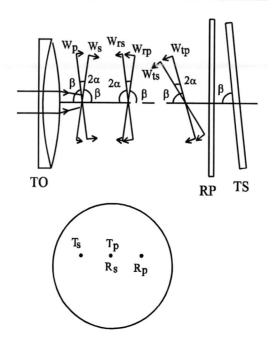

Figure 13. Incident and Fresnel reflected wave fronts. Relative positions of the focal spots are shown in the inset.

Nevertheless, it is possible to eliminate the effect of any residual aberration. The test surface is tilted by (-β) so as to retrace back the incoming p waves as shown in Figure 13. W_{tp} is made parallel to W_{rs} and the OPD between the interfering waves can be written as

$$\Delta_2 = [W_{tp}] - [W_{rs}] \qquad (9)$$

where $[W_{tp}] = [W_p] + [2W_t]$ and $[W_{rs}] = [W_s] + [2W_r]$ and we get

$$\Delta_2 = 2\,\{[W_t] - [W_r]\} - \{[W_s] - [W_p]\} \qquad (10)$$

$$\Delta = [W_t] - [W_r] = (\Delta_1 + \Delta_2)/4 \qquad (11)$$

To determine Δ_1 and/ or Δ_2 by phase shifting interferometry [35] a series of images are captured as the phase difference between the interfereing components is changed, in steps, by a constant amount by varying the angular of orientation of P (Figure 10). The intensity at a point (x, y) in the interference field can be written as [47]

$$I(x,y) = I_0(x,y)\,\{1 + V(x,y)\cos[\phi(x,y) + \alpha_j]\} \qquad (12)$$

where $I_0(x, y)$, $V(x, y)$, $\phi(x, y)$ are the mean intensity, fringe visibility and original phase difference between the two interfering beams at the point (x, y), respectively and $\alpha_j = (j-1)\alpha$, with j= 1,2, 3…and $0 < \alpha < \pi$., represents phase shift between the adjacent frames. We have used an algorithm [48], which uses five different values of intensity, I_1, I_2, I_3, I_4, I_5 corresponding to j=1, 2, 3, 4, 5 and $\alpha=\pi/2$ and ϕ is given by

$$\phi = \arctan[2(I_4-I_2)/(I_1+I_5-2I_3)] \tag{13}$$

For an error ε in phase shift, (so that $\alpha=\pi/2 +\varepsilon$, where ε is a small quantity) the error in ϕ is given by [48].

$$\Delta\phi = (\varepsilon^2/4)\sin 2\phi \tag{14}$$

which shows the insensitivity of this algorithm to phase shift miss-calibration type system error.

Taking absolute values of numerator and denominator of Eq. (13), ϕ is first calculated in modulo $\pi/2$ and then comparing signs of $\sin\phi$ and $\cos\phi$, [ref. 35, pp. 366] ϕ is transformed to modulo 2π.

The calculated phases are wrapped and unwrapping / phase integration is performed by comparing the phase difference between adjacent pixels [49]. Unwrapped phase values are least square fitted and deviation from a reference plane is obtained. Phase variations are converted to optical path variations by using the following relation

$$\delta = (\lambda/4\pi)\phi \tag{15}$$

where δ and λ represent optical path length and wavelength of light, respectively.

In the present technique, the phase shifts/steps are introduced by varying the angular orientation of P (Figure 10) in steps of 45° for introducing phase shift of $\pi/2$ between the frames and five frames are captured. Figure 14 shows a typical Fizeau fringes for good quality flat surface.

Advantage of the technique is that the separation between virtual point sources, which are used to produce a pair of collimated beams with orthogonal polarizations in the FI cavity, is variable and can be adjusted to smaller values so as to reduce the effect of off axis aberration to practically negligible values.

Nevertheless, it is possible to eliminate the effect of any residual aberration by following a two-step averaging method discussed in the preceding paragraphs.

3.5. Generation of White Light Haidinger Fringes

Fringes of equal inclination [50], which are also called Haidinger fringes [51], can be obtained by illuminating a transparent plane parallel glass plate with light from a monochromatic point source. The diverging beams of light Fresnel reflected from the plane surfaces of the near parallel plate interfere to form non localized Haidinger fringes. Applications of Haidinger fringes for the measurement of residual wedge angle of quasi parallel plates and deviation angle of prisms have been discussed in [52]. For a broad band light source, the Haidinger fringes can be obtained only with thin films whose thicknesses satisfy the coherence requirement.

Figure 14. shows the nature of typical Fizeau fringes.

In this section, a CPOC based technique for the production of Haidinger fringes with a low coherence broad band white light source is discussed. [15]. A schematic of the optical setup is shown in Figure 15. Expanded collimated light beam represented by plane wave front (ABC) passes through a plate type beam splitter (BS) and falls on BS1 at an angle of incidence of 45°. BS1 and plane mirrors M1, M2 form a CPOC. M1 and M2 are inclined at 45° to each other and are placed at equal distances with respect to BS. Thus the CPOC is set for zero lateral shear between the beam components produced due to splitting up of the incident beam by BS1. Due to the presence of a relatively long focal length lens (L), as shown in Figure 15, two converging beams emerge from either side of L. The converging emergent beams coming out along the input beam port of the BS1, are deflected by BS towards a rotating diffuser screen (RDS). In Figure 15, the beam paths are shown by central rays. Suppose, for a position (L1) of the converging lens, the focal spots coincide at F. A longitudinal displacement ($s_0/2$) of the converging lens from the symmetrical position L1 to another position L would produce a longitudinal separation s_0 between the focal spots denoted by (F1) and (F2). Rays of light diverging out of F1 and F2 interfere to produce non-localized circular equal inclination fringes on RDS. An imaging lens (IL) transfers the fringe image onto the plane of a two dimensional CCD array. The image is digitized and captured by a frame grabber and Personal computer (PC).

The optical path difference (OPD) (Δ) between the rays F2x and F1x is given by

$$\Delta = \{[F2x] + [s_0] - [F1x]\} \qquad (16)$$

Applying simple geometry, Eq. (16) can be written as

$$\Delta = s_0 - \{\sqrt{[(D+s_0)^2 + x^2]} + \sqrt{(D^2 + x^2)}\} \qquad (17)$$

Where D = F2J and x = Jx as shown in Figure 15.

Since Δ=0, for x=0, i. e., at the center of the interference field, as can be seen from Eq. (17), zero order bright fringes for all the wave lengths would coincide at the center of the field as shown in Figure 16.

Also the fringe order increases as one moves away from the center of the field. These behaviors are different from Haidinger fringes normally produced by using parallel plate or thin films, where the fringe order is highest at the center and decreases as one moves away from the center of the field.

Because of the coherence requirement, white light/ broad band Haidinger fringes can only be produced by thin film or by Michelson interferometer while a similar type fringes of equal inclination can be produced using the setup shown in Figure 15.

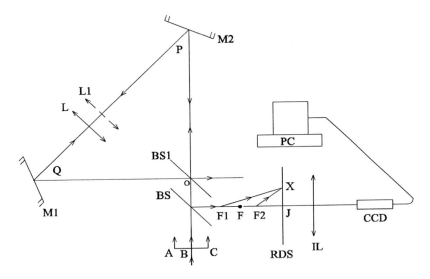

Figure 15. Schematic of the optical setup for the production white light Haidinger fringes.

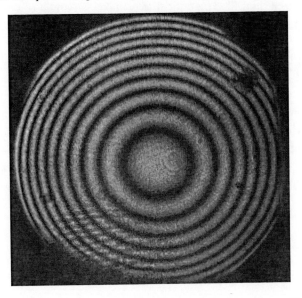

Figure 16. Fringes of equal inclination obtained with the setup shown in Figure 15.

The fringes formed by emergent beams from the output port of the BS has dark centre because of the extra p phase suffered by one of the interfering components. Also, because of the unequal number of reflections suffered by the interfering beams, the visibility depends on the R/T splitting ratio of BS.

The technique discussed has been applied for the measurement of centering error [Section 3.6] and focal length of lenses [Section 3.7].

3.6. Measurement of Centering Error of Lenses

The non-coincidence of optical axis of a lens, which is defined as the line joining the center of curvatures of the two surfaces, with its physical/mechanical axis is called centering error.

For a lens with circular shape, which is most common, the axis of the edge cylinder is considered as the mechanical axis of the lens. The parallel shift of the optical axis from the mechanical axis is referred to as de-center while the angular inclination between the optical axis and mechanical axis is called tilt.

In this section a technique for the determination of centering error of lenses is discussed. [16] Consider a positive lens, represented by cardinal points [53], and illuminated by expanded, collimated and counter-propagating He-Ne (632.8nm) laser beams as shown in Figure 17.

In general, centering error has both tilt and de-center components and the optical axis and mechanical axis can be skew lines in three dimensional spaces. For the sake of simplicity and ease of understanding, we can consider the optical axis in the plane of the figure and apply principles of paraxial optics. N1, N2 and N1H1, N2H2 represent nodal points and principal planes of the lens, respectively.

The focal planes are shown by lines which are normal to the optical axis (C1C2) of the lens at the focal points F and F'. The optical axis makes an angle ϕ with the direction of the laser beams represented by plane wave-fronts (A1B1C1) and (A2B2C2). The ray direction is shown by arrow.

The collimated beams come to focus at F1 and F2 on the respective focal planes. The positions of F2 and F1 can be obtained by considering rays (B1N1) and (B2N2) which are directed towards N1 and N2, respectively.

It is evident from the property of nodal points [54] that the ray BN1 comes out along the direction (N2B2), which is parallel to B1N1, after traveling through the lens and intersects the focal plane at F2. Similarly, the ray B2N2 travels along N1B1 and intersects the corresponding focal plane at F1. The lateral (in a direction perpendicular to the direction of propagation of the beams) separation between F1 and F2 can be written as

$$\tau = N1N2 \sin \phi \tag{18}$$

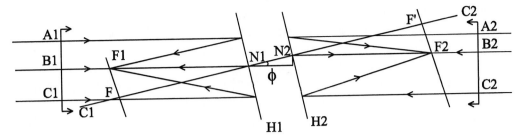

Figure 17. Lateral shift of focal points F1 and F2 of a lens, shown by cardinal points, due to a tilt (φ) in the optical axis.

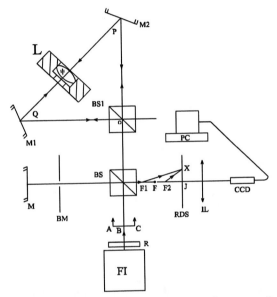

Figure 18. Schematic of the optical setup for measurement of centering error of a lens.

The counter-propagating collimated beams discussed in the preceding paragraphs are produced in a CPOC, formed by non-polarizing beam splitter (BS1) and plane mirrors (M1) and (M2), as shown in Figure 18. M1 and M2 are inclined at 45° to each other. Expanded collimated beam from a Fizeau interferometer (FI) is coupled to the CPOC via a beam splitter (BS). Only central ray paths of the beams are shown in Figure 18. (R) is a reference surface of the FI. Positions of M1 and M2 are adjusted with respect to BS1 to make the reflected and transmitted components of the parallel beam, as split by BS1, to travel the same triangular path (TP) in opposite directions and exit from the CPOC with zero optical path difference and zero lateral shear. The emergent beams retracing back along the incident beam but in opposite direction are only considered. The alignment for counter propagating beams, with zero lateral shear and zero optical path difference, which comes out along the incident beam direction can be done with interferometric precision by utilizing the Fizeau fringes formed between the emergent beams and the Fresnel reflected component of the incident collimated beam off R, which is pre-aligned normal to the incident parallel beam by comparing the light reflected from R with that retro-reflected from a corner cube prism [11], which is temporarily placed in the beam path. The test lens is placed in the CPOC, as shown in Figure 18 where test lens is represented by L, with its mechanical axis along the direction of the beams. The test lens focuses the counter-propagating beams at (say) F1 and F2. Diverging beams from F1 and F2 interfere and non-localized fringes of equal inclination, which are in the form of concentric circles on a plane normal to the direction of propagation of the beams, are formed [15]. The center of the circular fringe system would lie on the line joining F1 and F2. Suppose for a particular longitudinal position of the test lens in the TP, F1 and F2 would coincide at (F) and a longitudinal shift of (S/2) of the test lens from this position would introduce a longitudinal separation S between F1 and F2. Suppose, there is a centering error in the test lens and its optical axis makes an angle φ with the direction of the counter-propagating beams, the lateral separation between F1 and F2 is given by τ (Eq. 18) and the angle (α) between the line joining F1 and F2 and the central ray (CR) direction of the counter propagating beams, as shown in Figure 19, is given by

$$\operatorname{Tan}\alpha = \tau/S = M \operatorname{Sin} \phi \tag{19}$$

where M=(N1N2 / S)

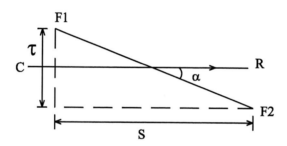

Figure 19. Positions of the focal points with respect to the central ray.

The centering error of the test lens may have both tilt and de-center components. A narrow collimated reference pencil beam which traverses the path of the central rays of the counter-propagating beams and also travels along the mechanical axis of the test lens is thus needed along with the diverging beams for the determination of the tilt and de-center of the test lens. As any reference collimated beam, traveling along the mechanical axis is transformed by test lens into converging beam, an equivalent reference pencil beam is sampled from the input collimated beam outside the CPOC by BS, plane mirror M and a circular beam mask (BM) with a central hole of 1.0mm diameter in the following way: During the initial setting up process a similar BM is temporarily placed in the input parallel beam from FI (between R and BS) to allow a central pencil beam to pass through and serves as input beam to CPOC. The mechanical axis of the lens holder, without test lens, is made to coincide with the counter-propagating pencil beams by placing an appropriate BM (with outer diameter same as the lens and with a central hole of 1.0mm diameter) in the lens holder and doing the necessary adjustments. M is adjusted for the retro-reflection of the pencil beam. It is evident that the pencil beam retro-reflected by M travels collinearly with that emerge out of the CPOC and travel towards a rotating diffuser screen (RDS) as shown in Figure 18, and thus serves as reference pencil beam. The BM between BS and M is adjusted for the passage of the pencil beams. The BM placed in the input beam is then removed. The RDS is placed normal to the direction of the reference pencil beam and an imaging lens (IL) is used to transfer the image formed on the RDS to the CCD detector plane. The images of the circular fringes of equal inclination along with the reference spot formed on the RDS due to the reference pencil beam are grabbed by Frame grabber and Personal computer (PC). As can be seen in Figure 18, the optical path difference (OPD) Δ, between the rays F2X and F1X is given by

$$\Delta = \{[F2X] + [S] - [F1X]\} \quad (20)$$

Applying simple geometry, Eq. (20) can be written as

$$\Delta = S - \{\sqrt{[(D+S)^2 + x^2]} + \sqrt{(D^2 + x^2)}\} \quad (21)$$

where D = F2J and x = JX are as shown in Figure 18.

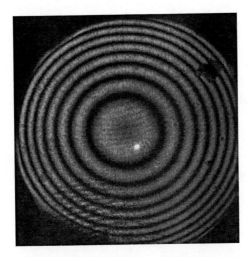

Figure 20. Grabbed fringes of equal inclination and the position of the reference pencil beam.

It is evident from Eq. (21) that for x=0, i. e., at the centre of the field, Δ is zero. The centre of the fringes of equal inclination as defined by Eq. (21) is found from the position co-ordinate of the centre of the circle obtained by fitting the position co-ordinate data of the fringe minimums on the first ring.

Both the tilt and the de-center components of the centering error can be determined by applying a null method. The lens holder is mounted on a high quality tip tilt mirror mount, which in turn is fitted on a three axes (x, y and z) linear translation stage. Also, a thin highly parallel plate, such as a pellicle beam splitter, is mounted on lens holder with its surface-normal along the mechanical axis of the lens holder.

During the initial setting up, test lens is kept in the lens holder with its mechanical axis along the direction of the reference pencil beam. Figure 20 shows grabbed fringes of equal inclination along with the laser spot for a lens with centering error. .

Lens holder is adjusted to make the optical axis of the test lens to coincide with the reference pencil beam direction as can be ensured from the coincidence of the center of the fringes of equal inclination with the center of the reference pencil beam for two different longitudinal positions of the RDS.

The resultant angular and/or lateral adjustments necessary to make the optical axis of the test lens to coincide with the reference pencil beam direction give both the tilt and the de-center components of the centering error. The tilt component of the centering error, which often is the predominant one, can also be determined with interferometric precision from the spacing and direction of the two-beam Fizeau fringes formed between the collimated beam reflected from the thin parallel plate mounted in lens holder and that from the reference surface R of the FI.

The accuracy of measurement depends on the accuracy of coincidence of the center of the circular fringes with the center of the reference beam. Thus in the limit, accuracy is limited by the effective dimension, Δr, of the pixel of the CCD array on RDS. We have

$$\Delta r = (d_l/d_c)\, \delta r \tag{22}$$

where d_l and d_c are the diameters of the first ring on the RD and CCD detector plane, respectively and δr is the actual dimension of the pixel of the CCD detector. Maximum value of d_c depends on the size of the CCD array. The linear separation (Δl) on the RDS between the center of the fringes of equal inclination and the center of the reference pencil beam, due to a residual tilt angle ϕ between optical axis and mechanical axis, can be written as

$$\Delta l = D\, M \sin \phi \tag{23}$$

It is evident that Δl should be greater than or equal to Δr and the minimum measurable tilt error (ϕ_m) is given by

$$\Delta l = \Delta r$$
$$D\, M \sin \phi_m = (d_l/d_c)\, \delta r$$

and
$$\sin \phi_m = [(S\, d_l)/ D]\, [\delta r/(N1N2\, d_c)] \tag{24}$$

The term in the second square bracket on the right hand side of Eq. (24) is a constant for a particular TL and CCD detector array while that within the first square bracket varies with S.

The technique discussed in the preceding paragraphs is non-contact and is insensitive to external vibrations as both the beams travel through the same optical circuit. Also, the contrast of the equal inclination fringes is good as the interfering beams suffer same number of reflections and transmissions. Main advantage of the technique is that there is no need for a high accuracy rotation spindle as normally required, for the measurement of centering error of lenses. Though in the present work only positive lens is considered, it is evident that the technique is equally applicable for negative lenses.

3.7 Determination of the Focal Length of Lenses

There are standard methods for measurement of focal length of lenses. A polarization CPOC based technique for the measurement of focal length of lenses is discussed in this section [17].

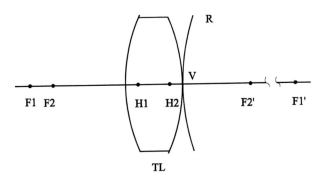

Figure 21. Optical schematic for illustration of the basic principle.

We consider Figure 21 for explanation of the basic principle of the method. (F1) and (F2) are two point sources situated on the axis of a test lens (TL). Suppose F2 coincides with the back focal point of TL and F1 is at a small separation x_1 from F2. Thus collimated beam would be produced by TL due to light from F2. Since x_1 is small, the real image due to F1 would be formed at long distance from TL, say at F1', as shown in Figure 21. The radius of curvature of the image forming wave front at the vertex (V) of TL is R. Suppose H1, H2 are the position of principal points and F2' is the front focal point of TL and H2F2' =F, the focal length of TL, F2'F1' = x_2, H2V=d. We get

$$x_2 = (R + d - F) = (R-F) \tag{25}$$

since (R-F)>> d.

According to Newton's formula, where object and image distances are measured from the focal points, the focal length is related to object and image distances by the following relation

$$F = \sqrt{(x_1 x_2)} = \sqrt{[x_1 (R-F)]} \tag{26}$$

Solving for F, we obtain

$$F = [-x_1 + \sqrt{(x_1^2 + 4Rx_1)}]/2 \tag{27}$$

A polarization CPOC is used to obtain two point sources separated by a small known separation on the axis of TL. The point sources are formed by light with linear orthogonal polarizations so that by adjusting the orientation of the linear polarizer in the incident beam to CPOC, one of the point sources can be selected. A wedge shear plate is used to measure the radius of curvature R of the nearly collimated beam corresponding to an object point situated at a small separation x_1 from another point object adjusted to be at the focal point of TL.

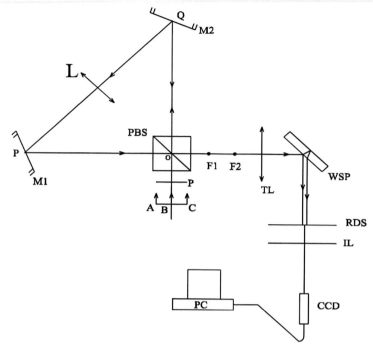

Figure 22. Schematic of the optical setup for the measurement of focal length of lenses.

The optical setup is shown schematically shown in Figure 22. Expanded collimated He-Ne (1=632.8nm) laser beam with linear polarization, represented by plane wave front (ABC), is allowed to fall on a polarization beam splitter (PBS) at an angle of incidence 45°. PBS along with plane mirrors M1 and M2, which are inclined at 45° to each other, forms a CPOC. PBS splits up the incident beam into reflected and transmitted components having orthogonal polarizations namely, p and s. The plane of polarization of the incident beam is set at appropriate orientation (45°). M1 and M2 are adjusted to be at equal distances to make the lateral shear zero. A convergent lens (L) with relatively longer focal length (greater than the cavity length of the CPOC) is placed in the hypotenuse arm of the CPOC at a separation of $x_1/2$ from the symmetrical position, as shown in Figure 22, so as to produce focal points F1 and F2 with separation x_1 between them. F1 and F2 are produced due to convergent beams with orthogonal polarizations. F1 is made off and F2 is set at the back focal point of TL. This is done by ensuring collimated beam output from TL. The beam collimation is checked by a wedge shear plate (WSP), which is kept with its wedge in a direction perpendicular to the direction of lateral shear of the WSP.[1] Separation between TL and F2 is adjusted to align the lateral shearing fringes in a direction normal to the wedge direction of the WSP, as shown in Figure 23. Now F2 is switched off and the radius of curvature of the nearly collimated converging beam forming image of F1 due to TL is found out from the lateral shear fringes formed on RDS, using the relation

$$R = \{[s\ d/(1\ \sin a)] + D\} \qquad (28)$$

where s, d and a are the amount of lateral shear, fringe spacing, and the angular orientation of the fringes with the horizontal, as illustrated in the Figure 24, respectively and D is the distance between RDS and the vertex of TL. Hence from the known value of x_1 and R, F can be found using Eq. (27).

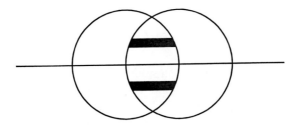

Figure 23. Lateral shear fringes with collimated beam.

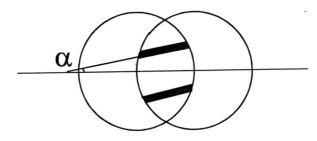

Figure 24. Lateral shear fringes with slightly non collimated beam.

The same technique can also be applied for the determination of focal length of negative lenses [17]. For negative lens, F1 and F2 should act as virtual objects.

CONCLUSION

CPOC based interferometric techniques for external measurement of dihedral right angle of optical components, simultaneous measurement of refractive index and wedge angle of optical windows, measurements of centering error and focal length of lenses have been discussed. Productions of Young's and Haidinger fringes with broad band sources, using CPOC based optical setup, have been described. The optical setup for producing Young's fringes can be used to estimate temporal coherence of low coherence sources and a similar setup is used for the implementation of polarization phase shifting Fizeau interferometer. The setup for producing Haidinger fringes has important applications such as measurement of centering error and focal length of lenses.

A CPOC based optical setup has the following advantages: 1] simple and easy to setup and align 2] practically insensitive to external mechanical vibrations 3] the interfering beams travel identical optical paths, hence monochromatic, quasi- monochromatic or broad band sources can be used

ACKNOWLEDGMENTS

I thank Pavan kumar for careful reading of the manuscript and Rishipal for drawing of the figures.

REFERENCES

[1] M. Strojnik, G. Paez, and M. Mantravadi, "Lateral shear interferometers "*Optical ShopTesting*,3rd ed., D. Malacara, Ed., pp. 140-142, Wiley, Hoboken, NJ (2007).
[2] S. Chatterjee and Y. P. Kumar, *Appl. Opt.* 2002, 41, 5857-5859.
[3] S. Chatterjee and Y. P. Kumar, *Opt. Eng.* 2010, 49, 053605.
[4] J A. Ferrari and E.Garbasi, *Appl. Opt.* 2005, 44, 4510-4512.
[5] Y. H. Lo, A. R. D. Somervell, T. H. Barnes, *Opt. Laser Eng.* 2005, 43, 33-41.
[6] M. P. Kothiyal and C. Delisle, *Appl. Opt.* 1985, 24, 4439-4442.
[7] D. Malacara "Radial, Rotational and Reversal shear interferometers "*Optical ShopTesting*,3rd ed., D. Malacara, Ed., pp. 189-197, Wiley, Hoboken, NJ (2007).
[8] P. Harharan and M. Roy ,*J. Mod. Opt.* 1992, 39, 1811-1815.
[9] C. C. Cheng and M. G. Raymer *J. Mod. Opt.* 2000, 47, 1237-1246.
[10] J. S. Darlin, M. P. Kothiyal and R. S. Sirohi, *J. Mod. Opt.* 1997, 44, 929-939.
[11] S. Chatterjee and Y. P. Kumar, *Appl. Opt.* 2009, 48, 1598-1605.
[12] Y. P. Kumar and S. Chatterjee, *Appl. Opt.* 2009, 48, 4756-4761.
[13] S. Chatterjee and Y. P. Kumar, *Appl. Opt,* 2008, 47 2956-2960.
[14] S. Chatterjee and Y. P. Kumar, *Opt. Eng.* 2009, 48, 115601(1-6).
[15] S. Chatterjee and Y. P. Kumar, *Opt. Lett.* 34 (8) 1291-1293.
[16] S. Chatterjee and Y. P. Kumar, *Opt. Eng.* 2010, 49, 043601(1-5).
[17] Y. P. Kumar and S. Chatterjee *Opt. Eng.* 2010, 49, 053604.
[18] M. V. R. K. Murty, "Newton, Fizeau and Haidinger inteferometers," in *Optical Shop Testing*, D. Malacara, ed. (Wiley, 1992), pp.22-23, pp.30-34.
[19] C. Ai and K. Smith *Appl. Opt.* 1992, 31, 519-527.
[20] D. Thomas and J. C. Wyant, *J.Opt. Soc. Am.* 1977, 67, 467-472.
[21] A. Saxena and L. Yeswanth, *Opt. Eng.* 1990, 29, 1516-1520.
[22] D. Malacara and R. F. Hernandez, *Proc. SPIE*-1332, 36-40 (1991).
[23] J. E. Ludman and C. Wards *Proc.SPIE-299*, 106-111 (1981).
[24] J. Burke, B. Oreb, B. Platt, and B. Nemati *Proc. SPIE* 5869, 225-235 (2005).
[25] B. Oreb, J. Burke, R. P. Netterfield, J. A. Seckold, A. Leistner, M. Gross, and S. Dligatch *Proc. SPIE* 6292, 1-13, (2006).
[26] Z. Ge and M. Takeda, *Appl. Opt.* 2003, 42, 6859-6868.
[27] S. Chatterjee and Y. P. Kumar, *Appl. Opt.* 2008, 47, 4900-4906.
[28] O. Kafri, K. M. Kerske and E. Keren, *Appl. Opt.* 1988, 27, 4602-4603.
[29] D. Bhattacharyya, A Ray, B. K. Dutta and P. N. Gosh, *Opt. Laser Technol.* 2002 34, 93-96.
[30] Y. R. Wang, X. M. Qu, L. Z. Cai, B. M. Ma and D. L. Sun, *J. Mod. Opt.* 1999, 46, 1369-1376.

[31] M. Born and E. Wolf, "Elements of the theory of interference and interferometers" in *Principles of Optics* (Pergamon, 1989), pp-260-265, pp- 268-271, pp-271-277.
[32] J. F. Barrera, F. F. Medina, and J. G. Swerquia, *Optik* 2007, 118, 402-406.
[33] M. Born and E. Wolf, *Principles of Optics*, p. 286-291, Pergamon Press, Oxford (1989).
[34] M. V. Mantravadi and D. Malacara, "Newton, Fizeau and Haidinger interferometers," in *Optical Shop Testing,* D. Malacara, Ed. p. 17-32, Wiley (2007).
[35] K. Creath, "Phase measurement interferometry techniques," in *Progress in Optics, xxviii*. E. Wolf, Ed. p. 349-393, North-Holland Amsterdam: (1988).
[36] J. Schwider, "Advanced evaluation technique in interferometry," in *Progress in Optics, xxviii*, E. Wolf, Ed. p.271-359, North-Holland Amsterdam: (1990).
[37] J. E. Greivenkamp and. J. H. Bruning, "Phase shifting interferometry," in *Optical Shop Testing,* D. Malacara, Ed. p. 547-655, Wiley (2007).
[38] C. Ai and J. C. Wyant, *Appl. Opt.* 1987 26, 1112-1116.
[39] P. Hariharan, "The geometric phase," in *Progress in Optics, xiviii*, E. Wolf. Ed. p.149-204, North Holland, Amsterdam (2006).
[40] S. Chatterjee, Y. P. Kumar, and B.Bhaduri *Opt. and Laser Tech.* 2007, 39, 268-274.
[41] K. Onuma, K. Tsukamoto, and S. Nakadate J. *Crystal Growth*. 1993, 129, 706-718.
[42] J. E. Millard, N. J. Brock, J. B. Hayes, M. B. North-Moris, M. Novak and J. C. Wyant *Proc. SPIE* 5531 304 (2004).
[43] B. Kimbrough, J. Millard, J. Wyant and J. B. Hayes *Proc. SPIE* 6292 62920F-1 (2006).
[44] J. C. Wyant "Dynamic interferometry" Opt. and Phot. News. 14(1) 38-41 (2003).
[45] J. C. Wyant, *Opt. and Photonics News* 2007, 18, 32-37.
[46] T. H. Barnes *Appl. Opt.* 1987 26, 2804-2809.
[47] P. Hariharan *Opt. Eng.* 2000, 39, 967-969.
[48] P. Hariharan, B. F. Oreb, T. Eiju, *Appl. Opt.*, 1987, 26, 2504-2505.
[49] D. Malacara, S. Malacara, and Z. Malacara, in *Interferogram Analysis for Optical Testing*, p. 248-55, Marcel Dekker, New York: (1998).
[50] M. Born and E. Wolf, "*Principles of Optics*," chap-VII, pp-281-285, Pergamon Press, sixth edition (1989).
[51] C. V. Raman and V. S. Rajagopalon J. *Opt. Soc. A.* 1939, 29, 413-416.
[52] M. V. Mantravadi and D. Malacara, "Newton, Fizeau and Haidinger Interferometers," Chap-1, 35-40, *Optical Shop Testing* (3rd edition) ed., D. Malacara, Wiley (2007).
[53] M. Born and E. Wolf, "*Principles of Optics*," , pp. 161-163, Pergamon Press, Sixth (corrected) edition (1989).
[54] F. A. Jenkins and H. E. White, "*Fundamentals of Optics*," Chap-5, pp. 71-78, McGraw-Hill Book Company, 2nd edition(1950).

In: Interferometry Principles and Applications
Editor: Mark E. Russo

ISBN 978-1-61209-347-5
© 2012 Nova Science Publishers, Inc.

Chapter 5

SINGLE-SHOT PHASE-GRATING PHASE-SHIFTING INTERFEROMETRY

Gustavo Rodriguez-Zurita[1,a] and Noel-Ivan Toto-Arellano[2,b]

[1]Benemérita Universidad Autónoma de Puebla, Facultad de Ciencias Físico-Matemáticas
Apartado Postal, Puebla, México

[2]Centro de Investigaciones en Óptica A.C., Col. Lomas del Campestre, León, Gto. México

ABSTRACT

Phase-shifting interferometry requires of several interferograms of the same optical field with similar characteristics but shifted by certain phase values to retrieve the optical phase. This task has been usually performed by stages with great success and requires of a series of sequential shots. However, time-varying phase distributions are excluded from this schema. Several efforts for single-shot phase-shifting interferometry have been tested successfully, but some of them require of non-standard components and they need to be modified in some important respects in order to get more than four interferograms. Two-windows grating interferometry, on the other hand, has been proved to be an attractive technique because of its mechanical stability as a common-path interferometer. Moreover, gratings can be used as convenient phase modulators because they introduce phase shifts through lateral displacements. In this regard, phase gratings offer more multiplexing capabilities than absorption gratings (more useful diffraction orders because higher diffraction efficiencies can be achieved). Furthermore, with two phase gratings with their vector gratings at 90° (grids) there appear even more useful diffraction orders. Modulation of polarization can be independently applied to each diffraction order to introduce a desired phase-shift in each interference pattern instead of using lateral translations. These properties combine to enable phase-shifting interferometric systems that require of only a single-shot, thus enabling phase inspection of moving subjects. Also, more than four interferograms can be acquired that way. A simple interferogram processing enables the use of interference fringes with different fringe modulations and intensities. In this chapter, the basic properties of two-windows interferometry with phase

---

[a] E-mail address: gzurita@fcfm.buap.mx. Tel.: (+52) 222 229 5500 Ext. 2109, Fax: (+52) 222 229 5636.
[b] E-mail address: ivantotoarellano@cio.mx.

gratings and modulation of polarization is reviewed on the basis of the far-field diffraction properties of phase gratings and grids. Phase shifts in the diffraction orders can be used as an advantage because they simplify the needed polarization filter distributions. Examples of experimental set-ups (such as basic configurations, lateral-shear and radial-shear) are shown and discussed. It is finally remarked, that these interferometers are compatible with interference fringes exhibiting spatial frequencies of relative low values and, therefore, no great loss of resolution is related with several interferograms when simultaneously using the same image field of the camera.

1. INTRODUCTION

Phase gratings have been employed as an optical element with more diffraction efficiency than absorption gratings to perform a variety of tasks. Among them are beam splitting for interferometry, intensity measuring [1] and optical shop testing [2,3]. The performance of phase gratings depends strongly upon their Fourier spectra. The case of sinusoidal phase grating has been discussed since long [4], so it is well known that its Fourier coefficients are Bessel functions of the first kind of integer order q, J_q [5-6]. Such functions are real valued and their values oscillate around zero, so they can introduce eventually π-phase shifts in a given grating Fourier spectrum. These shifts are of little relevance, if any, to applications where the power spectrum is the main concern, as is often the case in spectroscopy [7].

It is also known that a grating interferometer with two windows in the object plane performs as a common path interferometer [8]. Several advantages have been shown, as its mechanical stability [9, 10]. Moreover, in conjunction with a suitable modulation of polarization, single-shot phase-shifting interferometric systems can be implemented with phase gratings [11]. Two windows phase-grating interferometers (TWPGI) are based on the interference between neighboring diffraction orders [8]. Thus, the fringe modulation of each interference pattern can be affected when π-phase shifts are presented [11,12]. Furthermore, these phase shifts have to be taken into account for the overall performance of the system and their practical advantages can even influence its design.

This review is first aimed at the phase shifts between diffraction orders that have been observed in the Fourier spectra of a phase grating. An example of phase sinusoidal grating is calculated with a standard FFT routine. This would serve for interpretation of later experimental observations. Because some experimental consequences of phase shifts appear also in phase grids, corresponding discussion for sinusoidal phase gratings follows. Experimental observations in agreement with the previous discussions are then shown by using commercial phase gratings. Finally, as an application of the above, a system able to obtain four interferograms 90° phase-apart in only one shot with a phase grid is proposed and demonstrated. More that four interferograms can also be taken in one shot and the way they can be usel for phase-shifting interferometry are also described.

2. BASIC CONSIDERATIONS

A TWPGI is depicted in Figure 2.1. Basically, it consists of a 4f Fourier optical system, f being the focal length of each transforming lens. In an experiment, as used in following sections, the illumination can come from an YVO$_4$ laser operating at $\lambda = 532$ nm. A quarter-wave plate retarder Q in conjunction with a linear polarizer P allows adjustment of linear polarization at 45° with respect to the horizontal. Two windows (A and B) are placed in the object plane (x, y). For modulation of polarization, two retarders (Q_R, Q_L) are placed just before each window, so as to obtain nearly circular polarizations of opposite signs at the output of each one. A periodic phase-only transmittance $G(u/\lambda f, v/\lambda f)$ is placed in the frequency plane (u,v). Then, $\mu = u/\lambda f$ and $\varsigma = v/\lambda f$ are the frequency coordinates scaled to the wavelength λ and the focal length. In the plane (u,v), the period of G is denoted by d and its spatial frequency, by $\sigma = 1/d$. Two neighboring diffraction orders thus have a distance of $F_0 \equiv \lambda f / d$ in the image plane for a grating. Then, $\sigma \cdot u = F_0 \cdot \mu$ and F_0 can be used as a frequency. In the following sections, phase shifts in the image plane of this system due to the grating are discussed.

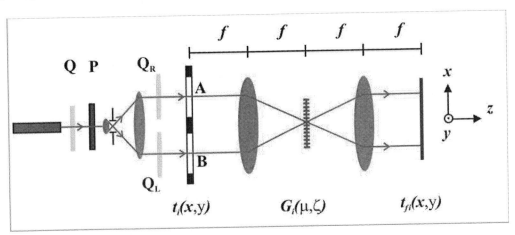

Figure 2.1. TWPGI with phase periodic element G_i of period d. $i = 1$ one-dimensional case (grating), $i = 2$ two-dimensional case (grid).

2.1.1. Sinusoidal Phase Gratings

For simplicity, a sinusoidal phase grating centered in the Fourier plane of a 4f system would be considered with no loss of generality. Its complex amplitude can be expressed as

$$G_1(\mu,\varsigma) = e^{i2\pi \cdot A_g \sin[2\pi F_0 \mu]} = \sum_{q=-\infty}^{\infty} J_q(2\pi A_g) e^{i2\pi \cdot q F_0 \mu} \quad (2.1)$$

with $2\pi A_g$ the grating's phase amplitude and J_q the Bessel function of the first kind of integer order q. The Fourier transform of Eq.1.1 is then given by

$$\tilde{G}_1(x,y) = \sum_{q=-\infty}^{\infty} J_q(2\pi A_g) \delta(x - qF_0, y) \quad (2.2)$$

with $\delta(x,y)$ the two-dimensional Dirac delta function. Thus, the Fourier spectrum of a sinusoidal phase grating comprises point-like diffraction orders of amplitude weighted by Bessel functions. Such spectrum can be detected at the image plane of the $4f$ system.

2.1.2. Phase Grids

A sinusoidal phase grid can be generated by multiplication of two sinusoidal phase gratings whose respective grating vectors are forming an angle of 90 degrees. Taking the rulings of one grating along the "μ" direction and the rulings of the second grating along the "ς" direction, the resulting centered phase grid can be written as

$$G_2(\mu,\varsigma) = e^{i2\pi A_g \sin[2\pi \cdot X_0 \mu]} e^{i2\pi A_g \sin[2\pi \cdot Y_0 \varsigma]} = \sum_{q=-\infty}^{\infty} J_q(2\pi A_g) e^{i2\pi \cdot qF_0 \mu} \sum_{r=-\infty}^{\infty} J_r(2\pi A_g) e^{i2\pi \cdot rF_0 \varsigma}$$
, (2.3)

where the frequencies along each axes directions are taken as $X_0 = Y_0 = F_0$. The Fourier transform of the phase grid becomes

$$\tilde{G}_2(x,y) = \sum_{q=-\infty}^{q=\infty} \sum_{r=-\infty}^{r=\infty} J_q(2\pi A_g) J_r(2\pi A_g) \delta(x-qF_0, y-rF_0)$$
, (2.4)

which consists of point-like diffraction orders distributed in the image plane on the nodes of a lattice with a period given by the value F_0.

2.2. Two-Window Phase-Grating Interferometry: Fringe Modulation

Phase grating interferometry is based on a phase grating placed as the pupil of a $4f$ Fourier optical system [8-12]. The use of two windows at the object plane in conjunction with phase grating interferometry allows interference between the optical fields associated to each window with higher diffraction efficiency [8-12, 13]. Such a system performs as a common path interferometer (Figure 2.1). A convenient window pair for a grating interferometer implies an amplitude transmittance given by

$$t_1(x,y) = w(x+\frac{x_0}{2}, y) + w'(x-\frac{x_0}{2}, y)$$
(2.5)

where x_0 is the separation of center to center between two windows. One rectangular aperture, $w(x,y)$, can be written as $w(x,y) = rect[x/a] \cdot rect[y/b]$ whereas the second one, written as $w'(x,y) = w(x,y) \exp\{i\phi(x,y)\}$, with an object phase function being described by $\phi(x,y)$. As shown in Figure 2.2, a and b represent the side lengths of each window (A and B). Placing a grating of spatial period $d = \lambda f / F_0$ in the Fourier plane, the corresponding transmittance is given by $G_1(\mu,\varsigma)$. The image formed by the system consists basically of replications of each window at distances F_0. This image is defined by $t_{f1}(x,y)$, that is, the convolution of $t_1(x,y)$ with the point spread function of the system, which is $\tilde{G}_1(x,y)$. This results in the following

$$t_{f1}(x,y) = t_1(x,y) * \tilde{G}_1(x,y)$$
$$= w\left(x+\frac{x_0}{2},y\right) * \sum_{q=-\infty}^{\infty} J_q(2\pi A_g)\delta(x-qF_0,y)$$
$$+ \left[w\left(x-\frac{x_0}{2},y\right)e^{i\phi(x-\frac{x_0}{2},y)}\right] * \sum_{q=-\infty}^{\infty} J_q(2\pi A_g)\delta(x-qF_0,y). \quad (2.6)$$

The symbol (*) denotes convolution. By adding the terms q and $q-1$ (both located within the same replicated window $w\left(x-qF_0+\frac{x_0}{2},y\right)$) and for the case of matching the windows positions with the diffraction order positions ($F_0 = x_0$), the previous equation simplifies to

$$t_{f1}(x,y) = \sum_{q=-\infty}^{\infty} \left[J_q(2\pi A_g) + J_{q-1}(2\pi A_g)e^{i\phi\left(x-x_0\left[q-\frac{1}{2}\right],y\right)}\right] w\left(x-x_0\left[q-\frac{1}{2}\right],y\right). \quad (2.7)$$

Thus, an interference pattern between fields associated to each window must appear within each replicated window. The fringe modulation m_q of each pattern would be of the form

$$m_q = \frac{2J_q J_{q-1}}{J_q^2 + J_{q-1}^2}. \quad (2.8)$$

Further phase-shifting techniques must be used in order to introduce proper additional shifts in each pattern. An appropriate displacement of the grating is a possible technique to be used [10]. An alternative approach is the use of polarization-induced phase shifting [11], which is the method than will be described later and the one depicted in Figure 2.1.

2.2.1. π-shifts of the Fourier Spectra of Sinusoidal Phase Gratings

According to Eq. (7), m_q and thus, the contrast, of each interference pattern depends on the signs of J_q and J_{q-1}. The interference fringe modulation is positive for one half of the diffraction orders if the grating's Fourier coefficients are all positive for $q > 0$, whereas the other half would show alternating fringe modulations due to the odd parity of J_{2q+1}. These results can also be depicted as in Figure 2.3 for a hypothetical case, where only the signs of the amplitudes of the diffraction orders from a given grating are separately shown displaced from the origin due to the respective displacement of windows A and B (first two plots from above). The case of $J_4 < 0$ is shown.

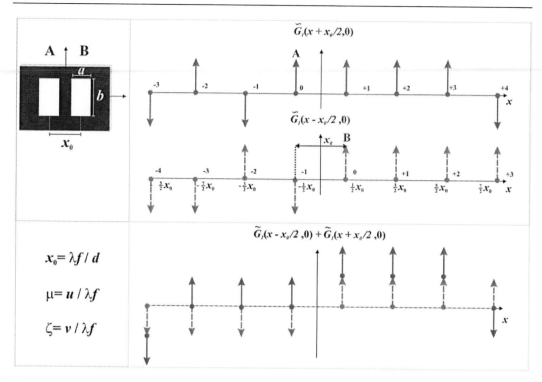

Figure 2.2. Amplitude signs of diffraction orders (hypothetical grating) in the image plane of a TWPGI resulting from windows displacement of $\pm x_0 / 2$. Upper left: windows configuration.

For simplicity, the replicated windows are not plotted. The third row in the graph exhibits the superposition of the two previous spectra. Fringe modulation changes must appear in both halves of the image plane. The expected signs of each fringe modulation in this case would be +,-,-,-,+,+,+,- from left to right. The figure then shows that odd diffraction orders have odd-order Bessel function parity, and conversely for even diffraction orders and even parity.

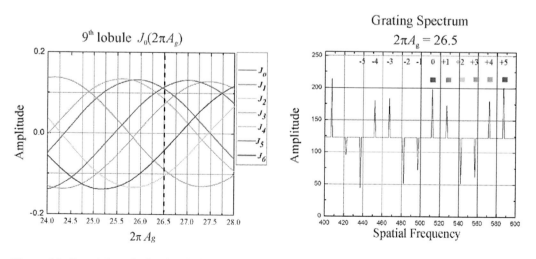

Figure 2.3. Bessel domain for the ninth lobule of J_0 (left). Corresponding Fourier spectrum of the phase grating (right).

Figure 2.3 shows a case of positive J_q-values for $q = 0,1,4,5$ and negative for $q = 2, 3, 6$ as an example (left). This situation corresponds to the value of $2\pi A_g = 26.5$, which belongs to a range in the Bessel domain within the 9th lobule of J_0. A phase grating of the type represented by Eq.2.1 (with the same amplitude $2\pi A_g$ as before) have a Fourier spectrum whose components show π-shifts accordingly (see the positive orders in the right of Figure 2.3). π-shifts such as these can be detected as changes in the signs of fringes modulations in the experimental interferometers discussed later.

2.2.2. π-shifts of the Fourier Spectra of Sinusoidal Phase Grids

A rectangular phase grid $G_2(\mu,\varsigma)$ can be generated with two phase gratings of equal spatial frequency. Figure 2.4(a) depicts the signs of the diffraction orders of a grid made up from two crossed gratings having spectra as the one of Figure 2.2. Positive signs are denoted with hollow circles, whereas negative signs are marked with crosses. The dashed lines form regions enclosing diffraction orders of index pairs 0,n or m,0. Then, the order 0,0 is found in the intersection of these regions. Two possible configurations of windows can be considered for a TWPGI. These configurations are shown in Figure 2.4(b). They are denoted by W_1 and W_2. Respective displacements of diffraction patterns are also indicated with displaced dashed lines. For the case of phase grids with windows in configuration W_2, the image can be written as:

$$t_{f2}(x,y) = t_2(x,y) * \tilde{G}_2(x,y)$$

$$= \sum_{q=-\infty}^{\infty} w\left(x + \frac{x_0}{2}, y + \frac{x_0}{2}\right) * J_q(2\pi A_g) J_r(2\pi A_g) \delta(x - qF_0, y - rF_0)$$

$$+ \sum_{q=-\infty}^{\infty} w\left(x - \frac{x_0}{2}, y - \frac{x_0}{2}\right) * J_q(2\pi A_g) J_r(2\pi A_g) \delta(x - qF_0, y - rF_0). \quad (2.9)$$

Again, with $F_0 = x_0$

$$t_{f2}(x,y) = \sum_{q=-\infty}^{\infty}\sum_{r=-\infty}^{\infty} J_q(2\pi A_g) J_r(2\pi A_g) w\left(x + \frac{x_0}{2}[1-2q], y + \frac{x_0}{2}[1-2r]\right) +$$

$$\sum_{q=-\infty}^{\infty}\sum_{r=-\infty}^{\infty} J_{q-1}(2\pi A_g) J_{r-1}(2\pi A_g) w\left(x - \frac{x_0}{2}[1-2q], y - \frac{x_0}{2}[1-2r]\right), \quad (2.10)$$

where $t_2(x,y) = w\left(x + \frac{x_0}{2}, y + \frac{x_0}{2}\right) + w\left(x - \frac{x_0}{2}, y - \frac{x_0}{2}\right)$.

This result simplifies to

$$t_{f2}(x,y) = \sum_{q=-\infty}^{\infty}\sum_{r=-\infty}^{\infty}\left[J_q(2\pi A_g)J_r(2\pi A_g) + J_{q-1}(2\pi A_g)J_{r-1}(2\pi A_g)e^{i\phi\left\{\left(x-x_0\left[q-\frac{1}{2}\right]\right),\left(y-x_0\left[r-\frac{1}{2}\right]\right)\right\}}\right]$$
$$\times w\left(x-x_0\left[q-\frac{1}{2}\right], y-x_0\left[r-\frac{1}{2}\right]\right).$$
(2.11)

Similarly as for gratings, the fringe modulation of the interference pattern within a window centered in $\left(x_0\left[q-\frac{1}{2}\right], x_0\left[r-\frac{1}{2}\right]\right)$ is

$$m_{qr} = \frac{2J_q J_{q-1} J_r J_{r-1}}{(J_q J_r)^2 + (J_{q-1} J_{r-1})^2}.$$
(2.12)

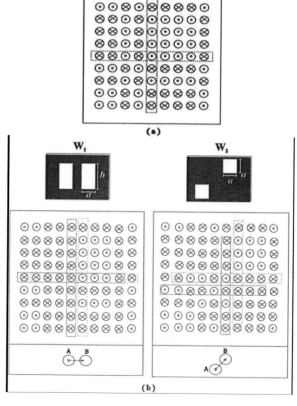

Figure 2.4. a) π-phase distribution of diffraction orders of grids. The dashed lines enclose diffraction orders of indexes 0,n or m,0. b) TWPGI order superposition: Configuration W₁: interference pattern signs for windows displaced along the horizontal axis. Configuration W₂: interference patterns signs for displaced windows along a line at 45°. Respective displacements of diffraction patterns are remarked in dashed lines. An explanation of dot and cross patterns to be found in the text.

Each corresponding fringe modulation depends on the relative phases between J_q, J_{q-1}, J_r and J_{r-1}. These relationships are discussed with the same example of Figure 2.4(a). Figure 2.4(b) shows the relative phases (and thus, expected signs of m_{qr}) of each diffraction order arising from configuration W_1 or W_2. If positive signs or negative ones coincide in a diffraction order, the fringe modulation will be positive. This is plotted with one symbol (cross or hollow dot). Only when a cross with a hollow dot appears, the modulation is expected to be negative. "Vertical" bands with equal sign of m_{qr} are thus expected in configuration W_1. Regions with equal sign of m_{qr} can be seen in configuration W_2.

2.3. Experimental Testing of the Phase-Shifts in Phase Gratings and Phase Grids

Figure 2.5 shows the superposition of Fourier amplitude spectra under windows configuration W_1 for a phase grating with $110\, lines/mm$. The separation between the centers of the windows was of about $9.4\, mm$. Other parameters used were $f \approx 160\, mm$, $a = 3\, mm$ and $b = 6\, mm$. The fringe modulation of the experimental interference patterns shown can be interpreted as if its first four Fourier coefficients had phase relations as the ones sketched in Figure 2.3. The fringes were obtained with the system of Figure 2.1 before placing the retardation plates Q_R, Q_L and the polarizing filters on the image plane. The patterns show the relative phases of the diffraction orders discussed in previous sections (Figure 2.2). The phase-shifted steps of the experimental fringe patterns can be calculated by applying the algorithm proposed by Kreis [15] after proper scaling process. The resulting mean values (in radians) are, from left to right, 0.000, 3.150, 3.131, 3.151, 0.005, 0.014, 0.008, 3.154. It can be seen from the values that they depart by small amounts from 0 or π.

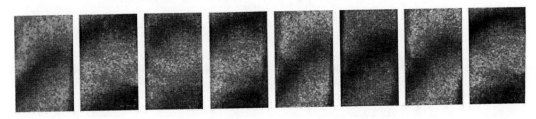

Figure 2.5. Experimental results with a phase-grating. Each image was subject to the same scaling process (from 0 to 255).

For the case of the diffractions orders belonging to a phase-grid constructed with two crossed gratings of equal frequency, the corresponding interference patterns are shown in Figure 2.6 for the windows configuration W_2. Each grating gives patterns as in Figure 2.4(b) (right) when placed alone in the system of Figure 2.1 with no plate retarders neither linear polarizing filters. The whole figure is a composite image because patterns of higher order have lower intensities. The fringe modulation signs are in agreement with the conclusions derived from Figure 2.4. The relative phase values of the 16 patterns within the square (drawn with dotted lines in the patterns of Figure 2.6) employing the method from Ref. 15 can be seen in Table 2.1. Configuration W_1 gives modulation signs in agreement with Figure 2.4(b) (the patterns are not shown). Any grating displacement on its plane only introduces a constant phase term in Eqs.1.6 and 1.8 which, in turn, only shifts each interference pattern by the same amount independent of the diffraction order [9,10]. Modulation of polarization employed to attain the needed shifts in each interferogram is described in the next sections taking only four of 16 interferograms.

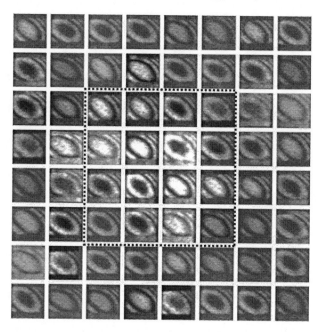

Figure 2.6. Experimental patters for a phase-grid (composite image, windows configuration W_2).

Table 2.1. Phase shifts of the 16 patterns within the dotted square of Figure 6, as measured by the method from Ref.14 (from the left, above)

	Shifts (rad)		
0.016	0.040	3.129	3.173
0.009	0.016	3.150	3.187
3.144	3.173	0.000	0.010
3.137	3.122	0.017	0.037

3. MODULATION OF POLARIZATION

The Figure 3.1 shows the arrangement of an ideal one-shot phase-shifting grating interferometer incorporating modulation of polarization. A combination of a quarter-wave plate Q and a linear polarizing filter P generates linearly polarized light at an appropriate azimuth angle (45°) entering the interferometer. Two quarter-wave plates (Q_L and Q_R) with their orthogonal fast axes are placed in front of the two windows of the common-path interferometer so as to generate left and right circularly polarized light as the corresponding beam leaves each window. A phase grating is placed at the system's Fourier plane as the pupil. In the image plane, superimposition of diffraction orders result, causing replicated images to interfere.

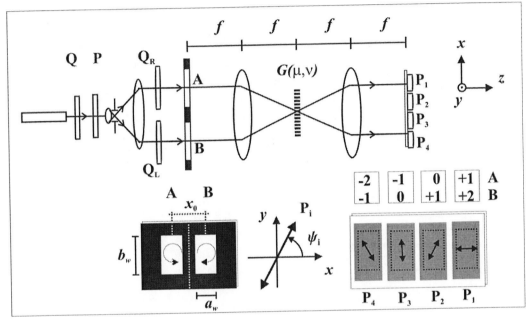

Figure 3.1. One-shot phase-shifting grating interferometer with modulation of polarization. A, B: windows. Side view: ψ_i: polarizing angles (with i = 1,2,3,4 to obtain phase-shifts ξ_i =0°, 90°, 180°, 270° respectively). Involved interference order superpositions are indicated from -2 to +2.

The phase shifting ξ_i, i= 1...4, results after placing a linear polarizer to each one of the interference patterns generated on each diffracting orders in the exit plane (P_1, P_2, P_3, P_4). Each polarizing filter transmission axis is adjusted at different angle ψ_i, so as to obtain the desired phase shift ξ_i for each pair of orders. For a 90° phase-shift ξ_i between interfering fields, the polarization angles ψ_i in each diffraction order must be 0°, 45°, 90° and 135° for the case of ideal quarter-wave retardation (α'= 90°). In the next sections, some particularities arising from the optical components available for our set-up are discussed. Among these, the calculation of ψ_i for the case of a non exact quarter-wave retardation is considered through an example.

3.1. Interference Patterns with Polarizing Filters and Retarding Plates

When using birefringent plates which do not perform exactly as quarter-wave plates for the wavelength employed, the polarization angles of the linear polarizing filters to obtain 90° phase-shifts must change.

To calculate the phase shifts induced in a more general polarization states by linear polarizers, consider two fields whose Jones vectors are described respectively by

$$\vec{J}_L(x,y) = \frac{1}{\sqrt{2}}\begin{pmatrix} 1 \\ e^{i\alpha'} \end{pmatrix} \quad \vec{J}_R(x,y) = \frac{1}{\sqrt{2}}\begin{pmatrix} 1 \\ e^{-i\alpha'} \end{pmatrix} e^{i\phi(x,y)}.$$

These vectors represent the polarization states of two beams emerging from a retarding plate with phase retardation $\pm\alpha'$. Each beam enters the plate with linear polarization at $\pm 45°$ with respect to the plate fast axis. Due to their orientations, the electric fields of the beams rotate in opposite directions, thereby the indices L and R. The beam with index R is supposed to carry a phase distribution $\phi(x,y)$. When each field is observed through a linear polarizing filter whose transmission axis is at an angle ψ, the new polarization states are

$$\vec{J}' = J_\psi^l \vec{J}_L \quad \vec{J}'' = J_\psi^l \vec{J}_R \tag{3.1}$$

where the spatial dependence has been dropped for simplicity and with the linear polarizing transmission matrix given by

$$J_\psi^l = \begin{pmatrix} \cos^2\psi & \sin\psi\cos\psi \\ \sin\psi\cos\psi & \sin^2\psi \end{pmatrix}. \tag{3.2}$$

If the new fields are left to interfere, the resulting irradiance can be written as

$$\begin{aligned}\left|\vec{J}_T\right|^2 &= \left|\vec{J}' + \vec{J}''\right|^2 \\ &= 1 + \cos\phi\cdot\cos^2\psi + \sin 2\psi\cdot\cos\alpha' + \sin 2\psi\cdot\cos[\alpha'-\phi(x,y)] \\ &\quad + \sin^2\psi\cdot\cos[2\alpha'-\phi(x,y)] \\ &= 1 + \sin 2\psi\cdot\cos\alpha' + A(\psi,\alpha')\cos[\xi(\psi,\alpha') - \phi(x,y)]\end{aligned} \tag{3.3}$$

where

$$\xi(\psi,\alpha') = \tan^{-1}\left[\frac{\sin 2\psi\cdot\sin\alpha' + \sin^2\psi\cdot\sin 2\alpha'}{\cos^2\psi + \sin^2\psi\cdot\cos 2\alpha' + \sin 2\psi\cdot\cos\alpha'}\right] \tag{3.4}$$

and

$$A(\psi,\alpha') = \left[\cos^4\psi + \sin^4\psi + \left(1+\tfrac{1}{2}\cos 2\alpha'\right)\cdot\sin^2 2\psi + 2\sin 2\psi\cdot\cos\alpha'\right]^{\frac{1}{2}}. \tag{3.5}$$

Plots of $\xi(\psi,\alpha')$ and $A(\psi,\alpha')$ are shown in Figure 3.2 for several values of α'. For the special ideal case of $\alpha'= \pi/2$ (quarter-wave plate), it is readily found that

$$\xi(\psi,\pi/2) = 2\psi, \quad A^2(\psi,\pi/2) = 1, \tag{3.6}$$

which can be verified in Figure 3.2. Then, the irradiance of Eq.3.3 reduces to

$$|J_T|^2 = 1+\cos[2\psi-\phi(x,y)], \tag{3.7}$$

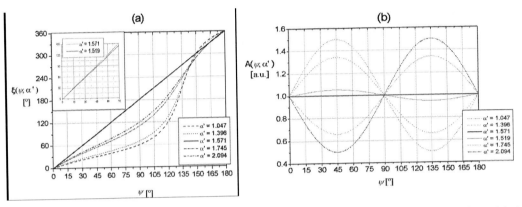

Figure 3.2. a) Phase shift $\xi(\psi,\alpha')$ as a function of ψ for several values of α'. Insert: α' for ideal retardation and experimental retardation. b) Amplitude $A(\psi,\alpha')$ as a function of ψ for several values of α'.

which is a well known expression used already for phase shifting [18]. In fact, by denoting four irradiances at four different angles as

$$|J_i|^2 = 1+\cos[2\psi_i-\phi(x,y)], \tag{3.8}$$

with $i = 1...4$, the relative phase can be calculated as [17]

$$\tan\phi = \frac{|J_1|^2-|J_3|^2}{|J_2|^2-|J_4|^2} \tag{3.9}$$

where $|J_1|^2$, $|J_2|^2$, $|J_3|^2$ and $|J_4|^2$ are the intensity measurements with the values of ψ given by $\psi_1 = 0, \psi_2 = \pi/4, \psi_3 = \pi/2, \psi_4 = 3\pi/4$. When the phase retardation is different from $\pi/2$, Eq.3.3 to Eq.3.5 must be used. In those cases, the value of ψ can be determined from Eq.3.4 looking for $\xi = 0, \pi/2, \pi, 3\pi/2$. For $\xi = 0$ it is easy to see from the Figure 3.2.a that $\psi_1 = 0$.

Cases $\xi = \pi/2, 3\pi/2$ lead to the condition

$$\cos^2\psi + \sin^2\psi \cdot \cos 2\alpha' + \sin 2\psi \cdot \cos\alpha' = 0, \qquad (3.10)$$

which can be transformed in the following second degree equation for $\cos^2\psi$

$$\{(1-\cos 2\alpha')^2 + 4\cos^2\alpha'\}\cos^4\psi + \{2(1-\cos 2\alpha')\cdot\cos 2\alpha' - 4\cos^2\alpha'\}\cos^2\psi + \cos^2 2\alpha' = 0 \qquad (3.11)$$

with two solutions $\psi_{a,b}$ given by

$$\cos^2(\psi) = \{2\cos^2\alpha' - (1-\cos 2\alpha')\cos 2\alpha'\}$$

$$\frac{\pm\sqrt{\{(1-\cos 2\alpha')\cos 2\alpha' - 2\cos^2\alpha'\}^2 - \{(1-\cos 2\alpha')^2 + 4\cos^2\alpha'\}\cos^2 2\alpha'}}{(1-\cos 2\alpha')^2 + 4\cos^2\alpha'} \qquad (3.12.a)$$

which enables the choosing of

$$\psi_2 = \psi_a, \quad \psi_4 = \psi_b + \pi, \qquad (3.12.b)$$

where ψ_a and ψ_b are two meaningful different solutions arising from Eq.3.12.a and $|\psi_b| < \psi_a$. For the case $\xi = \pi$, it is found that the following condition must be fulfilled

$$\sin 2\psi \cdot \sin\alpha' + \sin^2\psi \cdot \sin 2\alpha' = 0$$

which leads to

$$\psi_3 = n\pi + \arctan(-\sec(\alpha')). \qquad (3.12.c)$$

The value with $n = 1$ can be chosen.

3.2. Interferometer with Phase-Grating

The phase object under test is placed in one of the two windows in the object plane. Thus, the Jones vector in the object plane can be written as

$$\vec{O}(x,y) = \vec{J}_L(x+\tfrac{1}{2}x_0,y)\cdot w(x+\tfrac{1}{2}x_0,y) + \vec{J}_R(x-\tfrac{1}{2}x_0,y)\cdot w(x-\tfrac{1}{2}x_0,y) \qquad (3.13)$$

where x_0 is the separation of center to center between the two windows, the object phase being described with the function $\phi(x,y)$ included in $\vec{J}_R(x,y)$. Here, two different polarizations are related to the field of each window and the corresponding Jones vectors are denoted by $\vec{J}_i(x,y)$, $i = R, L$ as before. The rectangular aperture $w(x,y)$ can be written as $w(x,y) = rect[x/a_w] \cdot rect[y/b_w]$, where a_w and b_w represent the widths of the window.

A grating of spatial period $d = \lambda f / X_0$ is placed in the Fourier plane. Then, the corresponding transmittance is given by

$$G(\mu,\nu) = G_P(\mu) * \sum_{n=-\infty}^{\infty} \delta\left(\mu - \frac{n}{X_0}\right) \tag{3.14}$$

with $\delta(\mu)$ denoting the Dirac delta function and * the convolution operation. $\mu = u / \lambda f$ and $\nu = v / \lambda f$ are the frequency coordinates scaled to the relevant wavelength λ and the focal length f. The actual frequency coordinates are thus u and v. The grating's profile for a period is given by $G_P(\mu)$. The point spread function of a system with such a pupil can be obtained with the inverse Fourier transform of $G(\mu,\nu)$, which results in

$$\tilde{G}(x,y) = X_0 \cdot \sum_{n=-\infty}^{\infty} \tilde{G}_P(n \cdot X_0) \cdot \delta(x - n \cdot X_0, y) \tag{3.15}$$

The image formed by the system is the convolution of $\bar{O}(x,y)$ with $\tilde{G}(x,y)$, which is basically the replication of each window at distances X_0. The replications of each window are displaced by $\pm \frac{1}{2} x_0$ with respect to the origin, so they are superimposed if $Nx_0 = X_0$. Figure 2.3 shows the case of $N = 1$. The case of similar amplitudes for each spectral diffraction orders (from order -2 to order +2) is presented, a situation which can be found in phase gratings [13]. Also, in the Figure 3.3 one of the four orders shown, the +1, is depicted π out of phase with respect to the others. This effect can be obtained with phase gratings because odd order amplitudes are proportional to Bessel functions of odd orders, which have in turn odd parity, whereas even order amplitudes follow Bessel functions of even parity. Thus, diffraction orders as described are expected to be obtained with phase gratings due to their particular distribution $\tilde{G}(x,y)$.

Assuming that no additional changes of polarization occurs other than the one imposed by the retarding plates, the diffraction order superposition is expected to follow the description outlined in sec. 3.1. So, in order to obtain four interferograms with a phase shift of $\pi/2$, the previous sections justify the use of an interferometer consisting of two birefringent windows separated by x_0 in the plane object and a phase grating in the plane of Fourier. The interference of the fields of each window is obtained in the plane image when superposing itself the appropriate orders of diffraction. Linear polarizers in front of each order at the proper angle ψ would give the values of phase shifts according with Eq.3.3. In order to superpose orders +1 +2, 0 +1, -1 0 and -2 -1 it is necessary to fulfill the condition $d = \lambda f / x_0$.

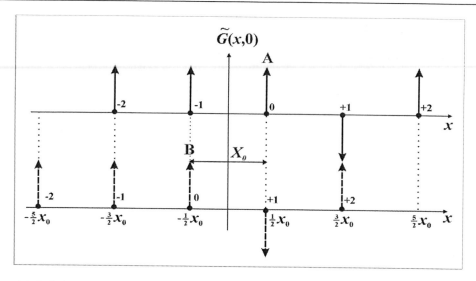

Figure 3.3. Relative positions of diffraction orders from windows A and B. The case of order +1 out of phase by π with respect to the others is shown. Two interferograms with inverse contrast result. $X_0 = x_0$.

3.3. Experimental Set-up

A green laser light with $\lambda = 532\ nm$ was employed to illuminate the system of Figure 2.1. Figure 3.4 shows four interferograms from the system as a preliminary observation. They were obtained before placing retardation plates and polarizing filters. The patterns show the relative phases of the diffraction orders as discussed in Sec.3.2. The two rightmost interferograms have the same fringe contrast. Such contrast appears to be the complementary one of the remaining leftmost pair of interferograms. The Fourier spectrum of the grating behaves as the one of Figure 3.3 (with a π phase difference between even and odd single orders). In fact, the contrast of the remaining patterns (not shown) follows changes that can be explained in agreement with the parity properties of the Bessel functions.

To complete the set-up, off-the-shelf retarding plates designed as quarter-wave plates for $\lambda_a = 514.5\ nm$ were used in the windows. As an example, a nominal retardation of

$$\alpha' = \frac{\pi}{2}\frac{\lambda_a}{\lambda} = 1.519\ rad \qquad (3.16)$$

is calculated. Introducing this value in the solutions described in sec.3.1, the results $\psi_1 = 0\quad \psi_2 = 46.577°\quad \psi_3 = 92.989°\quad \psi_4 = 136.42°$ were obtained. In the experimental set-up, these values must be changed to $\psi'_1, \psi'_2, \psi'_3,$ and ψ'_4 due to the additional 180° phase difference, as described later on.

Figure 3.4. Image plane from a system as depicted in Figure 1. Neither plate retardation plates nor polarizing filters were used. Two opposite fringe contrast can be seen. Compare with Figure 3.3.

The phase grating employed (110 ln/mm) generates five diffraction orders of similar but not equal average irradiance (Figure 3.4), as expected. Because the respective irradiances do vary due both to the diffraction order amplitude and the variations of pattern amplitude (Eq.3.5), each interferogram was subject to a normalization process to each maximum of its irradiance before using Eq. 3.9. The separation between window centers was of $x_0 \approx 10mm$. Other parameters used were focal lengths of $f \approx 160mm$, $a_w = 6mm$ and $b_w = 10mm$.

3.4. Experimental Results

An object phase has placed in window A and the window B is the reference. The transmission axes of the polarizing filters attached to interferograms of equal contrast were adjusted according with the first two values calculated as explained in previous sections ($\psi'_1 = \psi_1$ and $\psi'_2 = \psi_2$) to achieve mutual phase differences of $\Delta\xi = \xi(\psi'_2, \alpha') - \xi(\psi'_1, \alpha') = \pi/2$. The remaining two axes were adjusted taking into account the additional phase shift of 180°. Using the same angles $\psi'_3 = \psi_1$ and $\psi'_4 = \psi_2$, the two required phase differences $\Delta\xi = \xi(\psi'_3, \alpha') + \pi - \xi(\psi'_1, \alpha') = \pi$ and $\Delta\xi = \xi(\psi'_4, \alpha') + \pi - \xi(\psi'_1, \alpha') = 3\pi/2$ can be obtained.

3.4.1. Static Distributions

Two test objects were prepared evaporating magnesium fluoride (MgF$_2$) on a glass substrate: a disk or phase dot and a phase step. When each object was placed separately in one of the windows using the interferometer of Figure 2.1 with polarizers P$_1$, P$_2$, P$_3$ and P$_4$, using the previously calculated angles ψ'_1, ψ'_2, ψ'_3, and ψ'_4, the interferograms of Figure 2.5 were obtained. For each object, the four interferograms are shown together with the unwrapped phase calculated with Eq.3.9 at the right (in 256 grey levels). As examples, some typical raster lines for each unwrapped phase are shown in Figure 3.6 (in arbitrary phase units).

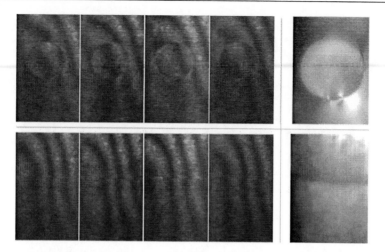

Figure 3.5. Upper row: phase dot. Four 90° phase-shifted interferograms and unwrapped phase. Lower row: phase step. Four 90° phase-shifted interferograms and unwrapped phase.

These observations suggest a further simplification for the polarizing filters array for the case of the phase shift of π from a phase grating. It consists of using only one filter big enough to cover the two central interferograms at the same angle of $\psi_2{}'$ ($\psi_2{}'=\psi_2=\psi_3$) instead of two separate filters at ψ_2, ψ_3 respectively. Thus, only three linear polarizing filters have to be used. The transmission axes of the filters P_1 and P_4 can be both horizontally oriented ($\psi_1{}'=\psi_1=\psi_4$), see Figure 3.7.

This only alters the order of the shifted interferograms. Because the departure from an ideal case for the experimental conditions is relatively small, this configuration could be experimentally tested with the described set-up resulting in qualitative good results.

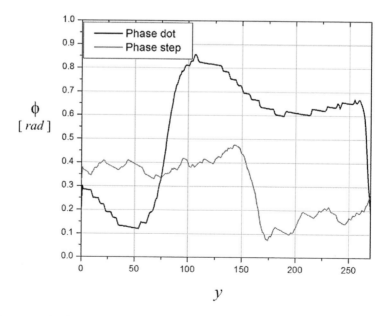

Figure 3.6. Unwrapped calculated phases along typical raster lines of each object of Figure 2.5. Scale factor : 0.405 rad.

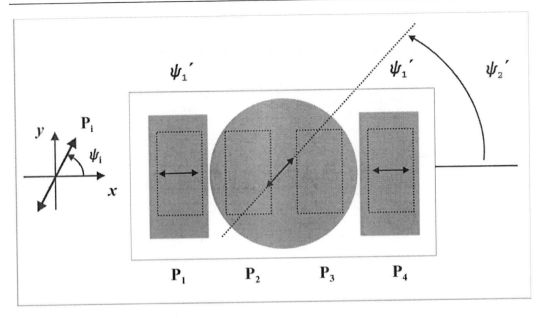

Figure 3.7. Simplification for the polarizing filters array.

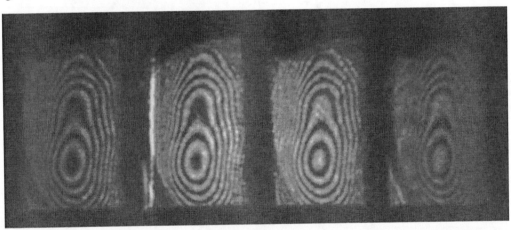

Figure 3.8. Typical four 90° phase-shifted interferograms from oil flowing (animation 1.24 Mb).

3.4.2. Moving Distributions

Immersion oil was applied to a glass microscope slide and allowed to flow under the effect of gravity by tilting the slide slightly. The slide was put in front of one of the object windows of the system of Figure 3.1. Figure 3.8 shows a typical sequence of four shifted interferograms from the oil flow in arbitrary units. Figure 3.9 shows the resulting unwrapped phase evolution of another oil flow.

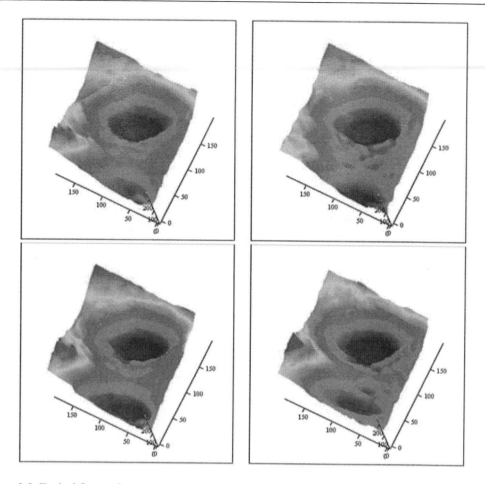

Figure 3.9. Typical frames from an unwrapped phase from interferograms of oil flowing.

3.5. Phase-Grid Interference Patterns with Modulation of Polarization

Incorporating modulation of polarization, a TWPGI can be used for dynamic interferometry. This system is able to obtain four interferograms 90° phase-apart with only one shot. Phase evolving in time can then be calculated and displayed on the basis of phase-shifting techniques with four interferograms. The system performs as previous proposals to attain four interferograms with a single shot [16,17]. In the following sections, a variant of a TWPGI able to capture four interferograms 90° shifted apart in one shot is described. It consists of the set-up shown in Figure 2.1 with configuration W_2 and a phase grid. The system uses a grid as a beam splitter in a way that resembles the well-known double-frequency shearing interferometer as proposed by Wyant [18], but the present proposal differs from it not only because of its modulation of polarization, the use of a single frequency and the use of two windows, but also in the phase steps our system introduces. Besides, the proposal is not a shearing interferometer of any type.

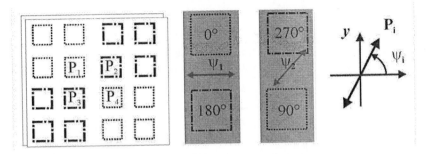

Figure 3.10. Polarizing filters array for 90° phase stepping.

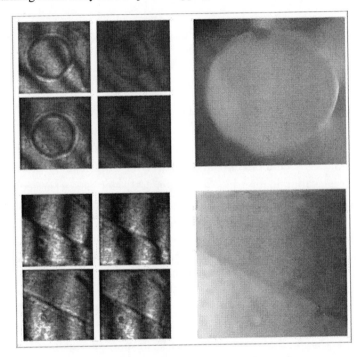

Figure 3.11. Upper row: phase dot. Four 90° phase-shifted interferograms and unwrapped phase. Lower row: phase step. Four 90° phase-shifted interferograms and unwrapped phase.

The Figure 2.1 shows the arrangement of a one-shot phase-shifting grid interferometer including modulation of polarization with retarders for the windows and linear polarizers on the image plane. The system generates several diffraction orders of similar irradiances in the average but not equal fringe modulations, as expected (Eq.2.12). In order to use Eq.3.9 properly, each interferogram image was scaled to the same values of grey levels (from 0 to 255). Previous reports show that a simplification for the polarizing filters array can be attained when using the phase shifts of π [11] to obtain values of ξ of 0, $\pi/2$, π and $3\pi/2$. For the configuration W_2, due to the π-shifts, only two linear polarizing filters have to be placed (instead of four filters, without the π-shifts). The transmission axes of the filter pairs P_1, P_3 and P_2, P_4 can be the same for each as long as they cover two patterns 180° phase apart (Figure 2.7). According with Sec.3.3, the needed values of ψ have to be of $\psi_1 = 0°$ and $\psi_2 = 45°$ with ideal quarter-wave retarders. But considering the retarders at disposal, it can be shown with Eq.2.11 that ψ can be of $\psi_1 = 0°$ and $\psi_2 = 46.577°$. They are sketched in Figure 3.10. The square enclosing the 16 windows replicas in the same figure is to be compared with the similar square of Figure 2.6 (dotted lines).

Two objects for testing are a phase disk and a phase step. When each object was placed separately in one of the windows using the TWPGI with the polarizers array, the interferograms of Figure 3.11 were obtained. For each object, the four interferograms are shown together with the calculated unwrapped phase. However, more than four interferograms could be used, whether for *N*-steps phase-shifting interferometry [19] or for averaging images with the same shift.

4. APPLICATION: LATERAL SHEARING INTERFEROMETRY

4.1. Lateral Shearing Interferometry

In lateral shearing interferometry, two mutually displaced versions of the same wavefront are brought to interfere [20]. The resulting interference pattern consists of fringes of equal wavefront slope with respect to the shear (shearograms), and they are to be interpreted approximately as lines of equal directional derivative for a sufficiently small shear Δs. They require low coherence illumination [3,21] or laser [20] and can act as either two beam or multiple beam interferometers [3,23]. Self-reference two-beam laser shearing interferometers have been used in many applications such as testing of optical components [24,25], the study of flow and diffusion phenomena in gases and liquids [26-27], beam collimation testing [30,31] and detection of optical phase singularity [32,33] among many others. To produce two versions of a given wavefront, a beamsplitter method is, in general, employed. One successful and versatile technique consists of the use of a spatially periodic diffractive element such as a grating [34, 35] or a double grating [18,36,37]. On the other hand, two-window phase grating interferometry (TWPGI) has been demonstrated to be useful as a simple beamsplitting method for single-shot phase-shifting interferometry [11, 38-40], so it seems worthy to combine shearing techniques with phase-shifting methods. That way, phase extraction of directional derivatives of wavefronts could be obtained with the typical accuracy of phase-shifting [41]. In phase-shifting interferometry (PSI), $n = (N+1)$ interferograms are obtained with a proper phase shift in between, where *N* is the number of shifts to be carried out [19, 41]. A system of equations permits calculation of wrapped phase [42] for further corresponding unwrapping. The case of $n = 4$ is widely used, especially when well-contrasted, low-noise interferograms are available [41]. To reduce errors, from noisy patterns for instance, algorithms with $n = 5$ or 7 can be used instead [18]. Of course, shearing interferometry has already incorporated PSI [42-45]. However, the reported systems make the capture of interferograms sequentially, which limit its application to static phase distributions as in their present form. In this review, a lateral-shear interferometer capable of generating *n* interferograms with proper shifts using only a single shot is described. This can be obtained with a TWPGI and modulation of polarization [11, 39]. A similar approach has been reported for a radial shearing interferometer to obtain several phase-shifted interferograms in one shot [40]. Aside from the inherent differences between lateral and linear shearing interferometers, the lateral shearing as we describe can perform in two configurations. Such configurations (denoted by A and B in the following sections) can be of practical significance for lateral shearing and, although realizable for other interferometers (for example, in the system described in [40]), such configurations has not be discussed in detail yet. Experimental results for the cases of $n = 4$, 5 and 7 are also shown.

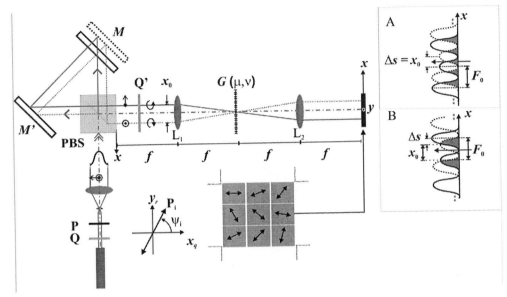

Figure 4.1. Variable lateral shear interferometer with phase grating and modulation of polarization. Q, Q' quarter-wave retarders, P, P_i: linear polarizers, ψ_i: transmission angle of polarization. Δs linear shear, x_0 beam separation from PBS, F_0 order separation. x_q, y_r: local coordinates (see text). Dot-dash ray tracing: axial ray for $x_0 = 0$; continuous and dotted ray tracings: transmitted and reflected rays in PBS ($x_0 \neq 0$). These last traces are associated with orthogonal polarizations (linear between PBS and Q'. circular after leaving Q' up to detection). A, diffraction orders of the same diffraction numerical order superimpose. B, diffraction orders of different numerical order superimpose.

Figure 4.1 shows the experimental set-up. Illumination comes from a polarized He-Ne laser operating at $\lambda = 632.8$ nm, where the expanded beam transversal section is $a = 8.6$ mm. Its polarization is arranged to be at 45° by using a quarter-wave retarding plate (Q) and a linear polarizer (P). The set-up consists of the coupling of two main systems or stages. The first stage is a cyclic shearing interferometer (CSI), which comprises a polarizing beamsplitter (PBS) and two mirrors (M, M') [39,45]. The second stage, a TWPGI, consists of a $4f$ Fourier imaging system with similar achromatic lenses (L_1 and L_2) of focal length $f \approx 20$ cm. and a pair of crossed phase gratings (grid) $G(\mu,v)$ as the system's pupil with spatial period d, where $\mu = u / \lambda f$ and $v = v / \lambda f$ are the frequency coordinates (u,v) scaled to the wavelength λ and the focal length f. Thus, the output of the CSI consists of two versions of the same wavefront, each one with mutually orthogonal linear polarizations (vertical and horizontal). A wave retardation plate of $\lambda/4$ (Q') is placed in front of the entrance lens, so as to achieve counter rotating circular polarization for each sheared wavefront (left and right, $\bar{J}_L$ and $\bar{J}_R$) with equal amplitudes. In Figure 4.1, plots A and B represent the superposition of amplitude spectra in the output of configuration A or B. Each configuration emerges according to the value of the shear Δs with respect to the beam side a (assumed to be of a squared section), as explained in the following sections. Similar traces in plot lines and ray tracing were used in the Figure 4.1.

4.2. Interference Pattern Replication and Modulation of Polarization

We assume for simplicity a beam section of the form $w(x,y) = rect[x/a] \cdot rect[y/a] \cdot \exp\{i\phi(x,y)\}$. Thus, the amplitude in front of the entrance lens would be given by

$$\vec{t}_i(x,y) = \vec{J}_L \cdot w(x - \frac{x_0}{2}, y) + \vec{J}_R w(x + \frac{x_0}{2}, y), \tag{4.1}$$

so the beam displacement x_0 results along the x-axes only. The displacement x_0 between the wavefronts can be adjusted by translating the mirror M of the CSI. To draw theoretical conclusions, we assume that the phase grid placed in the TWPGI Fourier plane is made up from two crossed phase gratings of equal frequencies [39]. In the image plane of the TWPGI, the amplitude can be written as

$$\vec{t}_o(x,y) = \vec{t}_i(x,y) * \widetilde{G}_2(x,y)$$
$$= \sum_{q=-\infty}^{\infty}\sum_{r=-\infty}^{\infty} J_q(2\pi A_g) J_r(2\pi A_g) \left\{ \vec{J}_L \cdot w(x - qF_0 - \frac{x_0}{2}, y - rF_0) + \vec{J}_R \cdot w(x - qF_0 + \frac{x_0}{2}, y - rF_0) \right\}$$
$$\tag{4.2}$$

where the order separation is $F_0 \equiv \lambda f/d$ and J_q denotes the Bessel function of the first kind and integer order q, and $\widetilde{G}_2(x,y)$ is the resulting Fourier transform of the centered phase grid. In the image plane of the system, a series of replicated beams can be observed. Because of their polarization, an interference pattern can be detected when a linear polarizing filter is placed before the detector at an angle $\psi = \psi_i$ (Figure 4.1).

4.2.1. Configuration A: Phase Shifts

Two configurations will be considered. In the first configuration (configuration A) the shear of choice $\Delta s = x_0$ is of a value smaller than the size a of the beam section, which in turn, is smaller than the order separation F0 (Figure 4.1). Thus, two beams with a shear x_0 enter the TWPGI close to the optical axes. At the image plane of the TWPGI, the superposition of the two beam copies with a mutual shear x_0 would appear around each diffraction order, each superposition isolated from the others, for this case. Under these conditions, the irradiance in each superposed copies would be proportional to $|J|^2$, or to the squared modulus of Eq. (4.2):

$$|J|^2 = 2J_q^2 J_r^2 \left(1 + \cos[\xi(\psi) - \Delta\phi(x_q, y_r)]\right), \tag{4.3}$$

where a translation of coordinates was used around the order position ($x_q = x - qF_0$ and $y_r = y - rF_0$), and $\Delta\phi(x,y) \equiv \phi(x + x_0/2, y) - \phi(x - x_0/2, y)$. The pattern would show a shift $\xi(\psi) = 2\psi$ [38] and fringe modulation of unity, without dependence of the orders. This configuration uses the CSI as beam divider and produces the shear (box A in Figure 4.1). The grid only makes copies of both superimposed beams, but they keep their circular polarizations to be detected through linear polarizers, each of them at different angle $\psi = \psi_i$ (Figure 4.1).

4.2.2. Configuration B: Phase-Shifts and π-Shifts

In the second configuration to discuss (configuration B) the beam displacement x_0 is greater than the beam size a, but such that only the first neighboring orders partially overlap. Thus, two windows with a given separation $x_0 > a$ apart enter the TWPGI, while their shear in the image plane is now $\Delta s = F_0 - x_0 < a$ (box B in Figure 4.1). Then, the isolated term is the superposition of orders qr and $(q+1)r$ in Eq.(4.2). Under similar conditions for detection as in configuration A, the corresponding irradiance results now proportional to

$$|J|^2 = (J_q J_r)^2 + (J_{q+1} J_r)^2 + 2 J_q J_r J_{q+1} J_r \cdot \cos[\xi(\psi) - \Delta\phi(x_q, y_r)] \quad (4.4)$$

so, its fringe modulation is

$$m_q = \frac{2 J_q J_{q+1}}{J_q^2 + J_{q+1}^2} \quad (4.5)$$

Each corresponding fringe modulation depends on the relative phases between the J_q. In configuration B, there are changes in the sign of m_q because of π-shifts of the Bessel functions in Eq. (4.2) [46]. This result simplifies the polarization filter array to be employed because it is possible to use only one filter to cover two shearograms shifted by 180°. The coefficient J_r contributes to the total irradiance but has no influence in m_q. In this second configuration, the CSI imposes orthogonal linear polarizations in two beam copies. The grid divides further but makes the shear also. Still, this shear can be adjusted with mirror M in the CSI through x_0.

Figure 4.2 shows the obtained patterns for both configurations A (Figure 4.2.a) and B (Figure 4.2.b) before shifting. To detect these patterns, a polarizer at the same angle of $\psi = 36°$ has been placed over the whole field containing all of the shown diffraction orders for each case. The phase shifts obtained in several patterns has been calculated following the method from Kreis with respect to the pattern in the upper left corner of the rectangle with dashed lines. These shifts are listed in table A and table B. For the case A, the phase shifts result close to zero, whereas for the case B, there are shifts close to π. From the pictures, it is possible to see that at least nine patterns can be useful for phase shifting. To obtain different phase shifts, a polarizing linear filter at a proper angle $\psi = \psi_i$ has to be placed on each replicated interferogram, as described below.

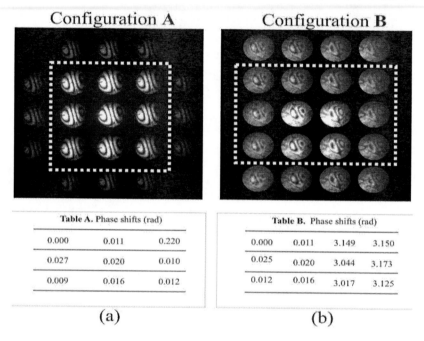

Figure 4.2. Interference patterns from configuration A or B. Phase-shifts with respect to the pattern in the upper left corner of the rectangle in dashed lines in each case. (a) Replicated interference patterns for configuration A. (b) Replicated interference patterns for configuration B showing some phase shifts of π.

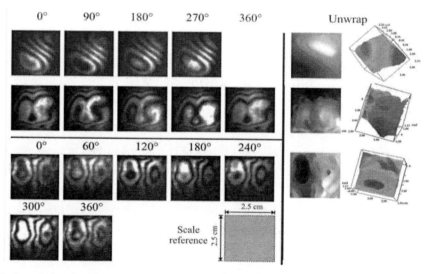

Figure 4.3. Interference patterns (shearograms, left) and phase unwrapped (right) in configuration A. For $n = 4$ (first row from above), the polarizing filter angles in degrees were: $\psi_1 = 0$, $\psi_2 = 46.58$, $\psi_3 = 92.99$, $\psi_4 = 136.40$, and for $n = 5$ (second row), $\psi_1 = 0$, $\psi_2 = 46.58$, $\psi_3 = 92.99$, $\psi_4 = 136.40$, $\psi_5 = 180$ (both sets for phase shift of 90°). For $n = 7$ (last two rows) and phase shifts of 60°, $\psi_1 = 0$, $\psi_2 = 30.80$, $\psi_3 = 62.33$, $\psi_4 = 92.98$, $\psi_5 = 122.16$, $\psi_6 = 150.69$, $\psi_7 = 180$. A square is included to show the scale.

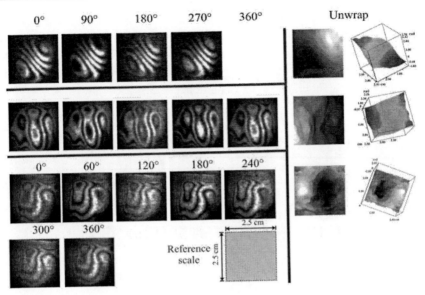

Figure 4.4. Interference patterns (shearograms) in configuration B. $n = 4$ (first row from above) and 5 (second row) with phase shift of 90° and $n = 7$ (last rows) with phase shift of 60°. First row (right): unwrapped phase showing the directional derivative from a wavefront affected with spherical aberration ($\psi_1 = \psi_2 = 0, \psi_3 = \psi_4 = 46.58$ in degrees). Second and lower rows: interference patterns and unwrapped phase (right) for a static oil drop. For $n=5$, $\psi_1 = \psi_2 = 0, \psi_3 = \psi_4 = 46.58$, $\psi_5 = 92,99$ and for the case of symmetrical seven (each phase step of 60°), $\psi_1 = \psi_2 = 0$, $\psi_3 = \psi_4 = 30.80$, $\psi_5 = \psi_6 = 62.33$ and $\psi_7 = 92.99$ (values in degrees).

4.2.3. Experimental Results

In the experimental set-up, the phase gratings do not follow exactly a sinusoidal distribution. However, its first diffraction orders which were used behave very similarly to a sinusoidal grating showing π-shifts between some of them.

Then, apart from the amplitude of diffraction orders, the relevant conclusions still remain. In the experimental set-up, $d = 9$ μm and $F_0 = 13.905$ mm for both configurations (A and B). Phase calculations were performed using well-known phase solutions for four-steps-in-cross algorithm and also for n-symmetrical step algorithms [19]. Before the phase calculations, each shearogram was subjected to the same scaling process (0-255 gray level for minimum to maximum of intensities) in order to overcome the differences in irradiance and fringe modulation. A low-pass filtering process was also applied prior to phase calculation. Figure 4.3 shows sets of n experimental shearograms as obtained with configuration A and the resulting unwrapped $\Delta\phi$. In the upper row, the case $n = 4$ is shown (wavefront affected with spherical aberration), with $x_0 = 0.8$ mm. In the middle and lower rows, the cases $n = 5, 7$ respectively for oil drops falling down on glass slide plates are shown ($x_0 = 0.6$ mm).

Figure 4.4 shows sets of 4, 5 and 7 shearograms as obtained with configuration B, as well as unwrapped phase correspondingly with $x_0 = 10.4$ mm. Note that for $x_0=F_0$, the configuration B reduces to a common-path TWPGI [39], i.e. with no shear ($\Delta s = 0$). In such a case, one beam can convey an imposed phase information (a transparency, for instance) of a given amplitude distribution while the other beam can serve as the reference. This type of system could be applied for measurement of surface deformations in fluids [27-28] or for measurement of concentration gradient profile of liquids [29]. Figure 4.5 shows the case of moving distributions (configuration B) corresponding to an immersion oil drop on a microscope slide. Figure 4.5(a) shows a typical sequence of four shearograms and Figure 4.5(b) shows the resulting unwrapped $\Delta\phi$ evolution.

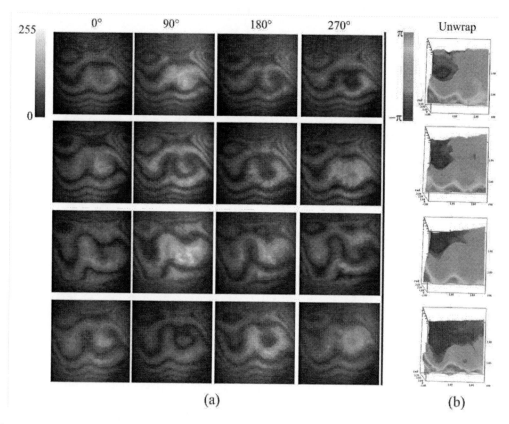

Figure 4.5. Moving refractive index distributions as observed under configuration B. (a) Shearograms from oil flowing.(b) Unwrapped $\Delta\phi$ of oil flowing.

4.2.4. Conclusions

A variable lateral shear interferometer with a phase grid to simultaneously achieve several shearograms for phase measurements using phase-shifting techniques has been demonstrated in two configurations. Configuration A works with beams entering the TWPGI close to the optical axis and rendering fringe modulations of unity at its image plane. Each interferogram has a theoretical modulation of unity. Higher, homogeneous apertures are not mandatory in this configuration. Configuration B is basically a common-path TWPGI, so the involved beams do not travel close to the optical axis and its fringe modulations is not unity, but can be close to one for phase gratings and phase grids of higher values in its phase amplitude modulation However, this configuration allows an independent manipulation of the beams before entering the second stage, so an eventual additional operation could be performed. In spite of using a stage additional to the TWPGI for both versions, such stage consists of a cyclic interferometer and does not affect the system's mechanical stability in an appreciable way. These configurations exemplify how other similar adaptation to a TWPGI can be carried out in order to apply single-shot phase-shifting techniques in phase measurements. Other grating types could be used, but order irradiances ratios and fringe modulations values might change, as well as the polarizing filter distributions.

4.3. Application: Radial Shearing Interferometry

In radial shearing interferometry (RSI), two versions of the same wavefront are brought to interfere, one of the versions contracted or expanded with respect to the other (different radial scaling) [48]. The radial shear R is defined as the ratio of the small (a') to the large (a'') diameters. The applications of RSI have been reported in aberrations measurements [49], optical testing [50], aspherical surfaces measurements [51,52], corneal topographic inspection [53,54], adaptive optics [55] (for example, mirror control [56], wavefront correction [57] and real-time blurred image restoration [58]), wavefront sensing [59,60] and beam characterization [61,62] among others. RSI systems have been implemented mainly with classical components [48,63-66], but also with gratings [67], zone plates [68], holography [69] and, in a very versatile way due to wavefront electronic storage capability, with speckle techniques [70-73]. RSI permits also phase reconstruction [73] both for the case of large magnification and for smaller magnification cases [75]. Many of these applications can receive benefits from the accuracy of the methods of phase-shifting interferometry (PSI) [76]. In PSI, $n = (N+1)$ interferograms are obtained with a proper phase shift in between, where N is the number of shifts to be carried out [41]. A system of equations permits calculation of wrapped phase for the further corresponding unwrapping. The case of $n = 4$ is widely used, especially when well-contrasted, low-noise interferograms are available. To reduce errors, as when having noisy patterns, algorithms with n = 5 or 7 might be used instead [41]. However, by performing the capture of interferograms sequentially, the PSI limits its application to static phase distributions. In this communication, we describe a radial-shear interferometer capable of generate n interferograms with proper shifts by using only a single shot. In this way, RSI techniques can be expanded to applications where phase evolves in time. We report a method to obtain single-shot PSI by employing two-windows phase-grating interferometry (TWPGI) and modulation of polarization [11,39] although other possibilities might be adapted as well [77]. Two main configurations (I and II) can be directly employed and both are presented. Experimental results for the cases of $n = 4$, 5 and 7 are also shown as examples.

4.3.1. Basic Considerations

Figure 4.6 shows the experimental set-up. It consists of the coupling of two main blocks or stages. Configurations differ only in the first stage, which can be whether the block I or the block II (left side of Figure 4.6). One block is attached to a TWPGI, which constitutes the second block.

Illumination comes from a polarized HeNe laser below operating at λ = 632.8 *nm*. Its polarization is arranged to be at 45° by using a quarter-wave retarding plate (Q) and a linear polarizer (P). The first stage in configuration I is a cyclic radial interferometer (CRI), which comprises a polarizing beamsplitter (PBS), two lenses (L_1, L_2) and two mirrors (M, M') [17]. The first stage of configuration II is a Mach-Zehnder radial-shear interferometer (MZR), which comprises a telescope in each arm [65]. These two blocks and their variants are widely used as part of systems in different applications (for example, CRI [56,57,62,76] and MZR [57,58,60,61,74,75]), so they can be considered as representative cases of potential adaptations for single-shot PSI. The second stage, a TWPGI, consists of a 4f Fourier imaging system with similar achromatic lenses of focal length $f \approx 20\,cm$ and a pair of crossed phase gratings (grid) $G(\mu,v)$ as the system's pupil with spatial period d. $\mu = u/\lambda f$ and $v = v/\lambda f$ are the frequency coordinates (u,v) scaled to the wavelength λ and the focal length f. Thus, the output of the CRI consists of two versions of the same wavefront, each one with mutually orthogonal linear polarizations (vertical and horizontal). A wave retardation plate of $\lambda/4$ (Q') is placed in front of the entrance lens, so as to achieve counter rotating circular polarization for each radial-sheared wavefront (left and right, $\vec{J}_L$ and $\vec{J}_R$) with equal amplitudes.

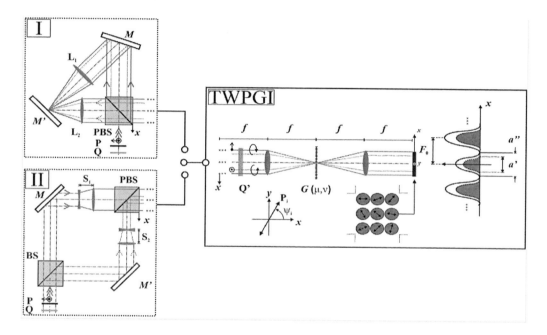

Figure 4.6. Setup: $d = 0.009090\,mm$, $\lambda = 632.8\,nm$, $f = 200\,mm$, $F_0 = 13.905\,mm$. Configuration I: a CRI and a TWPGI. Configuration II: a MZI and a TWPGI.

4.4. Phase Grid Interferometry with Modulation of Polarization

Assuming beam sections $w''(x,y) = circ[\rho/M_a] \cdot \exp\{i\phi(x/M_a, y/M_a)\}$ and $w'(x,y) = circ[\rho] \cdot \exp\{i\phi(x,y)\}$, the amplitude in front of the entrance of TWPGI would be given by

$$\vec{t}_i(x,y) = \vec{J}_L \cdot w'(x,y) + \vec{J}_R w''(x,y) \tag{4.6}$$

where $\rho = \sqrt{x^2 + y^2}$ and $M_a = 1/R$ denotes the relative magnification of the pupils. We assume that the phase grid is made up from two phase gratings of amplitude $2\pi A_g$ each and with orthogonal gratings vectors. The resulting Fourier transform of the centered phase grid can thus be written as

$$\tilde{G}_2(x,y) = \sum_{q=-\infty}^{q=\infty} \sum_{r=-\infty}^{r=\infty} J_q(2\pi A_g) J_r(2\pi A_g) \delta(x - qF_0, y - rF_0) \quad (4.7)$$

where the order separation is $F_0 \equiv \lambda f / d$ and J_q denotes the Bessel function of the first kind and integer order q [10]. In the image plane of the TWPGI, the amplitude can be written as

$$\tilde{t}_o(x,y) = \tilde{t}_i(x,y) * \tilde{G}_2(x,y)$$
$$= \sum_{q=-\infty}^{\infty} \sum_{r=-\infty}^{\infty} J_q(2\pi A_g) J_r(2\pi A_g) \{\vec{J}_L \cdot w'(x - qF_0, y - rF_0) + \vec{J}_R \cdot w''(x - qF_0, y - rF_0)\} \quad (4.8)$$

with ($*$) denoting convolution. In the image plane of the system, a series of replicated beams can be observed. Because of their polarization, an interference pattern can be detected when a linear polarizing filter is placed before detection. Using Jones calculus with

$$\vec{J}_L{}' = \mathbf{J}_\psi^L \vec{J}_L, \quad \vec{J}_R{}' = \mathbf{J}_\psi^L \vec{J}_R, \quad \mathbf{J}_\psi^L = \begin{pmatrix} \cos\psi & -\sin\psi \\ \sin\psi & \cos\psi \end{pmatrix}, \quad (4.9)$$

the pattern irradiance results proportional to the squared modulus of Eq. (4.8) in the general case. Thus, two beams with a certain R enter the TWPGI close to the optical axes. At the image plane of the TWPGI, the superposition of the two beam copies with R would appear around each diffraction order, each superposition isolated from the others if a'' are properly chosen. The two radial sheared beams around a given order qr will have counter rotating circular polarization. A diffraction order qr can pass through a filter designed to block out all of the remaining orders. Under these conditions, by detecting the irradiance with a linear polarizer at angle ψ with respect to the horizontal [11,39], only the contribution of an isolated term of order qr can be considered and its irradiance would be proportional to,

$$(J_q J_r)^2 + (J_q J_r)^2 + 2 J_q J_q J_r J_r \cdot \cos[\xi(\psi) - \Delta\phi(x_q, y_r)] = 2 J_q^2 \cdot J_r^2 (1 + \cos[\xi(\psi) - \Delta\phi(x_q, y_r)]) \quad (4.10)$$

where a translation of coordinates was used around the order position ($x_q = x - qF_0$ and $y_r = y - rF_0$), and $\Delta\phi(x,y) \equiv \phi(x,y) - \phi(x/M_a, y/M_a)$. The pattern would show a certain shift $\xi(\psi) = 2\psi$. Its fringe modulation would be of unity. These configurations use the CRI or MZR as beam dividers and produce the radial shear. The grid makes copies of both superimposed beams, but they keep its circular polarizations to be detected with the help of linear polarizers, each of them at different angle ψ. In the experimental set-up, the phase gratings do not follow exactly a sinusoidal distribution. However, at least its first diffraction orders, which are the ones used, behave very similarly to a sinusoidal grating showing π-shifts between some of them. So, apart from the amplitude of diffraction orders, the main conclusions still remain.

4.5. Configuration A: Phase Shifts

4.5.1. Experimental Results

In the experimental set-up, $d = 0.009090$ mm and $F_0 = 13.905$ mm for both configurations (I and II). Phase calculations were performed using standard phase solutions for four-steps-in-cross algorithm and for symmetrical-five and symmetrical-seven step algorithms [19]. Each interferogram was subject to the same scaling (256 gray levels) and low-pass filtering process before phase calculation. The Figure 4.7 shows sets of n experimental interferograms as obtained with configuration I and the resulting unwrapped $\Delta\phi$. In the upper row, the case $n = 4$ is shown (wavefront affected with spherical aberration), with $M_a = 1.7$. In the middle and lower rows, the cases $n = 5$ with $M_a = 1.2$ are shown.

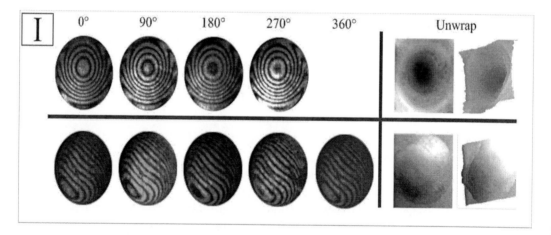

Figure 4.7. TWPGI in combination with a cyclic radial shear interferometer (I). $n = 4$ and 5 with phase shift of 90°. Upper row: Typical radial shearograms obtained with a slightly defocused lens ($f_1 = 110 mm, f_2 = 170 mm, a' = 7.0 mm, a'' = 8.6 mm$), of axis. Lower row: typical radial interferograms of a lens with spherical aberration, defocusing and tilt ($f_1 = 110 mm, f_2 = 130 mm, a' = 8.0 mm \ a'' = 8.6 mm$). Unwrapped phase at right.

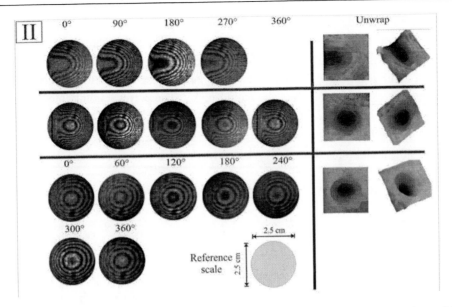

Figure 4.8. Telescopic system in a Mach-Zehnder interferometer to produce radial shear in combination with a TWPGI (II). ($f_1 = -30\,mm$, $f_1' = 120\,mm$, $f_2 = -30\,mm$, $f_2' = 140\,mm$, $a' = 12\,mm$, $a'' = 22\,mm$). Upper row: Typical radial interferograms obtained with a slightly defocused lens of axis. $n = 4$ and 5 with phase shift of 90° Lower row: typical radial interferograms of a lens with spherical aberration, defocusing and tilt. Unwrapped phase at right.

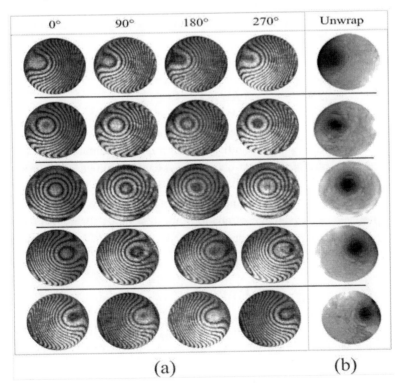

Figure 4.9. Moving phase distribution. (a) Typical interferograms obtained by moving mirror M at a constant speed in configuration I (Media 1). (b) Unwrapped $\Delta\phi$ at right.

Figure 4.8 shows sets of 4,5 and 7 interferograms as obtained with configuration II, as well as unwrapped phase correspondingly with $M_a = 1.2$. Note that if mirror M were so adjusted in both configurations, they would introduce a linear shear Δs ($\Delta s \neq 0$). In such a case, for $R = 1$, the resulting system serves as a variable linear shear interferometer. A typical phase distribution evolving in time is shown in Fig 4.9. The corresponding interferograms for successive captures of the CCD camera are show on Figure 4.9(a). The figure 4.9(b) shows the obtained unwrapped $\Delta \phi$.

These results are obtained by moving the mirror M at a constant speed in configuration I, with $M_a = 1.7$.

4.6. Conclusions

In conclusion, an adaptation of a radial shear interferometer with a phase grid to simultaneously achieve several shearograms for phase measurements using phase-shifting techniques has been demonstrated in two configurations. This system uses the CRI or MZR as beam divider and produces the radial shear. The grid makes copies of both superimposed beams, but they keep its circular polarizations to be detected with the help of linear polarizers, each of them at different angle ψ. The system does not demand of higher, homogeneous apertures. Configuration I is mechanically very stable and Configuration II shows less stability. This interferometeric system allows capture of phase-shifted shearograms evolving in time to be used for phase extraction. Other grating types could be used, but order irradiances ratios and fringe modulations values might change, as well as the polarizing filter distributions.

CONCLUSION

We have presented a series of interferometric systems based on a phase-grating and modulation of polarization , with the ability to obtain n-simultaneous phase shifts to process optical phase, because of this, dynamic phase object can be studied with these systems. The deployed systems can be reconfigured to obtain the phase or the directional derivative of the phase. Have been proposed for some applications in microscopy, tomography and singular optics, being analyzed the advantage of adapting these systems for use with opaque objects and high reflactance, also have been proposed to study the slope of a phase objects.

REFERENCES

[1] R. M. A. Azzam, "Division-of-Amplitude Photopolarimeter (DOAP) for the Simultaneous Measurement of all Four Stokes Parameters of Light," *Opt. Acta* **29**, 685-689 (1982).

[2] V. Ronchi, "Forty Years of History of a Grating Interferometer," *Appl. Opt.* **3**, 437-451 (1964).

[3] Cornejo-Rodríguez, "Ronchi Test," in *Optical Shop Testing*, D. Malacara editor, Wiley, New York, (1992).

[4] E. S. Barrekette, H. Freitag, "Diffraction by a Finite Sinusoidal Phase Grating," *IBM Journal*, 345-349 (1963).

[5] J. W. Goodman, Introduction to Fourier Optics, 2nd edition, McGraw-Hill, (1988).

[6] J. Li K. Fu, "Studying Diffractive Orders of the Phase Grating, Fiber and Integrated Optics, 29, 96-105(2010) .

[7] F. Kneubühl, "Diffraction Grating Spectroscopy," *Appl. Opt.* **8**, 505-519 (1969).

[8] V. Arrizón, D. Sánchez-De-La-Llave, "Common-Path Interferometry with One-Dimensional Periodic Filters," *Opt. Lett.* **29**, 141-143 (2004).

[9] C. Meneses-Fabian, G. Rodriguez-Zurita, and V. Arrizon, "Optical Tomography of Transparent Objects with Phase-Shifting Interferometry and Stepping-Wise Shifted Ronchi Ruling," *J. Opt. Soc. Am. A* **23**, 298-305 (2006).

[10] C. Meneses-Fabian, G. Rodriguez-Zurita, and V. Arrizon, "Common-Path Phase-Shifting Interferometer with Binary Grating," *Opt. Commun.* **264**, 13-17 (2006).

[11] G. Rodriguez-Zurita, C. Meneses-Fabian, N. Toto-Arellano, J. F. Vázquez-Castillo, and C. Robledo-Sánchez, "One-Shot Phase-Shifting Phase-Grating Interferometry with Modulation of Polarization: case of four interferograms," *Opt. Express* **16**, 7806-7817 (2008).

[12] D. A. Thomas and J. C. Wyant, "High Efficiency Grating Lateral Shear Interferometer," *Opt. Eng.*, **15**, 5, pp. 477 (1976).

[13] P. W. Ramijan, "Processing Stereo Photographs by Optical Subtraction," Ph. D. Thesis, University of Rochester (1978).

[14] J. Schwieder, R. Burow, K.-E. Elssner, J. Grzanna, R. Spolaczyk, K. Merkel. "Digital Wave-Front Measuring Interferometry: some systematic error sources," *Appl. Opt.* **22**, N.21, 3421-3432 (1983).

[15] T. Kreis, "Digital Holographic Interference-Phase Measurement Using the Fourier-Transform Method," *J. Opt. Soc. Am. A* **3**, 847-855 (1986).

[16] B. Barrientos-García, A. J. Moore, C. Pérez-López, L. Wang and T, Tschudi, "Spatial Phase-Stepped Interferometry using a Holographic Optical Element," *Opt. Eng.*. **38**, 2069-2074 (1999).

[17] M. Novak, J. Millerd, N. Brock, M. North-Morris, J. Hayes, and J. Wyant, "Analysis of a micropolarizer array-based simultaneous phase-shifting interferometer," *Appl. Opt.* **44**, 6861-6868 (2005).

[18] J.C. Wyant, "Double Frequency Grating Lateral Shear Interferometer," *Appl. Opt.* **12**, N.9, 2057-2060 (1973).

[19] D. Malacara, M. Servin, Z. Malacara, C.6 "Phase detection algorithms" in *Interferogram Analysis for Optical Testing*, Marcel Dekker (New York) (1998).

[20] M. V. Mantravadi, Lateral shearing interferometers, c.4 in Optical Shop Testing, 2nd ed., D. Malacara, ed. (Wiley, New York, 1992).

[21] J. C. Wyant, White light extended source shearing interferometer, *Appl. Opt.* **13**,200-202 (1974).

[22] M. V. R. K. Murty, The use of a single plane parallel plate as a lateral shearing interferometer with a visible gas laser source, *Appl. Opt.* **3**, 531-534 (1964).

[23] R. S. Sirohi, T. Eiju K. Matsuda, T. H. Barnes, Multiple-beam lateral shear interferometry for optical testing, *Appl. Opt.* **34**, 2864-2870 (1995).

[24] S. Yokoezeki, K. Ohnishi, Spherical aberration measurement with shearing interferometer using Fourier imaging and moiré method, *Appl. Opt.* **14**, 623-627 (1975).

[25] K. Matsuda, Y. Minami, T. Eiju, Novel holographic shearing interferometer for measuring lens lateral aberration, *Appl. Opt.* **31**,6603-6609 (1992).

[26] P. J. Gardner, M. C. Roggemann, B. M. Welsh, R. D. Bowersox, T. E. Luke, Statistical anisotropy in free turbulence for mixing layers at high Reynolds numbers ,*Appl. Opt.* **35**, 4879-4889 (1996).

[27] B. J. Pelliccia-Kraft, D. W. Watt, Three-dimensional imaging of a turbulent jet using shearing interferometry and optical tomography, *Experiments in Fluids* **29**, 573-581 (2000).

[28] N. Rashidnia ,R. Balasubramaniam ,J. Kuang, P. Petitjeans , T. Maxworthy , Measurement of the Diffusion Coefficient of Miscible Fluids Using Both Interferometry and Wiener's Method ,*International Journal of Thermophysics*, **22**, 547-555 (2001).

[29] Priti Singh, Chandra Shakher, Measurement of the temperature of a gaseous flame using a shearing plate, *Opt. Eng.* **42**, 80-85 (2003).

[30] K. V. Sriram, P. Senthilkumaran, M. P. Kothiyal, R. S. Sirohi, Double-wedge-plate interferometer for collimation testing: new configurations, *Appl. Opt.* **32**, 4199-4203 (1993).

[31] J. S. Darlin, K. V. Sriram, M. P. Kothiyal, R. S. Sirohi, Modified double-wedge-plate shearing interferometer for collimation testing, *Appl. Opt.* **34**, 2886-2887(1995).

[32] D.P. Ghai, .P Senthilkumaran, R.S. Sirohi, Shearograms of an optical phase singularity,*Opt. Commun.*, **281**, 1315-1322 (2008).

[33] D. P. Ghai, S. Vyas, P.Senthilkumaran, R. S. Sirohi, Shearograms of a singular beam using wedge plate lateral shear interferometer, *Opt. and Lasers in Engineer.* **46** 797-801 (2008).

[34] D. A. Thomas, J. C. Wyant, High efficiency grating lateral shear interferometer, *Opt. Engineer.* **15**, 477 (1976).

[35] H Schreiber, J Schwider, Lateral shearing interferometer based on two Ronchi phase gratings in series, *Appl. Opt.* **36**, 5321-5324 (1997).

[36] P. Hariharan, W. H. Steel, J. C. Wyant, Double grating interferometer with variable lateral shear, *Opt. Commun.* **11**, 317-320 (1974).

[37] M. P. Rimmer, J. C. Wyant, Evaluation of large aberrations using a lateral shear interferometer having variable shear, *Appl. Opt.* **14**, 142-150 (1975).

[38] T. Kiire, S. Nakadate, M. Shibuya, Simultaneous formation of four fringes by using a polarization quadrature phase-shifting interferometer with wave plates and a diffraction grating, *Appl. Opt.* **47**, 4787-4792 (2008).

[39] N. I. Toto-Arellano, G. Rodriguez-Zurita, C. Meneses-Fabian, J. F. Vázquez-Castillo, Phase shifts in the Fourier spectra of phase gratings and phase grids: an application for one-shot phase-shifting interferometry, *Opt. Express*, **16** , 19330-19341 (2008).

[40] N I Toto-Arellano, G. Rodriguez-Zurita, C. Meneses-Fabian, and J. F. Vazquez-Castillo, Adjustable lateral-shear single-shot phase-shifting interferometry for moving phase distributions **J. Opt. A: Pure Appl. Opt. 11** 045704 (2009).

[41] K. Creath, *Phase-measurement interferometry techniques*, vol 26 in Progress in Optics, E. Wolf, eds, 349-393 (North-Holland, 1998).

[42] M. Kothiyal, C. Delisle, Shearing interferometer for phase shifting interferometry with polarization phase shifter, *Appl. Opt.* **24**, 4439-4442 (1985).

[43] D. W. Griffin, Phase shifting shearing interferometry, *Opt. Lett.* **26**, 140-141 (2001).
[44] H.-H. Lee, J.-H. You, S.-H. Park, Phase-shifting lateral shearing interferometer with two pairs of wedge plates, *Opt. Lett.* **28**, 2243-2245 (2003).
[45] R. Xu, H. Liu, Z. Luan, L. Liu, A phase-shifting vectorial-shearing interferometer with wedge plate phase-shifter, *J. Opt. A: Pure Appl. Opt.* **7**, 617-623 (2005).
[46] P. Hariharan, D. Sen, Cyclic shearing interferometer, *J. Sci. Instrum.* **37**, 374 (1960).
[47] G. Rodriguez-Zurita, N I Toto-Arellano ,C. Meneses-Fabian, and J. F. Vazquez-Castillo, One-shot phase-shifting interferometry: five, seven, and nine interferograms ,*Opt. Letters* **33**, 2788-2790 (2008).
[48] M. V. Mantravadi, Radial rotational and reverse shear interferometers *Optical Shop Testing* ed Wiley (New York,) C.5. (1992).
[49] D. Malacara, Mathematical interpretation of radial shearing interferometers *Appl. Opt.* **13**, 1781-1784 (1974).
[50] W. H. Steel , A radial shear interferometer for testing microscope objectives *J. Sci. Instrum.*, **42**, 102-104 (1965).
[51] P. Hariharan, B. F. Oreb, Z. Wanzhi, Measurement of aspheric surfaces using a microcomputer controlled digital radial-shear interferometer, *Jour. Mod. Optics*, **31**, 989-999 (1984).
[52] D. Liu, Y. Yang, Y. Shen, J. Weng, Y. Zhuo, *Proc. SPIE System optimization of radial shearing interferometer for aspheric testing*, **6834** (2007) .
[53] W. W. Kowalik, B. E. Garncarz, H. T. Kasprzak, Corneal topography measurement by means of radial shearing interference: part I- theoretical considerations *Optik* **113** 39-45 (2002).
[54] W. W. Kowalik, B. E. Garncarz, H. T. Kasprzak, Corneal topography measurement by means of radial shearing interference: part II- measurements errors *Optik* **114** 199-206 (2003).
[55] R. Tansey, A. Phenis, K. Shu, Conf. Proc on *Use of a radial shear interferometer as a self reference interferometer in adaptive optics* The Advanced Maui Optical Space Surveillance Technologies The Maui Economic Development Board Wailea, Maui, Hawaii p 10-14 (2006).
[56] R. F. Horton,Design of a white light radial shear interferometer for segmented mirror control *Opt. Engineer.* **27**, 1063-1066 (1998).
[57] T. Shirai ,T. H. Barnes, T. G. Haskell,Adaptive wave-front correction by means of all-optical feedback interferometry *Opt. Lett.* **25** 773-775 (2000).
[58] T. Shirai, T. H. Barnes, T. G. Haskell,Real-time restoration of a blurred image with a liquid-crystal adaptive optics system based on all-optical feedback interferometry *Opt. Commun.* **118** 275-282 (2001).
[59] J. M. Geary, Wavefront sensors, C.IV in *Adaptive Optics Engineering Handbook* R. K. Tyson ed. p 123-150 (2000).
[60] R. A. Hutchin, Combined shearing interferometer and Hartmann wavefront sensor *Pat. Num.* 4518854 (1985).
[61] C. Hernandez-Gomez, J. L. Collier, S. J. Hawkes, C. N. Danson, C. B. Edwards, D. A. pepler, I. N. Ross, T. B. Winstone , Wave-front control of a large-aperture laser system by use of a static phase corrector, *Appl. Opt.*, **39**, 1954-1961 (2000).
[62] D. Liu, Y. Yang, L. Wang, Y. Zhuo, Real-time diagnosis of transient pulse laser with high repetition by radial shearing interferometer, *Appl. Opt.*, **46**, 8305-8314 (2007).

[63] P. Hariharan, D. Sen, Radial shearing interferometer, *J. Sci. Instrum.*, **38**, 428 (1961).

[64] D. S. Brown, Radial shear interferometry, *J. Sci. Intrum.*, **39** 71-72 (1962).

[65] M. V. R. K. Murty, A compact radial shearing interferometer based on the law of reflection, *Appl., Opt.* **3**, 853 (1964).

[66] M. V. R. K. Murty, Radial shearing interferometers using a laser source *Appl. Opt.* **12** 2765-2767 (1973).

[67] D. E. Silva, Talbot interferometer for radial and lateral derivatives, *Appl. Opt.*, **11**, 2613-2624 (1972).

[68] R. N. Smartt, Zone plate interferometer, *Appl. Opt.*, **13**, 1093-1099 (1974).

[69] J. C. Fouéré, D. Malacara, Holographic radial shear interferometer, *Appl. Opt.*, **19**, 2035-2039 (1974).

[70] R. K. Mohanty, C. Joenathan, R. S. Sirohi, High sensitivity tilt measurement by speckle shear interferometry, *Appl. Opt.*, **25**, 1661-1664 (1986).

[71] C. Joenathan, A. R. Ganesan, R. S. Sirohi, Fringe compensation in speckle interferometry: application to nondestructive testing, *Appl. Opt.*, **25**, 3781-3784 (1986).

[72] C. Joenathan, R. Torroba, Simple electronic speckle-shearing-pattern interferometer, *Opt. Lett.*, **15**, 1159-1161 (1986).

[73] A. R. Ganesan, D. K. Sharma, M. P. Kothiyal, Universal digital speckle shearing interferometer, *Appl. Opt.*, **27**, 4731-4734 (1988).

[74] D. R. Kohler, V. L. Gamiz, Interferogram reduction for radial-shear and local-reference holographic interferograms, *Appl. Opt.*, **25**, 1650-1652 (1986).

[75] E. López-Lago, R. de la Fuente, Amplitude and phase reconstruction by radial shearing interferometry, *Appl. Opt.* **47**, 372-376 (2008).

[76] M. P. Kothiyal, C. Delisle, Shearing interferometer for phase shfting interferometry with polarization phase shifter, *Appl. Opt.*, **24**, 4439-4442 (1985).

[77] T. Kiire, S. Nakadate, M. Shibuya, Simultaneous formation of four fringes by using a polarization quadrature phase-shifting interferometer with wave plates and a diffraction grating, *Appl. Opt.*, **47**, 4787-4792 (2008).

In: Interferometry Principles and Applications
Editor: Mark E. Russo

ISBN 978-1-61209-347-5
© 2012 Nova Science Publishers, Inc.

Chapter 6

SAR INTERFEROMETRY FUNDAMENTALS AND HISTORIC EVOLUTION IN TERRAIN MOVEMENTS APPLICATIONS

*Paz Fernández-Oliveras**
Associate Professor. University of Granada, Spain

ABSTRACT

This chapter includes an introduction about fundaments of the InSAR technique and its historical evolution focus on terrain movement applications. The fundamentals include the SAR image parameters and characteristics that are important to consider in the use of InSAR techniques in the study of different types of terrain movements (subsidence, landslides, volcanic activity, earthquakes...). The Chapter starts with a short introduction about SAR images wavelength and main advantages of active sensors against passive sensor in the images formation (solar radiation independence, cloud penetration). After this introduction, it includes an explanation about the SAR images geometric distortions, due to the sensor *line of sight* (LOS), and their influence in the terrain movement detection. The next section corresponds to the phase and amplitude components and their use, the definitions of Interferometry and Differential Interferometry and the DInSAR fundamental equations necessary to apply this technique to the terrain movement investigations. After that the chapter includes a review of the Advanced InSAR techniques (called A- InSAR or A-DInSAR) and Multi-interferogram techniques existing actually, that constituted a great advance in the use of Interferometric Image in the quantitative assessment of terrain movements.

Finally, the chapter conclude with a review of the historic evolution of the InSAR technique related to the use of SAR images to detect quantify and study the evolution of the different types of terrain movements, and in deep review analisis of subsidence and landslide applications.

---

* E-mail: pazferol@ugr.es.

X.1.- SAR AND DINSAR FUNDAMENTALS

Notions about RADAR Images

Interferometry technique Works with Synthetic Aperture RADAR images. These images are generated by an active sensor that transmits and receive its own radiation in the electromagnetic spectrum microwave band (wave length between 0.3 and 100 centimeters) (Figure X.1).

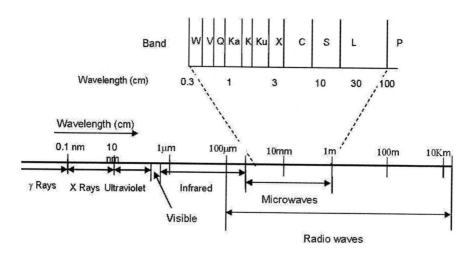

Figure X.1. Wavelength employed in Remote Sensing with microwaves ampliation.

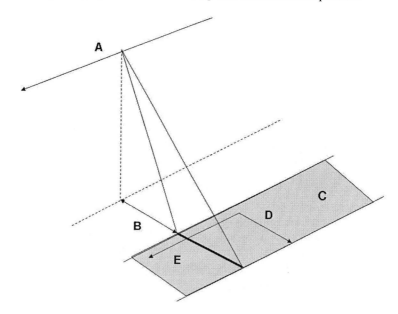

Figure X.2. RADAR images geometric acquisition: A sensor trajectory; B perpendicular sensor-ground surface distance; C: sensor capture area or path; D: range in the perpendicular sensor direction (across-track); E: range in the sensor movement direction (along-track).

Due to the sensor transmit and receive our own radiation, the images can been acquired in any time of the day, during the day or the night. This characteristic constitutes an advantage with respect to the passive sensor images. Another of the advantages is the Independence of the climate conditions in the acquisition time, because microwave can penetrate clouds, but could appear atmospheric effects that could be necessary to take into account.

SAR Images Characteristics: Distortions

RADAR images geometry is quite different to optics remote sensing images, because the sensors have lateral vision instead of vertical. Due to this, RADAR images are rectangular and have a decreasing resolution between closer (*near range*) and further (*far range*) sensor areas. (Figure X.2) This kind of acquisition does that the RADAR images have some deformations that conditioned their interpretation and the areas that could been analyzed. One of the consequences is that the relief seems to be "inclinated" in the images (Figure X.3). Surface deformations phenomena are *foreshortening*, *shadow* and *layover*.

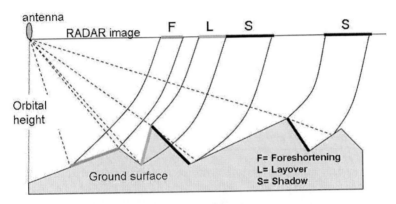

Figure X.3. Geometric distortions in SAR images. Modified of Crosetto et al, 2005.

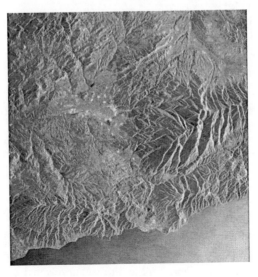

Figure X.4. Example of an ERS1 amplitude SAR image of the south of the Granada province (Spain) with 10 azimut times and 2 range times compression (Fernandez, 2009).

Each of these deformations is shown in Figure X.3. Is important to take into account that sensor have lateral vision, because all measures are obtained in the sensor line of sight (LOS). In the case of the ERS1 and ERS2 satellites sensor this inclination is about 23 grades respect to the vertical direction.

SAR Images Characteristics: Amplitude and Phase

Synthetic Aperture RADAR images (SAR) are complex images and has two components (real and imaginary) (Ferreti et al, 2007):

$I = A \cdot \cos\Phi$ Real Component
$Q = A \cdot \sin\Phi$ Imaginary Component

where A= amplitude and Φ = phase
With these components amplitude and phase images can be obtained:

$\Phi = \tan^{-1}\left(\dfrac{Q}{I}\right)$ Phase image

$A = \sqrt{I^2 + Q^2}$ Amplitude image

Amplitude image is the result of the sensor-ground interaction, e.g. is a function of the wave-ground interaction in the earth surface. An amplitude image example is shown in Figure X.4.

Amplitude images are quite noisy, e.g. noise as signal perturbation. The noise and, in consequence, the SAR images are condicionated by speckle phenomena, typical of the coherent acquisition systems and that is shown in the image as an *salt and pepper effect* (Rosen et al. 2000; Ferretti et al. 2007).

This effect, join with geometrical distortions, already commented, do difficult Amplitude SAR images interpretation.

From the interferometric point of view, amplitude images are useful to recognize in the space the area over interferometric studies will been done, because surface elements can be identified.

Phase image is the inteferometry fundamental part, because it contains the indispensable information to the InSAR ground surface study (Crosetto et al. 2005).

SAR Interferometry (InSAR) and Differential SAR Interferometry (DInSAR)

Interferometric techniques are focus in the study and exploitation of the phase images information of the SAR complex images (Hansen, 2001). Using images acquired from quite different points of view, surface topographic information can be obtained (in the case of Digital Elevation Models DEM) and ground vertical deformation displacements (in the case of the Differential Interferometric SAR, DInSAR)

InSAR and DInSAR techniques are both base on the explotation of the information contained in one or more interferograms (Klees and Massonnet, 1999).

An interferogram is a complex image that results of the multiplication of an SAR image and the conjugate complex of other one (Massonnet and Feigl, 1998; Rosen et al. 2000; Ferretti et al. 2007). The information contents on the interferogram phase is the phase difference between the two SAR images used and its values are from $-\pi$ and $+\pi$, that is called wrapped phase. The phase differences are coded with a color wheel in which a complete rotation corresponds to 2π or 360° phase difference (Klees and Massonnet, 1999) (Figure X.5). The two images used to form an interferogram are called *Master* and *Slave*. This images, with the exception of the tandem ERS1/2 mission in which the satellites are in the same spatial position with a 1 day time delay, correspond to quite different positions of the same sensor in different orbits and time. The vector that connects both orbits positions are called *baseline* and its projection in the LOS perpendicular direction are called *perpendicular baseline* (Rosen et al. 2000). The interferometric phase has basically two information types: one related with terrain topography, exploited for the InSAR techniques and other related with ground motion, that is exploited for the DInSAR techniques (Rosen et al. 2000).

Differential Interferometry techniques studies ground motion by means of differential interferograms. Interferograms represent the phase differences between the two images used to calculate them that can be related with terrain forms and their displacement (Klees and Massonnet, 1999).

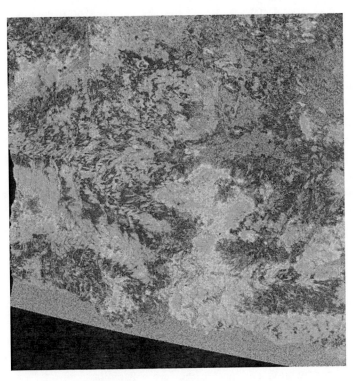

Figure X. 5. Wrapped differential interferogram of the Granada province (Spain) generated with two ERS1 images with 70 days temporal delay (Fernandez, 2009).

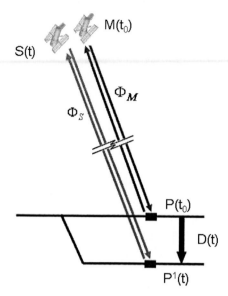

Figure X.6. Image acquisition diagram in a subsidence case between two satellite passes (Modified of Crosetto et al. 2005).

So, to study ground motion topographic information must be removed from the interferograms. To do that topographic information are introduced in the process by means of a Digital Elevation Model (DEM) (Klees and Massonnet, 1999). In this method a synthetic interferogram are simulated using the DEM and the orbital parameters of both images used to calculate the interferogram. Them the synthetic interferogram are subtracted to the real interferogram and the phase difference related with ground motion are obtained. This new interferogram are called Differential Interferogram (Gabriel et al. 1989).

The main limitation of this method is the topographic DEM error. In this sense the shorter images baseline used in the interferogram calculation, the lower differential interferogram sensibility to the DEM inaccuracy (Klees and Massonnet, 1999). Figure X.6 shows a simplification of the image adquisition process, considering only one ground point P (Crosseto et al. 2005).

In the case of the ground deformation detection where point P have been moved to point P^1 during time delay between two image acquisitions, the interferometric phase are affected for several parameters (Crosetto et al. 2005):

$$\Delta\Phi_{Int} = \Phi_S - \Phi_M = \frac{SP - MP}{\frac{\lambda}{4\cdot\pi}} + \frac{SP^1 - SP}{\frac{\lambda}{4\cdot\pi}} + \Phi_{Atm} + \Phi_{Noise} = \Phi_{Topo} + \Phi_{Mov} + \Phi_{Atm} + \Phi_{Noise}$$

Where:

Φ_S y Φ_M are the slave and master interferometric phases.

Φ_{Atm} is the atmospheric contribution,

Φ_{Noise} is the noise component,

Φ_{Topo} is terrain topographic component,

Φ_{Mov} is the ground motion component,

SP is the slave satellite- point distance, and

λ is the radar wavelength.

If topographic surface are known by means of a DEM, the Φ_{Topo} contribution can be extracted (Φ_{Topo_Sim}) of $\Delta\Phi_{Int}$, obtained the differential phase $\Delta\Phi_{D-Int}$,

$$\Delta\Phi_{D-Int} = \Delta\Phi_{Int} - \Phi_{Topo\_Sim} = \Phi_{Mov} + \Phi_{Atm} + \Phi_{Res\_Topo} + \Phi_{Noise}$$

where Φ_{Res_Topo} represents the residual DEM error.

Advanced Differential Interferometry (A-DInSAR) and the Multi-Interferogram Techniques

In the case of a studies that involves a wide time period, that is not possible to cover with an unique interferogram or with a few of them, it is necessary to consider a differential interferogram series that will processed together with an specific software, to obtain a movement average rate of the complete time period. Techniques that used a differential interferogram stack processed all together are called Multi-interferogram techniques or Advanced Differential Interferometry (A-DInSAR). Usually these techniques employed atmospheric corrections (Crosetto et al. 2005) and they achieve to correct classic DInSAR techniques deficits, obtaining:

- Temporal displacement evolution estimation
- Millimetric accuracy
- Topographic and atmospheric errors minimization
- Slow movements monitoring

A briefly review of the more useful Multi interferogram techniques are presented. See the mentioned publications for more details about each of them.

Small Baseline DInSAR Technique (SBAS)

This technique is base on an adequate differential interferograms combination generated with small baseline images (Berardino et al. 2002). The algorithm start from an unwrapped differential interferograms It includes topographic error estimation and atmospheric phase filters.

After phase unwrapping, the deformation temporal component (LP, low pass) and possible topographic effects are estimated together by minimum square estimation. Then the parameters calculated are subtracted from each differential interferogram, and another phase unwrapping is calculated.

One the residual phase unwrapping are being calculated, the temporal component LP are added to estimate the deformation. This deformation estimation can be affected by atmospheric effects. To remove them a doble filter are applied: a low pass spatial one and a high pass temporal one, due to atmospheric effects have high spatial correlation and low temporal correlation. Finally, the estimated phase is subtracted to the signal and the conversion from signal to displacement is done.

Interferometric Point Target Analisis Technique (IPTA)

Published by Werner et al. 2003, it is implemented in the commercial software Gamma. It uses vectorial interferogram information storage and management, to save storage space and be more efficient, and finally it does a data raster conversion.

The processing starts with the SLC corregistrated stack images and a SLC image pairs list to be considered to interferograms calculation, as well as the DEM to be used.

In the second step a pixel candidate list to be considered are calculated, based on low temporal retro scattered variability.

In the third step the SLC values of the pixels candidate are extracted in a punctual data stack and the initial interferometric baselines are calculated.

In the fourth step the differential interferograms are calculated.

In the fifth step the stack of differential interferograms calculated are analyzed, first in the temporal domain. A bidimensional regression phase analysis with the perpendicular baseline is done and another one with the phase and the temporal interval for each of the points selected. The quality criteria used to accept or discharge the points in the IPTA analysis is the phase standard deviation respects to the regression.

The results in this step are: altitude corrections, lineal deformation rate, a quality measurement, residual phases and unwrapped differential interferometric phase that are used as initial values for a new iteration of the process. The residual phases include atmospheric phase, the non linear deformation and the errors.

Then, IPTA results includes for each point analyzed, the altitudes, the linear deformation rates, the atmospheric phase, refine baselines, quality information and the non linear deformation.

Coherent Pixels Technique (CPT)

The algorithm used in this technique (Mora, 2004; Mora et al. 2003) estimates the linear and non linear displacement components, Dem error and atmospheric effects for un reduced number of interferograms that are processing together and the no need to have a master image in common. Pixels candidates are selected with a model using a coherence criterion. Only the pixels that have a coherence value higher than a selected value in all the interferograms will be considered in the analysis.

The main steps are the following:

a) Selection a couple of images that covers the temporal interval of the study,
b) Optimum images pairs formation, for the interferograms calculation, considering a maximum baseline
c) Pixels candidates identification with a criterion of their coherence stability considering their stability in the interferogram stack

d) Pixels selected triangulation to establish a phase relation between them an d the adjust of the linear deformation model and the DEM error.

When the linear deformation has being estimated the next step is the non linear deformation components estimation, isolated this component the atmospheric effects, using spatial and temporal filters successively. Finally the results are integrated and interpolated to generate the displacements maps.

Permanent Scatterers Technique

This patented technique (Ferretti et al. 2000, Ferretti et al. 2001) is one of the most widespread and is based on the velocity estimation in certain image points that are called Permanent Scatterers (PS). The points correspond to specific surface objects that remain invariables in all the images along the time period analysis considered. These points must have stable amplitude in all the images. From all the PS detected as candidates will be considered as correct PS the ones that have a certain coherence threshold. Moreover to do the analysis is necessary a minimum PS density of 25 PS per Km^2.

This technique needs at least 30 images all with the same master image, that in some cases are difficult. However the fact of that the PS size is smaller than the resolution cell, make possible the use of interferograms with big baseline.

With this technique lineal and non lineal displacement components and topographic error are calculated for each PS considering atmospheric effects.

Stable Point Network Tecnique, SPN

This technique expose in several publications (Arnaud et al. 2003, Duro et al. 2003; Duro et al. 2005, Crosetto et al. 2008) was the first one that can combine ERS and ENVISAT satellite data (Duro et al. 2005). The SPN software is based on the commercial interferometric chain DIAPASON for all the SAR processing, images corregistration and interferogram calculation. The basic elements are described in Crosetto et al. (2008) and are the following: 1) Pixel selection based on the SAR amplitude stability or the spectral coherence; 2) the use of multiple images 3) Modeling capability.

The technique estimates from DInSAR phase: the ground deformation, the topographic residual error and atmospheric effects separating these components to the phase noise.

The main products obtained are three: the first is the average deformation map, representing by the lineal velocity deformation rate for the whole study period in the sensor LOS and referenced to an stable point selected in the images. The accuracy that can reach the estimation is less than 1mm per year, similar to the PS technique. The second product is the residual topographic errors map and the third deformation temporal series for each stable point, that describes the temporal deformation evolution of each one (Crosetto et al. 2008).

X.3.-BACKGROUND

X.3.1. DInSAR Technique Origins and Historic Evolution

First RADAR images applications start with lateral RADAR airborne platforms SLAR and were used as an alternative to the aerial photographs for tropical areas studies that are frequently cover of clouds, due to the cloud penetration properties of RADAR images (Allan, 1983, Trevett, 1986). The main problem of these sensors was the low spatial resolution due to the small antenna diameter (Chuvieco, 1990). This inconvenient disappeared with satellite borne Synthetic Aperture RADAR, SAR. Its beginning was in 1978 with the SEASAR NASA satellite to oceanographic applications that was used successfully to geode oceanographic determination, swells, submerged sand banks and ocean currents (Elachi, 1982). It success did that two new missions with similar characteristics, SIR-A and B, were launched in 1981 and 1984 and they confirmed the possibility of using these images to other applications like vegetation thematic cartography. Growing interest for this kind of images and its potential applications did that new special projects included RADAR capabilities (Chuvieco, 1990) Table X.1. (Hanssen, 2001). First study that used SAR images to ground deformation detection was Gabriel et al. (1989), in which appeared DInSAR technique for the first time using SEASAT images to study ground deformation related with clays expansion in the Imperial Valley, California, USA. First SAR images experiments with European satellites images to surface ground detection were done in 1992 with ERS1 images by means of a European Spatial Agency (ESA), INS Stuttgart University and Milano Polytechnic project (Prati et al. 1992; Hartl et al. 1992).

In this experiment two artificial reflectors, called *corner reflectors*, were placed near Bonn (Germany), with 1 centimeter height difference between first and second image acquisition. These images were used to form an interferogram. A few millimeters change was detected in the processing (Gens and Genderen, 1996). Another more practical example of millimetric vertical ground deformation detection was proposed in van Halsema et al. (1995), on the application of SAR interferometry to accuracy ground subsidence measurements in Holland.

Table X.1. Satellites with SAR capabilities

Satellite	Launch	Operative to	Nationality
ERS-1	1991	2000	European
JERS-1	1992	1998	Japanese
ERS-2	1995	2005	European
RADARSAT-1	1995		Canadian
ENVISAT	2002		European
ALOS	2006		Japanese
TerraSAR-X	June 2007		German
RADASAT-2	December 2007		Canadian

Table X.2. First and most important events studied with RADAR Interferometry and main references (modified of Hanssen, 2001)

Location	References
Earthquakes	
Kobe, Japan	Ozawa et al, 1997
Landers, California, USA	Massonnet et al, 1993; Massonnet and Adragna, 1993; Zebker et al, 1994; Peltzer et al, 1994; Massonnet et el, 1994; Feigl et al, 1995; Peltzer et al, 1996; Massonnet et al, 1996b; Hernandez et al, 1997; Price and Sandwell, 1998; Michel et al, 1999.
Hector, California, USA	Sandwell et al, 2000
Manyi, Tibet	Peltzer et al, 1999
Izmit, Turkey	Barbieri et al, 1999; Hanssen et al, 2000; Reilinger et al, 2000.
Eureka Valley, California, USA	Massonnet and Feigl, 1995; Peltzer and Rosen, 1995.
Northridge, California, USA	Massonnet et el, 1996a; Murakami et al, 1996; Kawai and Shimada, 1994.
Kagoshima-kenhokuseibu, Japan	Fujiwara et al, 1998
Nuweiba, Aqaba Gulf	Baer et al, 1999; Klinger et al, 2000.
Grevena, Greece	Meyer et al, 1996; Clarke et al, 1996
Colfiorito, Umbria-Marche, Italy	Stramondo et al, 1999
San Andrés Fault Creeping, Parkfield, USA.	Rosen et al, 1998; Bürgmann et al, 2000.
Volcanoes	
Vatnajöküll, Iceland	Roth et al, 1997;Thiel et al, 1997
Krafla expansion, Iceland	Sigmundson et al 1997
Etna, Italy	Briole et al, 1997; Delacourt et al, 1997; Lanari et al, 1998; Williams and Wadge, 1998.
Iwo Jima, Japan	Ohkura, 1998
Izu Peninsula, Japan	Fuliwara et al, 1998
Katmai, Alaska	Lu et al, 1997; Lu and Freymueller, 1999.
Kilauea, Hawaii	Mouginis-Mark, 1995a; Rosen et al, 1996; Zebker et al, 1996.
Soufriere Hills, Monserrat	Wadge et al, 1999
Campi Flegrei, Italy	Usai et al, 1999; Avallone et al, 1999; Lundgren et al, 2001.
Yellowstone Caldera, USA	Wicks et al, 1998
Piton de la Fourniase, Reunión	Sigmundsson et al, 1999
Long Valley, California, USA	Thatcher and Massonnet, 1996
Unzen, Japan	Fujii et al, 1994
Galapagos	Mouginis-Mark, 1995b; Jonsson et al, 1999; Amelung et al, 2000.
Glaciar/ Ice movement	
Antártico/Patagonia	Goldstein et al, 1993; Hartl et al, 1994a, b; Rott and Siegel, 1997; Rott et al, 1998; Joughin et al, 1999;
Groenland	Kwok and Fahnestock, 1996; Joughin et al, 1996; Rignot et al, 1997; Joughin et al, 1997; Mohr et al, 1998; Hoen and Zebker, 2000.

To investigate the potentialities and limitations of the technique the FEL-TNO project was developed using ERS1 images supplied by ESA (Gens and Genderen, 1996). In the last 15 years, the DInSAR technique development had been related with its successful application to different ground deformation process. Seismic activity is one of the research fields with the highest number of published works. Massonet et al. (1993), Peltzer et al. (1996), Rosen et al. (1998) were the precursors of this application.

Published works had studied presisimic, cosismic and postsismic deformations, choosing to the interferogram calculation two images previous, one previous and another one posterior, or two posterior images respectively. As a recent example of this applications the first interferograms generation after the L'Aquila earthquake (Italy) in April 6th of 2009, did by Istituto per il Rilevamento Elettromagnetico dell' Ambiente (IREA-CNR) and their interpretation by the Istituto Nazionale di Geofisica e Vulcanologia (INGV), that combines one presismic ENVISAT- ASAR of February 1st 2009 with a postsismic one of April 12th 2009 (*Synthetic report of the activities of ASI-SIGRIS personnel and Earthquake Remote Sensing Group of the Remote Sensing Laboratory - National Earthquake Center, INGV – Rome, 2009*). In volcanology field DInSAR technique has been applied successfully in inflation and deflation volcanoes studies, as is shown in Massonet and Sigmundsson, (2000) and Amelung et al. (2000). Other application is the glaciological studies of Antarctic ice movement velocities and topographic ice layers measurements (Goldstein et al. 1993, Kwok and Fahnestock, 1996). Table X.2. includes first published works of each of these applications

Other two important DInSAR technique application fields related with ground deformation are subsidence and landslides that are treated in the following sections.

Moreover, first reviews about the DInSAR technique different applications are included in Massonet and Feigl (1998) and Hanssen (2001). There is also an internet link, inside the ESA webpage in which it is possible to consult last DInSAR studies based on ERS and ENVISAT images founded by ESA: http://www.esa.int/esaEO/.

In spite of the DInSAR applications proliferation in the nineties decade, the technique showed certain problems, which solution was considered lately. Those problems are related with noise and artifacts (systematic distortions) produced during the data acquisition and processing (Zhou et al. 2003).

Noise are generated by temporal decorrelation between acquisition time of the two images that are involved in the interferogram, and the spatial and temporal atmospheric variation generate artifacts that are difficult to detect in the ground deformation estimation (Zhou et al. 2003).

The appearance of DInSAR multi interferogram or Advanced A-DInSAR techniques (Ferretti et al. 2000; Berardino et al. 2002; Werner et al. 2003; Mora et al. 2003) had been an important advance in the DInSAR ground deformation studies, because they allow the quantitative deformation estimation, due to the data redundancy, an accuracy increasing to millimetric order values and a wider temporal analysis based on multiple interferograms and reducing the coherence loss due to temporal decorrelation (Crosetto et al. 2005). Their applications and commercial software implementation are now the present of the DInSAR technique.

X.3.2.-DInSAR Technique Application to Subsidence Studies

There are many publications that show DInSAR technique results in different world regions and contexts. Most of them are related with subsidence due to mining activities (Camec et al. 1996; Carnec and Delacourt, 2000; Strozzi et al. 2003; Raucoules et al. 2003; Crosetto et al. 2005b; Jung et al. 2007; Herrera et al. 2007; Mei et al. 2008) and water pumping or aquifer overexploitation (Carnec and Fabriol, 1999; Sneed et al. 2001; Zhou et al. 2003; Strozzi et al. 2003; Chang et al. 2004; Teatini et al. 2005; Tomás et al. 2005; Fruneau et al. 2005; Zhang et al. 2007; Stramondo et al. 2007, Finnegan et al. 2008; Fernandez et al, 2009). An in deep analysis of the most meaningful of the last ones is shown in Table X.3. They include study areas in Europe, America and Asia in different climatic environments.

The technique used in the works has changed during the time. First studies used InSAR technique, while in the last ones different ADInSAR algorithms are being considered. The spatial extension are in most of the cases are kilometer square order.

Most of the works have used ERS1-2 images and only a few have combined images of more than one satellite. The image number varies from less than 10 to more than 10 almost at equal parts, including more than 10 images in all the ADInSAR studies.

X.3.3.- DInSAR Technique Application to Landslides Studies

Landslide application is one of the most ambitious applications of the DInSAR technique. The lack of coherence areas in hilly environments, the difficult results interpretation due to the combination between LOS and geometric slope characteristics are some of the reasons.

Table X.3 shows an in deep analysis of the most relevant research publications in the landslides DInSAR application.

In the cases in which an article includes several cases, the information that corresponds to each landslide is included in the table with a number.

With respect to the technique, only the first reference, Fruneau et al. (1996) has used InSAR, obviously due to the research date, when ADInSAR techniques didn't exist yet. Seven of the investigations used PS ADInSAR technique that has advantages in non urban areas with respect to other ones. Most of the articles covers areas of square kilometers and ERS1-2 are the most images used, in some cases in combination with other satellites images. Temporal interval analyzed is commonly between 2 and 10 years, but 6 cases consider less than a year, corresponding with a landslide reactivation. The number of images considered is from 2 to 10 or more than 10 almost as equal parts, considering all the PS studies more than 10 images as is required to the application of this technique. Complementary studies are available in most of the DInSAR landslides application, generally previous to the DInSAR application. These studies are in situ measurements by several techniques, leveling and GPS basically. In 5 cases there were no additional information, and in 4 of them the studies part from the landslides movements knowing and their qualitative characteristics. Only in one case the landslide was discovered by DInSAR technique application. Finally landslides types studied are mostly complex and translational and very few of other types, like rotational landslides and rockfalls.

Table X.3. Summary of subsidence research published using SAR technique

Nº Ref.	Technique			Extension		Images used		Temporal interval				N. imag		Complementary Studies						Subsidence types					
	InSAR	DInSAR	ADInSAR	ha	Km²	ERS 1-2	Others	Month -1yr	1-2 years	2-10 years	2-10	+10	leveling	GPS	Extensom.	others	Water pumping	Mining	Tectonic	Non specified	Others
1	X				X	X			X		X					X	X				
2	X				X	X				X	X		X				X				
3	X				X					X	X			X			X				
4	X				X					X	X			X			X				
5	X				X					X	X			X			X				
6		X			X	X		X			X		X			X (geostadist)	X				
7	X(3)	X (1), (2)		X(2)	X(1) X(3)		X JERS	X(2), X(3)		X(1)	X(1, 2, 3)		X			X(ERS DInSAR)	X(1, 2)	X(3)			
8		X			X	X				X	X					X(subt water hole net)	X				
9			X (SBAS)	X		X				X	X	X	X								
10			X (IPTA)		X	X				X	X	X	X	X		X DInSAR	X				
11			X (CPT)		X	X				X	X	X	X		X	X(piez)	X				
12		X		X		X				X	X	X	X			X(piez)	X				
13			X (PS)		X		X RADARSAT			X		X					X (drenaje)				X(consolidation)
14		X			X	X	X ENVISAT			X	X		X								
15			X (SBAS)		X	X				X	X	X	X				X		X		
16		X			X	X	X RADARSAT1			X		X		X			X		X		
17			X(SPN-CPT)		X	X	X ENVISAT			X		X			X		X				
18			X (IPTA)		X	X	X ENVISAT			X		X				X(geotecnic inf.)					X(consolidation)

N°	Technique	Extension	Images used	Temporal interval	N. imag	Complementary Studies	Subsidence types		
N°	Technique	Extension	Images used	Temporal interval	N. imag	Complementary Studies	Subsidence types		
19	X(simplifi)	X	X ENVISAT		X			X(2)	X (geotecnic prop.)
20	X (PS)	X	X		X	X	X		
21	X(simplifi)	X	X		X		X		X(hidrotermal)

ARTICLES NUMBERS INCLUDED IN THE TABLE X.3:

1. Galloway D.L., Hudnut K.W., Ingebritsen S.E., Phillips S.P., Peltzer G., Rogez F., Rosen P.A. 1998. Detection of aquifer-system compaction and land subsidence using interferometric synthetic aperture radar, Antelope Valley, Mojave Desert, California. *Water Resour. Res.* 98 Vol. 34, N°. 10, p. 2573.
2. Amelung F., Galloway D. L., Bell J. W., Zebker H. A., and Laczniak R. J. 1999. Sensing the ups and downs of Las Vegas; InSAR reveals structural control of land subsidence and aquifer-system deformation. Geology; June 1999; v. 27; no. 6; p. 483-486. DOI: 10.1130/0091-7613(1999)027.
3. Hoffman, J., Zebker, H.A., Galloway D.L., Amelung F. (2001). Seasonal subsidence and rebound in Las Vegas Valley, Nevada, observed by synthetic aperture radar interferometry. *Water Resources Research*, Vol 37, N°. 6, 1551–1566, 2001
4. Sneed, M., Ikehara, M., Balloway, D., Amelung, F., 2001. Detection and measurement of land subsidence using global positioning system and interferometric synthetic aperture radar, Coachella Valley, California, 1996–1998. Water-Resources Investigations. Report 01-4193, US Geological Survey, USA.
5. Bell J. W., Ramelli A. R. and Blewitt G. 2002. Land Subsidence in Las Vegas, Nevada, 1935–2000: New Geodetic Data Show Evolution, Revised Spatial Patterns, and Reduced Rates. Environmental and Engineering Geoscience; August 2002; v. 8; no. 3; p. 155-174; doi: 10.2113/8.3.155.
6. Zhou Y., Stein A. and Molenaar M. (2003). Integrating interferometric SAR data with levelling measurements of land subsidence using geostatistics. *International Journal of Remote Sensing* 24: 18, 3547 – 3563. doi:10.1080/01431160210000023880.
7. Strozzi T., U. Wegmüller, C. L. Werner, A. Wiesmann, V. Spreckels, 2003. JERS SAR Interferometry for Land Subsidence Monitoring. *IEEE Transactions on Geoscience and Remote Sensing*, Vol. 41, No. 7, July 2003, pp 1702-1708.
8. Chang C.P., Chang T.Y., Wang C.T., Kuo C.H., Chen K.S. 2004. Land-surface deformation corresponding to seasonal ground-water fluctuation, determining by SAR interferometry in the SW Taiwan. *Mathematics and Computers in Simulation* 67 (2004) 351–359. doi:10.1016/j.matcom.2004.06.003
9. Lanari R., Zeni G., Manunta M., Guarino S., Berardino P. and Sansosti E. 2004. An integrated SAR/GIS approach for investigating urban deformation phenomena: a case of study of the city of Naples, Italy. *Int. J. Remote Sensing*, 20 July, 2004 vol 25, n°14, 2855-2862. doi: 10.1080/0143116031000164775O.

10. Teatini P., L. Tosi, T. Strozzi, L. Carbognin, U. Wegmüller, F. Rizzetto. 2005. Mapping regional land displacements in the Venice coastland by an integrated monitoring system. *Remote Sensing of Environment* 98 (2005) 403 – 413. doi:10.1016/j.rse.2005.08.002

11. Tomás R., Y. Márquez, J. M. Lopez-Sanchez, J. Delgado, P. Blanco, J.J. Mallorqui, M.Martínez, G. Herrera, J. Mulas. 2005. Mapping ground subsidence induced by aquifer overexploitation using advanced Differential SAR Interferometry: Vega Media of the Segura River (SE Spain) case study. *Remote Sensing of Environment* 98 (2005) 269 – 283. doi:10.1016/j.rse.2005.08.003.

12. Fruneau, B.; Deffontaines, B.; Rudant, J.P. and Le Parmentier A.M. (2005). monitoring vertical deformation due to water pumping in the city of Paris (France) with differential Interferometry. *C.R. Geoscience*, 337, 1173-1183. doi:10.1016/j.crte.2005.05.014

13. Dixon, T.H., Amelung, F., Ferretti A., Novali, F., Rocca, F., Dokka, R., Sellall, G., Kim, S.-W., Wdowinski, S., and Whitman, D. (2006). Subsidence and flooding in New Orleans. *Nature*, 441, 587-588. doi:10.1038/441587a

14. Zhang Q., Zhao C., Din X. and Peng, J. 2007. Monitoring Xian Land Subsidence Evolution by Differential SAR Interferometry, en *Geomatics Solutions for Disaster Management*. Edited by J. Li, S. Zlatanova, A. G. Fabbri. New York. Springer, 2007.

15. Stramondo S., M. Saroli, C. Tolomei, M. Moro, F. Doumaz, A. Pesci, F. Loddo, P. Baldi, E. Boschi. 2007. Surface movements in Bologna (Po Plain, Italy) detected by multitemporal DInSAR. *Remote Sensing of Environment*. 110 (2007) 304-316. doi:10.1016/j.rse.2007.02.023.

16. Finnegan, N.J., Pritchard, M. E., Lohman, R. B and Lundgren, P. R. 2008. Constrains on surface deformation in the Seattle, WA, urban corridor from satellite radar interferometry time –series analysis. *Geophysical Journal International*, 174, 29-41.

17. Herrera, G., Tomás R., Lopez-Sanchez J.M., Delgado J., Vicente, F., Mulas, J., Cooksley, G., Sanchez, M., Duro, J., Arnaud, A., Blanco, P., Duque S., Mallorqui J.J., De la Vega-Panizo, R., Monserrat, O. 2008. Validation and comparison of Advanced Differential Interferometric Techniques: Murcia metropolitan area case study. ISPRS Journal of Photogrammetry and Remote Sensing. Article in Press. doi:10.1016/j.isprsjprs.2008.09.008.

18. Stramondo S., M. Saroli, C. Tolomei, M. Moro, F. Doumaz, A. Pesci, F. Loddo, P. Baldi, E. Boschi. 2007. Surface movements in Bologna (Po Plain, Italy) detected by multitemporal DInSAR. *Remote Sensing of Environment*. 110 (2007) 304-316. doi:10.1016/j.rse.2007.02.023.

19. Vallone, P., Giammarinaro, M.S., Crosetto, M., Agudo, M., Biescas, E. (2008). Ground motion phenomena in Caltanissetta (Italy) investigated by InSAR and geological data integration. *Engineering Geology*, 98, 144-155

20. Vilardo, G., Ventura G., Terranova C., Matano F., Nardò S. 2009. Ground deformation due to tectonic, hydrothermal, gravity, hydrogeological, and anthropic processes in the Campania Region (Southern Italy) from Permanent Scatterers Synthetic Aperture Radar Interferometry. *Remote Sensing of Environment* 113, 197–212.

21. Fernandez, P., Irigaray C., Jimenez, J., El Hamdouni, R., Crosetto, M., Monserrat, O. and Chacon, J.2009. First delimitation of areas affected by ground deformations in the Guadalfeo River Valley and Granada metropolitan area (Spain) using the DInSAR technique. *Engineering Geology* 105 (2009) 84–101.

In the cases in which an article includes several cases, the information that corresponds to each case is included in the table with a number.

Table X.4. Summary of DInSAR landslides researches published in international journals

Art N°	Technique			Spatial order			Imag used		Temporal interval			Number imag			Complementary Studies		Landslide type				
	InSAR	DInSAR	PS	m²	ha	Km²	ERS 1-2 ENVISAT	JERS	Month-1yr	1-2 yrs	2-10 yrs	2	2-10	+10	No	Yes	Traslational	Rotacional	Complex	Non especified	Others
1	X			X			X		X				X			X	X				
2		X				X	X				X		X			X	X		X		
3		X			X		X			X			X			X			X		
4		X				X	X				X		X			X		X			
5		X	X			X	X							X		X		X		X	
6		X			X			X	X			X			X						
7		X				X	X				X		X				X				X
8		X	X			X	X				X		X			X	X				
9		X		X(3)	X(1)X(2)X(4)	X(1)X(3)X(4)	X(1)X(2)X(3)X(4)	X(1)X(2)X(4)	X(3)			X(1)X(2)X(3)X(4)		X(3)	X(1)X(2)X(4)	X(4)		X(1)X(2)X(3)			
10		X			X		X				X		X	X		X	X				
11		X			X				X				X			X		X			
12		X	X			X	X				X			X	X				X		
13		X	X		X		X			X				X		X	X				
14		X	X			X	X				X		X	X		X			X		
15		X	X			X	X				X		X	X	X(3)	X(1)X(4)			X		
16		X	X			X	X				X		X	X		X			X		
17		X			X		X				X		X	X		X				X	
18		X			X		X			X			X	X		X			X		
19		X		X			X				X		X	X		X	X				

ARTICLES NUMBERS INCLUDED IN TABLE X.4:

1. Fruneau, B., Achache, J., Delacourt, C. 1996. Observation and modelling of the Saint-Etienne de Tinee landslide using SAR interferometry. *Tectonophys*, 265 (3–4), 181-190.

2. Rott H., Scheuchl B. and Siegel A. 1999. Monitoring very slow slope movements by means of SAR interferometry: A case study from a mass waste above a reservoir in the Ötztal Alps, Austria. *Geophysical Research Letters* Vol. 26, No. 11, 1629-1632.

3. Rizo V. et al.. 2000. SAR Interferometry and Field Data of Randazzo Landslide (Eastern Sicily, Italy). *Physics and Chemistry of the Earth*, Part B: Hydrology, Vol. 25, No. 9, 771-780.

4. Berardino P., Constantini M., Franceschetti G., Iodice A., Pietranera L. and Rizzo V. 2003. Use of differential SAR interferometry in monitoring and modelling large slope instability at Maratea (Basilicata, Italy). *Engineering Geology*, 68, 31-51. doi:10.1016/S0013-7952(02)00196-5.

5. Colesanti, C., Ferretti, A., Prati, C., Rocca, F. 2003. Monitoring landslides and tectonic motions with the Permanent Scatterers Technique. *Engineering Geology*, 68, 3–14. doi:10.1016/S0013-7952(02)00195-3.

6. Catani F., Farina P., Moretti S., Nico G. and Strozzi T. On the application of SAR interferometry to geomorphological studies: estimation of landform attributes and mass movements. *Geomorphology*, v. 66, iss. 1-4 [SPECIAL ISSUE], 119-131. doi:10.1016/j.geomorph.2004.08.012

7. Singhroy V. and Molch K. 2004. Characterizing and monitoring rockslides from SAR techniques. *Advances in Space Research* Volume 33, Issue 3, 290-295.

8. Hilley, G.E., Bürgmann, R., Ferretti, A., Novali, F., Rocca, F..2004. Dynamics of Slow-Moving Landslides from Permanent Scatterer Analysis. *Science*, 304, 1952-1955.

9. Strozzi, T., P. Farina, A. Corsini, C. Ambrosi, M. Thüring, J. Zilger, A. Wiesmann, U. Wegmüller and C. Werner. 2005. Survey and monitoring of landslide displacements by means of L-band satellite SAR interferometry. *Landslides* 2 (3), 193-201. doi:10.1007/s10346-005-0003-2.

10. Squarzoni C., Delacourt C., Allemand, P. 2003. Nine years of spatial and temporal evolution of the La Valette landslide observed by SAR interferometry. *Engineering Geology*, 68, 53-66.

11. Kimura, H., Yamaguchi, Y., 2000. Detection of landslide areas using satellite radar interferometry., *Photogramm Eng Rem S*, 66 (3), 337-344.

12. Bovenga, F.; Nutricato, R.; Refice, A.; Wasowski, J. 2006. Application of multi-temporal differential interferometry to slope instability detection in urban/peri-urban areas. *Engineering Geology*, 2006/12/15, 88, 3–4, 218-239. doi:10.1016/j.enggeo.2006.09.015.

13. Colesanti, C. and Wasowski, J., 2006. Investigating landslides with space-borne Synthetic Aperture Radar (SAR) interferometry. *Engineering Geology*, 88, 3–4, 173–199. doi:10.1016/j.enggeo.2006.09.013.

14. P. Farina, D. Colombo, A. Fumagalli, F. Marks, S. Moretti. 2006. Permanent Scatterers for landslide investigations: outcomes from the ESA-SLAM project. *Engineering Geology* 88, 200–217. doi:10.1016/j.enggeo.2006.09.007.

15. C. Meisina, F. Zucca, D. Fossati, M. Ceriani, J. Allievi. 2006. Ground deformation monitoring by using the Permanent Scatterers Technique: The example of the Oltrepo Pavese (Lombardia, Italy). *Engineering Geology* 88, 240–259. doi:10.1016/j.enggeo.2006.09.010.

16. C. Meisina, F. Zucca, F. Conconia, F. Verria, D. Fossatib, M. Ceriani, J. Allievi. 2007. Use of Permanent Scatterers technique for large-scale mass movement investigation. *Quaternary International* 171-172, 90-107. doi:10.1016/j.quaint.2006.12.011.

17. Corsini; P. Farina; G. Antonello; M. Barbieri; N. Casagli; F. Coren; L.Guerri; F. Ronchetti; P. Sterzai; D. Tarchi. 2006. Space-borne and ground-based SAR interferometry as tools for landslide hazard management in civil protection. *International Journal of Remote Sensing*, 27:12, 2351-2369. doi:10.1080/01431160600554405.

18. Peyret, M., Djamour Y., Rizza M., Ritz J.-F., Hurtrez J.-E., Goudarzi M.A., Nankali H., Chéry J., Le Dortz K., Uri F. 2008. Monitoring of the large slow Kahrod landslide in Alborz mountain range (Iran) by GPS and SAR interferometry. Engineering Geology 100, 131–141.

19. Fernandez, P., Irigaray C., Jimenez, J., El Hamdouni, R., Crosetto, M., Monserrat, O. and Chacon, J.2009. First delimitation of areas affected by ground deformations in the Guadalfeo River Valley and Granada metropolitan area (Spain) using the DInSAR technique. *Engineering Geology* 105 (2009) 84–101.

REFERENCES

Allan, T.D. Ed.1983.*Satellite Microwave Remote Sensing*. Ellis Howard. New York.

Amelung, F., Jonson, S. Zebker, H. A. and Segall, P. 2000. Widespread uplift and "trapdoor" faulting on Galapagos volcanoes observed with radar interferometry. *Nature* 407, 993-996.

Arnaud, A., Adam, N., Hanssen, R., Inglada, J., Duro, J., Closa, J., Eineder, M. 2003. ASAR ERS Interferometric phase continuity. *Proceedings International Geoscience and Remote Sensing Symposium*, Toulouse, 21-25 July, pp. 1133-1135.

Berardino, P., Fornaro, G., Lanari, R. and Sansosti, E. 2002. A new algorithm for surface deformation monitoring based on Small Baseline Differential SAR Interferograms. *Transactions on Geoscience and Remote Sensing*, 40, 2375–2383.

Carnec, C., Massonnet, D., King, C. 1996. Two examples of the use of SAR interferometry on displacement fields of small spatial extent. *Geophysical Research Letters*, 23 (24), pp. 3579-3582.

Carnec, C., Delacourt, C., 2000. Three years of mining subsidence monitored by SAR interferometry, near Gardanne, France. *Journal of Applied Geophysics* 43, 43-54.

Carnec, C., Fabriol, H. 1999. Monitoring and modeling land subsidence at the Cerro Prieto geothermal field, Baja California, Mexico, using SAR interferometry. *Geophysical Research Letters*, 26 (9), pp. 1211-1214.

Chang C.P., Chang T.Y., Wang C.T., Kuo C.H., Chen K.S. 2004. Land-surface deformation corresponding to seasonal ground-water fluctuation, determining by SAR interferometry in the SW Taiwan. *Mathematics and Computers in Simulation* 67 (2004) 351–359. doi:10.1016/j.matcom.2004.06.003.

Chuvieco, E. 1990. *Fundamentos de Teledetección Espacial*. Rialp. Madrid.

Crosetto, M., Crippa, B., Biescas, E., Monserrat, O. Agudo, M. Fernández, P. 2005. "Land deformation measurement using SAR interferometry: state-of-the-art", *Phogrammetrie Fernerkundung Geoinformation*, 06-2005, 497-510. Stuttgart (Germany) ISSN: 1432-8364/05/2005/0497.

Crosetto M., Crippa B. and Biescas E., 2005b. Early detection and in-depth analysis of deformation phenomena by radar interferometry. *Engineering Geology*, Volume 79, Issues 1-2, 3 June 2005, 81-91. doi:10.1016/j.enggeo.2004.10.016.

Crosetto, M., Biescas, E., Duro, J., Closa, J., Arnaud, A. 2008. Generation of advanced ERS and ENVISAT interferometric SAR products using stable point network technique. *Photogrammetric Engineering and Remote Sensing* 74 (4), 443-451.

Duro, J., Inglada, J., Closa, J., Adam, N., Arnaud, A. 2003. High Resolution Differential Interferometry using time series of ERS and ENVISAT SAR data. *Proc. of FRINGE 2003 Workshop,* Frascati, Italy,1–5 December 2003 (ESA SP-550, June 2004).

Duro, J., Closa, J., Biescas, E., Crosetto, M., Arnaud, A. 2005. High Resolution Differential Interferometry using time series of ERS and ENVISAT SAR data. *Proc. 6th Geomatic Week*, Barcelona, 8-11 February (CDROM).

Elachi, C. 1982. Radar Images of the Earth from Space. *Scientific American*, vol 247, 46-53.

Fernandez,Oliveras P. 2009. Determinación de movimientos verticales del terreno mediante técnicas de interferometria RADAR DInSAR. Thesis Doctoral. Universidad de Granada, Spain.

Fernandez, P., Irigaray C., Jimenez, J., El Hamdouni, R., Crosetto, M., Monserrat, O. and Chacon, J. 2009. First delimitation of areas affected by ground deformations in the Guadalfeo River Valley and Granada metropolitan area (Spain) using the DInSAR technique. *Engineering Geology* 105 (2009) 84–101.

Ferretti, A., Prati, C. and Rocca, F. 2000. Non linear subsidence rate estimation using Permanent Scatterers in Differential SAR Interferometry. *Transactions on Geoscience and Remote Sensing*, 38, nº5, 2202–2212.

Ferretti, A., Prati, C. and Rocca, F. 2001. Permanent Scatterers in SAR interferometry. *Transactions on Geoscience and Remote Sensing*, 39, nº1, 8–20.

Ferretti, A., Monti-Guarnieri, A., Prati, C. Rocca, F. and Massonnet, D. 2007. *InSAR Principles: Guidelines for SAR Interferometry Procesing and Interpretation.* ESA Publications TM-19. The Netherlands.

Finnegan, N.J., Pritchard, M. E., Lohman, R. B and Lundgren, P. R. 2008. Constrains on surface deformation in the Seattle, WA, urban corridor from satellite radar interferometry time –series analysis. *Geophysical Journal International,* 174, 29-41.

Fruneau, B.; Deffontaines, B.; Rudant, J.P. and Le Parmentier A.M. 2005. Monitoring vertical deformation due to water pumping in the city of Paris (France) with differential Interferometry. *C.R. Geoscience*, 337, 1173-1183. doi:10.1016/j.crte.2005.05.014.

Gabriel, A. K., R. M. Goldstein, and Zebker H. A. 1989. Mapping Small Elevation Changes Over Large Areas: Differential Radar Interferometry, *J. Geophys. Res.*, 94(B7), 9183–9191.

Gens, R. and Genderen J. L. van. 1996. SAR interferometry: issues, techniques, applications. *Int. J. Remote Sensing*, vol 17, nº10, pp1803-1835.

Halsema, D. van, Kooij, M. W. A. van der, Groenewoud, W., Huising, J., Ambrosious, B.A.C., and Klees, R. 1995. SAR interferometrie in Netherlands. *Remote sensing Nieuwsbrief,* Junio, pp. 31-34.

Hanssen, R. 2001. *Radar Interferometry. Data interpretation and error analysis.* Kluwer academic publishers. The Netherlands.

Kampes, B. 2006. *Radar Interferometry. Persistent Scatterer Technique.* Springer. The Netherlands.

Harlt, Ph., Reich, M, Thiel, K. H. and Xia, Y. 1992. SAR-interferometry applying ERS1: some preliminary test results. *Proceedings of the first ERS-1Symposium.* ESA Paris, 219-222.Goldstein et al. 1993.

Herrera G., Tomás R., Lopez-Sanchez J.M., Delgado J., Mallorqui J.J., Duque S., Mulas J. 2007. Advanced DInSAR analysis on mining areas: La Union case study (Murcia, SE Spain). *Engineering Geology* 90 (2007) 148–159. doi:10.1016/j.enggeo.2007.01.001.

INGV, 2009. *Synthetic report of the activities of ASI-SIGRIS personnel and Earthquake Remote Sensing Group of the Remote Sensing Laboratory - National Earthquake Center, INGV – Rome, 2009.* http://portale.ingv.it/primo-piano-1/news-archive/2009-news/april-6-earthquake/sar-preliminary-results

Jung H.C., Kim S-W, Jung H-S, Min K.D, Won J-S. 2007. Satellite observation of coal mining subsidence by persistent scatterer analysis. *Engineering Geology* 92 (2007) 1–13. doi:10.1016/j.enggeo.2007.02.007.

Klees, R. and Massonnet, D. 1999. Deformation measurements using SAR interferometry: potential and limitations. *Geologie en Mijnbouw* 77: 161-176.

Kwok, R. and Fahnestock, M. A. 1996. Interferometric Estimation of Three-Dimensional Ice-Flow Using Ascending and Descending Passes, *IEEE Transactions on Geoscience and Remote Sensing*, 36(1):25– 37.

Massonnet, D; Rossi, M.; Carmona, C; Adragna, F.; Pelzer, G.; Feigl K. And Rabaute, T. 1993. The displacement field of the Landers earthquake mapped by radar interferometry, *Nature*, 364 (8): 138-142.

Massonnet D. and Adragna F., 1993. A full scale validation of Radar interferometrywith ERS-1: the Landers earthquake. *Earth Observation Quarterly*, 41.

Massonnet, D. and Feigl, K. L., 1998. Radar interferometry and its application to changes in the Earth's surface. *Reviews of Geophysics, 36(4), 441-500.*

Massonet, D. and Sigmundsson, F. 2000. Remote sensing volcano deformation by radar interferometry from various satellites. In: Mouginis- Mark et al (eds). *Remote Sensing of active volcanism. Geophysical Monographs* 116, American Geophysical Union, 207-221.

Mei, S., Poncos, V., Froese, C. 2008. Mapping millimetre-scale ground deformation over the underground coal mines in the Frank Slide area, Alberta, Canada, using spaceborne InSAR technology. *Canadian Journal of Remote Sensing*, 34 (1-2), pp. 113-134.

Mora, O., 2004. *Advanced differential SAR techniques for detection of terrain and building displacements.* Thesis Doctoral, Universidad Politécnica de Cataluña, 182 pp.

Mora, O., Mallorquí, J.J., Broquetas, A., 2003. Linear and nonlinear terrain deformation maps from a reduced set of interferometric SAR images. *Transactions on Geoscience and Remote Sensing*, 41, 2243–2253.

Peltzer, G.;Rosen, P; Rogez, F. and Hudnut, K. 1996. Postseismic rebound in fault step-overs caused by pore fluid flow. *Science*, 273: 1202-1204.

Prati, C, Rocca, F and Monti-guarnieri, A. 1992. SAR interferometry experiments with ERS-1. *Proceedings of the first ERS-1Symposium*. ESA Paris. pp. 211-218.

Raucoules D., C. Maisons, C. Carnec, S. Le Mouelic, C. King, S. Hosford. 2003. Monitoring of slow ground deformation by ERS radar interferometry on the Vauvert salt mine (France). Comparison with ground-based measurement. *Remote Sensing of Environment* 88 (2003) 468–478. doi:10.1016/j.rse.2003.09.005.

Rosen, P.; Werner, C.; Fielding, E.; Hensley, S.; and Vincent, S.B.P. 1998. Aseismic creep along the San Andreas fault northwestof Parkfield, CA, measured by radar interferometry, *Geophysical Research Letters,* 25 (6). 825-828.

Rosen, P.A, Hensley S., Joughin, I.R., Li, F.K., Madsen, S. N., Rodriguez, E. and Goldstein, R. M. 2000. Synthetic Aperture Radar Interferometry. *Proceedings of the IEEE. Vol 88, (3), 333-382. Invited paper.*

Sneed, M., Ikehara, M., Balloway, D., Amelung, F., 2001. Detection and measurement of land subsidence using global positioning system and interferometric synthetic aperture radar, Coachella Valley, California, 1996–1998. Water-Resources Investigations. Report 01-4193, US Geological Survey, USA.

Stramondo S., M. Saroli, C. Tolomei, M. Moro, F. Doumaz, A. Pesci, F. Loddo, P. Baldi, E. Boschi. 2007. Surface movements in Bologna (Po Plain, Italy) detected by multitemporal DInSAR. *Remote Sensing of Environment*. 110 (2007) 304-316. doi:10.1016/j.rse.2007.02.023.

Strozzi T., U. Wegmüller, C. L. Werner, A. Wiesmann, V. Spreckels, 2003. JERS SAR Interferometry for Land Subsidence Monitoring. *IEEE Transactions on Geoscience and Remote Sensing*, Vol. 41, No. 7, July 2003, pp 1702-1708.

Tomás R., Y. Márquez, J. M. Lopez-Sanchez, J. Delgado, P. Blanco, J.J. Mallorquí, M.Martínez, G. Herrera, J. Mulas. 2005. Mapping ground subsidence induced by aquifer overexploitation using advanced Differential SAR Interferometry: Vega Media of the Segura River (SE Spain) case study. *Remote Sensing of Environment* 98 (2005) 269 – 283. doi:10.1016/j.rse.2005.08.003.

Teatini P., L. Tosi, T. Strozzi, L. Carbognin, U. Wegmüller, F. Rizzetto. 2005. Mapping regional land displacements in the Venice coastland by an integrated monitoring system. *Remote Sensing of Environment* 98 (2005) 403 – 413. doi:10.1016/j.rse.2005.08.002.

Trevett, J.W. 1986. *Imaging Radar for Resources Surveys*. Chapman and Hall. London.

Werner, C., Wegmüller, U., Strozzi, T. and Wiesmann, A., 2003, Interferometric point target analysis for deformation mapping. *Proceedings of IGARSS 2003*, 21–25 July 2003, Toulouse, France.

Zhang Q., Zhao C., Din X. and Peng, J. 2007. Monitoring Xian Land Subsidence Evolution by Differential SAR Interferometry, en *Geomatics Solutions for Disaster Management*. Edited by J. Li, S. Zlatanova, A. G. Fabbri. New York. Springer, 2007.

Zhou Y., Stein A. and Molenaar M. 2003. Integrating interferometric SAR data with levelling measurements of land subsidence using geostatistics. *International Journal of Remote Sensing* 24: 18, 3547 – 3563. doi:10.1080/0143116021000023880.

REFERENCES TABLE X.2

Amelung, F., Jonson, S. Zebker, H. A. and Segall, P. 2000. Widespread uplift and "trapdoor" faulting on Galapagos volcanoes observed with radar interferometry. *Nature* 407, 993-996.

Avallone, A., Zollo, A., Briole, P., Delacourt, C. and Beauducel, F. 1999. Subsidence of Campi flegrei (Italy) detected by SAR interferometry. *Geophysical Research Letters*, 26 (15): 2303-2306.

Baer, G., Sandwell, D., Williams, S. Bock, Y. and Shamir, G. 1999. Coseismic deformation associated with the november 1995 Mw=7.1 Nuweiba earthquake, Gulf of Elat (Aqaba), detected by synthetic aperture radar interferometry. *Journal of Geophysical Research*, 104 (B11): 25221-25232.

Barbieri, M., Lichtenegger, J. and Calabresi, G. 1999. The Izmit Earthquake: A quick Post-seismic Analysis with satellite observations, *ESA Bulletin*, 100: 107-110.

Briole, P. Massonnet, D. and Delacourt, C. 1997. Post eruptive deformation associated with the 1986-87 and 1989 lava flows of Etna detected by radar interferometry. *Geophysical Research Letters*, 24: 37-40.

Bürgmann, R., Schmidt, D., Nadeau, R. M., d'Alessio, M., Fielding, E. J., Manaker, D., McEvilly, T. V. and Murray, M. H. 2000. Earthquake Potential Along the Northern Hayward Fault, California, Science, 289:1178–1182.

Clarke, P. J., Paradissis, D., Briole, P., England, P. C., Parsons, B. E., Billiris, H., Veis, G., and Ruegg, J.-C. 1996. Geodetic investigation of the 13 May 1995 Kozani-Grevena (Greece) earthquake, *Geophysical Research Letters*, 24:707–710.

Delacourt, C., Briole, P., Achache, J., Fruneau, B. and Carnec, C. 1997. Correction of the tropospheric delay in SAR interferometry and application to 1991-93 eruption of Etna volcano, Italy, in: AGU Fall meeting, December 8-12, San Francisco, USA,.

Feigl, K. L, Sergent, A. And Jacq, D. 1995. Estimation of an earthquake focal mechanism from a satellite radar interferogram: Aplication to the December 4, 1992 Landers aftershock. *Geophysical research Letters*, 22 (9): 1037-1040.

Fujii, N., T Nakano, T. O. and Yamaoka, K. 1994. Detection of Ground Deformations by the Multi– pass Differential SAR– interferometry. Examples of the Active Volcanic Area, in: paper presented at the *1st workshop on SAR interferometry*, Tokyo, Japan, December 1994, NASDA.

Fujiwara, S., Yarai, H., Ozawa, S., Tobita, M., Murakami, M., Nakagawa, H., Nitta, K., Rosen, P. A. and Werner, C. L. 1998. Surface displacement of the March 26, 1997 Kagoshima-kenhokuseibu earthquake in Japan from synthetic aperture radar interferometry, *Geophysical Research Letters*, 25(24):4541–4544.

Goldstein, R. M., Engelhardt, H., Kamp, B. and Frolich, R. M. 1993. Satellite radar interferometry for monitoring ice sheet motion: Application to an Antarctic ice stream, *Science*, 262:1525–1530.

Hanssen, R., Vermeersen, B., Scharroo, R., Kampes, B., Usai, S., Gens, R. and Klees, R. 2000. Deformatiepatroon van de aardbeving van 17 augustus 1999 in *Turkije gemeten met satelliet radar interferometrie, Remote Sensing Nieuwsbrief*, 90:42–44, In Dutch.

Hartl, P., Thiel, K.-H. and Wu, X. 1994a. Information extraction from ERS-1 SAR data by means of INSAR and D-INSAR techniques in Antarctic research, in: Second ERS1 Symposium— Space at the Service of our Environment, Hamburg, Germany, 11–14 October 1993, ESA SP-361, pp. 697–701.

Hartl, P., Thiel, K. H., Wu, X., Doake, C. and Sievers, J. 1994b. Application of SAR Interferometry with ERS-1 in the Antarctic, Earth Observation Quarterly, 43:1–4.

Hernandez, B., Cotton, F., Campillo, M. and Massonnet, D. 1997. A Comaprison between short term (co-seismic) and long term (one year) slip for the Landers earthquake : Measurements from strong motion and SAR interferometry. *Geophysical Research Letters*, 24(13): 1579-1582.

Hoen, E. W. and Zebker, H. A. 2000. Penetration Depths Inferred from Interferometric Volume Decorrelation Observed over the Greenland Ice Sheet, *IEEE Transactions on Geoscience and Remote Sensing*, 38(6):2571–2583.

Jonsson, S., Zebker, H., Cervelli, P., Segall, P., Garbeil, H., Mouginis-Mark, P. and Rowland, S. 1999. A Shallow-Dipping Dike fed the 1995 Flank Eruption at Fernandina Volcano, Galapagos, Observed by Satellite Radar Interferometry, *Geophysical Research Letters*, 26(8): 1077–1080.

Joughin, I. R., Winebrenner, D., Fahnestock, M., Kwok, R. and Krabill, W. 1996. Measurement of ice-sheet topography using satellite-radar interferometry, *Journal of Glaciology*, 42(140):10–22.

Joughin, I. R., Fahnestock, M., Ekholm, S. and Kwok, R. 1997. Balance Velocities of the Greenland ice sheet, Geophysical Research Letters, 24(23) :3045–3048. Joughin, I. R., Kwok, R. and Fahnestock, M. A. (1998. Interferometric Estimation of Three-

Dimensional Ice-Flow Using Ascending and Descending Passes, *IEEE Transactions on Geoscience and Remote Sensing*, 36(1):25– 37.

Joughin, L, Gray, L., Bindschadler, R., Price, S., Morse, D., Hulbe, C., Mattar, K. and Werner, C. 1999. Tributaries of West Antarctic Ice Streams Revealed by RADARSAT Interferometry, *Science*, 286(5438):283– 286.

Kawai, S. and Shimada, M. 1994. Detections of Earth Surface Deformation Change by means of INSAR technique, in: paper presented at the *1st workshop on SAR interferometry*, Tokyo, Japan, December 1994, NASDA.

Klinger, Y., Michel, R. and Avouac, J.-P. 2000. Co-seismic deformation during the Mw 7.3 Aqaba earthquake (1995) from ERS-SAR interferometry, Geophysical Research Letters, 27(22):3651– 3655.

Kwok, R. and Fahnestock, M. A. 1996. Ice Sheet Motion and Topography from Radar Interferometry, *IEEE Transactions on Geoscience and Remote Sensing*, 34(1): 189– 200.

Lanari, R., Lundgren, P. and Sansosti, E. 1998. Dynamic deformation of Etna volcano observed by satellite radar interferometry, *Geophysical Research Letters*, 25:1541– 1544.

Lu, Z., Fatland, R., Wyss, M., Li, S., Eichelberger, J., Dean, K. and Freymueller, J. 1997. Deformation of New Trident Volcano measured by ERS 1 SAR interferometry, Katmai National Park, Alaska, *Geophysical Research Letters*, 24(6):695– 698.

Lu, Z. and Freymueller, J. T. 1999. Synthetic aperture radar interferometry coherence analysis over Katmai volcano group, Alaska, *Journal of Geophysical Research*, 103(B12):29887– 29894.

Lundgren, P., Usai, S., Sansosti, E., Lanari, R., Tesauro, M., Fornaro, G. and Berardino, P., 2001. Modeling surface deformation observed with synthetic aperture radar interferometry at Campi Flegrei caldera, *J. Geophys.Res.*, 106, 19 355–19 366.

Massonnet D. and Adragna F., 1993. A full scale validation of Radar interferometrywith ERS-1: the Landers earthquake. *Earth Observation Quarterly*, 41.

Massonnet D. and Feigl, K. 1995. Satellite radar interferometry map of a coseismic deformation field of the M=6.1 Eureka Valley, California earthquake of May 17, 1993. *Geophysical Research Letters*, 22 (12):1541-1544.

Massonnet, D., Feigl K; Rossi, M and Adragna F. 1994. Radar interferometric mapping of defromationa year after the Landers earthquake. *Nature* 369: 227-230.

Massonnet, D; Rossi, M.; Carmona, C; Adragna, F.; Pelzer, G.; Feigl K. And Rabaute, T. 1993. The displacement field of the Landers earthquake mapped by radar interferometry, *Nature*, 364 (8): 138-142.

Massonnet,D; Feigl, K; Vadon, H. and Rossi, M. 1996a. Coseismic deformation field of the M=6.7 Northridge, Califormia earthquake of January, 17, 1994 recorder by two rdar satellites using radar interfeometry. *Geophysical Research Letters*, 23 (9): 969-972.

Massonnet, D.; Thatcher, W and Vadon, H. 1996b. Detection of postseismic fault zone collapse following the Landers earthquake. *Nature,* 382:489-497.

Meyer, B., Armijo, R., Massonnet, D., de Chabalier, J. B., Delacourt, C., Ruegg, J. C., Achache, J., Briole, P. and Panastassiou, D. 1996. The 1995 Grevena (Northern Greece) earthquake: fault model constrained with tectonic observations and SAR interferometry, *Geophysical Research Letters,* 23:2677– 2680.

Michel, R.; Avouac, J. P. and Taboury, J. 1999. Measuring ground displacement from SAR amplitude images:application to the Landers earthquake. *Geophysical Research Letters*, 26 (7): 875-878.

Mohr, J. J., Reeh, N. and Madsen, S. N. 1998. Three-dimensional glacial flow and surface elevation measured with radar interferometry, *Nature*, 291:273– 276.

Mouginis-Mark, P. J. 1995a. Analysis of volcanic hazards using Radar interferometry, *Earth Observation Quarterly*, 47:6– 10.

Mouginis-Mark, P. J. 1995b. Preliminary Observations of Volcanoes with the Radar, *IEEE Transactions on Geoscience and Remote Sensing*, 33(4):934– 941.

Murakami, M; Tobita, M; Fujiwara, S and Saito, T. 1996. Coseismic crustal deformations of the 1994 Northridge, California, earthquake detected by interferometric JERS-1 Syntetic aperture radar. *Journal of Geophysical Research*, 101 B4: 8605-8614.

Ohkura, H. 1998. Applications of SAR data to monitoring Earth surface changes and displacements, *Advances in Space Research*, 21(3):485– 492.

Ozawa, S; Murakami, M.; Fujiwara, S. and Tobita, M. 1997. Syntetic aperture radar interferogram of the 1995 Kobe earthquake and its geodetic invesion. *Geophysical research Letters*, 24 (18), 2327-2330.

Peltzer, G. and Rosen, P. 1995. Surface Displacement of the 17 May 1993 Eureka Valley, California Earthquake Observed by SAR Interferometry, *Science*, 268:1333– 1336.

Peltzer, G., Crampé, G. and King, G. 1999. Evidence of non linear Elasticity of the Crust from Mw 7.6 Manyi (Tibet) Earthquake. *Science*, 286 (5438): 272-276.

Peltzer, G.; Hudnut, K. W. and Feigl K. L. 1994. Analisis of coseismic surface displacements gradients using radar interferometry: New insights into the Landers earthquake. *Journal of Geophysical Research*, 99 B11: 21971- 21981.

Peltzer, G.;Rosen, P; Rogez, F. and Hudnut, K. 1996. Postseismic rebound in fault stepovers caused by pore fluid flow. *Science*, 273: 1202-1204.

Price, E.J. and Sandwell, D. T. 1998. Small scale deformations associated with the 1992 Landers, California, earthquake mapped by synthetic aperture radar interferometry phase gradients. *Journal of Geophysical Research*, 103(B11):27001-27016.

Reilinger, R. E., Ergintav, S., Bürgmann, R., McClusky, S., Lenk, O., Barka, A., Gurkan, O., Hearn, L., Feigl, K. L., Cakmak, R., Aktug, B., Ozener, H. and Töksoz, M. N. 2000. Coseismic and Postseismic Fault Slip for the 17 August 1999, M = 7.5, Izmit, Turkey Earthquake, *Science*, 289(5484):1519– 1524.

Rignot, E. J., Gogineni, S. P., Krabill, W. B. and Ekholm, S. 1997. North and northeast Greenland ice discharges from satellite radar interferometry, *Science*, 276:934– 937.

Rosen, P., Werner, C., Fielding, E., Hensley, S. and Vincent, S. B. P. 1998. Aseismic creep along the San Andreas fault northwest of Parkfield, CA measured by radar interferometry, *Geophysical Research Letters*, 25(6):825– 828.

Rosen, P. A., Hensley, S., Zebker, H. A., Webb, F. H. and Fielding, E. J. 1996. Surface deformation and coherence measurements of Kilauea Volcano, Hawaii, from SIR-C radar interferometry, *Journal of Geophysical Research*, 101(E10):23109– 23125.

Roth, A., Adam, N., Schwäbisch, M., Müschen, B., Böhm, C. and Lang, O. 1997. Observation of the effects of the subglacial volcano eruption underneath the Vatnajöküll glacier in Iceland with ERS-SAR data, in: *Proc. of the third ERS symposium*, Florence, Italy, 17-20 March 1997.

Rott, H. and Siegel, A. 1997. Glaciological Studies in the Alps and in Antarctica Using ERS Interferometric SAR, in: 'FRINGE 96' workshop on ERS SAR Interferometry, Zürich, Switzerland, 30 Sep– 2 October 1996, pp. 149– 159, ESA SP-406, Vol II.

Rott, H., Stuefer, M., Siegel, A., Skvarca, P. and Eckstaller, A. 1998. Mass fluxes and dynamics of Moreno Glacier, Southern Patagonia Icefield, *Geophysical Research Letters*, 25(9):1407– 1410.

Sandwell, D. T., Sichoix, L., Agnew, D, Bock, Y. and Minster, J. B. 2000. Near realtime radar interferometry of the Mw 7.1 Hector mine Earthquake, *Geophysical Research Letters*, 27 (19): 3101-3104.

Sigmundsson, F., Vadon, H. and Massonnet, D. 1997. Readjustment of the Krafla spreading segment to crustal rifting measured by Satellite Radar Interferometry, *Geophysical Research Letters*, 24(15): 1843– 1846.

Sigmundsson, F., Durand, P. and Massonnet, D. 1999. Opening of an eruptive fissure and seaward displacement at Piton de la Fournaise volcano measured by RADARSAT satellite radar interferometry, *Geophysical Research Letters*, 26(5):533.

Thatcher, W. and Massonnet, D. 1996. Crustal deformation at Long Valley Caldera. eastern California, 1992-1996 inferred from satellite radar interferometry, *Geophysical Research Letters*, 24(20):2519– 2522.

Thiel, K.-H., Wu, X. and Hartl, P. 1997. ERS-tandem-interferometric obsertvation of volcanic activities in Iceland, in: Third ERS Symposium— Space at the Service of our Environment, Florence, Italy, 17– 21 March 1997, ESA SP-414, pp. 475– 480.

Usai, S. and Klees, R. 1999. SAR Interferometry On Very Long Time Scale: A Study of the Interferometric Characteristics Of Man-Made Features, *IEEE Transactions on Geoscience and Remote Sensing*, 37(4):2118– 2123.

Wadge, G., Scheuchl, B. and Stevens, N. F. 1999. Spaceborne radar measurements of the eruption of Soufriere Hills Volcano, Montserrat during 1996-99, *J. Volcanology and Geothermal Research*.

Wicks, Jr, C., Thatcher, W. and Dzurisin, D. 1998. Migration of Fluids Beneath Yellowstone Caldera Inferred from Satellite Radar Interferometry, *Science*, 282:458– 462.

Williams, C. A. and Wadge, G. 1998. The effects of topography on magma chamber deformation models: Application to Mt. Etna and radar interferometry, *Geophysical Research Letters*, 25(10):1549– 1552.

Zebker, H.A., Rosen, P. A., Goldstein, R. M, Gabriel, A. And Webner, C.L. 1994. On the derivation of coseismic displacement fields using differential radar interferometry: The Landers earthquake. *Journal of Geophysical Research*, 99, B10: 19617-19634.

Zebker, H., Rosen, P., Hensley, S. and Mouginis-Mark, P. 1996. Analysis of active lava flows on Kilauea volcano, Hawaii, using SIR-C radar correlation measurements, *Geology*, 24:495– 498.

In: Interferometry Principles and Applications
Editor: Mark E. Russo

Chapter 7

HIGH CONTRAST SCHLIEREN DIFFRACTION INTERFEROMETRY

Raj Kumar[*]

Central Scientific Instruments Organisation, Chandigarh, India
Institute for Plasma Research, Bhat, Gandhinagar, India

ABSTRACT

Schlieren techniques are among the simplest and oldest known optical methods for visualizing refractive index gradients in transparent media. Conventional schlieren methods are generally used to obtain first hand qualitative information about the test field. To obtain quantitative information these techniques are transformed into interferometers.

In schlieren diffraction interferometry, the position of schlieren diaphragm/ diffracting element is adjusted in such a way that it diffracts a part of the incident unperturbed geometrical light. This diaphragm diffracted light serves as reference beam while other part of geometrical light modulated with test media serves as object beam.

Interference of these beams generates the schlieren interferogram which could be used to get required information about the test media. In this chapter, methods for enhancing contrast and sensitivity of schlieren diffraction interferometer are described. A comparative study of various schlieren diffracting elements is presented.

Since schlieren, shadowgraphy and interferometry provides information related to different aspects of the test media, thus, a combined system, using holographic optical elements, is described which could simultaneously provide information related to systems involving a wide range of index or density gradients. This combined system is particularly useful for studying the highly transient phenomena such as plasma. Applications of these methods in conventional and new emerging fields are discussed.

---

[*] E-mail: raj_csio@yahoo.com.

1. INTRODUCTION

Optical methods, which are non-invasive, non-contact, highly accurate and sensitive as well as suitable for parallel processing instead of point-to-point testing, are finding newer applications in various established and emerging fields of science and technology [1]. These methods can be broadly classified in accordance to four basic principles: thermal radiation, light scattering from tracer particles, interaction of fluid flow with a solid surface, and methods relying on refractive index changes in the test field. The development of optical interferometry [2], which falls in the last category, has led to an extraordinary advancement in metrology. Conventionally used interferometric schemes, such as Mach–Zehnder interferometer, provide excellent results on phase objects, e.g. flow visualization. The main disadvantage related to these interferometric schemes is the requirement of a large number of good quality optical elements and spatially separated arms (object beam and reference beam) which make the interferometer more costly and prone to mechanical vibrations as well as environmental fluctuations related to temperature variations and pressure changes. In order to minimize these problems, various single beam interferometric schemes have been reported [3]. Among these, schlieren systems [4, 5] are the oldest and simplest known methods for making density gradients visible in transparent media. The German word "schliere" means "streak" or "flaw" and it refers to optical distortions caused by inhomogeneous refractive index in the otherwise homogeneous test fields. These in-homogeneities remain invisible to human eye and system used to make these visible is known as schlieren system. Schlieren techniques rely on the principle of deflection of light beam from in-homogeneities during its passage through the test media/field. Index gradients in the test field cause the incoming light rays to undergo angular deviations. This deflection of light beam from its original path due to interaction with test field provides information about the spatial gradients of refractive index in the test field, integrated along the optical path. Depending upon the type and shape of the schlieren element used, these angular deviations could be subsequently visualized as gradients in irradiance, color, fringe position, contrast, or a combination of these. Robert Hooke was the first to demonstrate schlieren effect before the Royal Society of London in 1672, where he showed in an experiment that besides the flame and some smoke of a candle there is a continual stream rising up from it, distinct from the air. However, Hooke's work was forgotten and Leon Foucault and August Toepler are generally regarded as the discoverers of this group of techniques. During the course of time a number of variations of the technique have been developed for specific and general purposes [4-7]. These techniques are widely used for performing various types of test studies on transparent objects such as depicting deviations in light beams induced by density, temperature or refractive index gradients in optical shop testing, combustion research, laminar and turbulent fluid flow, shock and detonation waves, plasma diagnostics, acoustic studies and other steep refractive index gradients associated with heat and mass transfer, or pressure changes, but not confined to these. As an instrument, a schlieren apparatus is suitable to measure the slope of index gradients and is sensitive to transverse refractive index gradients in the test field. The relative magnitude of index gradients is estimated by observing the intensity patterns generated in the observation plane. The first area to become dark has the positive gradients, followed by the flat areas, and finally the negative gradients. Various possible configurations like systems using single lens/mirror or two lenses/mirrors and requirement of opto-mechanical components for realization of these systems have been well documented in the literature [4, 5]. It may be noted that schlieren configurations using single lens/mirror are suitable for qualitative observations only but for quantitative analysis a collimated beam of light is required to traverse the test field instead of a diverging or converging beam. For this reason either in-line configurations using two lenses or Z-type configurations using two mirrors are generally used in schlieren systems instead of a single mirror/lens configuration. The deflection of light beam at the schlieren element from its original path due to interaction with index gradients of the test media, in case of two lenses/mirrors system, is $x = \delta f$, where δ is angle of deflection due to inhomogeneity in the test field and f is focal length of the focusing lens/mirror. This deviation is independent on the position of inhomogeneity within the parallel/collimated beam of light between the lenses/mirrors.

A conventional schlieren system in two lens configuration is schematically shown in Figure 1. It consists of a point or slit light source S, a collimating lens L_1, and another lens L_2 which focuses this collimated beam on the schlieren element, for example, knife-edge. The test object, to be investigated, is inserted in the collimated beam between lenses L_1 and L_2. The schlieren element placed in the focal plane of second lens/mirror (source image plane) is used to block/cutoff the source image at various positions. The light deflected from the test field either misses the knife-edge or is blocked by it thereby producing change in illumination at observation plane. The shape of this shadow-light pattern is widely used in optical shop testing to get first hand information about the optics aberrations. For the case of air or other gases, the change in optical path length may be expressed in terms of gas density ρ by Dale-Gladstone law $n - 1 = k\rho$ where n the index of refraction and k is a constant known as Dale-Gladstone constant.

Conventional schlieren methods are generally used to obtain first hand qualitative information about the test field. To obtain quantitative information these techniques are transformed into interferometers [4, 8, 9]. In schlieren interferometry a single beam of light is divided into two beams after passing through the test media and further superposition of these beams results in the schlieren interferogram. Requirement of a single beam makes schlieren interferometry not only simple to use and cost effective but also offer benefits like rendering it comparatively insensitive to environmental effects such as mechanical vibrations, thermal fluctuations etc. in comparison with conventional interferometers. One of the schlieren system based interferometric scheme known as schlieren diffraction interferometer [4] is described in next section.

2. SCHLIEREN DIFFRACTION INTERFEROMETER

It is known that diffraction is an integral part of the schlieren image-formation process. Following are the two main diffraction effects observed in schlieren systems [4, 5]. One is diffraction of direct light (source image) at the schlieren element, which limits the sensitivity and resolution of schlieren systems. The second type is the diffraction of light, deflected from the test object at the schlieren diffracting element. This second type of diffraction degrades the quality of schlieren results. Apart from these diffraction effects, diffraction also takes place at object edges in the test area, creating halos in the image. Due to this type of diffraction, dust particles on the optical elements and their in-homogeneities as well as sharp edges of the objects become visible, when the direct beam is blocked. These diffraction halos are not a very serious problem for schlieren analysis because these just outline object edges with thin white lines/fringes and usually do not obscure the schlieren image [5]. Gayhert and Prescost [10] reported the first schlieren based interferometric scheme and demonstrated that the diffraction of incident light at schlieren element could effectively be used for quantitative measurements. They adjusted the width of the source slit to several microns and observed the formation of interference bands in the schlieren images of the convective flows around a heated plate and a heated pipe. The narrowing of source image resulted in diffraction of incident beam and thereby generating a fringe pattern which is used for quantitative analysis of phase objects. In this configuration the geometrical light which is modulated by the test object but passes to the observation plane without being diffracted by the schlieren element serves as the object beam while the diffracted light which diverges out from the schlieren element serves as the built-in reference beam. Superposition of these two beams generates the interference pattern which could be used for quantitative measurements on phase objects. Temple [11] presented a detailed theoretical and experimental analysis of the schlieren diffraction interferometer using the theory of Fourier optics.

In the conventionally used Fourier optics approach to explain schlieren process, the focal plane filter, also known as schlieren stop or diffracting element modifies in one or other way the spatial frequency spectrum of the test field, letting zero order frequency undisturbed (either blocked or allowed to pass). It is well known that the zero order spatial frequency of the input plane information corresponds to the central disk of the Airy pattern, known as Airy disk that accommodates around 84% of the incident light. In conventional schlieren diffraction interferometry, explained by Temple [11] the Airy disk is blocked with schlieren diffracting element and interference takes place between a strong object beam (light deflected/ refracted from the test object) and the weak light diffracted from schlieren element, serving as reference beam. Upon recombination, these two beams interfere with each other, and one can observe a distribution in intensity that is related to the phase shift which has occurred in the object beam while passing through the test field. Though it proved a simple and cost effect interferometer for getting quantitative information but its applications remain limited due to low contrast of the interferogram because of the difference in amplitudes of the interfering strong object beam with a weak reference beam.

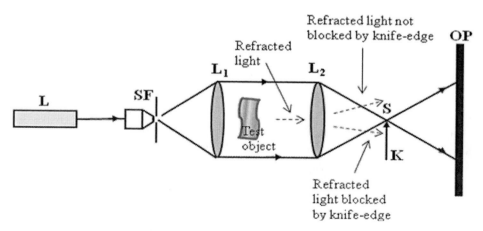

Figure 1. Schematic representation of a schlieren setup.

3. EFFECT OF AIRY DISK ON THE CONTRAST OF SCHLIEREN DIFFRACTION INTERFEROMETER

As we have seen above that in conventional schlieren diffraction interferometry the schlieren element (knife-edge, wire, etc.) is used to block the Airy disk, which contains about 84% of the incident light. This procedure is being practiced because of the fear that the strong background light from the Airy disk could result in weak contrast of the schlieren interferogram and it could also mask phase information of the test object. It is also considered that diffraction of geometrical light at the schlieren element will add extra noise in the results thereby making the analysis more difficult and inaccurate. To avoid these diffraction effects and to obtain optimum amount of information a criterion has also been reported for the size and location of obstacles/schlieren elements in the focal plane of a schlieren diffraction interferometer [11, 12]. This configuration of schlieren diffraction interferometer (Airy disk blocked) was used for a long time but it produces low contrast interferograms due to unequal amplitudes of the interfering beams. In an effort to improve contrast in schlieren diffraction interferogram experiments on the diffraction of light from the knife-edge were revisited [13]. It was found that a conventional knife-edge diffraction pattern formed by a converging or diverging beam incident on the knife-edge can be manipulated in such a manner that a single diffraction fringe covers the whole field of view. In other words an infinite fringe mode knife-edge diffraction pattern is obtained in the observation plane (OP). In this infinite fringe mode condition knife-edge diffraction pattern acts as a two beam interferometer similar to that of schlieren diffraction interferometer and could be used for performing test studies on phase objects. Here the knife-edge diffracted wave serves as a reference beam while the geometrical wave passing undisturbed by the knife-edge serves as object beam. It is known that an infinite fringe mode condition is achieved only when both the interfering beams travel collinear i.e. the two sources of light (focus and illuminated part of knife-edge) coincide with each other. Thus in this configuration knife-edge diffracts light from Airy disk instead of blocking it. The amplitude of knife-edge diffracted light at the point of observation P can be expressed using the boundary wave diffraction theory [14],

$$U(P) = U^g(P) + U^d(P) \tag{1}$$

Here U^g propagates according to the laws of geometrical optics and is known as the geometrical wave while U^d is generated from every point of the illuminated boundary of the knife-edge and is called the boundary diffraction wave. Field distributions of these waves can be represented as:

$$U^g(P) = \frac{A\exp(jkR)}{R} \quad \text{when P is in the direct beam}$$
$$= 0 \quad \text{when P is in the geometrically shadow region} \tag{2}$$

and

$$U^d(P) = \frac{A}{4\pi} \int_\Gamma \frac{\exp(jkr)}{r} \frac{\exp(jks)}{s} \frac{\cos(\vec{n},\vec{s})}{[1+\cos(\vec{s},\vec{r})]} \sin(\vec{r},\vec{dl})dl \qquad (3)$$

where R is the distance from source S (focus in our case) to the point of observation P; s is the distance between a typical point on the knife-edge K and P; Γ denotes the boundary of illuminated part of K; dl is an infinitesimal element situated on Γ; $\vec{n}$ is unit vector perpendicular to the edge of the aperture and to the incident light, $\vec{s}$ is unit vector from observation point P towards the point Q on Γ; $\vec{r}$ is unit vector in the direction of incident light and j = √-1. The geometrical representation of diffraction of light from the knife-edge according to boundary diffraction wave theory is schematically shown in Figure 2. The first factor in the integrand, exp(jkr)/r, of Eq. (3) represents the amplitude and phase of the wave incident on the knife-edge, the second factor, exp(jks)/s, corresponds to the amplitude and phase of the elementary spherical boundary diffraction wave starting from a point on the aperture edge Γ at the observation point P and the third factor, cos($\vec{n}$, $\vec{s}$) sin($\vec{r}$, $\vec{dl}$)/[1+ cos ($\vec{s}$, $\vec{r}$)], determines the angular dependence of the boundary diffraction wave. Equation (3) shows that amplitude of boundary diffraction wave is directly proportional to the amplitude of geometrical light received by the knife-edge for diffraction. Thus, its value will be maximum when knife-edge diffracts light from the Airy disk. The experimental results showing conventional finite fringe mode knife-edge diffraction pattern, an infinite fringe mode knife-edge diffraction pattern and test results on an optical glass plate in this infinite fringe mode knife-edge diffraction pattern are shown in Figure 3. These results show that in the infinite fringe mode condition the knife-edge diffraction pattern could effectively be used as a schlieren diffraction interferometer.

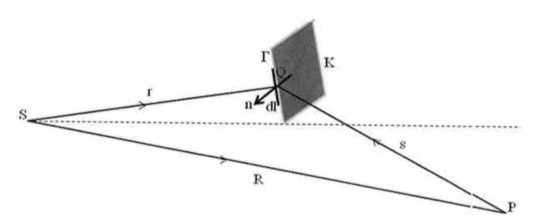

Figure 2. Geometrical representation of knife-edge diffraction according to Maggi-Rubinowicz boundary diffraction wave theory.

In Figure 3 (d) results are shown on the same glass plate but it has been rotated by 180° along the vertical axis so that the light refracted by the glass plate falls on the knife-edge and hence is blocked by it thereby no information about optical path length variations in the glass plate can be obtained corresponding to these index gradients.

These results [Figure 3 (c) and Figure 3 (d)] show that using knife-edge as schlieren diffracting element only information related to index gradients which deflect light in a direction away from the knife-edge can be obtained.

In order to demonstrate experimentally that contrast of schlieren diffraction interferogram becomes maximum when schlieren element diffracts light from the Airy disk instead of blocking it, the position of knife-edge was adjusted corresponding to these two situations. These situations are schematically shown in Figure 4; where in position I knife-edge diffracts light from the Airy disk whereas in position II it has blocked the Airy disk.

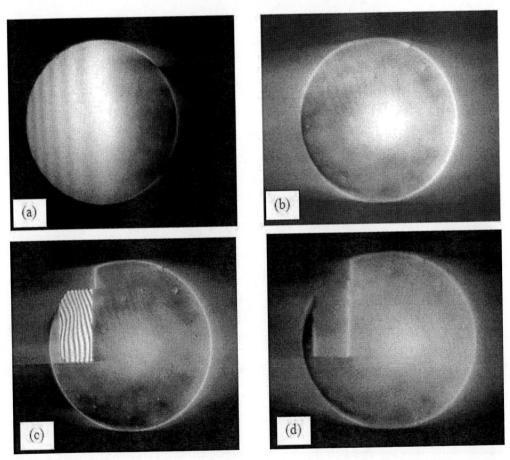

Figure 3. Experimental results on knife-edge diffraction pattern, (a) conventional finite fringe mode diffraction pattern, (b) infinite fringe mode diffraction pattern, (c) test results on an optical glass plate in infinite fringe mode diffraction pattern, and (d) results on the same glass plate, rotated by 180° about its vertical axis so that light deflected from the glass plate is blocked by the knife-edge and hence no information is available about phase/refractive index variations.

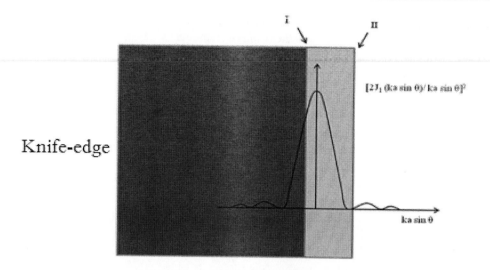

Figure 4. Schematic representation of two positions of knife-edge in the Airy pattern, in position I knife-edge diffracts light from the Airy disk, and in position II knife-edge has blocked the Airy disk.

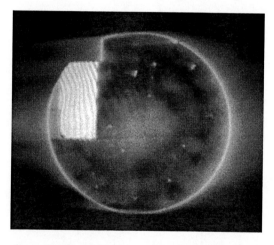

Figure 5. Experimental results on the same glass plate when Airy disk is blocked with the knife-edge.

Experimental results on an optical glass plate corresponding to position I and position II of the knife-edge (a good quality razor blade) of Figure 4 are shown in Figure 3 (c) and Figure 5 respectively. A comparison of these results clearly shows an increase in contrast of schlieren diffraction interferogram when schlieren element diffracts light from the Airy disk instead of when schlieren element has blocked it.

To demonstrate the effect of the amplitude of incident light on the contrast of schlieren diffraction interferogram, an experiment was performed where light from different parts of the geometrical beam was blocked using a screen in the collimated beam between lenses L_1 and L_2 as reported in reference [13]. This blocking of geometrical light shows decrease in contrast of schlieren diffraction interferogram due to corresponding decrease in amplitude of boundary diffraction wave from the knife-edge. Thus instead of blocking the Airy disk, diffracting light from it enhances contrast of the interferogram. To demonstrate that in infinite fringe mode condition knife-edge indeed diffracts light from the Airy disk, an another experiment was performed in which through a series of experimental results it is shown that the increase in contrast of schlieren interferogram is due to an increase in amplitude of the boundary diffraction wave and ultimately contrast becomes maximum when schlieren element diffracts light from the Airy disk [15]. To observe the change in amplitude of diffracted light, with respect to position of schlieren element in the Airy pattern, the system was first adjusted in the conventional schlieren position (Airy disk blocked, resulting in a dark fringe at observation plane) and than F-number of the optical system was increased. This results finally in a situation where diffracted light becomes quite strong giving again bright fringe in observation plane. In this situation the amplitude of boundary diffraction wave becomes maximum and it can easily be seen outside the geometrically illuminated region, demonstrating that at this position schlieren element diffracts light from the Airy disk.

It is known that the sensitivity $S \sim f/w$ of schlieren diffraction interferometer [5] is directly dependent on focal length 'f' and inversely proportional to unobstructed width 'w' of the source image. The ultimate lower limit on 'w' is reached when schlieren element diffracts light from the Airy disk, where due to complex nature of focus/caustic it becomes immaterial that from which part of the Airy disk light is being diffracted. Thus diffraction of light from Airy disk also improves sensitivity of schlieren diffraction interferometer as compared to the case when Airy disk is blocked. For large F-number the system becomes more sensitive but if lens aperture is small then light refracted from the test medium at larger angles will miss the focusing lens and information related to these index gradients will be lost. Thus a trade-off is required between focal length and the lens aperture, i.e. an optimum F-number should be chosen to accommodate the desired information. The lenses used for schlieren diffraction interferometry must be aberration free otherwise lens aberrations will modulate the test object information, and information retrieved will not be correct. Effect of various types of aberrations in schlieren optics on the information retrieved has been discussed in reference [16]. It may be noted that this schlieren diffraction interferometer is suitable for the case when test object does not cover the whole width of the collimated geometrical beam. For the case when test object covers whole width of the collimated beam, due to refraction effects the position of the focused spot (Airy disk) will be shifted from its original position and diffracting aperture could not receive any light for diffraction and hence no reference beam will be available for interferometric purposes. In this case system will provide information analogous to that which is retrievable with conventional schlieren system.

4. CONTRAST ENHANCEMENT BY USING MIRROR-EDGE AS SCHLIEREN ELEMENT

In previous section we saw that how effective positioning of schlieren diffracting element in the Airy pattern could improve the contrast of schlieren diffraction interferometer. But still there is possibility of further improvement in contrast because here also amplitude of boundary diffraction wave, which serves as reference beam in the schlieren diffraction interferometer, is much less than the amplitude of geometrical object beam.

Since contrast of a two-beam interferometer is represented by

$$V = \frac{I_{max} - I_{min}}{I_{max} + I_{min}}$$

$$= \frac{2a_1 a_2}{a_1^2 + a_2^2} \quad (4)$$

where I_{max} and I_{min} are the intensities of light waves in the interference maxima (constructive interference) and interference minima (destructive interference) respectively of the interference pattern and a_1 and a_2 are amplitudes of the two interfering beams. Thus, contrast, also known as visibility of the interference pattern, will be maximum with a value of one when the amplitudes of the interfering beams (reference beam and object beam) are equal i.e. $a_1 = a_2$. Thus, in order to further enhance contrast in schlieren diffraction interferometer it is required to increase the amplitude of boundary diffraction wave. One method is to use a properly fabricated $\lambda/2$ phase plate [4]. To fabricate this plate refined technology is required and the phase plate is wavelength dependent also. Other method to increase the amplitude of boundary diffraction wave is use of mirror-edge as schlieren diffracting element [17] as shown schematically in Figure 6.

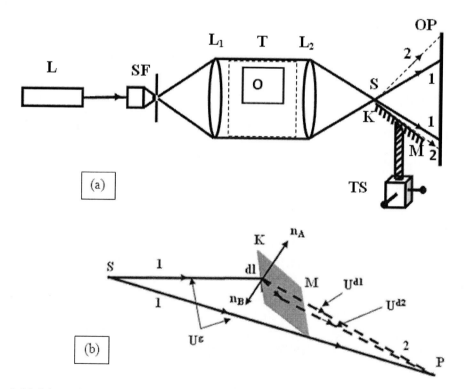

Figure 6. (a) Schematic representation of mirror-edge as schlieren diffracting element; 1 represents geometrical waves and 2 represent the boundary diffraction wave and in (b) diffraction process at mirror-edge is shown for clarity.

In this configuration a Lloyd's mirror (20 mm × 40 mm × 2 mm, SiO2-protected, front surface silver-coated, reflectivity ~ 94%) is adjusted in such a way that conventional finite fringe mode Lloyd's mirror interferogram approaches the infinite fringe mode condition. In this situation a portion of the geometrical light starts striking the leading edge of the Lloyd's mirror and gets diffracted there. This condition is generally avoided in conventional Lloyd's mirror interferometer [18] because this diffracted light spoils/distorts the test results. In our case this mirror-edge diffracted light (boundary diffraction wave) has been effectively used as a stronger reference beam for the realization of schlieren diffraction interferometer. The process of increasing the amplitude of boundary diffraction wave, serving as reference beam in the schlieren diffraction interferometer, can be easily visualized with experiments performed on the knife-edge diffracted wavefront using a Lloyd's mirror in the newly reported diffraction Lloyd's mirror interferometer [19]. These experiments show that two boundary diffraction waves (one starting from the knife-edge and another from virtual image of the knife-edge formed by the Lloyd's mirror) could interfere to generate the interference pattern. Further, fringe width of these fringes can be varied by varying distance between the knife-edge and the Lloyd's mirror, and one could obtain an infinite fringe mode situation when distance between two sources (knife-edge and its mirror image) approaches zero. The concept of infinite fringe mode condition for interference fringes formed by superposition of two boundary diffraction waves can physically be realized using the mirror-edge as diffracting element. In this case when incident light is diffracted from the leading edge of the mirror, a portion of this diffracted light is reflected back from the front surface of the same mirror. In this system two sources of boundary diffraction waves (mirror-edge and its image formed by reflecting surface of the mirror) coincide with each other and thus one gets an infinite fringe mode condition. Using boundary diffraction wave theory, the total field in the observation plane will be,

$$U(P) = U^{g1}(P) + U^{g2}(P) + U^{d}(P), \tag{5}$$

where

$$U^{g1}(P) = \frac{A\exp(jkR)}{R} \quad \text{when } P \text{ is in the direct beam};$$
$$= 0 \quad \text{when } P \text{ is in geometrically shadow region} \tag{6}$$

is the geometrical wave directly reaching to the observation plane, and $U^{g2}(P)$ is the geometrical wave reflected off the mirror surface and then reaching the observation plane. Furthermore,

$$U^{d}(P) = U^{d1}(P) + U^{d2}(P)$$

$$= \frac{A}{4\pi}[1-\exp(jk\Delta)]\int_{\Sigma} \frac{e^{jk(r+s)}}{rs} \frac{\cos(\vec{n}_A,\vec{s})}{[1+\cos(\vec{r},\vec{s})]} \sin(\vec{r},\vec{dl})dl \tag{7}$$

where

$$U^{d1}(P) = \frac{A}{4\pi}\int_{\Sigma} \frac{e^{jk(r+s)}}{rs} \frac{\cos(\vec{n}_A,\vec{s})}{[1+\cos(\vec{r},\vec{s})]}\sin(\vec{r},\vec{dl})dl\,;$$

$$U^{d2}(P) = \frac{A}{4\pi}\int_{\Sigma} \frac{e^{jk(r+s+\Delta)}}{rs} \frac{\cos(\vec{n}_B,\vec{s})}{[1+\cos(\vec{r},\vec{s})]}\sin(\vec{r},\vec{dl})dl$$

and

$$\cos(\vec{n}_A,\vec{s}) = -\cos(\vec{n}_B,\vec{s})\,. \tag{8}$$

Here $U^d(P)$ is the total amplitude of boundary diffraction wave starting from the mirror-edge K and reaching the observation plane; U^{d1} is mirror-edge diffracted wave directly proceeding towards the observation plane; U^{d2} is mirror-edge diffracted wave which is reflected from the mirror surface and then reaches to the observation plane and Δ is additional optical path length introduced in boundary diffraction wave U^{d2} corresponding to a phase change due to reflection from the mirror surface. This is clear from Eq. (7) that total amplitude of boundary diffraction wave depends upon the phase difference between two interfering boundary diffraction waves. The amplitude of boundary diffraction wave $U^d(P)$ will become maximum (twice as compared to that from the solid knife-edge) whenever the phase introduced in the reflected wave is $(2n + 1)\pi$ (bright fringe) and its value becomes zero corresponding to a phase change of $2n\pi$ (dark fringe), where n denotes an integer. In this infinite fringe mode condition, two boundary diffraction waves U^{d1} and U^{d2} combine to give total boundary diffraction wave U^d. This total diffracted field then interferes with geometrical wave U^g giving total field of schlieren diffraction interferometer at the observation plane represented by Eq. (5). The intensity distribution at the observation point corresponding to Eq. (5) is given by

$$I_2 = U(P)\,U(P)^* = |U^{g1}(P) + U^{g2}(P) + U^d(P)|^2 \tag{9}$$

Equation (9) shows that the presence of boundary diffraction wave modifies the intensity distribution of the conventional Lloyd mirror interferometer. By giving proper translation and tilt to the Lloyd's mirror its position can be adjusted such that its leading edge diffracts light from the Airy disk and its surface reflects only the boundary diffraction wave to reinforce total amplitude of the boundary diffraction wave but it does not folds/reflects any portion of the geometrical wave. This process of diffraction at mirror-edge and realization of folding mirror diffraction interferometer is described in detail in reference [20]. Dropping the term corresponding to geometrical folded beam, Eq. (9) gives

$$I_2(P) = |U^{g1}(P) + U^d(P)|^2 \tag{10}$$

which represents intensity distribution in the interferogram due to schlieren diffraction interferometer. It may be noted that at grazing incidence a phase change of π radians gets introduced in the beam reflected off the mirror surface automatically thereby doubling the amplitude of boundary diffraction wave as compared to that available with the solid knife-edge and thus giving higher fringe contrast. Experimental results obtained by using mirror-edge as schlieren diffracting element are shown in Figure 7, where contrast of interference fringes formed due to optical path length variation in the object beam because of the test object is better as compared to that results presented in Fig 3 and Fig 5 obtained with a solid knife-edge for two different situations.

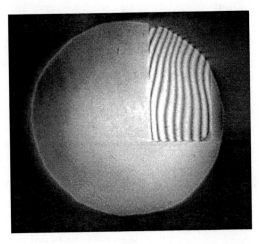

Figure 7. Experimental results on the glass plate using mirror-edge as schlieren diffracting element.

It is known that the amplitude of boundary diffraction wave is maximum near the geometrically illuminated to geometrically shadowed transition boundary where its value is approximately equal to half of the incident light [21]. Thus in the infinite fringe mode condition where a diffraction fringe near the geometrical shadow boundary is broadened to cover the field of view, the amplitude of boundary diffraction wave can be considered uniform over the fringe and its value may be approximated to that near the geometrically shadow boundary which is $\sim U_g/2$. In case of infinite fringe mode condition obtained with mirror-edge diffraction total amplitude of the boundary diffraction wave for constructive interference will be $U_g/2 + U_g/2 = U_g$ i.e. its value becomes equal to that of amplitude of geometrical wave. Thus interferogram obtained by using mirror-edge as schlieren diffracting element could have contrast equal to unity which is similar to what one would get using a well designed and properly fabricated $\lambda/2$ phase knife-edge [4, 22] or a good contrast two beam interferometer like conventional Lloyd's mirror interferometer. Use of mirror as a diffracting element in schlieren diffraction interferometer has twofold advantages over the conventional Lloyd's mirror interferometer. First, it provides a larger test path, and second, it become suitable to carry out thermal studies and flow visualization, which was not easy to be performed in the conventional Lloyd's mirror configuration due to proximity of test and reference beams.

5. COMPARISON OF VARIOUS SCHLIEREN DIFFRACTING ELEMENTS

Choice of proper schlieren diffracting element is an important characteristic of schlieren diffracting element for retrieving the required information about the test field/test object as the amount of information available using different diffracting elements is not the same. A conventional solid knife-edge is easy to get (a simple razor blade works well) but it blocks half of the information corresponding to light refracted towards the knife, as can be seen in the results shown in Figure 3(d). Also a portion of deflected light which falls on the knife-edge gets diffracted there, thereby reducing the quality of the interferogram. Thus knife-edge is suitable to provide information related only to transverse refractive index or density variations which deflect light away from the knife. In this case contrast of the schlieren interferogram is poor due to difference in amplitude of boundary diffraction wave (reference beam) and the geometrical wave (object beam). The contrast can be improved using a mirror-edge as an alternative of the knife-edge but amount of information retrievable still remains same. A phase knife-edge lets the both side deflected light to pass to the observation plane and for a proper selection of wavelength and thickness of the phase element the contrast in schlieren interferogram could become twice to that of solid knife-edge [4]. Experimental results showing information obtainable with a phase knife-edge for both side deflected light are shown in Figure 8. It may be noted that in case of a properly fabricated phase knife-edge, though, the contrast becomes twice to that obtainable with solid knife-edge but light passing through the phase plate gets an additional phase shift as compared to the light passing over it (free space). Here also only information related to index or density gradients transverse to the phase knife-edge can be retrieved. Extra efforts are also required to fabricate the properly finished phase knife-edge. As an alternative to phase knife-edge a thin wire can also be used as schlieren diffracting element to get information related to two side deflected light but use of wire results in loss of a portion of the information related to low spatial frequencies which are blocked by the wire.

Thus a wire of diameter less than Airy disk is required to get high contrast interferogram providing information related to both side deflecting transverse index gradients in the test field. Use of wire or phase knife-edge as schlieren diffracting element has advantage over the solid knife-edge, in that it provides more information about test object.

Additionally, phase knife-edge has an advantage over a wire as schlieren element in providing fuller information as it does not block any light even as that deflected from weak optical in-homogeneities. As discussed in section 2, that if light deflected from the gradients of the test field strikes the schlieren element it will gets diffracted there and could spoil the results.

In order to show the effect of diffraction of light deflected from the test field, experimental results on a burning candle using tip of a thin optical fiber as schlieren diffracting element are shown in Figure 9. These results show that a portion of light deflected from the candle flame and striking on the optical fiber gets diffracted there while information related to light deflected in other three directions become available in the schlieren interferogram. It may be seen in Figure 9 that diffraction of test field deflected light at tip of optical fiber has added extra noise in the results and information related to this deflected-diffracted light is propagating parallel to the shadow of the candle up to deep inside the geometrically shadow region. The problem related to diffraction of test field deflected light at the schlieren element remains in almost all the schlieren diffracting elements as described in reference [23].

In that work the amount of information retrievable using different schlieren diffracting elements including phase knife-edge, square phase aperture and tip of an optical fiber was experimentally studied with reference to information available (not blocked by the schlieren element) and diffraction of test field deflected light at the schlieren element. An optical glass plate having refractive index gradients in a particular direction was chosen and information related to light deflected into different directions using these schlieren diffracting elements was studied by rotating the glass plate in different directions.

A comparison of information retrievable using various schlieren diffracting elements shows that a properly fabricated tip of a thin optical fiber as schlieren diffracting element could provide more information as compared to the solid knife-edge, wire, phase knife-edge or corner of a square phase aperture because in this case only light deflected towards the optical fiber and striking it is diffracted from the schlieren diffracting element thereby providing information related to the gradients deflecting light in other three directions. However, it may be noted that use of solid knife-edge, phase knife-edge or a wire as schlieren diffracting element provides information about transverse phase variations only. The two-dimensional information may be obtained if the schlieren diffracting element also has two dimensions [11] i.e. either two crossed-wires, one oriented in horizontal direction and the other has its orientation in vertical direction or a small square phase aperture. Our investigations show that the use of corner of a square phase aperture as schlieren diffracting element could provide two-dimensional information of the test object but the diffraction of light deflected from test object at the schlieren diffracting element noticeably affects the test results [23], same will also occur when using two crossed-wires or square phase apertures as schlieren diffracting elements unless diffracting aperture has dimensions small compared to the deflection of light produced by the low order spatial frequencies in the test field.

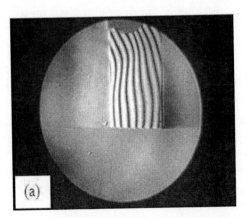 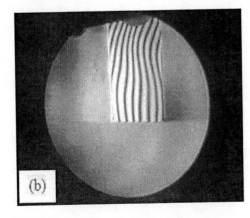

Figure 8. Experimental results on the glass plate using a $\lambda/2$ phase plate as schlieren diffracting element. Here information become available related to light deflected in both the directions i.e. away from knife-edge (a) and towards the knife-edge (b).

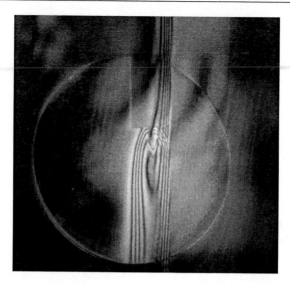

Figure 9. Experimental results on a burning candle using tip of an optical fiber as schlieren diffracting element. Here effects of diffraction of the light deflected from the flame and falling on tip of optical fiber become obvious.

In order to avoid diffraction of test object deflected light at the schlieren element and to obtain complete information of the test field one has to use a schlieren element which has dimensions less than the diameter of Airy disk. In this configuration light deflected from the test field in any direction will pass to the observation plane without going diffracted from the schlieren element. Such a system has been elegantly demonstrated in the well known point diffraction interferometer [24] which can be considered as an extension of the work of Brackenridge and Gilbert [25] who suggested use of a circular disk for getting two-dimensional information. For further enhancing contrast of the schlieren interferogram a clear hole, small compared with the size of Airy disk, is done in an absorbing layer on a clear substrate. Alternatively an opaque disk of similar size can also be used in place of the clear aperture. In this case a diffracted spherical wave interferes with the object wave (modified by test field) that has passed through the absorbing layer. Transmittance of the absorbing layer has to be chosen such that amplitudes of both the interfering beams become almost equal to get maximum contrast in the interferogram. Since the size of Airy disk corresponds to numerical aperture at which the interferometer is being used and the size of diffracting aperture (hole or disk) must be small than Airy disk so that it acts as an unresolved point that diffracts light as a uniform spherical wave. Thus, diffracting apertures of different sizes are required to suite different numerical apertures. Also selection of plates is required, to suite different numerical apertures, and, for each, a range of transmittances to suits different tilts. Point diffraction interferometer has benefits over mirror-edge in terms that it provide full information of the test objects but still mirror-edge could be used to get useful information related to axis-symmetric systems without difficulties related to different numerical aperture of the system.

6. COMBINED SCHLIEREN, SHADOWGRAPH AND INTERFEROMETRIC SYSTEM

Schlieren, shadowgraph and interferometric methods are able to perform non-contact and non invasive diagnostics on various transparent objects. These techniques have their own advantages and shortcomings and their ranges of providing information are also different. Interferometry is highly sensitive, accurate and it relates phase variations linearly to optical path length variations but it does not provide accurate information where index gradients are high, which produces path-length variations greater than that the wavelength of light used. Schlieren techniques can be used to get information related to high index gradients.

Schlieren technique provides information related to first order derivative of the index gradients. If index gradients are still high, than shadowgraph methods are useful that provide information related to second order derivative of the index gradients. Interferometry is used for obtaining quantitative information while schlieren and shadowgraph methods are generally used to get the first hand qualitative information about the test field.

In problems of physical interest generally it is not possible to have a single kind of density or index gradients and the system involve a range of index gradients. In such systems it is not possible to get complete information using a single optical method and thus it become desirable to use different methods to retrieve complete and accurate information of the test field. Choice of a particular system depends upon the nature of test field and information required along with other parameters like source of light, space availability, etc. Because of the complementarity nature of schlieren, shadowgraph and interferometric methods and the overlap of their ranges of usefulness, combination of these methods is particularly valuable for situations involving large range of refractive index gradients. For the case of static objects having more than one type of index gradients, information corresponding to different optical methods can be retrieved sequentially using the same single beam schlieren system by proper alignment of the schlieren element in the Airy pattern. But for the case of transient phenomena, where characteristics of test field changes with time, for example gases and plasmas, simultaneous information instead of sequential information retrievable with different techniques become necessary. Thus a system which can effectively combine more than one or all of these techniques could overcome inherent shortcomings of individual methods. Combined systems involving a large number of conventional optics with different optical arrangements have been reported in the literature [26-28].

An alternative and simple method for combining these techniques in a single system is use of the concept of holography [29, 30]. A properly recorded and reconstructed holographic optical element could provide the same information that is obtainable with conventional optics and additionally fabrication of HOE's is simple and cost effective than conventional optics. Use of HOE's for realization of interferometers makes their alignment easy as compared to conventional optics based interferometers.

Also, other advantages offered by HOE's like light weight, multiple optical functions in a single HOE, etc. make HOE based systems more compact and rigid [29-32]. Using multiplexing property of holographic optical elements (HOE) multiple converging beams can be captured with a collimated reference beam [33].

After chemical processing of the recorded hologram it is illuminated with a collimated beam which reconstructs the recorded converging beams for realization of combined schlieren, shadowgraphic and interferometric system.

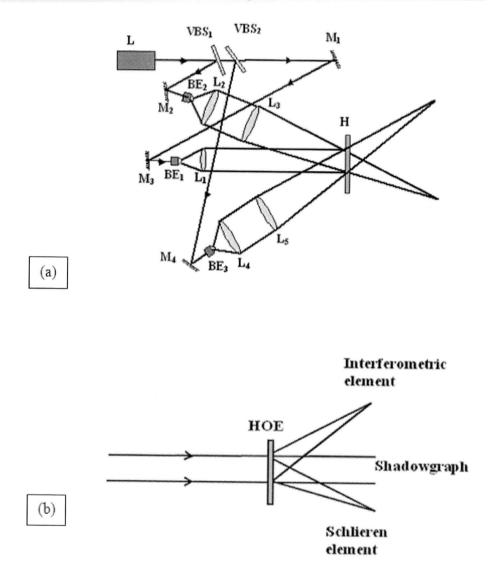

Figure 10. Schematic of experimental arrangement for (a) formation of multiplexed holographic optical element and (b) realization of combined systems using this multiplexed HOE.

The experimental arrangement for the formation of HOE enabling combined operation of these techniques is schematically shown in Figure 10 (a). A helium–neon laser (Coherent Inc., USA 35 mW) was used as the coherent light source for recording the HOE and for the realization of the system. Here a collimated reference beam R, generated using a collimating lens L_1 in conjunction with the beam expander assembly BE_1, is combined with two converging beams O_1 and O_2 propagating in different directions, in two separate holographic exposures on the same recording plate H. The convergent beams O_1 and O_2 were created using good quality telescopic set-up of lenses L_2 and L_3 and lenses L_4 and L_5 in conjunction with the beam expander assemblies BE$_2$ and BE$_3$ respectively to generate aberration-free focus points. After chemical processing, the plate H is illuminated with same reference beam R, which reconstructs the converging beams in their respective original directions. The test object is interposed in the beam R and schlieren elements [Figure 10(b)], i.e. knife-edge, positioned at the reconstructed focused spots generate there respective test results while the directly transmitting collimated beam provides the shadowgraphic results. The shear plate interferometric technique was used to ensure the optical quality of the collimated beams and Ronchi method was used for optical correction of the converging beam for astigmatism and coma, which would otherwise be introduced by the off-axis arrangement. Agfa-Gevaert 8E75HD plates were used for hologram recording and standard Kodak D-19 developer and R-9 bleach bath solutions were used for chemical processing. The results presented here have been captured with a Canon S-50 Power Shot digital camera (1024 × 768 pixels) in white-balance settings.

Consider the complex amplitude distributions of object beams and the reference beam as:

$$O_1 = (O_{o1}/r_1) \exp(-jk\vec{n}_1 \cdot \vec{r}_1),$$
$$O_2 = (O_{o2}/r_2) \exp(-jk\vec{n}_2 \cdot \vec{r}_2),$$
$$R = O_r \exp(jk\vec{n}_r \cdot \vec{r}), \tag{11}$$

where $\vec{n}_1$, $\vec{n}_2$ and $\vec{n}_r$ are unit vectors along the direction of propagation of the object beams O_1, O_2 and reference beam O_r respectively; $\vec{r}_1, \vec{r}_2$ and $\vec{r}$ are corresponding displacement vectors; $k = 2\pi/\lambda$, where λ is wavelength of the light used and $j = \sqrt{-1}$. O_{o1}, O_{o2} and O_r are the amplitude distributions of the beams. The amplitude transmittance of the processed H is given by [32]:

$$t_1 \sim |O_1 + R|^2 + |O_2 + R|^2 \tag{12}$$

The complex amplitude of the transmitted field from H, upon illumination with reference beam R, is:

$$U_1 = Rt_1 \sim R|O_1|^2 + R|R|^2 + O_1|R|^2 + O_1{}^*R^2 + R|O_2|^2 + R|R|^2 + O_2|R|^2 + O_2{}^*R^2 \tag{13}$$

We can consider $|R|^2$ to be constant across H, as a plane reference beam R is used for formation and reconstruction/illumination of H. Thus, only third and seventh terms on the right-hand side of Eq. (13) are of interest to us as these represent reconstructed beams O_1 and O_2. Now, if a phase object $S = \exp(j\varphi)$ is inserted in the reconstructing collimated beam R, the beam will be phase modulated accordingly and this phase modulated beam $R' = O_r \exp\{j(k\vec{n}_r \cdot \vec{r} + \varphi)\}$ will illuminate the processed H instead of beam R, giving (keeping only terms of interest)

$$U_1' = R't_1 \sim O_1 R^* R' + O_2 R^* R'$$

$$\sim [(O_{o1}/r_1) \exp(-j(k\vec{n}_1 \cdot \vec{r}_1 - \varphi) + (O_{o2}/r_2) \exp(-j(k\vec{n}_2 \cdot \vec{r}_2 - \varphi)] |R|^2 \qquad (14)$$

It becomes clear from Eq. (14) that reconstruction of the HOE with a phase modulated reference beam could transmit the information of the phase object modulating the reference beam into the reconstructed objects beams. These reconstructed object beams can further be manipulated with different schlieren/interferometric elements to convert their phase variations into intensity variations to carry out optical investigations.

Experimental results obtained on the flame of a burning candle using the holographic optical element instead of conventional optics are shown in Figure 11. Results presented in Figure 11 (a) are obtained in the directly transmitted collimated beam through the HOE and provide shadowgarphic results while results presented in Figure 11 (b) are obtained by using a small aperture on a phase plate i.e. a point diffraction interferometer. Though these results are not taken simultaneously due to availability of only one recording camera but still could provide an insight into the difference in nature of information available by using different techniques namely shadowgraphic and interferometric methods.

By applying multiple techniques, the inherent shortcomings of individual methods can be overcome and the risk of overlooking or misinterpreting certain features about the test field is reduced. The described combined system is suitable for industrial applications also because a single collimated beam incident perpendicularly on the holographic optical element serves as reference as well as test beam, making the system relatively insensitive to external vibrations. Also alignment of the proposed scheme is easy compared to other HOE based interferometers [32] due to the fact that, by observing and properly adjusting the back reflected light from the HOE, its position could easily be aligned. This was tested by removing the HOE from the set-up and again aligning it at different locations. It was observed that if the HOE had some misalignment, then the reconstructed focus spot will not be true and it will not be possible to obtain true information about the test object for precise quantitative measurements, but phase visualization could still be performed.

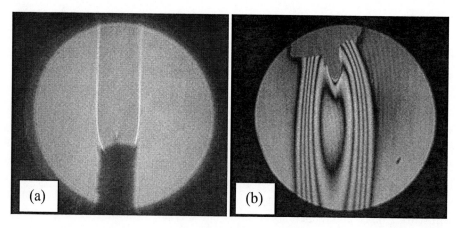

Figure 11. Experimental results on a burning candle with (a) shadowgraphy, and (b) point diffraction interferometer.

The error in retrieved information depends on the amount of misalignment in the repositioning of the HOE. It is known that recording of HOE on silver-halide plates introduces noise in the reconstructed wavefront. This noise could be reduced using dichromatic gelatin (DCG) as the recording material, as the grain size in this case is negligible and DCG gives better efficiency compared to silver-halide recording materials, thereby enhancing amplitude of the reconstructed wavefront.

Further, the only required HOE for realization of the combined schlieren, shadowgraphic and interferometric schemes could be produced in great numbers using the cost-effective hologram copying methods [29].

7. APPLICATIONS

Schlieren methods are frequently used for study of diffusion in liquids, combustion in flames, compressible flow in wind tunnels, electron density in plasmas and other conventional applications like optical shop testing etc. Recently this method has shown to have potential application as a diagnostic tool for the newly emerging fields and is used for real-time optoacoustic wave imaging [34], detection of nano-mechanical displacements [35], study of instabilities in plasma [36], detection of magnetization directions of individual nano-particles [37], etc. Schlieren diffraction interferometry has been used for gas density measurement [11], measurements of the thermal distributions near a heated metallic objects [12], and velocity measurement of charged particles [38] etc. Schlieren diffraction interferometer has been suitably modified for carrying out investigations on the diffraction of light [19]. This modified setup has been used for demonstrating existence of boundary diffraction wave in the 4π regions, its dependence on obliquity factor, and separate existence of boundary diffraction wave than the geometrical wave after the diffraction aperture [39]. Further it has been demonstrated that boundary diffraction wave is continuous in nature and does not have any discontinuity as required by the theory [40]. These results are consistence with other uniform theories of the boundary diffraction wave [41, 42]. Use of schlieren diffraction interferometer has also been reported for enhancing anti-counterfeiting characteristics of security holograms [43] which are widely used for protecting various important documents against forgery and important products against duplication. Investigations on the phenomenon of extraordinary optical transmission through sub-wavelength apertures [44, 45], which has potential applications in diverse areas of scientific and technology importance, has also been reported using this modified schlieren diffraction interferometer [46]. Recent progress in optoelectronics, for example, development of extremely sensitive and high speed cameras with frame-rates of several million images per second has open up new avenues to carryout investigations on highly transient processes with excellent temporal resolution. The combined system, involving more than one optical method, can be used for study of transient phenomena involving a wider range of refractive index or density variations. Other possible applications of schlieren diffraction interferometer include the areas where system under investigations becomes inaccessible for other interferometers or where setup/system vibrations and environmental effects make other interferometers unsuitable/complex, for example, in fusion devices: tokomaks and stellarators.

Conclusion

In this chapter, simple techniques are described to enhance contrast and sensitivity of schlieren diffraction interferometer. Schlieren diffraction interferometer has many advantages over the conventional two-beam interferometers: it uses a single light beam; it requires simple alignment; it is relatively inexpensive and is quite insensitive to environmental effects like vibrations and thermal fluctuations. It is demonstrated experimentally as well as theoretically that contrast of schlieren diffraction interferogram becomes maximum when schlieren element diffracts light from the Airy disk formed by the focusing lens instead of blocking it. Further, use of mirror-edge is proposed as schlieren diffracting element which could effectively double the contrast of schlieren interferogram as compared to the case of conventional solid knife-edge edge. Though use of mirror-edge could not provide complete information of the test field which can be obtained using a point diffraction interferometer but fabrication of mirror is easy as compared to phase plate used in point diffraction interferometer. Also same mirror can be used with different numerical apertures of the setup which is not possible in case of point diffraction interferometer. Use of holographic optical element is proposed for the realization of a combined schlieren, shadowgraph and interferometric setup which could be of great importance for simultaneously retrieving information related to different aspects of the index gradients in highly transient and complex systems like plasmas. It may be noted that the schlieren diffraction interferometer like other interferometric schemes provides integrated information about the whole depth of the measurement volume. In order to get point-to-point information about the test field with high spatial resolution light scattering techniques, for example, Thomson scattering, which is generally used for tokamak plasma diagnostics, can be worthwhile.

Acknowledgments

Author is grateful to Dr. A. K. Aggarwal and Mr. D. P. Chhachhia for fruitful discussion during the course of this work.

References

[1] Mayinger, F.; Feldmann, O. *Optical Measurements Techniques and Applications*; Ed.; Springer: Berlin, 2001.
[2] Hariharan, P. *Optical Interferometry*; Academic Press: CA, 2003.
[3] Malacara, D. *Optical Shop Testing*; Ed.; Wiley: NY, 1992.
[4] Vasiliev, L. A. Schlieren Methods; Baruch, A.; Trans.; Israel Program for Scientific Translations: NY, 1971.
[5] Settles, G. S. Schlieren and Shadowgraph Techniques; Springer: NY, 2001.
[6] Howes, W. L. *Appl. Opt.* 1984, 23, 2449-2460.
[7] Richard, H.; Raffel, M. *Meas. Sci. Technol.* 2001, 12, 1576-1585.
[8] Small, R. D.; Sernas, V. A.; Page, R. H. *Appl. Opt.* 1972, 11, 858-862.
[9] Stricker J.; Rosenblatt, F. *Opt. Lett.* 2003, 28, 1427-1429.

[10] Gayhart, E. L.; Prescott, *R. J. Opt. Soc. Am.* 1949, 39, 546-550.
[11] Temple, E. B. *J. Opt. Soc. Am.* 1957, 47, 91-100.
[12] Brackenridge, J. B.; Peterka, *J. Appl. Opt.* 1967, 6, 731-735.
[13] Kumar, R.; Kaura, S. K.; Sharma, A. K.; Chhachhia, D. P.; Aggarwal, A. K. *Opt. Laser Technol.* 2007, 39, 256-261.
[14] Born, M.; Wolf, E. *Principles of Optics*; Pergamon: Oxford, 1991; pp 370-592.
[15] Kumar, R. *Optik. Int. J. Light Electron Opt.* 2011, 122, 105-109.
[16] Hosch J. W.; Walters, *J. P. Appl. Opt.* 1977, 16, 473- 482.
[17] Kumar, R.; Chhachhia, D. P.; Aggarwal, A. K. *Appl. Opt.* 2006, 45, 6708-6711.
[18] Pollock, N.; *J. Phys. E: Sci. Instrum.* 1980, 13, 1062 - 1066.
[19] Kumar, R. *J. Opt.* (India) 2010, 39, 90-101.
[20] Kumar, R. *Appl. Phys. B - Lasers Opt.* 2008, 93, 415-420.
[21] Rubinowicz, A. *Nature* 1957, 180, 160–162.
[22] Kumar, R.; Mohan, D.; Kaura, S. K.; Chhachhia, D. P.; Aggarwal, A. K. *Pramana J. Phys.* 2007, 68, 581-589.
[23] Kumar, R., Kaura, S. K.; Chhachhia, D. P.; Mohan, D.; Aggarwal, A. K. *Pramana J. Phys.* 2008, 70, 121-129.
[24] Smartt, R. N.; Steel, W. H. *Jpn. J. Appl. Phys.* 1975, 14, 351–356.
[25] Brackenridge, J. B.; Gilbert, W. *P. Appl. Opt.* 1965, 4, 819-821.
[26] Kafri, O.; Kreske, K. *Appl. Opt.* 1988, 23, 4941–4946.
[27] Gregory-Smith, D. G.; Gilchrist, A. R.; Senior, P. *Meas. Sci. Technol.* 1990, 1, 419–424.
[28] Kleine, H.; Gronig, H.; Takayama, K. *Opt. Lasers Eng.* 2006, 44, 170–189.
[29] Caulfield, H. J. *Handbook of Optical Holography*; Ed.; Academic Press: NY, 1979.
[30] Mehta, P. C.; Rampal, V. V. *Lasers and Holography*; World Scientific: Singapore, 1993.
[31] Mohan, N. K.; Islam, Q. T.; Rastogi, P. K. *Opt. Lasers Engg.* 2006, 44, 871–880.
[32] Aggarwal, A. K.; Kaura, S. K.; Chhachhia, D. P.; Sharma, A. K. *Opt. Laser Technol.* 2004, 36, 545–549.
[33] Kumar, R.; Kaura, S. K., Chhachhia, D. P.; Mohan, D.; Aggarwal, A. K. *Curr. Sci.* 2008, 94, 184-188.
[34] Niederhauser, J. J.; Frauchiger, D.; Weber, H. P.; Frenza, M. *Appl. Phys. Lett.* 2002, 81, 571–573.
[35] Karabacak, D.; Kouh, T.; Huang, C. C.; Ekinci, K. L. *Appl. Phys. Lett.* 2006, 88, 193122.
[36] Moore, A. S.; Gumbrell, E. T.; Lazarus, J.; Hohenberger, M.; Robinson, J. S.; Smith, R. A.; Plant, T. J. A.; Symes, D. R.; Dunne, *M. Phys. Rev. Lett.* 2008, 100, 055001.
[37] Majetich, S.A.; Jin, Y. *Science* 1999, 284, 470–473.
[38] Schwart, M. J. R.; Thong, K. C.; Weinberg, F. J. *J. Phys. D: Appl. Phys.* 1970, 3, 1962-1966.
[39] Kumar, R.; Kaura, S. K.; Chhachhia, D. P.; Aggarwal, A. K. *Opt. Commun.* 2007, 276, 54-57.
[40] Kumar, R. *Appl. Phys. B -Lasers Opt.* 2008, 90, 379-382.
[41] Kouyoumjian, R. G.; Pathak, P. H. *Proc. IEEE* 1974, 62, 1448-1461.
[42] Umul, Y. Z. *J. Opt. Soc. Am. A* 2010, 27, 1613 - 1619.
[43] Kumar, R.; Aggarwal, A. K. *Indian J. Pure Appl. Phys.* 2007, 45, 429 - 433.

[44] Ebbesen, T. W.; Lezec, H. J.; Ghaemi, H. F.; Thio, T.; Wolff, P. A. *Nature* 1998, 391, 667 - 669.
[45] Garcia-Vidal, F. J.; Martin-Moreno, L.; Ebbesen, T. W.; Kuipers, L. *Rev. Mod. Phys.* 2010, 82, 729 - 787.
[46] Kumar, R. *Opt. Appl.* 2010, 40, 491- 499.

Chapter 8

DIFFRACTED BEAM INTERFEROMETRY

Elena López Lago,[*] Héctor González Núñez and Raúl de la Fuente
Departamento de Física Aplicada, Escuela Universitaria de Óptica y Optometría,
Campus Vida, Universidade of Santiago de Compostela, Galicia, Spain

ABSTRACT

Diffracted beam interferometry (DBI) is a self referenced characterization technique which was originally thought to reconstruct the phase of a beam starting from the interference data between the beam and its diffracted copy. The phase is recovered indirectly by means of an iterative algorithm relating the irradiances of the interfering beams and the phase difference. The first experimental demonstration of DBI was implemented on a Mach-Zehnder interferometer which incorporated an afocal imaging system in each arm, in order to form an image of a common object in different planes at the output of the interferometer. The irradiance and phase difference data were picked up from one of the image planes and entered into the iterative algorithm. Later modifications of the iterative algorithm made DBI able to characterize both the phase and the amplitude simultaneously. This new algorithm allows faster data acquisition which makes the method less influenced by environmental disturbances.

I. INTRODUCTION

Most optical applications only need information on the irradiance of an optical field, but there are some special tasks, with very diverse technological applications in particular in optics and astronomy, also requiring the knowledge of the phase. Some examples are laser beam diagnosis, beam control and beam shaping, optical testing, aberration correction, recovery of blurred images displacement and position sensing, active and adaptive control of optical systems, and several others.

---

[*] Departamento de Física Aplicada, Escuela Universitaria de Óptica y Optometría, Campus Vida, Universidade of Santiago de Compostela, E-15782 Santiago de Compostela, Galicia, Spain. E-mail: elena.lopez.lago@usc.es.

At the present there are various well consolidated methods for wavefront reconstruction. They mainly fall into three groups: (a) interferometric methods [1] which are based on the superposition of two beams with a well defined phase relationship; (b) methods based on the measurement of the wavefront slope or wavefront curvature, such as the Hartmann–Shack sensor [2] or techniques based upon the resolution of the irradiance transport equation [3]; (c) methods based on the acquisition of one or more non-interferometric images followed by the application of an iterative phase retrieval algorithm, such as phase retrieval [4, 5] and phase diversity methods [6].

Among these, interferometry is the method offering the greatest resolution in phase reconstruction [1]. It is based on the coherent superimposition of a light wave on a reference wave with a known phase, which is normally constant. The wavefront of the light wave is retrieved from the interference pattern. Unfortunately, wavefront interferometry presents some drawbacks due to high cost, sensitivity to mechanical vibrations and the critical need to control the reference wave. However, these drawbacks can be overcome by using self-referred interferometers that benefit from the unknown beam as its own reference. Examples of such interferometers are lateral shearing [7-13] and radial shearing interferometers [14-19] among others.

This chapter considers an alternative self-referencing interferometric method initially designed for wavefront measurement and known as diffracted beam interferometry (henceforth DBI) [20,21]. Unlike shearing interferometers, which use a copy of the test beam displaced or modified in some way transversally to the beam propagation direction, DBI is based on the coherent superposition of a light wave with a diffracted copy of itself. DBI can be regarded as a technique that applies an axial or longitudinal shearing to one of the interfering beams. But in contrast to shearing interferometers, the sheared beam is not an image of the test beam because diffraction affects to the amplitude and phase of the beam.

The key to DBI is that the essential information for wavefront reconstruction is obtained from the interference pattern and the intensity of both the light wave and its diffracted copy. However, these data do not provide direct information on the wavefront (only information about the phase difference between the light wave and its diffracted copy is obtained from the interference pattern); therefore an iterative numerical algorithm has to be applied to estimate the wavefront.

Sometimes, the objective is the simultaneous reconstruction of both the phase and the amplitude of an optical beam and some methods have been developed for this purpose over the last few years [19, 22-24]. Some of these methods are multi-shot and rely on phase-shifting interferometry [22] or on convolution of the test beam with suitable phase masks [23]. Single-shot methods [19, 24] are based on radial shearing interferometry. Obviously, the beam amplitude can be measured directly with an array detector, but the advantage of radial shearing techniques is that they provide both amplitude and phase, in a single measurement. In this context, our research revealed that slight modifications of the original DBI algorithm also allow us to infer phase and the beam amplitude simultaneously in such a way that the image data are not required [25].

In this chapter we present a review of the DBI technique from its origin as a phase-only characterization algorithm to the current amplitude and phase characterization algorithm. The chapter is organized as follows: in section II, we present some general observations concerning diffracted beam interferometers; in section III.A, we describe the iterative algorithm designed for the phase-only characterization algorithm and demonstrate the capabilities and limitations of the algorithm through numerical test. In section III.B, we present the modifications introduced in the code to extend the applications to characterize the amplitude of the signal beam too. In section IV, we deal with the design of the interferometer, paying particular attention to the parameters of the optical elements and to retrieval of the algorithm input data. In section V, we show some relevant experiments related to local or global characterization of optical elements. In sections VI, we briefly address to the advantages of DBI over other characterization methods, respectively. Finally, in section VII we present some conclusions and observations.

II. THE DIFFRACTED BEAM INTERFEROMETER: GENERAL OBSERVATIONS

Fig.1 shows a scheme of a generalized diffracted beam interferometer with the aim of clarifying how the technique works. In this diagram, the object plane refers to the plane where the beam is to be characterized. The beam from this plane, called object beam, is split into two replicas. These two replicas propagate along two different imaging systems and then they are recombined and superimposed on the observation plane. A magnified image of the object beam, given by one copy, is formed in this plane (called the image beam and which acts as a signal beam) and a diffracted beam given by the other copy (called diffracted beam and that usually acts as a reference beam).

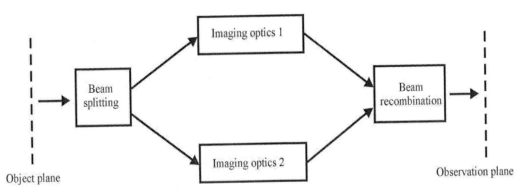

Figure 1. General scheme of a diffracted beam interferometer.

Some configurations compatible with this scheme are, for example, a Mach-Zehnder type interferometer with a different lens in each arm or a Michelson type interferometer with concave reflective optics. In all the research activities related to DBI we used a Mach-Zehnder interferometer. In all of the devices, the position and focal length of each imaging elements has to be adjusted to obtain the image of the object plane and a diffracted copy with appropiate relative magnification on the observation plane. It must be noted, that although different optical systems are inserted into each arm, a coherent superposition can be achieved if the optical path difference between the two "central rays" is less than the coherence length of the light source.

The reader must take into account that, when developing the algorithm, the imaging systems entering in the interferometer can induce phase changes in the beams which modify the interferogram: for example a simple lens introduces a quadratic phase in the image plane. However, this unwanted phase can be easily incorporated in the algorithm if it is known beforehand. In all the experiments, we use afocal lens systems for imaging, because in this case the optical field in the image plane reproduces the amplitude and also the phase of the object beam except for a magnification factor. In this chapter, we will show how the lens parameters must be chosen as a function of the characteristics of the element to be tested, in order to fulfill two requirements of DBI: one of the beams must overlap the other completely and the diffraction effects on the waves under propagation between the planes P_1 and P_2 must be significant to establish ligatures strong enough to manage the convergence of the algorithm to the actual phase.

III. THE ITERATIVE ALGORITHM: BASIS AND PROCEDURE

This section describes the physical fundaments and the operation of the algorithm. For the sake of simplicity, the study concentrates on a specific DBI interferometer based on a Mach-Zehnder configuration. In section III.A, we detail the structure of the original algorithm for wavefront reconstruction; in section III.B, we explain the light modifications the algorithm requires to spread its application to the full characterization of the beam, and mainly related to the initial conditions and to the reconstruction error. Finally in section III.B there are some numerical examples which aid understanding of the performance and what the technique might achieve.

In order to explain the physical principle of DBI we will refer, for simplicity, to a Mach-Zehnder configuration as the depicted in Fig.2 where the two telescopes, L_0L_1 and L_0L_2, share the first lens [21] placed before the interferometer. The interferometer arm holding lens L_1 and L_2 is called arm 1 and arm 2, respectively. The image formed by arm 1 of the interferometer is located on plane P_1 whereas the image formed by arm 2 is located on plane P_2. The distance between these two planes is equal to twice the difference between the back focal distances of lens L_1 and L_2. The quotient between these focal lengths gives the magnification ratio between the two images.

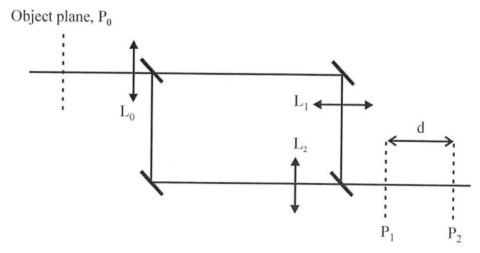

Figure 2. Scheme of a possible Mach-Zehnder interferometer for DBI characterization.

Any plane after the interferometer can be chosen as the observation plane. However, the choice of an image plane (i.e. P_1 or P_2) as the observation plane simplifies the phase retrieval of the object beam. From now on, plane P_2 will be the observation plane. On this plane, arm 2 of the interferometer provides a magnified image of the object beam while arm 1 provides a diffracted beam. We can relate these two beams by the Fresnel diffraction integral modified with a suitable scaling factor. This relation between beams will be used in the numerical algorithm for phase retrieval.

Let $u_0(x,y)$ be the complex amplitude of the object beam and let $u_1(x,y)$ and $u_2(x,y)$ be the complex amplitudes of the image beams at P_1 and P_2, respectively. They are related by a scaling factor:

$$u_i(x,y) = -\frac{f_0}{f_i} u_0\left(-\frac{f_0}{f_i}x, -\frac{f_0}{f_i}y\right) \quad (i=1,2) \tag{1}$$

Given that knowing any image field means the knowledge of the object beam, the phase retrieval of $u_2(x,y)$ can be set as the objective.

If $u_{1d}(x,y)$ refers to the complex amplitude of the beam coming from arm 1 at P_2. Using the scaling factor, it can be calculated from $u_2(x,y)$ by

$$u_{1d}(x,y) = C \iint u_2\left(-\frac{f_2}{f_1}x', -\frac{f_2}{f_1}y'\right) \times \exp\left\{\frac{ik[(x-x')^2 + ik(y-y')^2]}{2d}\right\} dx'\,dy' \tag{2}$$

where k is the wave number, d is the defocusing distance (distance between the image planes P_1 and P_2) and C is a complex constant. The irradiance distribution of an interferogram on the plane P_2 is given by

$$I(x,y) = I_2(x,y) + I_{1d}(x,y) + 2\sqrt{I_2(x,y)I_{1d}(x,y)}\cos(\Delta\phi(x,y)) \tag{3}$$

where I_2, I_{1d} are the irradiances of the beams in the observation plane and $\Delta\phi = \phi_2 - \phi_{1d}$ is the phase difference between them.

III.A. Phase-Only Characterization

Let us now center our attention on the iterative algorithm which has been developed to retrieve the phase of the image beam u_2. The input data of the algorithm are I_2, I_{1d} and the phase difference $\Delta\phi$ (which is obtained from the measured interferograms) and it uses the relation between amplitudes given by the Fresnel diffraction formula, explained in eq. (2). One iteration of the algorithm has the following structure (see also Fig. 3.):

i) In the k iteration an initial estimate of the complex amplitude of the image beam $u_2(x,y)$ is made starting from its real irradiance measured directly in the observation plane and from phase ϕ_2 which is obtained in the previous iteration:

$$u_2^{(ie)} = \sqrt{I_2}\,\exp(i\phi_2) \tag{4}$$

To make an initial guess in the first iteration, the real irradiance and an arbitrary phase, usually a constant phase, are taken, but simulations revealed that this initial phase is not a determining factor in the convergence of the algorithm.

ii) The complex amplitude of the diffracted beam $u_{1d}(x,y)$ is estimated by applying the Fresnel diffraction integral to $u_2^{(ie)}$ after correcting the magnification factor introduced by the imaging systems. To implement the diffraction integral, the angular spectrum formula [26] is used:

$$u_{1d}^{(e)}(x,y) = IFFT\{FFT[u_2^{(ie)}(x',y')]\exp[i\pi\lambda d(\eta_x^2 + \eta_y^2)]\} \qquad (5)$$

where $x' = -xf_2/f_1$, $y' = -yf_2/f_1$ are the rescaled spatial coordinates and $\eta_x = x/\lambda d$, $\eta_y = y/\lambda d$ are the spatial frequencies. The direct (FFT) and inverse (IFFT) Fourier Transform are calculated using a Fast Fourier Transform algorithm.

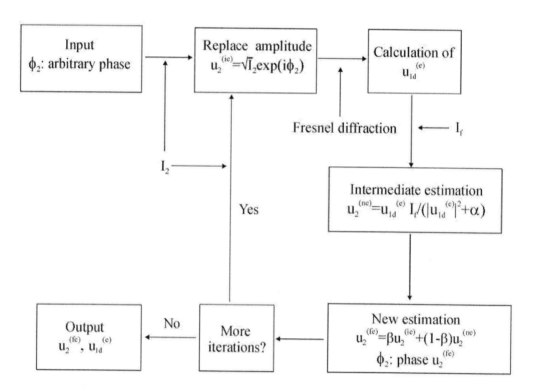

Figure 3. Flow chart of the DBI algorithm.

iii) Let us define $I_f = \sqrt{I_2 I_{1d}}\, e^{i\Delta\phi} = u_2 u_{1d}^*$. This magnitude is built from the experimental data. The complex amplitude of beam 2 can be written as $u_2 = u_{1d} I_f / |u_{1d}|^2$. If the current estimate of beam 1, that is $u_{1d}^{(e)}$, is inserted in this formula, another approximate expression for the complex amplitude of beam 2 can be calculated as

$$u_2^{(ne)} = u_{1d}^{(ie)} I_f / \left(\left|u_{1d}^{(ie)}\right|^2 + \alpha\right) \qquad (6)$$

where α is a small parameter inserted to avoid division by zero.

iv) A new estimate of the amplitude of the image field is made starting from the previous estimates obtained in steps 1 and 3. This is the final estimate in iteration k:

$$u_2^{(fe)} = \beta u_2^{ie} + (1-\beta) u_2^{(ne)} \qquad (7)$$

In this expression β is a parameter which takes a value from 0 to 1 which for convenience is set at the beginning of the algorithm. Lower values of β make the algorithm to converge faster, but larger values of β strengthen convergence.

Care must be taken with the relative phase of $u_2^{(ie)}$ and $u_2^{(ne)}$ in Eq. (7). In any experiment, it is difficult to adjust the two arms of the interferometer to the same length, so the phase of these two estimates can differ in a simple constant even when the algorithm tends to converge. In a hypothetical but unlucky situation where the two estimates have a phase difference close to π, they tend to cancel each other out and convergence fails. To avoid this problem, before calculating $u_2^{(fe)}$ from Eq. (7), $u_2^{(ie)}$ and $u_2^{(ne)}$ are put '*in phase*' by subtracting a constant phase from each one: the phase value taken at the center of the corresponding image map.

v) To judge the fitness of the algorithm, a comparison is made between the experimental the numerical data. This comparison can use the irradiance of the image beam, the irradiance of the diffracted beam and the phase difference between them (or any combination of these). After performing numerous numerical simulations we noted that when the algorithm converges, the major difference is between the experimental and numerical irradiances of the diffracted beam. Therefore, convergence is quantified by computing the rms error between the two irradiances:

$$Q = \frac{1}{N} \sqrt{\sum_{i,j=1}^{N} \left(I_d(x_i, y_j) \right) - \left| u_d^{(e)}(x_i, y_j) \right|^2} \qquad (8)$$

When the algorithm converges, the value of Q decreases until it stabilizes at a small value. Typically this occurs after ten iterations.

III.B. Amplitude and Phase Characterization

In section III.A, the input data starting the algorithm were the real amplitudes of the two fields at P_2, that is $|u_2|$ and $|u_{1d}|$, and the phase difference between them, $\Delta\phi$, obtained from the interference term. With these data, the magnitude I_f, given by $u_2 u_{1d}^*$, is built. In this section, we show how the phase and the amplitude can also be retrieved simultaneously by introducing simple modifications of the algorithm discussed in the previous section. That means that the full beam reconstruction can be made starting only from the product of complex amplitudes, I_f. (In [25] I_f was called complex amplitude of modulation) with the record of the two irradiance data being unnecessary.

Starting from the definition of I_f, the signal beam amplitude can be expressed as:

$$u_2(x,y) = \frac{I_f(x,y)}{u_{1d}^*(x,y)} \qquad (9)$$

The use of equations (2) and (9) together offers a simple way to determine u_2 by means of an iterative procedure. In the first step, u_{1d} is taken as a constant and is inserted in equation (9) to obtain $u_2=I_f$. Secondly, an estimate of u_{1d} is calculated by means of equation (2). In the third step, equation (9) is used again to calculate a new estimate of u_2 and equation (2) to calculate a new estimation of u_{1d}, and so on. After a few iterations, the estimated signal beam converges to the real signal beam.

Test can be carried out to quantify the error reconstruction. Some examples are a comparison between the retrieved and the original interfering beams, a comparison between both modulus and phase of the original and retrieved I_f term, or a comparison between the original and retrieved background interference term [24].

III.C. Numerical Examples

The aim of this section is to use numerical simulations to show the capabilities of the method in relation to determine the accuracy in the phase reconstruction and also identify critical situations to give possible indications to help overcome them. We also try to give an idea of the robustness of the method in the noise presence from different origin.

III.C.1. The Direct Algorithm and the Inverse Algorithm

One of the questions is what to do if the requirement of an overlap fails, which means that the size of the image beam is greater than the size of the diffracted beam. In this case, an alternative solution to changing the optics elements in the interferometer, or even changing the observation plane, is to interchange the role of the image beam and the diffracted beam. This way, if the original algorithm builds an approach of the image beam, it can be modified to build an approach to the diffracted beam and, starting from there, it is possible to get to the object beam.

Another question is related to the ambiguity of the sign of the phase difference $\Delta\phi$ and how it affects the phase retrieval. In this case, if the sign of $\Delta\phi$ is wrong the algorithm converges to a phase with opposite concavity and the numerical irradiances do not reflect the experimental ones and it also appears that the image reflects to the diffracted beam and viceversa. This situation is easy to identify, except when dealing with small phases (for example direct laser beam wavefronts); if this is the case, an inspection of the Q reconstruction error usually displays higher values for wrong values of the sign of $\Delta\phi$.

Fig. 4 shows some numerical results achieved with this algorithm. In these simulations, the following values were assigned to the focal lengths of the lens: $F_0 = 100$ mm, $F_1 = 200$ mm and $F_2 = 250$ mm. With these values, the distance between the image planes P_1 and P_2 is 100 mm. The object plane is illuminated by an He-Ne laser. All plots display an arbitrary radial direction in the observation plane. The first column corresponds to the phase of the image beam, ϕ_2, the second column to its irradiance, I_2, and the third column to the diffracted beam irradiance, I_{1d}. In the first example (Fig. 4(a), 4(b) and 4(c)), there is a simulation of the phase reconstruction of a defocused Gaussian beam with spherical aberration. As seen in the plots, the algorithm retrieves the real phase with great accuracy and, a rms phase error less than 5 10^{-5} waves is obtained. Furthermore, the retrieved and real irradiances are practically indistinguishable. The Q value achieved in this simple case is 2.6 10^{-4}. The convergence of the algorithm is very fast: a Q value less than 10^{-3} is obtained after 30 iterations.

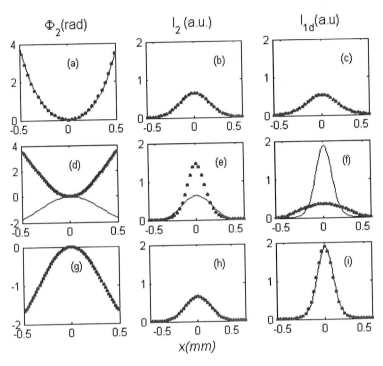

Figure 4. Results obtained with the numerical examples explained in the text.

In Fig. 4(d), 4(e) and 4(f) we show an example where the algorithm fails, an unsuccessful result which is expected. With respect to the first example, the sign of defocus has to be changed, so that the diffraction beam in the observation plane has a smaller spatial area than the image beam. Thus, the algorithm in step 3 makes an approximation of a broad beam starting from a narrow one (see Fig. 4(e) and Fig. 4(f)). In this case, possible divisions by very small numbers can result in erroneous values in Eq. (6) and the algorithm stagnates. To prevent stagnation, the role of the image and the diffracted beam in our algorithm can be interchanged. This 'inverted' algorithm starts with a first guess at the diffracted beam amplitude which now acts as a signal wave. This estimate is used to calculate the amplitude of the image beam by means of the Fresnel diffraction integral, which also serves to refine the amplitude of the diffracted field. This simple change avoids division by too small numbers in step 3 of the algorithm. To avoid confusions, from now on we will refer to this algorithm as the inverse algorithm, and the algorithm presented above will be called the direct algorithm. We used several numerical simulations to check that the application of the inverse algorithm helps convergence to the real phase when the diffracted beam is the narrow one. The last three plots (Fig. 4(g), 4(h) and 4(i)) display the results achieved with the inverse algorithm for the same object beam as in example two which is when the phase is retrieved. Note that an alternative solution to avoid stagnation in similar situations could be to change the observation plane to another one in which the diffracted beam overlaps the image beam.

III.C.2. Noisy Signals

In Fig. 5 we present further numerical simulations showing the performance of the algorithm. In these examples (corresponding to the same inputs as in Fig. 4) we added noise to the test interferograms. The maximum detected signal has a value of 16000 (arbitrary units) and a multiplicative Poisson noise is applied to each interferogram. We also added additive Gaussian noise with a mean of zero and standard deviation of 40. Finally, 1024 gray levels are simulated. The Q values achieved for these noisy images with 128x128 data points are about 0.009 in both examples. This Q value can be two orders of magnitude greater than the value attained in the corresponding ideal case; in spite of this and since the numerical irradiances resemble the experimental irradiances, it can be taken as a good value. At the same time, the retrieved phase also maintains good similitude to the real phase. In both examples a rms phase error of about 0.006 waves is obtained. It must be noted that, in order to calculate the phase error, we used a circular pupil to discard the contribution of data points with a small irradiance (in such points the phase has no significance).

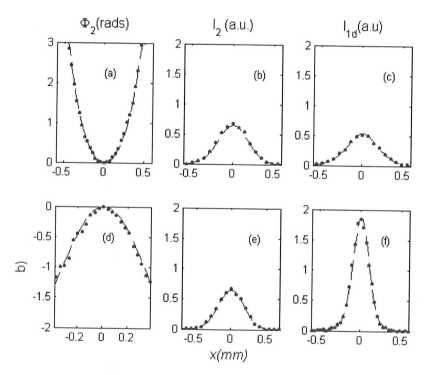

Figure 5. Same as Fig. 4 but with noisy signals.

Another set of simulations evaluated the capabilities of the algorithm when u_{1d} is the Fourier transform of the object beam, situation which represents the strongest diffraction effects. For a simulated object beam, it is first calculated the corresponding diffracted beam and the interference term, and after that it is applied a 1% multiplicative and additive noise to the images. A 256 grey level quantization is applied to simulate an eight-bit camera, too. The first example is for an aberrated wavefront with defocusing, spherical aberration, coma and astigmatism. In Fig. 3(a) the original wavefront is plotted. Figs. 6(b) and (c) show the reconstructed wavefront assuming Gaussian illumination and nearly uniform illumination, respectively. It can be seen that the retrieved wavefronts are almost indistinguishable from the correct ones. Rms. error values of 0.0097 for Gaussian illumination and 0.0163 for uniform illumination after convergence are obtained. For comparison, we noted that, without noise, the same wavefront was estimated with an rms error of 10^{-4} after 30 iterations. Figs. 6(d)–(f) show the results of wavefront reconstruction for an object with a cosinusoidal phase. In this case the rms errors were 0.0077 for Gaussian illumination and 0.0166 for uniform illumination. All of these wavefront estimates were obtained after 15–20 iterations with 256 × 256 data points (this means less than 30 s in a Matlab program running on a Pentium II 300 MHz computer). It should be noted that the results presented here are similar to those involving defocusing planes out of the Fourier plane.

We conducted several numerical simulations to test the validity and the accuracy of the simultaneous amplitude and phase characterization algorithm as done with the phase-only characterization algorithm, and found similar conclusions. Only two questions deserve special mention: one related to astigmatic signals and the other related to very noisy signals which will be visualized in section V.B supported by experiments.

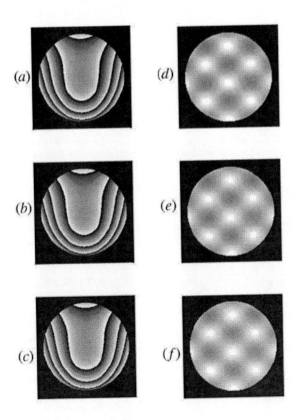

Figure 6. (a)–(c) Contour plots for a wavefront $\phi(x, y) = 8x2 + 2y2 - 3.3y(x2 + y2) + 0.4(x2 + y2)2$. (a) Actual wavefront. (b) Estimated wavefront with Gaussian illumination. (c) Estimated wavefront with nearly uniform illumination. (d)–(f) Same as (a)–(c) but $\phi(x, y) = \cos(4x) \cos(5y + 1)$.

IV. Design of the Interferometer

The design of the DBI interferometer must take into account three important aspects. Two of them are related to the lens parameters and the third is related to the extraction of I_f from the interference data. The lens parameters, that is, their focal length and position, must be chosen so that two conditions are satisfied. One of them, mentioned previously, is to ensure that one of the beams overlaps the other completely, and the second is that the diffraction effects affecting the beam between P_1 and P_2 should be significant. If the diffractive effects are low the interferometer approaches a radial shearing device and the algorithm proves less efficient [25]. The opposite occurs when the diffracted beam corresponds to the Fourier transform of the object beam. In this case, it is not unusual that the (imaged) Fourier transform has less spatial extent than the object beam. This can be taken into account in the experimental design using appropriate magnifying optics. Alternatively, as stated in the previous section the role of the object and diffracted beams in the numerical algorithm can be inverted. That is, we can begin with an estimate of the diffracted field, and use the object field to improve the first. At the end of the algorithm, the object field is determined from the retrieved diffracted one.

It is also important to pay attention to the detection system. Our experiments revealed that the most practical configuration is for a Badal telescope to image the observation plane on the CCD sensor. The lens parameters of this telescope are chosen so that the size of the beam matches the requirements of the CCD and also optimizes the irradiance levels of the recorded images to exploit the dynamic range of the camera.

IV.A. The Interference Term I_f

At the time of designing the DBI interferometer one of the relevant questions is how to get the I_f term, which is the product $u_2 u_{1d}^*$ (and can be rewritten as $|u_2||u_{1d}|e^{i\Delta\phi}$). Phase measurement interferometric techniques can be used to measure the phase difference between the interference beam $\Delta\phi$ and also can be used to extract the product $|u_2||u_{1d}|$. They are based on introducing a known phase change between the two beams and a later processing of the changes produced in the interferogram. These techniques can be divided into two categories: those that take the phase data simultaneously (spatial phase measurement data) and those that take the phase data sequentially (temporal phase measurement data)[27].

Among the temporal phase measurement techniques, phase stepping techniques are the best known. These consist in introducing a constant phase step between two intensity measurements.

The size of the phase step will depend on the number of intensity measurements to be taken. Three-step, Four-step or Carré methods are examples of algorithms that can be used to retrieve the phase difference and the modulation term [28].

A phase shift can be induced by moving a mirror, tilting a glass plate, rotating a half wave plate or analyzer, and so on. These elements are usually placed in an arm. In an interferometer with polarization isolation, the object and reference beams have orthogonal linear or circular polarizations. A rotating half-wave plate at the output of the interferometer will produce a frequency shift of twice its rotation frequency. Likewise, a rotating analyzer will produce a phase modulation at twice the rotation frequency [29].

In the spatial phase methods, the phase change is introduced spatially in the same interferogram [27]. These can be divided in two categories: phase-step methods and spatial carrier methods. In spatial carrier methods, a linear phase characterized by a spatial frequency is introduced in the interferogram, which is generated by tilting the reference wave. The information is processed in the frequency domain by Fourier transform methods that can provide I_f directly. Nevertheless, the use of this technique requires $\Delta\phi$ to verify some requirements. One of them is that the greatest gradient of the object phase has to be less that the spatial carrier phase, the background and contrast terms of the interferogram are slowly varying compared with the carrier frequency and the carrier frequency must fulfill the Nyquist conditions.

If the characteristics of the object beam allows the spatial carrier method to be applied, it simplifies the data acquisition process since only one measurement is needed to fully characterize the beam, and three measurements are needed for phase only characterization (in this case in addition to the interferogram, one frame for u_2 and another for u_{1d} have to be recorded). On the other hand, phase stepping techniques require a minimum of three interferograms to retrieve the I_f term, and two frames more if the two irradiance data are needed.

V. EXPERIMENTS

In this section we describe the most relevant results obtained with a DBI interferometer.

V.A. Phase-Only Characterization

The experimental results described in this section were obtained with a Mach-Zehnder interferometer with two afocal systems as shown in Fig. 7. The object plane, P_0, corresponds to the object focal plane of the first lens L_0 (an achromatic doublet of back focal length measuring 100 mm). This lens is located before the interferometer so it belongs to the path of the two beams generated in the interferometer. Each arm in the interferometer contains another lens (L_1 in arm 1 with a back focal length of 200 mm, and L_2 in arm 2 with a back focal length of 250 mm) which completes the corresponding afocal system. The object plane is directly illuminated with a linear polarizing He-Ne laser. A half-wave plate can change the polarization direction of light at the input of the interferometer, and its rotation allows the energy from one arm of the interferometer to be balanced with the other. Data were acquired by using a 12 bits CCD camera with 7.4 μm square pixels, placed at the observation plane, P_1 or P_2.

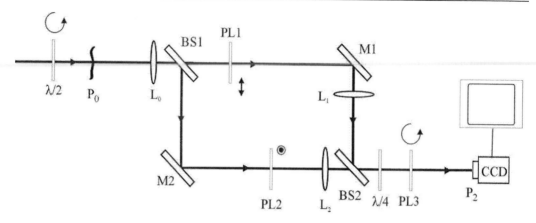

Figure 7. Experimental set-up. L0, L1, L2, achromatic doublets; PL1, PL2, PL3 linear polarizers; BS1, BS2 beam splitters, M1, M2 mirrors.

In order to retrieve the phase difference between the image and the diffracted beams, we recorded a set of four interferograms with a phase-shift $\pi/2$ produced by means of a rotating polarizing technique [29-30]. To apply this technique we insert a linear polarizer in each arm of the interferometer. The axes of these two polarizers are crossed so that, at the output there are two linearly polarized beams with orthogonal polarizations. The two linearly polarized beams are converted into right and left handed circularly polarized beams by means of a quarter wave plate with its axis at 45° to the polarization axes of the beams. A final polarizer set in a rotation mount transforms the beams back to linearly polarized beams. Rotation of this last polarizer at an angle θ now adds a phase-shift of 2θ to the phase difference induced in the interferometer, as it can be determined by simple polarimetric calculus [30]. Therefore, successive rotations of equal steps of $\theta = \pi/4$ give the desired phase shifts of $\pi/2$. The phase difference $\Delta\phi$ is obtained after processing the measured four interferograms with a four point phase stepping algorithm, in this case the Carré technique [28].

The technique is applied to the local characterization of several spherical and astigmatic lenses placed on the object plane and illuminated directly by the laser beam. In Fig. 8 we show the results for a convergent spherical lens with a nominal back focal length of 88.7 mm. The detection plane is P_2, i.e. the image plane for the beam that travels along arm 2 of the interferometer. In this case the image beam is broader than the diffracted beam so the inverse algorithm is used to retrieve the phase. The algorithm converges after 15 iterations with a Q value of 0.011. Fig. 8(e) plots the phase of the image beam after subtracting the phase of the laser (which was also determined with DBI). A simple quadratic fit determines the focal length of the lens. The value obtained was 86 ± 2 mm, close to the nominal value.

Measurements were also taken at plane P_1, the image plane for the beam propagated along arm 1 of the interferometer with similar results as in plane P_2. Fig. 9 gives a comparison of both measurements. Fig.9(a.1) and Fig.9(a.2) plot the beam irradiance in the object plane determined from its images taken at planes P_1 and P_2, respectively. Fig. 9(b.1) and Fig. 9(b.2) also show the beam irradiances in the object plane, but this time they are calculated from the beam irradiances obtained numerically by applying our algorithm in planes P_1 and P_2, respectively. The rms difference between the object irradiances determined directly from the experimental measurements is 0.0085. This value can be regarded as a lower limit to the Q value which gives back the DBI algorithm (0.011 in both planes) and also to the rms difference between the numerical object irradiances plotted in Figs 9(b.1) and 9(b.2) (this value is 0.012). Fig. 9(c.1) and Fig. 9(c.2) correspond to the object phase retrieved from the measurements taken in planes P_1 and P_2, respectively. The rms difference between these phases is 0.045 waves.

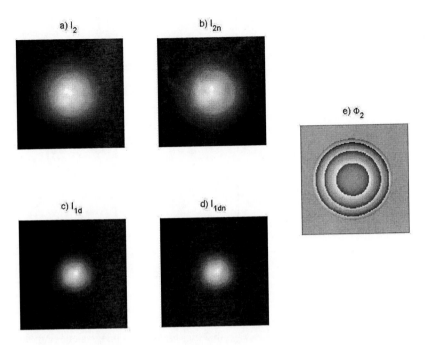

Figure 8. Experimental results for a spherical lens.

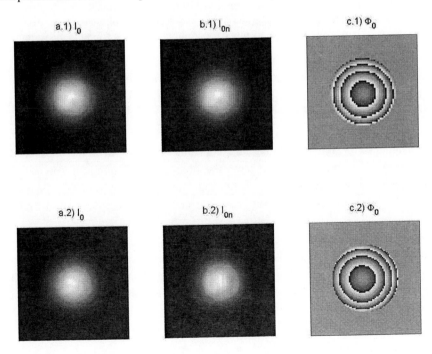

Figure 9. Comparison between measurements performed in planes P_1 and P_2 (see text).

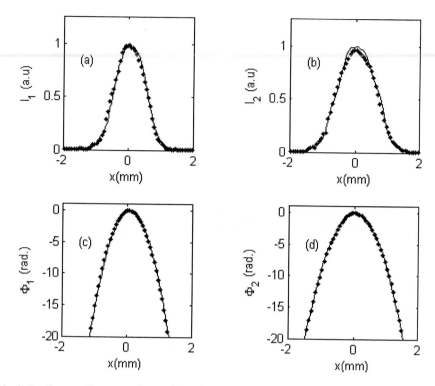

Figure 10. a) Continuous line: experimental irradiance of beam 1 measured at its image plane (P$_1$); dotted line: irradiance of beam 1 on its image plane calculated from its numerical irradiance at plane P$_2$. b) Continuous line: experimental irradiance of beam 2 measured on its image plane (P$_2$); dotted line: irradiance of the beam 2 on its image plane calculated from its numerical irradiance at plane P$_1$. c) Numerical phase of beam 1 on its image plane obtained from the measurement at this plane (continuous line) and from the measurement at plane P$_2$ (dotted line). d) Numerical phase of beam 2 on its image plane obtained from the measurement at this plane (continuous line) and from the measurement at plane P$_1$ (dotted line).

Moreover, we determined the complex amplitude of the image beams in plane P$_1$(u$_1$) and in plane P$_2$(u$_2$) starting from the complex amplitude of the diffracted beam in plane P$_2$ (u$_{1d}$) given by the DBI algorithm (in the first case), and starting from the complex amplitude of the diffracted beam given by the DBI algorithm in plane P$_1$ (u$_{2d}$). The dots in Fig. 10(a) and 10(b) show the irradiance of the calculated image beams together with the measured ones (in continuous line). The continuous line plots the phases of the image beams on each plane provided directly by the numerical algorithm and the dotted line plots the phase calculated after propagating the diffracted beams from the corresponding plane of diffraction to its image plane.

Fig. 11 presents the results obtained by using a negative cylindrical lens located at the object plane. The Q value is 0.0099. The retrieved phase is plotted along the axis of the cylinder and along the orthogonal direction. While a quadratic polynomial fits the phase along one axis, the phase along the orthogonal axis is practically constant. We note that in this case the phase returned by the numerical algorithm exhibits a small slope which can be attributed to some misalignment in the position of the lens. The slop is not included in the plots.

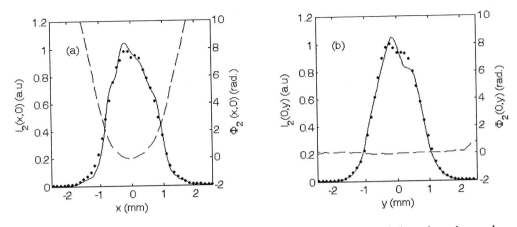

Figure 11. Results for the cylindrical lenses along the axes of the cylinder and along the orthogonal direction. Continuous line: measured irradiances; dotted line: numerical irradiances; dashed line: numerical phases.

V.B. Simultaneous Amplitude and Phase Characterization

In this case the experimental set-up, plotted in Fig.12, is similar to the one described in the previous section but with independent afocal imaging systems. At the output of the interferometer, a Badal system images one of the two mages planes, P_1 or P_2, onto a CCD detector. Let us suppose that this plane corresponds to plane P_1.

Moving the mirror along arm 1 tilts the beam u_1 in the image plane and a spatial frequency carrier is generated on the interferogram resultings from the superimposing the two beams. The irradiance distribution corresponding to this interferogram is written as

$$I(x,y) = |u_1(x,y)|^2 + |u_{2d}(x,y)|^2 + u_1(x,y)u_{2d}^*(x,y)\,e^{i2\pi Fx} + u_1^*(x,y)u_{2d}(x,y)\,e^{-i2\pi Fx} \qquad (9)$$

with u_1 and u_{2d} being the complex field amplitudes of the image and diffracted beams, respectively, and F is the induced spatial frequency. The distance between P_1 and P_2 is given by $d=2(f_1+f_2-f_3-f_4)$.

In a first experiment, a spatially filtered He-Ne laser beam illuminates an astigmatic ophthalmic lens located in the object plane P_0. The magnification of the telescopes in the interferometer arms is chosen to satisfy the two requirements of DBI, which are satisfied in this experiment with a diffraction distance $d=21$cm and a relative magnification between the two images $M_2/M_1=2.5$.

Fig.13 shows the results for a positive astigmatic lens with +0.25 dioptres and +0.50 dioptres along each principal meridian (+0.25sph+0.25cyl) and in Fig. 14 the results correspond to a lens with +0.25/-0.25 dioptres (+0.25sph-0.5cyl). For comparison, we also plotted the image beam and its diffracted copy obtained in separate measurements.

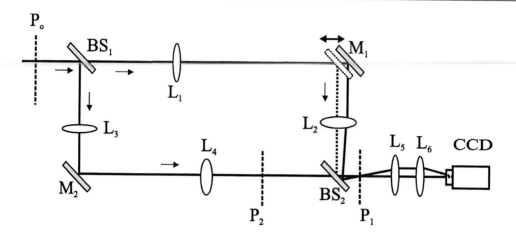

Figure 12. Experimental set-up. P_0, object plane; P_1, P_2 image planes. The upper arm of the interferometer corresponds to arm 1 and the lower arm to arm 2.

In both cases, the diffracted beams exhibit an elliptical shape as a result of lens astigmatism. The retrieved reference beams reproduce this shape with the proper orientation. It is highly unlikely that this would happen if the retrieved phase was wrong. For example, if the sign of the I_f term is reversed, the algorithm will converge into a beam with the opposite phase curvature. However, the irradiance of the reconstructed diffracted beam has a different and incorrect ellipticity and orientation. It is well-known that many phase-reconstruction methods do not give the sign of phase curvature. By using a diffracted replica of the object beam as a reference, DBI avoids this ambiguity.

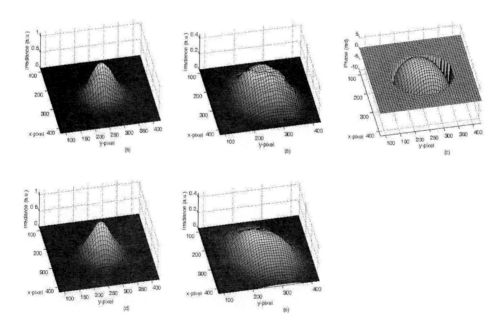

Figure 13. Amplitude and phase reconstruction for the astigmatic lens +0,25sph+0.25cyl. (a) image beam irradiance (b) diffracted beam irradiance (c) image phase. For comparison, we also show the measured image (d) and diffracted (e) beam irradiance.

A second set of experiments was designed to provide a global characterization of optical elements. In the present example the same laser beam illuminates a phase plate recorded in photoresist [31]. The phase of the plate reproduces a Zernike polynomial that simulates a secondary trefoil aberration. In this case, the diffraction distance was $d=5cm$ and the relative magnification between the images was $M_2/M_1=1.562$. We noted that the plate manufactured in photoresist was not perfect, so the phase object does not correspond exactly to the secondary trefoil but presents some other residual aberration. Furthermore, plate irregularities induce noise on the amplitude of the image beam. The results of reconstruction are shown in Fig. 15. It is observed that the reconstructed beam irradiances are noisier than the original beams. However, this has little effect on the retrieved phase that appears to be a secondary trefoil.

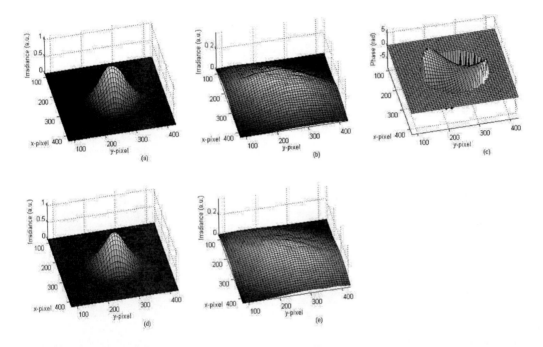

Figure 14. Amplitude and phase reconstruction for the astigmatic lens -0.25sph+0.5cyl. The plots are the same as in Fig. 13.

This experiment was repeated with $d=21cm$ and $M_2/M_1=2.5$. This configuration corresponds to complex experimental conditions because diffraction drastically alters the shape of the reference beam with respect to the signal beam. Fig. 16 shows the reconstructed and measured diffracted beams. The fringe pattern appearing in the measured reference beam is due to the diffraction caused by the sharp edge of the phase object acting as a circular pupil. The Fourier processing of the spatially modulated interferogram smooths the I_f term so that the retrieved diffracted beam does not present fringes. However, it displays a shape similar to the measured beam. Moreover, in spite of the complex shape of the diffracted beam and the I_f, the retrieved phase is quite correct. In fact, a comparison with a numerically-simulated secondary trefoil shows that the retrieved phase improves the one obtained in the previous experiment (see Fig. 17).

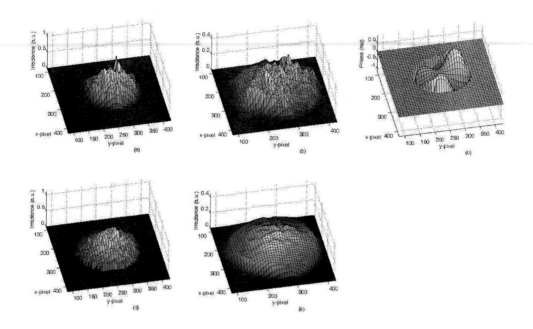

Figure 15. Amplitude and phase reconstruction for the phase plate obtained with a diffraction distance d = 5 cm and a relative magnification M_2/M_1=1.5625. The plots are the same as in Fig. 13.

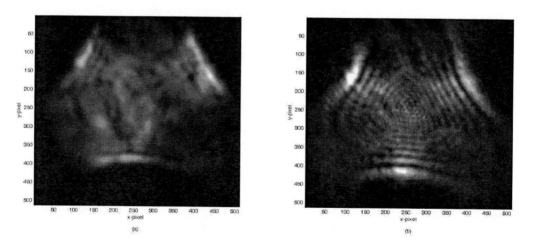

Figure 16. Reconstructed (a) and measured (b) irradiances of the diffracted beam for the phase plate obtained with d = 21 cm and M_2/M_1=2.5.

A possible reason for this is given below. The first experiment corresponds to a low diffraction limit where the reference beam is almost a magnified copy of the signal beam. This configuration approaches that of a radial shearing interferometer. In the second experiment the diffraction beam is highly diffracted, and consequently the I_f term contains information on the object beam and its evolution. This imposes severe conditions on the algorithm convergence.

Fig.18 shows the results of applying the procedure to characterize another phase plate whose phase reproduces a Zernicke polynomial. Note that in spite of the noise present in the signal amplitude, the algorithm accurately retrieves the phase if compared with the simulated one, except where the irradiance drops to zero.

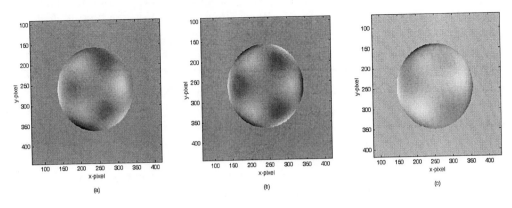

Figure 17. Reconstructed phase map of the signal beam for the phase plate (a) d = 5 cm and M_2/M_1=1.5625 (c) d = 21 cm, M_2/M_1=2.5. At the center, we show a numerical simulation of secondary trefoil aberration.

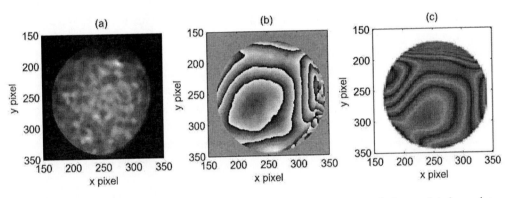

Figure 18. Signal beam amplitude (a) and phase reconstruction (b) of a second photoresist phase plate. The parameters are the same as in Fig. 17. (c) synthetic phase.

VI. DBI VERSUS OTHER CHARACTERIZATION TECHNIQUES

In this section we would like to stress the advantages of DBI over other interferometric and non interferometric characterization techniques.

Although the proposed technique is interferometric, the reader can see that it is also related to non-interferometric techniques such as phase retrieval from two intensity measurements and phase diversity. Phase retrieval methods are based on the acquisition of the image of the object under test and a diffracted replica (usually its Fourier transform). In phase diversity methods two images are taken of the object (usually in the Fourier plane) with and without phase diversity. In both methods the relation between the phase and the intensity data is known, and this relation is used in an iterative algorithm to retrieve the phase. Our method requires the image of the object, a diffracted image and the interference pattern. As in the non-interferometric methods, the relation between experimental measurements and the phase to be retrieved is also known, but there is also access to the phase difference between both images. It must noted that this additional information is crucial to enabling the method to perform well. This was confirmed by numerical simulations carried out both with and without taking into account data on this phase difference. However, if our method has to deal with a non-interferometric method, it needs a more complex experimental set-up

Readers may also be interested in the advantages of this method as opposed to other self-referencing interferometric techniques, such as lateral shearing interferometry or radial shearing interferometry. We would like to highlight two aspects. On the one hand, whereas lateral shearing interferometry is able to reconstruct the phase in the overlapping region of the two beams, our method retrieves the phase, at least numerically, even in the area where one of the beams goes to zero. On the other hand, lateral shearing interferometry allows us to retrieve the phase directly, but the final result can be altered by cumulative errors. Special software [32] allows us to minimize these errors, but the resulting algorithm is more complicated and time-consuming. Although the software in our method appears to be more complex, it seems to be robust in the presence of noise, so the phase retrieval can be done in several seconds.

Radial shearing can be thought of as a limit of the diffracted beam interferometry when the diffraction effects between the images planes tends to zero. If the procedure of the amplitude and phase characterization algorithm is compared with the procedure related in [24] for a radial shearing algorithm, although they are similar, there is an important difference between them. In the radial shearing algorithm the complex relation given by equation (2) relating the complex amplitudes of the image and its diffracted copy, can be split into two independent relations that allows the amplitude and the phase to be retrieved separately. This is not possible with the DBI algorithm: there is no way of retrieving the phase of the beam by using only the beam, and the same applies for the amplitude. This property, which may initially be seen as a drawback, can be very useful to give a quantitative idea of the quality of the phase obtained.

CONCLUSION

In this chapter, we review the main results obtained with DBI. The principal feature of this technique as opposed to other interferometric techniques is that each point of the observation plane provides information about the phase of the whole beam. This results from the fact that DBI measures the phase difference of an image of the test beam with a diffracted copy. This diffracted beam carries information about the phase evolution. There are other wavefront sensing techniques that provides information of beam evolution [5,6,33,34], but they are non interferometric and, therefore do not use the phase measurements as DBI does.

As with any self referenced technique, DBI uses a numerical algorithm to infer the phase which, in this case, consists of an iterative procedure. Numerical simulations showed that the method is robust and fast even with noisy signals. A rms. phase error less than 0.01 waves is easily achieved. We also identified situations where the algorithm fails and have proposed solutions to overcome such situations.

As explained in section II the DBI algorithm only works when the diffracted beam at the observation plane completely overlaps the image beam. If the image beam on observation plane has a larger spatial extent than the diffracted beam, their roles can be interchanged in equation (6) and (9) and at the end of the algorithm, the image beam is calculated from the diffracted one. Another solution is to use a suitable intermediate plane between planes P_1 and P_2 as the image plane. In this case, the relation between amplitudes (equation (2)) has to be generalized. Furthermore, choosing afocal systems with an adequate magnification can also overcome this limitation. Another completely different solution lies in replacing equations (6) and (9) by a nonlinear optimization-based routine that estimates the signal field amplitude from the I_f term.

We presented some experimental results to validate the DBI technique. To estimate the quality of the results we used a merit function (the Q function) to measure the difference between numerical and experimental beam irradiances. The Q values obtained in our experiments were as low as 0.01, similar to the Q values attained with numerical noisy signals. We also carried out several tests that indirectly confirmed that the phase is retrieved with good accuracy. We also showed that DBI enables simultaneous amplitude and phase reconstruction. Furthermore, the combination of DBI with spatial phase modulation techniques allows this task to be performed in a single shot. This allows the complete beam reconstruction to be carried out with a single measurement, thus avoiding the detrimental effects of environmental disturbances which are inherent in multi-shot interferometric techniques. Moreover, the application of the method on moving signals is enabled. We believe that this interferometric technique can be applied successfully to the diagnostics and control of high-power laser beams in real time.

Finally, we have also shown that if focus is only on the phase reconstruction, the signal and reference beam amplitudes can be used as a powerful test of reconstruction validity. A comparison between the reconstructed and measured beam irradiances provides qualitative evidence on the reconstruction quality.

We note that the related experiments not only show the feasibility of the DBI technique but that they also illustrate the importance of making the phase constraint given by eq. (6) to characterize the wavefront. In particular for the same cases as described in section V.A we tried to retrieve the phase by using an iterative algorithm that only uses the irradiances of the image and the diffracted beam and we were not successful.

ACKNOWLEDGMENTS

The authors would like to express their acknowledgment to the Spanish Ministry of Education and Science for providing financial support (contract FIS2007-63123). Héctor González Núñez acknowledges support to the Xunta de Galicia by María Barbeito Program. We also thank S Bará and M Gómez-García from the University of Santiago de Compostela for providing us with the phase plates used in the experiments. The authors acknowledge the Optical Society of America and the Institute of Physics for the inclusion of figures and extracts of texts of their intellectual property published in the following articles: López-Lago ,E; de la Fuente,R. "Wavefront sensing by diffracted beam interferometry" *J. Opt. A: Pure Appl. Opt.*2002, **4**, 299-302, (doi: 10.1088/1464-4258/4/3/314); López-Lago ,E; de la Fuente,

R, "Mach Zehnder diffracted beam interferometer", *Opt. Express* 2007, **15**, 3876-3887 (doi: 10.1364/OE.15.003876); López Lago, E.;de la Fuente ,R. "Single-shott amplitude and phase reconstruction by diffracted beam interferometry", *J.Opt. A:Pure Appl. Opt.* 2009, **11**, 125703 (6pp) (doi: 10.1088/1464-4258/11/125703).

REFERENCES

[1] Malacara, D; Servín M and Malacara Z 1998 *Interferogram Analysis for Optical Testing,* Marcel Dekker: New York NY, 1998.
[2] Shack, R. V.; Platt B. C. *J. Opt. Soc. Am.* 1971, 61, [3] Roddier, F. *Appl. Opt.* 1990, 29, 1402–1403
[3] Gershberg, R. W.; Saxton W .U. *Optik* 1972, 35,237–246
[4] Fienup, J. R . *Appl. Opt.* 1982, 21,2758–2769
[5] Gonsalves, R. A . *Opt. Eng.* 1982, 21, 829–832
[6] Bates ,W. J. *Proc. Phys. Soc. London* 1947, 59, 940-952
[7] Wyant, J. C. *Appl. Opt.* 1975, 14, 2622-2626
[8] Wyant, J. C.; Smith, F. D ,*Appl. Opt.* 1975, 14, 1607-1612.
[9] Liang, P.; Ding, J.; Jin, Z.; Guo, C. -S.; Wang, H. –T. *Opt. Express* 2006, 14, 625-634
[10] Dubra, A.; Paterson, C.; Dainty, C. *Opt. Express* 2004, 12, 6278-6288
[11] Ronchi, V. *Riv. Ottica Mecc. Precis.* 1923, 2,9–35
[12] Murty, M V R K *Appl. Opt.* 1964, 3, 531–51
[13] Hariharan, P; Sen, D *J. Sci. Instrum.* 1961,11, 428-432
[14] Brown , D. S. *J. Sci. Instrum.*1962, 39, 71-72
[15] Steel, W. H. *J. Sci. Instrum.* 1965, 42,102-104
[16] Li, D.; Chen, H.; Chen Z. *Opt. Eng.* 41, 1893-1898 (2002).
[17] Li, M ; Wang, P.; Li, X.; Yang, H.; Chen, H. *Opt. Lett.* 2005,30, 492-494.
[18] Chung, C. -Y. ; Cho, K. -C. ; Chang,;C. -C. , Lin, C. -H. ; Yen, W. -C. ; Chen, S. -J. *Appl. Opt.*2006, 45, 3409-3414.
[19] López-Lago, E; de la Fuente, R. *J. Opt. A: Pure Appl. Opt.*2002, 4, 299-302.
[20] López-Lago, E; de la Fuente, R, *Opt. Express* 2007, 15, 3876-3887
[21] Deutsch,B.; Hillenbrand, R.; Novotny, L *Opt. Express* 2008, 16 ,494–501
[22] Juanola-Parramon, R.; Gonzalez, N.; Molina-Terriza, G. *Opt. Express* 2008, 16, 4471–4478
[23] López Lago, E.; de la Fuente, R. *Appl. Opt.* 2008, 47 372–376
[24] López Lago, E.; de la Fuente, R. *J.Opt. A:Pure Appl. Opt.* 2009, 11, 125703 (6pp)
[25] Mendlovic, D.; Zalevsky, Z.; Konforti, N. *J. of Mod. Opt.* 1997, 44, 407-414.
[26] Creath, K., Kujawinska, M. Interferogram analysis: digital fringe pattern measurement techniques, Robinson, D.W; Reid, G.T.,0-7503-0197-X, Institute of Physics: Bristol, 1993,94-192.
[27] Creath, K. *Progress in Optics XXVI*; E. Wolf Ed., Elsevier Science, 1988, 349-393
[28] Hu, H. Z. *Appl. Opt* 1983, 22, 2052-2056
[29] Frins, E. M; Dultz, W.; Ferrari, J. A. *Pure Appl. Opt.* 1998, 7, 53-60
[30] Navarro, R.; Moreno-Barriuso, E.; Bará, S.; Mancebo, T. *Opt. Lett.*2000, 25, 236–238
[31] Elster, C.; Weingärner, I. *Appl. Opt.* 1999, 38, 5024–5031 and references therein

[32] Roddier, F.; Roddier, C.; Roddier, N. *Proc. Soc. Photo-Opt Instrum. Eng.* 1988, 976, 203-209
[33] Brady, G. R.; Fienup, J. R. *Opt. Express* 2006, 14, 474-486.

In: Interferometry Principles and Applications
Editor: Mark E. Russo

ISBN 978-1-61209-347-5
© 2012 Nova Science Publishers, Inc.

Chapter 9

BINARY GRATING INTERFEROMETRY WITH TWO WINDOWS

*Gustavo Rodriguez-Zurita** *and Cruz Meneses-Fabian*
Benemérita Universidad Autónoma de Puebla, Facultad de Ciencias Físico-Matemáticas
Apartado Postal, Puebla, México

INTRODUCTION

Along this review, by grating interferometry we will understand an interferometer which includes at least one grating (*i.e.*, a periodic amplitude transmittance) as an essential component of the system. Limiting the scope to the case of monochromatic or quasi-monochromatic cases, the task performed by such gratings includes at least one of the following: beam division, beam combination, beam replication (multiplexer) and phase shifter.

Sometimes, more than one of these tasks is performed by only one grating at the same time. In this review, we will also refer to absorption or phase amplitude transmittances as separate cases and exclude, for the sake of simplicity, more elaborate trasmittances, as anisotropic gratings for instance. In particular, we will consider interferometric systems using binary absorptive gratings with low to moderate spatial frequencies (Ronchi gratings are a good example of these) due to their wide use in the systems we review. There is a wide variety of interferometers which employ gratings as outlined. Interferometers which use plane gratings consisting of periodically spaced stripes with spatial frequencies ranging roughly from 20 to 2000 lines per inch have been widely proposed and used. In the Ronchi test of optical components, for example, partial superposition of diffraction orders generated by a binary grating performs as a lateral shear interferometer [1-4]. This technique of beam-splitting with a grating can be applied to samples other than optical elements, as metallic plates [5]. Such gratings can not only serve as beam dividers as in the previous cases, but also they can be used for beam recombination, as the all-gratings interferometers that have been demonstrated under several illumination conditions [6] with two, three or four gratings in

---

* Tel.: (+52) 222 229 5500 Ext. 2109, Fax: (+52) 222 229 5636, E-mail: gzurita@fcfm.buap.mx.

cascade [7]. In these examples, the resulting systems perform as two beam interferometers with space-invariant properties, and even polychromatic or incoherent illumination can be used [8].

Further systems use displacements within the range of some microns of properly placed gratings in order to obtain a given shift of the phases between the interfering beams [9, 10]. In these cases, the grating also works as phase shifter and not only as a beam splitter or as a beam recombiner. The incorporation of two windows and one grating is a special case of grating interferometers which forms a common-path interferometer. It has been presented a Fourier theory describing the image-formation process of a common-path interferometer which uses grating displacements [11]. The interferometer consists of a telecentric, $4f$-Fourier imaging system with two windows in the object plane and a binary ruling as a spatial filter. By comparison with the previous systems, it is possible to state that this interferometer comprises a wave-front divider (as in the Young interference experiment) and a grating as both a beam combiner and a phase shifter. It is different from them, therefore, in that it uses two displaced windows as an input. The following sections are devoted to develop and demonstrate the properties of this interferometer. Two windows placed in the object plane of an imaging optical system were earlier employed in character recognition applications, where a Young interference fringe pattern were to be expected in the frequency plane when the fields within each of the windows in the object plane have the same relative distribution. Inspection of the resulting fringes of two similar or dissimilar fields after convenient average has been reported to give useful information [12]. Placing a ruling with its fundamental spatial frequency matched with the frequency of this Young interference pattern would form a moiré pattern, as in the interferometer reported in ref.10 for testing flat plainness. In this case, shifting of such moiré fringes can be achieved by displacements of the grating and can be used for tuning the reference phase [10]. Two windows are used in the optical joint transform correlator as well [13].

In this case, however, a hologram is recorded with the two characters to be compared and placed in the Fourier plane of some lens for reconstruction. The purpose of this review is to present the basic properties of a simplified implementation of the interferometer proposed in ref.11 and related systems. One of such simplifications consists of using a Ronchi ruling (i.e. a binary absorption grating) as a spatial filter. Because this interferometer performs as a common-path interferometer, it seems to present some practical advantages as a good mechanical stability. Because these properties have been observed in tomographic applications [14], a discussion on some of related experiments are included. Furthermore, it is also shown how the aforementioned stability can be combined with its relative easiness to introducing phase steps which are useful to extract phase by phase-stepping interferometry techniques. Finally, in connection with phase-stepping methods, some efforts to reduce the number of shots needed to obtain four interferograms by incorporating linear polarization modulation are described.

1. BASIC CONSIDERATIONS

Let us consider a 4-f coherent imaging system with a binary grating (Ronchi ruling) in its Fourier plane (Figure 1.a). Two windows (L, F) are placed in its object plane.

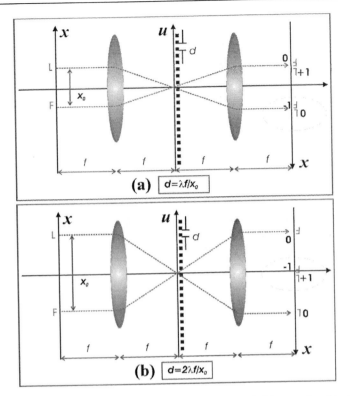

Figure 1.1. Superposition between diffraction orders of the optical fields associated to two windows (L, F) in the object plane at two different values of the distance between their centers, x_0. a) Fulfillment of condition $d = \lambda f / x_0$ leads to the superposition between (0,+1) and (-1,0) orders. b) Fulfillment of condition $d = 2\lambda f / x_0$ leads to the superposition between (-1,+1) orders.

Denoting with λ the wavelength of the illuminating monochromatic optical radiation and with f the focal length of the transforming lenses, the sketch is a modification of the common-path interferometer as reported in ref.[11], where only a cosine grating and a binary phase grating were considered.

In the image plane of this system, two series of diffraction orders appear (*i.e.* two discrete spectra), each series consisting of replications of each window falling apart by distances that are multipliers of $\lambda f / d$, d being the grating period. On the other hand, the two spectra are mutually displaced according to the spacing x_0 of the windows in the object plane. Then, by proper choice of the distance between windows with respect to the grating period d of the ruling, a convenient superposition of diffraction orders can be achieved. Under these conditions, interference patterns centered on each diffraction order can be obtained due to the coincidence of the replicated fields from the windows.

Figure 1.1.a. shows superposition of orders -1 with 0 and 0 with +1 (case of $d = \lambda f / x_0$) while Figure1.1.b. shows superposition of orders +1 with -1 (case of $d = 2\lambda f / x_0$.) In the following section, the corresponding Fourier analysis is presented.

1.1. Fourier Analysis

For a Ronchi ruling of spatial period $d = \lambda f / X_0$, with clear bar spatial width given by $A_w = \lambda f a_w$ and displaced by an amount $u_0 = \lambda f \mu_0$, its amplitude transmittance can be written in terms of spatial frequencies coordinates as

$$F(\mu,\zeta) = rect\left(\frac{\mu}{a_w}\right) * \sum_{n=-\infty}^{\infty} \delta(\mu - \mu_0 - n/X_0), \tag{1}$$

where $\mu = u/\lambda f$, $\zeta = v/\lambda f$, with u,v the actual coordinates of the Fourier plane, λ the wavelength of the illuminating optical beam, f the focal length of each Fourier transform lens of the telecentric system, and the symbol $*$ denoting convolution. The grating spatial period is denoted by X_0. If the imaging system has the Ronchi ruling as its pupil function, the corresponding impulse response can be written as the following

$$\tilde{F}(x,y) = \mathfrak{S}^{-1}\{F(\mu,\zeta)\} = a_w \cdot X_0 \text{sinc}(a_w x) \cdot \exp\{i2\pi\mu_0 x\} \sum_{n=-\infty}^{\infty} \delta(x - nX_0, y), \tag{2}$$

which reduces to

$$\tilde{F}(x,y) = a_w \cdot X_0 \sum_{n=-\infty}^{\infty} \exp\{i2\pi \cdot n\mu_0 X_0\} \cdot \text{sinc}(na_w X_0) \cdot \delta(x - nX_0, y). \tag{3}$$

For the case of a ruling with equal widths in clear and dark bars, $a_w = 1/2X_0$ and there will be no even diffraction orders (missing orders of order 2). Because the rulings used in the experiments usually do not have exactly equal widths in clear and dark bars, there are no missing orders in general. The transmittance of two windows apart one from each other by x_0 in the object plane (Figure1.1.a) is given by

$$t(x,y) = w(x + \frac{1}{2}x_0, y) + w(x - \frac{1}{2}x_0, y) \cdot \exp\left[i\varphi(x - \frac{1}{2}x_0, y)\right], \tag{4}$$

so the corresponding amplitude in the image plane can be determined by $t_f(x,y) = t(x,y) * \tilde{F}(x,y)$, which gives

$$t_f(x,y) =$$
$$a_w X_0 \sum_{n=-\infty}^{\infty} \exp\{i2\pi \cdot n\mu_0 X_0\} \cdot \text{sinc}(na_w X_0) \cdot [w(x + \frac{1}{2}x_0 - nX_0, y) + w(x - \frac{1}{2}x_0 - nX_0, y) \cdot \exp\{i\varphi(x - \frac{1}{2}x_0 - nX_0, y)\}]$$
$$\tag{5}$$

Thus, in the image plane there are two symmetrically displaced ruling with diffraction orders replicating whether one window or the other. The condition for overlapping orders n and $n+N$ is $NX_0 = x_0$, where N is an integer. A useful case of overlapping replicated windows is the case $N = 2$ as shown in ref.[11] for cosinusoidal absorptive gratings and phase gratings. Overlapping occurs for terms with orders $n = +1$ and $n = -1$. Another case of interest occurs when $N = 1$, in which overlapping occurs for the terms with $n = \pm 1$ and $n = 0$. Each term adopts the following form respectively

$$t_{0f}(x, y) = a_w x_0 \cdot [w(x + \frac{1}{2}x_0, y) + w(x - \frac{1}{2}x_0, y) \cdot \exp\{i\varphi(x - \frac{1}{2}x_0, y)\}] \quad (6.a)$$

$$t_{+1f}(x, y) = a_w x_0 \exp\{i2\pi \cdot \mu_0 x_0\} \cdot \mathrm{sinc}(a_w x_0) \cdot [w(x - \frac{1}{2}x_0, y) + w(x - \frac{3}{2}x_0, y) \cdot \exp\{i\varphi(x - \frac{3}{2}x_0, y)\}] \quad (6.b)$$

and

$$t_{-1f}(x, y) = a_w x_0 \exp\{-i2\pi \cdot \mu_0 x_0\} \cdot \mathrm{sinc}(a_w x_0) \cdot [w(x + \frac{3}{2}x_0, y) + w(x + \frac{1}{2}x_0, y) \cdot \exp\{i\varphi(x + \frac{1}{2}x_0, y)\}] \quad (6.c)$$

Then, there are overlapping of contributions of replicated windows at each diffraction order. Considering overlapping of pair of orders 0,1 or 0,-1 without further contribution of replicated windows of higher order, there would be two-beam interference patterns at points $(\pm x_0/2, 0)$. Within the region of a window width $w(x - x_0/2, y)$, the irradiance $I_{0,+1}(x, y)$ for orders 0 and 1 is given by

$$I_{0,+1}(x, y) = w\left(x - \frac{1}{2}x_0, y\right) \cdot |t_{0f}(x, y) + t_{+1f}(x, y)|^2$$
$$= 1 + \mathrm{sinc}^2(a_w x_0) + 2\mathrm{sinc}(a_w x_0)\cos[\varphi(x, y) + 2\pi \cdot \mu_0 x_0] \quad (7)$$

and similarly for 0 and -1. The introduction of a phase step can be performed with the displacement by an amount μ_0. The fringe visibility of these patterns is not unity. It is given by the following function of a_w

$$V(a_w) = \frac{2 \cdot \mathrm{sinc}(a_w x_0)}{1 + \mathrm{sinc}^2(a_w x_0)}. \quad (8)$$

Under experimental conditions, however, it appears a fringe contrast good enough for measurements for this case.

1.2. A Typical Experimental Set-up

As remarked earlier, Figure 1.a. shows superposition of orders -1 with 0 and 0 with +1 ($d = \lambda f / x_0$) while Figure 1.b. shows superposition of orders +1 with -1 ($d = 2\lambda f / x_0$.) Noteworthy, because the amplitudes of the ± 1 orders are theoretically equal for symmetrical grating spectra, the fringe visibility of their interference is expected to be unity (absolute value) for many types of gratings in the second case (Figure 1.1.b). In contrast, for the case of the superposition of non-symmetrical orders (as in Figure 1.1.a), this could be not the case and the corresponding fringe visibility should be less than unity in absolute value. However, this can be of no important consequence for many practical applications provided the resulting fringe visibility is not too low for a linear detection and for a later image processing as described in next sections. In addition, the fringes of these patterns can be easily shifted by translating the grating along its grating vector. These fringe shifts occur because a given displacement of the filter in the frequency plane introduces a relative phase between the interfering fields in the image plane which is proportional to the filter displacement. As in the described experiments, an actuator can drive the grating but the use of a liquid crystal display for displaying shifted gratings has also been suggested [11]. It can be thus recognized that such a system enables the introduction of phase values into diffraction orders as desired, thereby offering the possibility of shifting interference patterns as required in phase-shifting interferometry [9]. A slight departure from conditions of Figures.1.1.a. or 1.1.b, would lead to a linear shift between diffraction orders, thus forming lateral shearing interference patterns. This simple analysis skips the description of the moiré pattern generated in the frequency plane, but describes the interference patterns rather directly in doing so. Figure 1.2. shows a typical experimental set-up, which is basically the telecentric, double Fourier -transform spatial filtering imaging system already analyzed ($f = 48\, cm$). The object O under inspection is a transparent object or optical field placed just before the x-y object plane. The optical system can use a clean, vertically polarized and collimated He-Ne laser as the light source ($\lambda = 632.8\, nm$.) The x-y plane comprises two rectangular windows of sides a and b with a mutual separation of x_0 (Figure 1.2.a), being their respective positions given by coordinates (- $x_0 / 2$, 0) and ($x_0 / 2$, 0) respectively.

One of the windows allows the light emerging from the object to enter into the system, while the second window transmits light than has not traveled through the object, thus acting as a reference. In the frequency plane of laboratory coordinates (u, v), a Ronchi ruling R with a nominal spatial frequency of 1000 lines per inch (a corresponding period of $d = 25.4\, \mu m$) and a fill factor of roughly 0.5 is placed as a spatial filter with its rulings parallel to the vertical line. Spatial frequencies can be thus written as $u / \lambda f, v / \lambda f$. An actuator DC can drive horizontally the carrier which supports the Ronchi ruling. A translation $u = u_0$ of the ruling introduces a shift $2\pi u_0 / d$ in the phase difference of the diffraction orders [11, 14].

A mirror M attached to the carrier is a part of one arm of a Michelson interferometer for calibration and monitoring of the grating displacements u_0 (Figure 1.2). In the originally reported experiment, the illumination of this system is independent from the one used in the grating interferometer ($\lambda' = 543.5\, nm.$) A PIN detector at the field center of the Michelson interferometer output delivers an electrical signal to an oscilloscope for fringe motion detection and characterization.

A variable pulse generator was constructed to adjust both the voltage level and the pulse temporal width of the pulse applied to the actuator separately. The voltage level of the pulse controls the actuator action speed while its driving distance can be determined by the pulse temporal width.

At a constant voltage level, the pulse temporal width needed to obtain a desired interference patterns shift can be found. Typical displacements of the mirror M in fringe numbers as a function of the pulse width applied to the actuator employed were very linear for several voltages of the pulse (4-9 V) within a range from 2 to 120 ms (Figure 1.3).

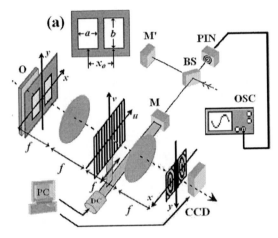

Figure 1.2. Common-path experimental grating interferometer: O, object sample; DC, actuator to drive the grating; CCD, camera receiving one of two possible interference patterns; PC, computer control to grab the pattern, display and processing. Michelson interferometer for calibration of the grating shift in dependence of the voltage pulse width applied to the actuator: M (attached to the driver displacement), M', mirrors; BS, beam splitter; PIN, point detector at Michelson pattern center; OSC, oscilloscope. a) Windows geometry (object plane.) The illumination of the Michelson interferometer is independent from the one used in the grating interferometer.

1.3. Experimental Results

The condition depicted in Figure 1.1.a, namely $d \equiv \lambda f / x_0$, is used in the following. This condition is convenient in order to keep x_0 as small as possible, so as to remain within the proper space range offered by the transforming lens aperture. The CCD-camera is aligned with one interferogram resulting from overlapping of orders 0 and +1. Figure 1.4. shows a sequence of five interferograms captured while observing Canada balsam placed on one of the faces of a microscope slide. The slide faces normal are horizontal during the observation, so the oil flows down. As the thickness of the oil layer changes, so the fringes modify themselves accordingly. Fringe visibility results good enough for data capture and visual observations.

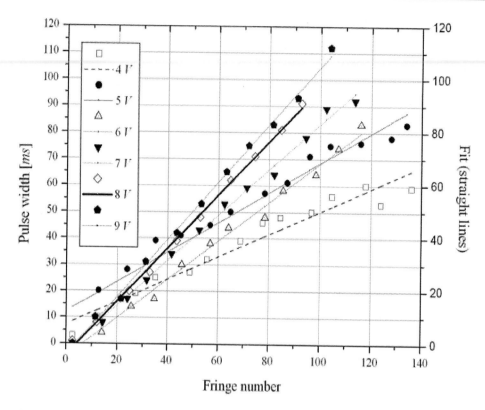

Figure 1.3. Relationship between pulse width on the actuator driving the grating and the observed fringe shifts.

It was chosen phase-shifting interferometry with shifts of 90° steps [9]. To achieve these shifts, first a phase shift of 180° was found by looking complementary interference patterns as obtained with a proper pulse width applied on the actuator. According with the experimental curves resulting from the Michelson interferometer fringe shifts, a fraction of this pulse width (obtained as described for a 180° shift) gives a fraction of 180° in the same direction of displacement. This linearity was used to find the desired shift of 90°. Such displacement of 180° amounts to be of $d/2$ in the grating plane (frequency) and it is equivalent to introduce a shift of $(d/2)/\lambda f = 1/2x_0$ in the image plane [11]. To verify the resulting grating shift, the traveling distance of the mirror M driven by the actuator was measured with the Michelson interferometer. It was not find contradiction in the resulting values.

Figure 1.4. A sequence of interference patterns from Canada balsam sliding down over a fixed glass slice in trans-illumination.

Binary Grating Interferometry with Two Windows

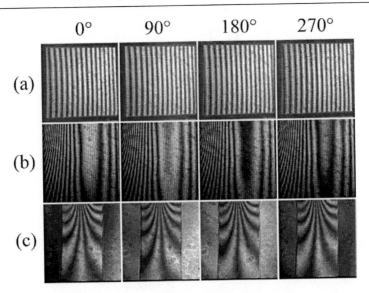

Figure 1.5. Four interferograms with phase shifts of 90° in between for three samples. a) A glass wedge. b) A folded piece of acetate foil. c) Another piece of acetate fold.

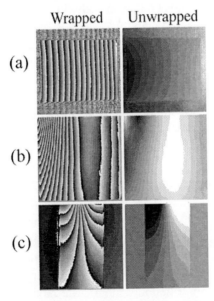

Figure 1.6. Resulting wrapped (256 gray steps for phase values between $-\pi$ and π) and unwrapped phases from the previous interference pattern sets shown in the Figure 4. a) From the glass wedge. Unwrapped phase range of about 31π. b) From the folded piece of acetate foil. Unwrapped phase range of about 34π. c) From the other piece of acetate fold. Unwrapped phase range of about 18π.

Experimental shifted interferograms are shown in Figure 1.5. in arrays of horizontal rows. In the upper row (Figure 1.5.a), the four interferograms mutually shifted by 90° phase and corresponding to a commercial glass wedge are displayed. In the middle row (Figure 1.5.b), the shifted interferograms of an acetate foil are shown, while at the lower row (Figure 1.5.c), the respective interferograms of another similar foil with a different bending shape are to be seen. Complementary contrast in interferograms pairs 0°-180° and 90°-270° can be seen. In Figure 1.6, wrapped and unwrapped phases for the samples previously shown in

Figure 1.5. are plotted in gray tones (8 bits) in arrays of two respective columns. A group of corresponding four interferograms was exported to a proper routine for wrapped phase calculation. The resulting wrapped phase image file was followed by another routine for phase unwrapping. From left to right, the wrapped phases are first shown. Then, the unwrapped phase plots in gray levels follow. No supplementary smoothing was employed in these results. In spite of the loss of aperture size and homogeneity in illumination, this system performs with reliability, as recently reported tomographic applications suggest [14]. It shows also great stability.

2. APPLICATION: OPTICAL TOMOGRAPHY

Tomography is a technique to reconstruct slices of the interior of samples by processing detected projections in a region in the outside of the sample. A projection is the spatial distribution of the output wave emerging from the sample to be inspected after an input wave enters it. Parallel projections refer to the case of projections which can be represented as a bundle of rays, each one of them suffering no deviations as they travels through the object. X rays behave in this way as they travel a human body, for example. It can be thought that each X ray suffers a gradual decrement in amplitude along its way to the detectors at the sample's output. The final amplitude distribution reached on the detection plane can then be seen as a parallel projection in this case. By using a number of parallel projections, each one taken at different angles (projection angles), properly processing with known algorithms (as the backprojection algorithm) allows an inner slice to be reconstructed. For the case of X ray tomography, the resulting slice is interpreted as an absorption distribution. In the following, we refer to tomographic techniques in connection with optical phase detection.

2.1. Liquid Gate for Low Refraction

Optical tomography refers to slice reconstruction of inner refractive index distributions instead of absorption within the visible region. Because the optical phase of a given ray traversing a sample changes its value along its way, the measurements of total changes demand of interferometric techniques. Simplicity of the reconstruction process makes desirable the use of parallel projection algorithms. But having perfect parallel projections in real samples means zero refraction, which is a very difficult condition to obtain. Transparent objects immersed within a gate filled with an immersion oil of a suitable refractive index have been suggested as a method to alleviate undesirable refraction effects [15]. The oil optically matches the sample interface with its surroundings, while the gate geometry, often a cuvette of a rectangular cross section, allows only normal incidence from the air at any sample's position. This way, a reasonable low refraction condition can be attained in experiments.

A gate performance can be visually appreciated in the Figure 2.1, where a glass block has been immersed within a liquid gate. The block, with length sides of some *cm*, is inspected under a two-beam interferometer and four phase shifts are applied in the reference beam. Three projections at a fixed angle are shown (a,b,c). At the sides, where the optical path changes linearly, fairly linear fringes can be seen. In the middle, where the optical path is almost constant, only one fringe occupies the field. These fringe distributions agree with a cross section of rectangular shape, as shown in Figure 2.1.d.

Figure 2.1. Interferograms from a glass block immersed within a liquid gate. a), b) and c), projections at a fixed projection angle, but with three different phase shifts. d), schema of a block cross section as seen from above.

2.2. An Experimental Set-up for Optical Tomography

Figure 2.2. shows the experimental set-up, which is basically a telecentric, double Fourier-transform spatial filtering imaging system (an application of Figure 1.1). The object O under inspection is a transparent rotating object placed just before the x-y object plane.

The optical system uses a clean, collimated, linear-polarized He-Ne laser as the light source $(\lambda = 632.8 nm)$. Each focal length was $f = 50 cm$. The x-y plane comprises two rectangular windows of sides a and b with a mutual separation of x_o, being their respective positions given by coordinates $(-x_0/2, 0)$ and $(x_0/2, 0)$ respectively (Figure 1.a). One of the windows allows the light emerging from the object to pass into the system, while the second window transmits light than has not traveled through the object, thus acting as a reference. The object remains immersed within a liquid gate filled with immersion oil, but it is vertically attached to the axis of a stepping motor in order to rotate from 0° to 360° at 0.9° steps. In the frequency plane of the output plane, a Ronchi ruling with a nominal frequency of 1000 lines per inch (the spatial period correspondent is $25.4 \mu m$) is placed as a spatial filter with its rulings parallel to the vertical line. Given the above parameters, the value of x_0 is about 1.2 cm. A DC actuator can drive horizontally the carrier which supports the Ronchi ruling. A mirror M attached to the carrier is a part of one arm of a Michelson interferometer (not shown) for calibration and monitoring of the slide displacements. A PIN detector (Newport CMA series) at the field center of the Michelson interferometer output delivers a signal to an oscilloscope for motion detection and characterization. Typical displacements of the mirror M in fringe numbers as a function of the pulse width applied to the actuator are very linear for several voltages of the pulse within a range from 2 to 120 ms as shown in Figure 1.4. To obtain a phase shift of 180° with a proper pulse width applied on the actuator in order to obtain $\mu_0 = 1/2x_0$, complementary interference patterns were first achieved. According with experimental curves resulting from the Michelson interferometer, a fraction of the pulse width gives a fraction of 180° in the same direction of displacement, there is a lineal depended in the local zone of the operation (Figure 2.3).

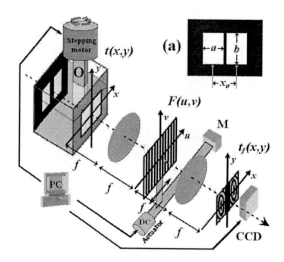

Figure 2.2. Experimental set-up. a) Windows pair.

For the presented experiments, pulses of 7.4 V were used with a temporal width of 32 ms for a phase shift of 90° (about 23 fringes counted in the Michelson interferometer, for which the wavelength was 543.5 nm). It was chosen to work with four shifts of values 0°, 90°, 180° and 270°.

2.3. Some Experimental Results

The CCD-camera is aligned with one interferogram resulting from overlapping of orders 0 and +1 (Figure 1.1.a). For data acquisition, a line of scanning coordinate $p = x$ is selected from the image received from the CCD-camera at a constant value of, say, $y = h$. This procedure defines an irradiance $I_\phi(p) = I_{01}(p,h)$ taken at a projection angle ϕ, which is stored (a row of 400 data). The object rotates for another irradiance capture. This procedure is repeated at 0.9° steps to cover the range of angles $0° \leq \phi \leq 360°$. The values of $I_\phi(p)$ are arranged as rows according to ϕ (an image with size of 400 x 400 pixels). Experimental interferograms $I_\phi(p)$ construct a composite interferogram (interfero-sinogram) over the plane $p - \phi$, resulting in an interferogram instead of a traditional sinogram [16, 17]. To obtain a sinogram, it is assumed that the parallel projection of phase $\tilde{f}_\phi(p) = \varphi(p,h)$ of the object slice with $y = h$ can be extracted from $I_\phi(p)$ by known techniques, such as phase-shifting interferometry [9]. Experimental results for a glass microscope slice as a sample was used, it was chosen the method of four shifts with 90° steps, such as is explained above section. With this procedure, the four shifts generating the interferograms shown in Figure 2.3. (a-d) were obtained. Composite continuous fringes are obtained in spite of the fact that each row has been taken separately. Note that the symmetry about $\phi = 180°$ is to be recognized with enough approximation. There are also continuous fringes of low frequency even in the background. Also, the induced phase step shifts the fringes in the sinogram plane as a whole. For each phase-shift, a complete turn of the object had to be done, but the uncertainty in the reproducibility of the zero position of the stepping motor after a complete turn does not appreciable affect the composite interferograms. The interference-fringes frequencies change on a very wide range. In particular, the borders of the object generate the highest frequency fringes values. In the lower row of Figure 2.3, the wrapped phase is first shown as obtained with a standard phase-stepping routine, plot (e). The unwrapped phase in a three-dimensional plot (f), which is to be taken as an estimation of the sinogram $\tilde{f}_\phi(p) = \Re\{f(x_h, z_h)\}$, is then subject to a standard filtered backprojection routine [16,17] to give the object slice reconstruction $f(x_h, z_h)$. $\Re$ denote the Radon transform [16] while $f(x_h, z_h)$ is the unknown phase slice distribution at level h using some coordinates x_h, z_h to describe the slice plane. The corresponding reconstruction is shown in Figure 3.g-h, as a gray tone plot (g) and as a three-dimensional plot (h). This particular reconstruction is calculated with all of the projections within the range [0°, 360°]. The dimensions of the slice object are 1mm and 8mm. The geometric proportions of the resulting slice are as expected, and the proportional factor (8 between its sides) can be approximately verifiable.

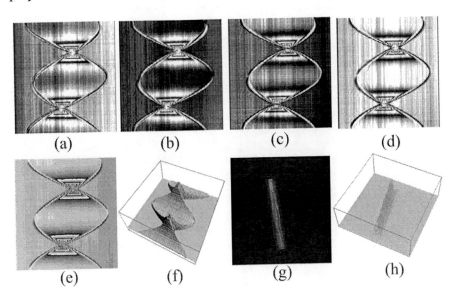

Figure 2.3. (a)-(d) $I_\phi(p) = I_{01}(p, h)$ as functions of p (horizontal axis) and ϕ (vertical axis) as constructed from phase-shifted interferograms. They are referred to as interfero-sinograms. The object is a microscope slice. (e) Wrapped phase, (f) unwrapped phase, on the same p-ϕ plane. (g) and (h) reconstructed object, level gray and three-dimensional plots, respectively.

Further results of using the samples shown in Figure 2.4. (glass plate, curved acetate foil, folded acetate foil) are shown in Figure 2.5. Acetate foils were cut in small pieces and folded to obtain arbitrary shapes as shown in Figure 4. Although the general slice form can be identified, there is some loss in borders. This can be due to the high frequency of the associated fringes, which not only fall out of the resolution range of the CCD-camera which was used, but also are too high for the phase shifting technique employed.

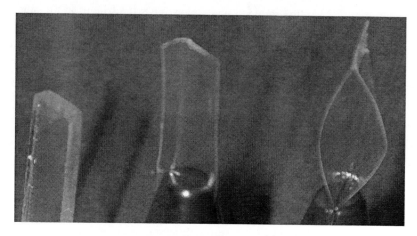

Figure 2.4. Transparent samples for experimental tomographic inspection.

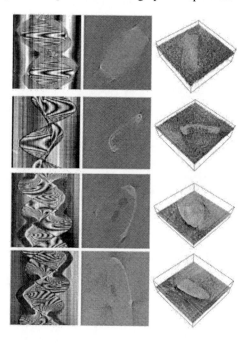

Figure 2.5. Interfero-sinograms (p, horizontal axis and ϕ, vertical axis) at a given phase shift (left column) and reconstructions (middle column: gray tone images, right column: three-dimensional plot) of some sections of the samples of Figure 4. From above: glass block, foil with "L"-shaped section, and two different sections of the third sample.

A common-path interferometer using two windows and a shifting grating for data acquisition in order to perform tomographic reconstruction of transparent objects in the visible range is thus possible. In spite of the loss of aperture size and homogeneity in illumination, this system performs with reliability. In particular, the appearance of interferosinograms consisting of continuous fringes is indication of the good stability of the system.

3. APPLICATION: LINEAR POLARIZATION MODULATION

Phase-shifting interferometry retrieves phase distributions from a certain number n of interferograms [18-9]. Each interferogram I_k must result from phase displacements by certain phase amounts f_k ($k = 0...n-1$) in order to form a solvable system of equations [9-19]. Because one of these phase amounts can be taken as reference, say $f_0 = 0$, it is possible to use the corresponding phase shifts, each denoted by $\alpha_i = f_i - f_{i-1}$, ($i = 1...n-1$). Among several possibilities, the case of $n = 4$ interferograms and $n-1 = 3$ equal shifts $\alpha_1 = \alpha_2 = \alpha_3 = \alpha$, $\alpha = 90°$, has been demonstrated to be very useful [20-21], especially for well contrasted interferograms [22]. To obtain phase-shifted interferograms, a number of procedures have been demonstrated but many of them needs of n shots to capture all of those interferograms. Thus, a simplification is desirable in order to reduce the time of capture. Single-shot interferometers capturing all needed interferograms simultaneously are good examples of this [23, 24]. On the other side, phase shifts can be induced by mechanical shifts of a proper element, as a piezoelectric stack [25] or a grating [22]. Modulation of polarization is another useful technique [24]. In grating interferometry, a grating can be transversally displaced by a quarter of a period to obtain shifts of $\alpha = 90°$ [26], for example. But in order to obtain several values, the same number of displacements is required [26-28]. Besides, when using gratings as phase shifters, the grating displacement must be carried out with sufficient precision. The higher the grating frequency, the smaller the grating displacement required. Thus, the use of high frequency rulings could compromise the precision of the phase shift.

As it has been described above, the use of binary gratings in conjunction with two windows gives several identical interference patterns. Previously, it has been remarked the use of only one pattern because the remaining ones contain the same information. In order to take advantage of at least a pair of interference patterns simultaneously, one of them have to carry some additional information. An independent phase can be introduced in each pattern by means of modulation of polarization, being the case of linear polarization the simplest one. In the following, such technique would be described in relation with the acquisition of four interferograms with only one transversal grating translation between two pairs of interferograms. This enables the introduction of phase stepping techniques for phase extraction for the case of four interferograms, but reducing the number of phase shifts needed. In addition, the value of the grating translation does not need to be known precisely, because the incorporation of a technique developed by Kreis [29] allows its calculation.

3.1. Phase Shifts by Modulation of Linear Polarization

Figure 3.1. shows the experimental setup. It comprises a $4f$ Fourier transform system under monochromatic illumination at wavelength λ. The transforming lenses have a focal length of f. Linear polarizer P_0 have its transmission axis at 45° and the linear polarizers P_1 and P_2, over windows A and B, have its transmission axis at 0° and 90° respectively.

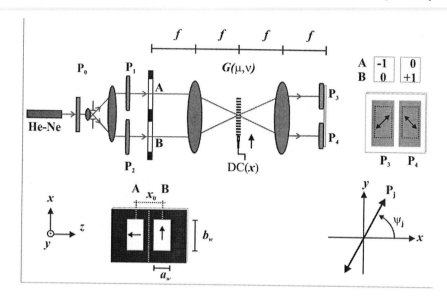

Figure 3.1. Experimental setup. P_j, $j = 0...4$: linear polarizers. A, B: rectangular windows in object plane. $G(\mu,\nu)$: Ronchi grating (μ,ν the spatial frequencies escalated by λf). DC(x): actuator. f: focal length. Side views of windows (lower left) and polarizers with $\psi = \psi_j$ in image plane (lower right) are also sketched.

The object plane (input plane) consists of two similar rectangular windows A and B, each of sides a_w and b_w. The windows centers are separated by the distance x_0. In general, an amplitude distribution of the form $A(x,y)$ can be considered in the window A as a reference wave, while $B(x,y)\exp[i\phi(x,y)]$ can represent the amplitude distribution in the window B (a test object, for instance). Then, the input transmittance can be expressed by

$$\mathbf{t}(x,y) = w(x+\tfrac{1}{2}x_0, y)A\left(x+\tfrac{1}{2}x_0, y\right)\mathbf{J}_A + w(x-\tfrac{1}{2}x_0, y)B\left(x-\tfrac{1}{2}x_0, y\right)\exp\left[i\phi\left(x-\tfrac{1}{2}x_0, y\right)\right]\mathbf{J}_B, \quad (3.1)$$

where $\mathbf{J}_A = \begin{pmatrix} 1 \\ 0 \end{pmatrix}$, $\mathbf{J}_B = \begin{pmatrix} 0 \\ 1 \end{pmatrix}$ are Jones vectors corresponding to orthogonal linear polarization states and the window function is written as $w(x,y) = rect(x/a_w) \cdot rect(y/b_w)$. A binary absorptive grating $G(\mu,\nu)$ with spatial period u_d and bright-band width u_w is placed in the frequency plane (Fourier plane, Figure 3.1). An actuator (DC) can translate the grating longitudinally through a given distance u_0. The grating (a Ronchi grating) can be written as

$$G(\mu,\nu) = rect\left(\frac{\mu - \mu_0}{\mu_w}\right) * \sum_{n=-\infty}^{\infty} \delta(\mu - n\mu_d), \quad (3.2)$$

with the spatial frequency coordinates are given by $(\mu,\nu)=(u/\lambda f, v/\lambda f)$, where (u,v) are the actual spatial coordinates, and $\mu_k = u_k/\lambda f$ with $k=0,w,d$ as a label for displacement, bright-band width and grating period respectively. The symbol $*$ means convolution.

In the image plane, the amplitude distribution can be written as the convolution between the amplitude of the object and the impulse response of the system, i.e., $t_O(x,y) = t(x,y) * \mathfrak{I}^{-1}\{G(\mu,\nu)\}$, with $\mathfrak{I}^{-1}$ the inverse Fourier transform operation performed by the second transforming lens as a convention taken in this work in accordance with an inversion in the image coordinates. Using Eq. (2), the convolution results in

$$t_O(x,y) = \sum_{n=-\infty}^{\infty} C_n \left\{ w_A\left[x-\left(\frac{n}{N_0}-\frac{1}{2}\right)x_0, y\right] \mathbf{J}_A + w_B\left[x-\left(\frac{n}{N_0}+\frac{1}{2}\right)x_0, y\right] \exp\left[i\phi\left(x-\left(\frac{n}{N_0}+\frac{1}{2}\right)x_0, y\right)\right] \mathbf{J}_B \right\}, \quad (3.3)$$

where it has been defined $w_K(x,y) = w(x,y) K(x,y)$ for $K = A, B$, $\mathfrak{I}^{-1}\{G(\mu,\nu)\} = \sum_{n=-\infty}^{\infty} C_n \delta(x - n/\mu_d)$ with $C_n = \frac{1}{2}\mathrm{sinc}(n/2)\exp(i2\pi n u_0/u_d)$ for $\mu_w = \frac{1}{2}\mu_d$, and assuming that x_0 equals some multiple integer N_0 of the period, in other words, $x_0 = N_0/\mu_d = N_0(\lambda f/u_d)$, (diffraction orders matching condition). According to Eq. (3), the amplitude in the image plane consists of a row of copies of the entrance transmittance, each copy separated by $1/\mu_d = \lambda f/u_d$ from the first neighbors. By adjusting the distance x_0 between windows such that $N_0/\mu_d = x_0$ and also assuring that the inequality $a_w \leq x_0$ is satisfied, the field amplitude $t_w(x,y)$ of window $w[x-(n/N_0 - 1/2)x_0, y]$ (as observed through a linear polarizer with transmission axis at angle ψ with respect to the horizontal) can be described as

$$t_w(x,y) = \{C_n A(x,y)\cos(\psi) + C_{n-N_0} B(x,y)\sin(\psi)\exp[i\phi(x,y)]\}\mathbf{J}_\psi. \quad (3.4)$$

where $\mathbf{J}_\psi^L \mathbf{J}_A = \cos(\psi)\mathbf{J}_\psi$, $\mathbf{J}_\psi^L \mathbf{J}_B = \sin(\psi)\mathbf{J}_\psi$, and

$$\mathbf{J}_\psi^L = \begin{pmatrix} \cos^2(\psi) & \sin(\psi)\cos(\psi) \\ \sin(\psi)\cos(\psi) & \sin^2(\psi) \end{pmatrix}; \quad \mathbf{J}_\psi = \begin{pmatrix} \cos(\psi) \\ \sin(\psi) \end{pmatrix}. \quad (3.5)$$

In particular, for the case $N_0 = 1$ and knowing that the irradiance results proportional to the square modulus of the field amplitude, $I(x,y) = |t_w(x,y)|^2$, the corresponding interference pattern is given by

$$I(x,y) = a(x,y) + b(x,y)\cos\left[\phi(x,y) - 2\pi\frac{u_0}{u_d}\right]. \quad (3.6.a)$$

where

$$a(x,y) = \frac{1}{4}\text{sinc}^2\left(\frac{1}{2}n\right)\cos^2(\psi)I_A(x,y) + \text{sinc}^2\left(\frac{1}{2}(n-1)\right)\sin^2(\psi)I_B(x,y), \qquad (3.6.b)$$

$$b(x,y) = \frac{1}{4}\sin(2\psi)\text{sinc}\left(\frac{1}{2}n\right)\text{sinc}\left[\frac{1}{2}(n-1)\right]\sqrt{I_A(x,y)I_B(x,y)}, \qquad (3.6.c)$$

with $I_K(x,y) = |K(x,y)|^2$ for $K = A, B$. This pattern results modulated by the functions $\sin(2\psi)$ and $\text{sinc}(n/2)$. Note that a and b are independent of position if the illumination is uniform in each object window. Otherwise, there can be smooth functions of the position and, in such a case there must give rise to corresponding spectra of a given extension, however small.

3.2. Phase Extraction

Selecting P_3 at angle $\psi = \psi_3 = 45°$ for $n = 0$ and P_4 at angle $\psi = \psi_4 = -45°$ for $n = 1$ with no grating displacement, $u_0 = 0$, two complementary patterns are first obtained at the image plane within replication regions given by $w[x + x_0/2, y]$ and $w[x - x_0/2, y]$. Such patterns can be written as

$$I_0(x,y) = a(x,y) + b(x,y)\cos[\phi(x,y)], \qquad (3.7.a)$$
$$I_1(x,y) = a(x,y) - b(x,y)\cos[\phi(x,y)]. \qquad (3.7.b)$$

This corresponds to patterns with a phase shift of $\alpha_1 = \pi$. Secondly, by performing an arbitrary translation of value $0 < u_0 < u_d/2$, the introduced phase shift is less than π radians. Then, another two interferograms result. Each one can be expressed as follows

$$I_2(x,y) = a(x,y) + b(x,y)\cos[\phi(x,y) - \Delta\phi], \qquad (3.7.c)$$
$$I_3(x,y) = a(x,y) - b(x,y)\cos[\phi(x,y) - \Delta\phi]. \qquad (3.7.d)$$

The corresponding phase shift for them is $\alpha_3 = \pi$. Considering patterns I_2, I_1 they differ by a phase shift of $\alpha_2 = \Delta\phi - \pi$, where $\Delta\phi = 2\pi \cdot u_0/u_d$. With this procedure, four interferograms with phase displacements of $f_0 = 0$, $f_1 = \pi$, $f_2 = \Delta\phi$, $f_3 = \Delta\phi + \pi$ can be obtained using only an unknown grating shift and, thus, two camera shots. The desired phase distribution can be calculated from

$$\phi_w(x,y) = \arctan\left\{\frac{I_2(x,y) - I_3(x,y) - [I_0(x,y) - I_1(x,y)]\cos(\Delta\phi)}{[I_0(x,y) - I_1(x,y)]\sin(\Delta\phi)}\right\}, \qquad (3.8)$$

where ϕ_w denotes the wrapped phase to be unwrapped further. From Eq.8, for the case of $\Delta\phi = 90°$, the well-known formula for four shifts can be obtained. Eq. (8) requires, of course, the knowledge of the value $\Delta\phi$ to be useful. In order to calculate $\Delta\phi$ from the same captured interferograms, the procedure suggested by T. Kreis [29] can be applied. This procedure is based on the Fourier transform analysis of fringes and a variant of it is proposed in the following sections to conceal it with the desired phase extraction.

3.3. Determination of $\Delta\phi$

Subtraction of Ec. (7b) from Ec. (7a) and Ec. (7d) from Ec. (7c) gives

$$g_1(x,y) = I_0(x,y) - I_1(x,y) = 2b(x,y)\cos[\phi(x,y)], \qquad (3.9.a)$$

$$g_2(x,y) = I_2(x,y) - I_3(x,y) = 2b(x,y)\cos[\phi(x,y) - \Delta\phi], \qquad (3.9.b)$$

procedure which eliminates a. In Eqs. (9.a) and (9.b), some dependence on position has been considered to include effects such as non uniform illumination, non linear detection or imperfections in the optical components. These subtractions avoid the Fourier transformation and the spatial filtering usually performed for the same purpose of eliminating a [11]. It is remarked that the Fourier transform procedure to eliminate a introduces an error due to the fact that, in general, the spectra from a and $b\cos\phi$ can be found mixed one with each other over the Fourier plane.

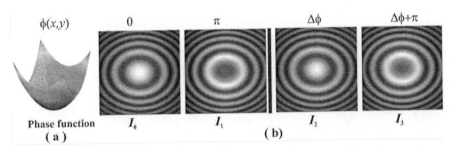

Figure 3.2. Simulated phase distribution ϕ (left) and corresponding shifted interference patterns.

Therefore, filtering out the a-spectrum around the zero frequency excludes also low frequencies from $b\cos\phi$ and, as a consequence, there is a corresponding loss of information related to ϕ and $\Delta\phi$. To calculate $\Delta\phi$, the method introduced by Kreis [29] is employed (an alternative can be seen in [30]).

An advantage of the variant that is proposed in this work consists of the elimination of a by subtracting two patterns. This way, there is no loss of information due to frequency suppression in the Fourier plane, as is the case of the method as proposed by Kreis. The proposed technique is illustrated in the example of the Figure 3.2. In addition, the technique is valid even when the phase function is more complex [29]. The 3-D plot at the left is the phase distribution $\phi(x,y)$, while the other plots are phase-shifted interferograms calculated from

$\phi(x,y)$. Interferograms I_0 and I_1 are mutually shifted by π rads, as well as I_2 and I_3, but between I_0 and I_2, and I_1 and I_3 is the same arbitrary phase of $\Delta\phi = \pi/7 \approx 0.39269908$ rads. This situation illustrates the kind of results that can be obtained with the setup of Figure1. Then, the problem is to find $\Delta\phi$ from the four interference patterns assuming that its value is not known. The solution is illustrated in Figure 3. The plots included are shown as an array of rows (seven letters) and columns (two numbers).

According to Eqs. (9.a) and (9.b), Figures 3.3-a1 and 3.3-a2 show the subtractions $g_1(x,y)$ and $g_2(x,y)$ respectively, where the four irradiances I_k were taken from the same interferograms of Figure 3.2. Next, Figure 3-b1 shows the Fourier spectrum of 3-a1 only ($g_1(x,y)$), while Figure 3-b2 depicts its resulting filtered spectrum in accordance with the method of Kreis. The used filter is a unit step [29], so the suppression of the left half of the spectrum is achieved. The following stage of the procedure to find $\Delta\phi$ consists of extracting the wrapped phase from the inverse Fourier transform of the already filtered $g_1(x,y)$, which is shown in Figure 3-c1. Figure 3-c2 shows the same procedure as applied to Figure 3.3-2a (i.e., $g_2(x,y)$). At this stage, the phases of the inverse Fourier Transform of each filtered interference pattern $g_1(x,y)$ and $g_2(x,y)$ are obtained. Therefore, these phases are wrapped, modified phases whose respective numerical integration results in two unwrapped, modified phases (Figures 3.3-d1 and 3.3-d2).

These unwrapped phases result in monotonous functions which are different from the desired phase ϕ, but their difference in each point (x,y) gives the modified phase difference $\Delta\phi'$ (Figures 3.3-e). This modified phase difference has to be constant for all interference pattern points, so an average over some range can be sufficient to calculate $\Delta\phi$ with a good approximation.

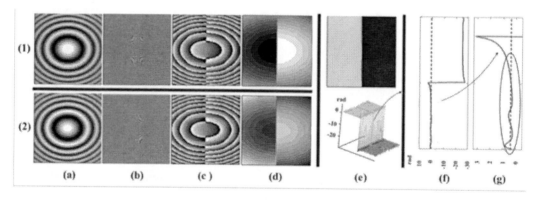

Figure 3.3. Variant of the method from Kreis (see text).

Figure3.3-f shows a line of Figure3.3-e1, and Figure 3.3-g shows a section of the Figure 3.3-f, where the used region to measure phase difference is indicated with an elliptic trace. The dots show the ideal phase difference $\Delta\phi$, while the red line shows the modified phase difference $\Delta\phi'$. The resulting value over the entire lower region of Figure 3.3-e2 was of $\overline{\Delta\phi'} = 0.39273364$, where the bar means average, so its difference with respect to the initial induced phase $\Delta\phi$ is of the order of 3.4556×10^{-5} rads. Taken $\overline{\Delta\phi'}$ as $\Delta\phi$, the wrapped phase distribution ϕ_w can be determined with Eq. (8) and the desired phase distribution ϕ can be identified with ϕ_u, the phase calculated from ϕ_w with standard unwrapping algorithm. The results of these stages are shown in Figure 3.4.

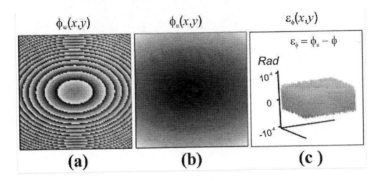

Figure 3.4. Phase extraction with Eq. (9) (left), unwrapped phase (center) and calculated error (right).

3.4. Experimental Results

The experimental setup follows closely the sketch of Figure 3.1, with $f = 479mm$, a laser He-Ne emitting at $\lambda = 632.8nm$, with a linear polarization at 45° by using P_0. $a_w = 10mm$, $b_w = 13mm$, and $x_0 = 12.45mm$, so the condition $a_w < x_0$ is fulfilled. The period of the Ronchi grating was $u_d = 25.4\mu m$. The grating was mounted on an actuator (Newport CMA-25CC). Note that the diffraction-order matching condition $x_0 = N_0(\lambda f / u_d)$ for $N_0 = 1$ is satisfied. The CCD camera (COHU 4815) is adjusted to capture the images of two interference patterns ($w[x + x_0/2, y]$ and $w[x - x_0/2, y]$) simultaneously. The fringe modulations of this interferogram pair are mutually complementary. By one actuator displacement, another pair of complementary interferograms can be obtained. This is shown in Figure 3.5, where the calculated value of the introduced phase was of $\Delta\phi = 14.4° = 0.2513274 rads$. The unwrapped phase is also shown. Each captured interferogram was subject to the same processing so as to get images with gray levels ranging from 0 to 255 before the use of Eq. (8). The sets of interferograms from three more objects are shown in Figure 6. Two of these objects were prepared by evaporating magnesium fluoride (MgF$_2$) on glass substrates (a phase dot, upper row, and a phase step, center row). The third object was oil deposited on a glass plate (bottom row). Each object was placed separately in one of the windows using the interferometer of Figure 3.1. Each set of four interferograms were obtained with a grating displacement between two of them as described. En each example, the phase-shifts induced by polarization can be visually identified from the complementary modulation contrasts of each pair. The calculated unwrapped phase for each object is shown in the rightmost column. Two typical profiles for some unwrapped phase are shown in Figure 3.7. The resulting unwrapping phases do not display discontinuities, as is the case when the wrong $\Delta\phi$ is taken in Eq.8.

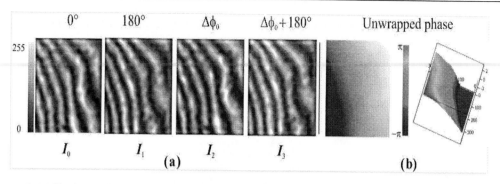

Figure 3.5. Tilted wavefront for testing. (a) Interference patterns. (b) Unwrapped phase. Phase shift measured according to the Kreis method: $\Delta\phi = 14.4° = 0.2513274 \ rads$.

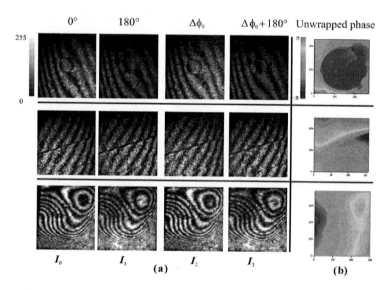

Figure 3.6. Test objects. (a) Interference patterns (b) Unwrapped phase. Upper row: phase dot, $\Delta\phi = 50.007° = 0.874 \ rad$. Center row: phase step, $\Delta\phi = 25.726° = 0.449 \ rad$. Lower row: still oil, $\Delta\phi = 60.700° = 1.059 \ rad$.

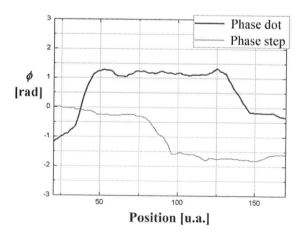

Figure 3.7. Typical slice profiles of the phase dot and the phase step along arbitrary directions.

The proposed phase-shifting interferometer is able to capture four useful interferograms with only one grating displacement and two shots. This grating displacement only requires being smaller than a quarter of period, but this condition is possible to verify by observing that the fringes do not shift enough to adopt a complementary fringe modulation. Because the phase shifts are achieved either by modulation of polarization or by grating displacement, not all of them results necessarily of the same value. The requirements of the arrangement are not very restrictive because it uses only very basic optical components, such as linear polarizers at angles of 0°, 45°, or 90°. A Ronchi grating from 500 to 1500 lines per inch can be used. As remarked before, the higher the grating frequency, the larger the distance between windows and more space to place samples becomes available, but the grating displacement has to be smaller. This last feature does not represent an impediment in this method because the actuator responsible for the displacement does not need of a calibration within a certain range, neither a precise displacement at a given prescribed value. Only one unknown displacement is needed and its value can be calculated each time it is employed. As for the fringe modulation, it is close to unity when using the 0 and ±1 diffraction orders for a typical Ronchi grating. Also, the two used patterns have the same fringe modulation because, as long as the two diffraction orders which superpose are of equal value, the involved amplitudes are the same. These features to extract static phase distributions make this proposal competitive as compared to the existing ones.

CONCLUSION

A common-path interferometer using two windows and a shifting grating for data acquisition is reliable. Although the interferograms involved are of different intensities and/or fringe contrast, a simple normalization procedure is enough to overcome these differences in order to apply phase-shifting techniques properly. The system is also able to perform tomographic reconstruction of transparent objects in the visible range using phase-shifting interferometry. In spite of the loss of aperture size and homogeneity in illumination, this system performs with reliability. In particular, the appearance of interferosinograms consisting of continuous fringes is an indication of the good stability of the system.

This system can be implemented to perform as a phase-shifting interferometer able to capture four useful interferograms with only one grating displacement and two shots. This grating displacement only requires being smaller than a quarter of period, but this condition is possible to verify by observing that the fringes do not shift enough to adopt a complementary fringe modulation. The requirements of the arrangement are not very restrictive because it uses only very basic optical components, such as linear polarizers at angles of 0°, 45°, or 90°. A Ronchi grating from 500 to 1500 lines per inch can be used. As remarked before, the higher the grating frequency, the larger the distance between windows and more space to place samples becomes available, but the grating displacement has to be smaller. This last feature does not represent an impediment in this method because the actuator responsible for the displacement does not need of a calibration within a certain range, neither a precise displacement at a given prescribed value. Only one unknown displacement is needed and its value can be calculated each time it is employed. As for the fringe modulation, it is close to unity when using the 0 and ±1 diffraction orders for a typical Ronchi grating. Also, the two used patterns have the same fringe modulation because, as long as the two diffraction orders which superpose are of equal value, the involved amplitudes are the same. These features to extract static phase distributions make this last proposal competitive as compared to the existing ones.

REFERENCES

[1] V. Ronchi, "Forty Years of History of a Grating Interferometer," *Appl. Opt.* 3, (1964) 437-451.
[2] R. Barakat, "General Diffraction Theory of Optical Aberration Tests, from the Point of View of Spatial Filtering," *J. Opt. Soc. Am.* 59 (1969) 1432.
[3] A. Cornejo-Rodriguez, *Ronchi test*, in: *Optical Shop Testing*, D. Malacara (Ed.), Wiley, New York (1992) pp. 321-365.
[4] M. V. Mantravadi, *Lateral shearing interferometers*, in: *Optical Shop Testing*, D. Malacara (Ed.), Wiley, New York (1992) pp. 123-172.
[5] A. Asssa, A. A. Betser, and J. Politch, "Recording slope and curvature contours of flexed plates using a grating shearing interferometer," *Appl. Opt.* 16 (1977) 2504-2513.
[6] E. N. Leith and B. J. Chang, "Space-Invariant Holography With Quasi-Coherent Light," *Appl. Opt.* 12 (1973). 1957-1963.
[7] B. J. Chang, R. Alferness, and E. N. Leith, "Space-invariant achromatic grating interferometers: theory," *Appl. Opt.* 14 (1975) 1592-1600.
[8] Yih-Shyang Cheng, "Fringe formation in incoherent light with a two-grating interferometer," *Appl. Opt.* 23 (1984) 3057-3059.
[9] K. Creith, *Phase-measurement interferometry techniques*, in: E. Wolf, (Ed.), *Progress in Optics*, vol. 26, North-Holland, Amsterdam, 1998.
[10] J. Schwider, R. Burow, K.-E. Elssner, J. Grzanna, and R. Spolaczyk, "Semiconductor wafer and technical flat planeness testing interferometer," *Appl. Opt.* 25 (1986) 1117-1121.
[11] V. Arrizón and D. Sánchez-de-la-Llave, "Common-path interferometry with one-dimensional periodic filters," *Opt. Lett.* 29 (2004) 141-143.
[12] H. Weinberger and U. Almi, "Interference Method for Pattern Comparison," *Appl. Opt.* 10 (1971) 2482-2487.
[13] J. W. Goodman, *Introduction to Fourier Optics*, McGraw-Hill, New York, 1988 pp. 243-246.
[14] C. Meneses-Fabian, G. Rodríguez-Zurita, and V. Arrizón, "Optical tomography of transparent objects with phase-shifting interferometry and stepwise-shifted Ronchi ruling," *J.O.S.A. A.* Vol.23, N.2, 2006, 298-305.
[15] S. R. Dean, *The Radon transform and some of its applications*, (Wiley, New York, 1983), pp.42-48, 128-147.
[16] A. C. Kak, M. Slaney, *Principles of computarized tomographic imaging*, (IEEE Press, New York, 1987).
[17] T. K. Gaylord, C. C. Guest, Optical interferometric liquid gate plate positioner, *Rev. Sci. Instrum.*, 55, n.6, 866-868 (1984).

[18] D. Malacara and S. Mallik, *C.3 Common-Path Interferometers*, in *Optical Shop Testing*, D. Malacara Ed., John Wiley and Sons, New York (2007) 97-118.

[19] J. E. Millerd, N. J. Brock, "Methods and apparatus for splitting, imaging, and measuring wavefronts in interferometry," U.S. Patent 20030053071A1 (2003).

[20] H. Schreiber, J.H. Bruning, *C.14 Phase Shifting Interferometry*, in *Optical Shop Testing*, D. Malacara Ed., Wiley and Sons, New York (2007) 547-655.

[21] D. Malacara, M. Servín, Z. Malacara, *C.5, Interferogram Analysis for Optical Testing*, Taylor and Francis, New York (2005) 159-254.

[22] J. Schwider, *Advanced evaluation techniques in interferometry*, Progress in Optics, E. Wolf, Ed., North-Holland (1990) 271-359.

[23] B. Barrientos-García, A. J. Moore, C. Perez-Lopez, L. Wang and T. Tshudi, *Appl. Opt.* 38 (1999) 5944-5947.

[24] M. Novak, J. Millerd, N. Brock, M. North-Morris, J. Hayes and J. Wyant, "Analysis of a micropolarizer array-based simultaneous phase-shifting interferometer," *Appl. Opt.* 44 (2005) 6861-6868.

[25] J. H. Bruning, D. R. Herriott, J. E. Gallagher, D. P. Rosenfeld, A. D. White, and D. J. Brangaccio, "Digital Wavefront Measuring Interferometer for Testing Optical Surfaces and Lenses," *Appl. Opt.* 13 (1974) 2693-2703.

[26] Hariharan, B.F. Obrem and T. Eiju, "Digital phase-shifting interferometry: a simple error-compensating phase calculation algorithm," *Appl. Opt.* 26 (1987) 2504-2506.

[27] Jirí Novák, Pavel Novák, Antonín Mikš, "Multi-step phase-shifting algorithms insensitive to linear phase shift errors," *Opt. Commun.* 281 (2008) 5302–5309.

[28] X. F. Xu, L. Z. Cai, Y. R. Wang, X. F. Meng, W. J. Sun, "Simple direct extraction of unknown phase shift and wavefront reconstruction in generalized phase-shifting interferometry: algorithm and experiments," *Opt. Letters* 33 (2008) 776-778.

[29] T. Kreis, "Digital holographic interference-phase measurement using the Fourier-transform method," *J. Opt. Soc. Am. A* 3 (1986) 847-855.

[30] X.F. Meng, L.Z. Cai, Y.R. Wang, X.L. Yang, X.F. Xu, G.Y. Dong, X.X. Shen, and X.C. Cheng, "Wavefront reconstruction by two-step generalized phase-shifting interferometry," *Opt. Commun.* 281 (2008) 5701–5705.

In: Interferometry Principles and Applications
Editor: Mark E. Russo

ISBN 978-1-61209-347-5
© 2012 Nova Science Publishers, Inc.

Chapter 10

ELECTRONIC SPECKLE PATTERN INTERFEROMETRY: PRINCIPLES AND APPLICATIONS

Jiong-Shiun Hsu,[1] Chi-Hung Hwang[2] and Wei-Chung Wang[3]

[1]Department of Power Mechanical Engineering, National Formosa University,
Taiwan, Republic of China
[2]Instrument Technology Research Center,
National Applied Research Laboratories, Taiwan, Republic of China
[3]Department of Power Mechanical Engineering and Dean, Office of International Affairs,
National Tsing Hua University, Taiwan, Republic of China

ABSTRACT

Since Electronic speckle pattern interferometry (ESPI) has the attractive merits such as non-contact, full-field, highly sensitive, etc., it has been a powerful tool for the measurement in practice, especially for the object with diffuse surface. In this chapter, the optical theories of ESPI including static and vibration measurements will first be introduced. Then the different optical arrangements in ESPI will be described and their advantages and disadvantages will be compared. Finally, some examples of ESPI respectively applied in-plane, out-of-plane and vibration measurements will be given.

1. INTRODUCTION

For engineering practice, mechanical behaviors and physical properties of engineering components need to be determined. Among the well-established measurement techniques, optical interferometry is a useful tool due to its real-time, whole-field, non-contact and highly sensitive characteristics. The surface of engineering components can be either diffuse or specular. Both the diffuse and specular surface characteristics can be measured by using the appropriate techniques of optical interferometry. In engineering practice, components with diffuse surface are more encountered e.g. vehicles, electronic packages, civil engineering structures, bio-membranes and so on; therefore, the demand of appropriate measurement techniques to measure and/or analyze diffuse surfaces are drastically increased.

Speckle phenomenon can be observed if a high coherent light, e.g. laser, is utilized to illuminate the surface of the diffuse object. Speckle distribution can be easily recorded by a charge coupled device (CCD) camera. Roosted with the speckle phenomenon and an interferometer, electronic speckle pattern interferometry (ESPI) was proposed in 1970s because of the use of photo-detectors. In contrast to the traditional holography [1], it is unnecessary to prepare a hologram with tedious procedures because the speckle image in ESPI can not only quickly be recorded but also promptly be updated by means of the digital image processing techniques. Moreover, in-plane, out-of-plane and 3-dimensional displacements of the object subjected to static or dynamic loading can be respectively measured with proper optical arrangements [2-4]. Therefore, ESPI is one of the best optical methods to be employed in the investigation of the deformation of diffuse objects.

ESPI was first proposed by Butters and Leendertz [5] in 1971. Since ESPI is a real-time, whole-field, non-contact and highly sensitive experimental technique, it has been widely used in engineering applications. Moore and Tyrer [6] adopted ESPI to extract fracture parameters around a crack. Liu and Shang [7] used the single mode optical fiber to establish ESPI in the non-destructive detection for composite materials containing defects. Bendek et al. [8] used ESPI to evaluate the measurement uncertainty of residual stress in plates by the layer removal technique. Dai et al. [9] established out-of-plane ESPI system based on Michelson interferometer to measure the thickness change of transparent plates subjected to tensile loading. In addition to aforementioned studies using ESPI in the static measurements, ESPI was also employed in the vibration measurement. Malmo and Vikhagen [10] used ESPI to measure the vibration of a car body. Slangen et al. [11] used ESPI to investigate the vibration characteristics of a cantilever aluminum plate and compared the results to those of finite element method. Jin et al. [12] utilized ESPI to evaluate the quality of microphones and loudspeakers from the observation of their vibration mode shapes. Spagnolo et al. [13] respectively adopted ESPI and local speckle correlation to observe the mode shapes of a plate.

The experiments in these studies relating to the vibration measurement of ESPI, namely traditional ESPI, are to sequentially grab two speckle images before and after the objects subjected to harmonic excitation. Then the fringe patterns indicating the model shape are obtained through subtracting these two speckle images.

However, the traditional ESPI has two drawbacks in the vibration measurement. The first is its fringe pattern is not clear and the other is that its sensitivity to displacement has to be improved. To simultaneously improve these two shortcomings without adding any optical components to that of traditional EPSI system, Hwang [14] and Wang, et al. [15] proposed amplitude fluctuation ESPI (AF-ESPI) based on the amplitude fluctuation of object under the fixed harmonic excitation. The AF-ESPI was employed to investigate problems in different areas [16-20].

The aim of this chapter intends to introduce the development of ESPI. The optical arrangements of ESPI for in-plane and out-of-plane displacement measurements will be described. Then the principles of EPSI including the static and vibration measurements will be given. Finally, applications of ESPI in evaluation of patching efficiency, nondestructive inspection of the defects inside composite materials and evaluating dynamic SIFs of structures containing cracks will be summarized.

2. OPTICAL ARRANGEMENTS OF ESPI

ESPI can be used to measure the in-plane and out-of-plane displacements with the proper optical arrangements. Typical optical arrangements of ESPI for in-plane and out-of-plane displacement measurements are depicted in Figure 2.1.(a) and Figure 2.1.(b), respectively.

For the optical arrangements of in-plane displacement measurement, a polarized laser is adopted as the light source and the power of laser used depends on the area of object to be measured. The laser beam is first divided into two beams by a non-polarized beam splitter (NPBS).

According to the optics theory [21], the contrast of fringe pattern interfered by two beams achieves maximum if the optical intensities of these two beam are the same. This is the reason that NPBS is employed herein. Two laser beams individually pass through two mirrors and continue to be expanded with noise being filtered out by spatial filters. Two expanded laser beams then illuminate the measured object and finally reflect into CCD camera. The speckle image is grabbed by the image processing board and displayed on the monitor.

For the optical arrangements of out-of-plane displacement measurement, a laser beam coming from a polarized laser is divided into two beams. One beam passes through a mirror and expands with noise being filtered out by a spatial filter to illuminate the surface of object. This beam is next reflected by the object into CCD camera and it is the so-called object beam.

On the other hand, the other beam is expanded with noise being filtered out and following the passes of three mirrors and a polarizer. Then the expanded beam strikes another beam splitter and is finally reflected into CCD camera after passing through a ground glass. Because this beam doesn't meet with the object, it is named as the reference beam. The placement of a ground glass on the path of reference beam is to increase the contrast of fringe pattern.

Based on the coherence theory [21], the optium lengths of optical paths of two beams have to be the same such that the obtained fringe pattern is much clearer. For this reason, travel distances of object and reference beams must be carefully checked.

Moreover, it was mentioned previously that intensities of two beams have to be the same to achieve the maximum contrast of fringe pattern. It should be noted that the intensity here means the intensity displays on the monitor rather than the intensity observed by eyes. Because the traveling distance after expanding object beam is longer than that of reference beam, the optical intensity of reference beam observed by eyes is significantly smaller than that of object beam.

To increase the experimental convenience, a polarizer is therefore placed on the path of reference beam so that optical intensity of reference beam can be continuously adjustable. This is also the reason that a polarized laser is employed herein. An attenuator is an alternative choice to adjust the optical intensity of reference beam but adopting a polarizer is strongly recommended.

Although the intensity of reference beam can also be continuously changed through rotating the attenuator, this usually leads the reference beam to tilt. Because a spatial filter comprises an objective and a pinhole, the intensity of reference beam may be significantly decreased due to the rotation of the attenuator and thus the spatial filter has to be iteratively adjusted until the intensities of object and reference beams on the monitor are sufficiently close.

Besides, the optical arrangements in Figure 2.1.(b) have both advantage and disadvantage. The advantage is the arrangements are suitable for the measurement of out-of-plane displacement of large size object whereas the disadvantage is that the stability of fringe pattern is easily influenced by the air disturbance.

If the area of measured object is small, it is recommended to establish the ESPI system based on the Michelson interferometer depicted in Figure 2.2. because the stability of its fringe pattern is less sensitive to air disturbance.

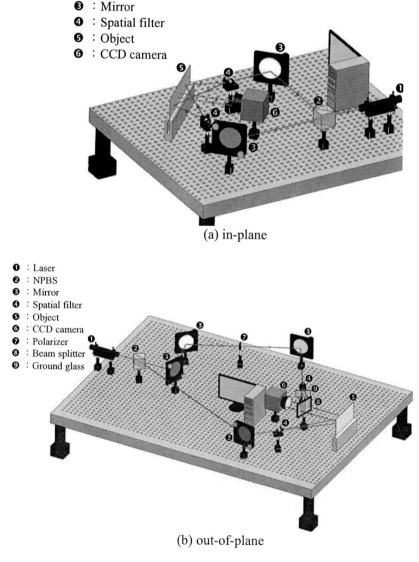

Figure 2.1. In-pane and out-of-plane ESPI setups.

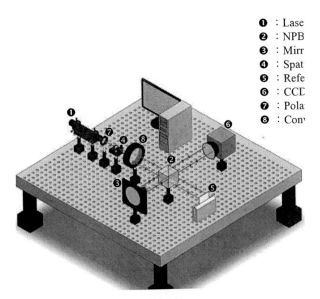

❶ : Lase
❷ : NPB
❸ : Mirr
❹ : Spat
❺ : Refe
❻ : CCL
❼ : Pola
❽ : Con

Figure 2.2. The out-of-plane optical arrangements of ESPI based on Michelson interferometer.

3. OPTICAL THEORIES OF ESPI [5, 14, 15, 22-24]

With the proper optical arrangements, ESPI can be employed to respectively measure the in-plane and out-of-plane displacements of objects subjected to either static or vibration loading. Their optical theories will be provided in this section. The ESPI theory of static measurement will first be introduced. Based on the theory of static measurement, traditional ESPI theory of vibration measurement will then be given. Finally, the theory of AF-ESPI will be given.

3.1. Principle of ESPI for Static Measurement

There are two common points of in-plane and out-of-plane ESPI static measurements. One is the need of two beams, object and reference beams, overlapping together in a zone to generate the interference. The other is to successively grab two speckle images before and after deformation and then subtract two images between each other to obtain the ESPI fringe pattern. Therefore, the intensity governing equation of ESPI fringe pattern has to be the same but the difference is the meaning between the fringe pattern and the corresponding displacement component. Prior to deformation of the object, the reference (u_r) and object (u_o) beams can be expressed as

$$ \tag{3.1.1}$$

$$ \tag{3.1.2}$$

where A_r and A_o respectively denote the amplitudes of reference and object beams; $\phi_r = \frac{2\pi}{\lambda} R_r$ and $\phi_o = \frac{2\pi}{\lambda} R_o$ are phases of reference and object beams, respectively; λ is the wavelength of laser; R_r and R_o are distances that reference and object beam respectively travels. The intensity before deformation, I_1, is

$$I_1 = I_o + I_r + 2\sqrt{I_o I_r} \cos\phi \qquad (3.1.3)$$

where ϕ is the phase difference between reference and object beams before deformation. After deformation, the intensity, I_2, is

$$I_2 = I_o + I_r + 2\sqrt{I_o I_r} \cos(\phi + \varphi) \qquad (3.1.4)$$

where φ is the phase difference between reference and object beams caused by applied loading. Subtract the image before deformation from the image after deformation and convert the negative signals to positive to avoid missing the deformation information. Thus, the intensity of fringe pattern, I, is

$$\begin{aligned} I &= \left| 2\sqrt{I_o I_r} \cos(\phi - \varphi) - \cos\phi \right| \\ &= \left| -4\sqrt{I_o I_r} \sin(\phi + \frac{\varphi}{2}) \sin\frac{\varphi}{2} \right| \end{aligned} \qquad (3.1.5)$$

Equation (3.1.5) consists of three terms, i.e. $4\sqrt{I_o I_r}$, $\sin(\phi + \frac{\varphi}{2})$, and $\sin\frac{\varphi}{2}$. However, the first term maintains a constant and the variation frequency of the second term is too high to be observed. Therefore, the fringe pattern is dominant by the third term of Eqn. (3.1.5).

3.1.1. In-plane ESPI

Suppose the deformation of object be rather small and the geometrical configurations of reference and object beams for in-plane ESPI before and after deformation of objects are illustrated in Figure 3.1, respectively. Let a point P on the surface of object moves to point P' after deformation. As can be seen from Figure 3.1, the increases of optical paths of reference, δ_r, and object, δ_o, beams resulting from deformation are

$$\delta_r = u\sin\alpha + w\cos\alpha \qquad (3.1.6)$$

$$\delta_o = -u\sin\beta + w\cos\beta \qquad (3.1.7)$$

where u is in-plane displacement; w is out-of-plane displacement; α is the incident angle of reference beam; β is the incident angle of object angle. Suppose the incident angles of reference and object beams both be equal to θ. The phase φ can be

$$\varphi = \frac{2\pi\delta}{\lambda} = \frac{4\pi u \sin\theta}{\lambda} \tag{3.1.8}$$

where $\delta = \delta_r - \delta_o$ is the difference of optical paths of reference and object beams introduced by deformation. In Eqn. (3.1.5), $I = 0$ when

$$\varphi = 2N\pi \ (N = 0, 1, 2, 3 \cdots) \tag{3.1.9}$$

Thus, the in-plane displacement is

$$u = \frac{N\lambda}{2\sin\theta} \ (N = 0, 1, 2, 3 \cdots) \tag{3.1.10}$$

where N is the fringe order.

3.2.2. Out-of-Plane

The geometrical configuration of out-of-plane ESPI is shown in Figure 3.2. in which a point P located on the object surface displaces to point P' after deformation. In contrast to the change of object beam in in-plane ESPI, the optical path of reference beam remains the same for out-of-plane ESPI arrangement during the deformation of object. As can be derived from Figure 3.2, the increase of optical path between object and reference beams, δ, is

$$\delta = w(\cos\alpha + \cos\beta) + u(\sin\alpha - \sin\beta) \tag{3.1.11}$$

where α and β are the incident and reflection angles of object beam, respectively. In addition to the small deformation assumption, the in-plane displacement, u, is assumed to be significantly small in comparison with the out-of-plane displacement, w, and it gives

$$\delta = w(\cos\alpha + \cos\beta) \tag{3.1.12}$$

It means that the corresponding phase difference between object and reference beams, φ, is

$$\varphi = \frac{2\pi\delta}{\lambda} = \frac{2\pi(\cos\alpha + \cos\beta)}{\lambda} \tag{3.1.13}$$

Substitute Eqn. (3.1.13) into Eqn. (3.1.5) and let $I = 0$ in Eqn. (3.1.5) yields

$$w = \frac{N\lambda}{\cos\alpha + \cos\beta} \quad (N =, 1, 2, 3 \ldots) \tag{3.1.14}$$

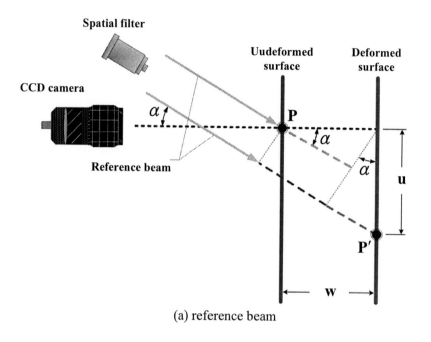

(a) reference beam

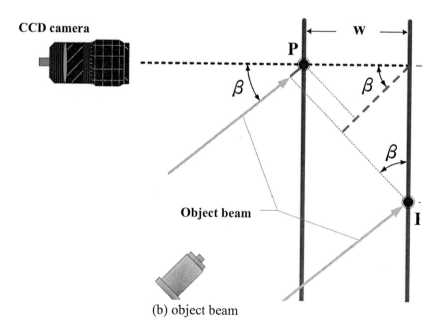

(b) object beam

Figure 3.1. The geometrical configuration of reference and object beams of in-plane ESPI.

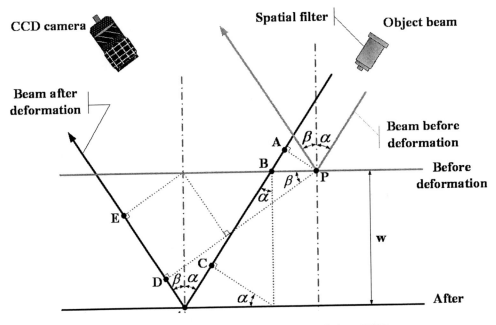

Figure 3.2. The geometrical configuration of object beam of out-of-plane ESPI.

3.2. Traditional ESPI Theory of Vibration Measurement

Regarding periodic vibration measurement, typically, there are three techniques which can be used with optical methods for vibration measurement, i.e. (a) time-averaged (TA) method; (b) stroboscopic method and (c) the use of a pulsed laser. The stroboscopic method and the pulsed laser application control the CCD shutter open frequency or lasing frequency to make the vibration object looks like at "static" state. Therefore, as the ESPI method combines with these two measurement techniques is applied to vibration measurement, the measurement principle is the same as the static case discussed in previous sections.

The time-averaged technique is the easiest one to implement with different optical methods, especially when measuring periodic vibration is requested. When the ESPI method combines with the time-averaged technique is termed as time-averaged ESPI method or abbreviated as TA ESPI. When the object is under periodic motion, the light intensity detected by a CCD camera at time t can be expressed as

$$I(\tilde{x},t) = I_o + I_r + 2\sqrt{I_o I_r}\cos[\varphi + \Delta(\tilde{d},t)] \qquad (3.2.1)$$

where $\Delta \equiv 2\pi\varsigma\,\tilde{n}\bullet(\tilde{d}\cos\omega t)/\lambda$; "$\bullet$" is the inner product operator ; ς is the optical arrangement parameter ; $\tilde{n}$ is the normal unit vector along observation direction ; $\tilde{d}$ is the vibration displacement vector and ω is the vibration frequency.

Since the CCD camera itself will accumulate the incoming light conducted electrons during shutter opening time interval, τ, then convert the light conducted electrons with readout circuit into output voltage. The output voltage signal becomes

$$V_{vib} = \alpha \int_0^\tau \left[I_o + I_r + 2\sqrt{I_o I_r} \cos(\varphi + \Delta) \right] dt \qquad (3.2.2)$$

where α is the slope of the CCD camera's sensitivity curve to indicate the relation between incoming light intensity and the output voltage signal. Assuming τ is set to be a complete integral vibration period to avoid possible mismatch between the observed fringe patterns and mathematical model. Then the output voltage signal, becomes

$$V_{vib} = \alpha \tau \left[I_o + I_r + 2\sqrt{I_o I_r} J_o(k\tilde{n} \bullet \tilde{d}) \cos\varphi \right] \qquad (3.2.3)$$

where $k \equiv 2\pi\varsigma/\lambda$. In most cases, the output voltage signal will be divided by shutter opening time interval, τ, to emphasize that the image is captured at a certain shot, then Eqn. (3.2.3) can be modified into

$$V_{TA} = \alpha \left[I_o + I_r + 2\sqrt{I_o I_r} J_o(k\tilde{n} \bullet \tilde{d}) \cos\varphi \right] \qquad (3.2.4)$$

For convenience, Eqn. (3.2.4) is defined as the TA vibration ESPI image. The vibration history of the object is continuously recorded by the CCD. This ESPI method is named as image-signal-addition ESPI method (ISAM ESPI). Eqn. (3.2.4) shows that the obtained ISAM ESPI image is modulated by the zero-order Bessel function with the sum of the reference and object light intensities as the background. Because of this background, the major disadvantage of ISAM ESPI is the poor fringe visibility. To enhance the ISAM ESPI fringe visibility, the first two terms of Eqn. (3.2.4) have to be blocked out. Operating the TA vibration ESPI image with high pass filter, the ISAM ESPI can be expressed as

$$V_{ISAM} = 4\alpha I_o I_r J_o^2(k\tilde{n} \bullet \tilde{d}) \cos\varphi \qquad (3.2.5)$$

The ISAM ESPI image shows dark fringe with a set of can null Eqn.(3.2.5). Then the vibration amplitude can be evaluated by

$$|\tilde{d}| = \frac{(\xi_{ISAM})_i}{k \cos(\tilde{n}, \tilde{d})} \qquad (3.2.6)$$

The other way to eliminate the background light intensity is to subtract an image with background from TA vibration ESPI image. The reference ESPI image, V_{TA-R}, is defined as an ESPI image captured under the same optical setup and the same shutter opening interval for the same object but without vibration.

Considering an object at stationary state is taken; then from Eqn. (3.2.1), the reference ESPI image captured by the CCD can be expressed as

$$V_{TA-R} = \alpha\left[I_o + I_r + 2\sqrt{I_o I_r}\cos\varphi\right] \quad (3.2.5)$$

Subtracting the reference ESPI image from the TA vibration ESPI image, the background is removed. This method is named as "image-signal-subtraction method (ISSM)". The output signal of the ISSM ESPI, V_{ISSM}, can be described by

$$V_{ISSM} = |V_{TA} - V_{TA-R}| = 2\alpha\tau\sqrt{I_o I_r}\left[1 - J_o(k\tilde{n}\bullet\tilde{d})\right]\cos\varphi \quad (3.2.5)$$

Mathematically, the ISSM ESPI fringes are modulated by function $\left[1 - J_o(k\tilde{n}\bullet\tilde{d})\right]$, and the function converges toward 1 and is always positive. Therefore, instead nulling Eqn. (3.2.5), the locations of the fringes are formulated by a set of ξ_{ISSM} that can locally minimize the values of Eqn. (3.2.5). The vibration amplitude $|\tilde{d}|$ can then be determined by

$$|\tilde{d}| = \frac{(\xi_{ISSM})_i}{k\cos(\tilde{n},\tilde{d})} \quad (3.2.6)$$

3.3. AF-ESPI Theory

For ISSM ESPI method, the reference ESPI image is captured as the object is at static state. However, the reference ESPI image can be acquired when the object experiences the same nominal vibration force as the TA vibration ESPI image. During vibration measurement, the variations of vibration driving force are generally introduced by environmental or electronic disturbance. When the disturbance is large, correlation between the reference ESPI image and TA vibration ESPI image is lost and cannot be used for measurement. On the contrary, when the disturbance is negligible, the disturbance works like modulation force and the vibration amplitudes can be assumed to change from $\tilde{d}$ into $\tilde{d} + \Delta\tilde{d}$. The output signal of the reference ESPI image now becomes

$$V_{TA-disturbed} = \alpha\left[I_o + I_r + 2\sqrt{I_o I_r}J_o(k\tilde{n}\bullet[\tilde{d}+\Delta\tilde{d}])\cos\varphi\right] \quad (3.3.1)$$

Subtracting the reference ESPI image from TA vibration ESPI image with disturbance considered, the output signal can be expressed as

$$V_{AF} = 2\alpha\sqrt{I_o I_r}\left\{J_o(k\tilde{n}\bullet[\tilde{d}+\Delta\tilde{d}]) - J_o(k\tilde{n}\bullet\tilde{d})\right\}\cos\varphi \quad (3.3.2)$$

In fact, in Eqn.(3.3.2) $\Delta \tilde{d}$ is one of the undetermined parameters and is difficult to control. As indicated in Eqn. (3.3.2), the output voltage V_{AF} is expressed by the zero-order Bessel function. The $\Delta \tilde{d}$ must be small enough to maintain the correlation between TA vibration ESPI image and reference ESPI image. Dividing each side of Eqn. (3.3.2) by $k \cos(\tilde{n}, \Delta \tilde{d})$ and taking limit of both sides, Eqn. (3.3.2) is simplified and can be expressed as a function of vibration amplitude only

$$\frac{\partial V_{AF}}{\partial (|\tilde{d}|)} = -2\alpha \left[\sqrt{I_o I_r} J_1(k\tilde{n} \bullet \tilde{d}) \cos\varphi \right] \tag{3.3.3}$$

When the output signal is transformed into image gray level; the negative voltage value must be rectified into the positive value, the output signal V_{AF} becomes

$$V_{AF} = 2\alpha \sqrt{I_o I_r} \left[J_1^2(k\tilde{n} \bullet \tilde{d}) \cos^2\varphi \right]^{1/2} \tag{3.3.4}$$

Based on Eqn. (3.3.4), the vibration amplitude can be determined by a set of $k\tilde{n} \bullet \tilde{d} = (\xi_{AF})_i$ which can null Eqn. (3.3.4); then the vibration amplitude is described by

$$|\tilde{d}| = \frac{(\zeta_{AF})_i}{k \cos(\tilde{n}, \tilde{d})} \tag{3.3.5}$$

4. APPLICATIONS OF ESPI

4.1. Evaluation of Patching Efficiency [25, 26-28]

Defects are occasionally present in structures or components during the manufacturing process and damages may be unavoidably produced in service. When defects or damages existing within the structures or components are not critical enough to endanger their safety, bonded repair provides a convenient and safe way to avoid replacing the structures or components. Besides, composite materials have advantages such as light weight, high strength-to-weight and stiffness ratios, etc; they have been widely utilized in various applications.

Therefore, the patching efficiency of aluminum alloy sheet containing a central crack repaired by a composite patching was investigated. In-plane ESPI was employed to measure the full-field in-plane displacement of cracked aluminum sheets subjected to tensile loadings before and after repaired by composite patching. The hybrid method combining ESPI displacement data and finite element method was then adopted to evaluate the repaired efficiency of composite patching with different fiber staking orientations.

Neglecting the effects of body force, the horizontal (u), and vertical (v) displacements at a point P for an isotropic, homogeneous and linear-elastic plate containing a central crack subjected to a tensile loading are

$$u = \frac{1+v}{E}\left\{\sum_{n=0}^{\infty}\frac{A_n}{(n-\frac{1}{2})}r^{(n+\frac{1}{2})}\left[\frac{(1-2v)}{(n+\frac{1}{2})}\cos(n+\frac{1}{2})\theta - \sin\theta\sin(n-\frac{1}{2})\theta\right] + \sum_{p=0}^{\infty}C_p r^{(p+1)}\left[\sin\theta\sin p\theta - \frac{2(1+v)}{(p+1)}\cos(p+1)\theta\right]\right\} \quad (4.1.1)$$

$$v = \frac{1+v}{E}\left\{\sum_{n=0}^{\infty}\frac{A_n}{(n-\frac{1}{2})}r^{(n+\frac{1}{2})}\left[\frac{2(1-v)}{(n+\frac{1}{2})}\sin(n+\frac{1}{2})\theta - \sin\theta\cos(n-\frac{1}{2})\theta\right] + \sum_{p=0}^{\infty}C_p r^{(p+1)}\left[\frac{\sin(p+1)\theta}{(p+1)} - (1-2v)\sin\theta\cos p\theta\right]\right\} \quad (4.1.2)$$

where E is the Young's modulus of plate; v is the Poisson's ratio; r is the distance between the crack tip to point P; θ is the azimuth angle of point P; A_i and C_i are coefficients. For mode I dominated problem, the stress intensity factor (SIF), K_I, and the far-filed stress, $(\sigma_{ox})_I$ are

$$K_I = -2\sqrt{2\pi}A_0 \quad (4.1.3)$$

$$(\sigma_{ox})_I = 2C_0 \quad (4.1.4)$$

where A_0 and C_0 are the first coefficient of the first and second power-series terms in the parenthesis of Eqn. (4.1.1) and Eqn. (4.1.2). Baker [27] showed that the modified equivalent stiffness, $E_{equ} t_{equ}$, on the patched area is

$$E_{equ}t_{equ} = E_s t_s + E_p t_p \quad (4.1.5)$$

where E_s and t_s are respectively the Young's modulus and thickness of a cracked plate; E_p and t_p are respectively the Young's modulus and thickness of a patching; $t_{equ} = t_s + t_p$ is the total thickness on the patching area. The generally orthotropic lamina in off-axis coordinates is shown in Figure 4.1.

The effective Young's modulus, E_x, for a single lamina is

$$E_x = \cfrac{1}{\cfrac{1}{E_1}\cos^4\theta + \left[-\cfrac{2\nu_{12}}{E_1} + \cfrac{1}{G_{12}}\right]\cos^2\theta\sin^2\theta + \cfrac{1}{E_2}\sin^4\theta} \qquad (4.1.6)$$

where E_1 is the longitudinal modulus of elasticity associated with the 1 direction; E_2 is the longitudinal modulus of elasticity associated with the 2 direction; ν_{12} is Poisson's ratio associated with the 1-2 plane; G_{12} is shear modulus associated with the 1-2 plane. The effective Young's modulus along the direction defined in Figure 4.1. can be substituted into E_p in Eqn. (4.1.5) to obtain the equivalent Young's modulus of the composite patching.

The geometrical configuration of the sheet containing a central crack is shown in Figure 4.2.(a) and its constituent material is aluminum alloy 6061-T6. Young's modulus and Poisson's ratio are 68.95 GPa and 0.33, respectively. The composite patching was prepared by stacking the prepregs (Toho Rayon Co., Japan) in a hot-press machine. The geometrical configuration is also shown in Figure 4.2.(b). The ESPI fringe patterns for different stacking orientations of with 3 layer double-sided patching are shown in Figure 4.3. It can be observed that the fringe patterns vary with stacking orientations of composite patchings. The corresponding SIF can be extracted from the aforementioned description and the repair efficiency for the composite patching can be evaluated.

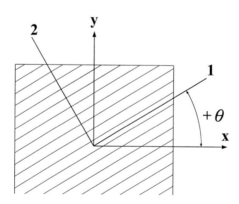

Figure 4.1. The generally orthotropic lamina in off-axis coordinates.

4.2. Nondestructive Inspection of the Defects inside Composite Material [29, 30]

The other application of ESPI method is the nondestructive inspection of the inner defects of composite structures made of carbon fiber reinforced plastic (CFRP). The CFRP is widely adopted to high-end bicycles, aeronautic vehicles, wind turbines, etc.; Voids, defects, delaminations and cracks might be introduced into CFRP laminates and degrads the structural performance.

To demonstrate the detection capability of inner defects of the ESPI method, as shown in Figure 4.4, a defect was embedded into a 6-ply and 0.95 mm thick CFRP plate. The inner defect contained CFRP specimen was manufactured by the thermal-press vacuum-bag method. To simulate the defect, an elliptical air-pocket was prepared by stacking two pieces of elliptical-shape nonporous polymer film and sealed by sealant tape along the edge then embedded in-between the selected plies.

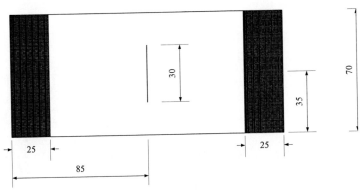

(a) the sheet containing a central crack

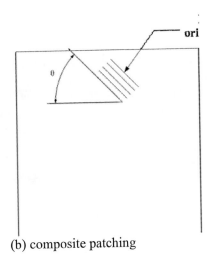

(b) composite patching

Figure 4.2. The geometrical configuration of the sheet containing a central crack and the composite patching (unit:mm).

The thickness of the air-pocket is about 0.3 mm. The elliptical ratio of the embedded air-pocket is 1.51 with 15.0 mm major-semi axis length. The air-pocket was embedded in-between the 4th and 5th ply. The specimen was then held on a fixed panel and put into a specially designed vacuum chamber as shown in Figure 4.5. By controlling the valves of the vacuum chamber, pressure difference can be introduced between the embedded artificial defect and the ambient pressure of the vacuum chamber. The specimen's surface directly above the defect bulged out slightly and the corresponding out-of-plane displacement could be detected by the ESPI method. If there is no defect inside the CFRP, the ESPI image shows no fringe pattern because there is no pressure difference between the inner specimen and the chamber. An out-of-plan ESPI fringe pattern obtained at 180.0 mmHg pressure difference is shown in Figure 4.6.(a). The ESPI fringe pattern shows the defect can be detected even the defect is located in-between 4th and 5th ply. However, the location, shape and size of the defect are not easily identified from the obtained ESPI fringe pattern. Complex image processing procedure has to be implemented to extract the out-of-plane displacement field and the displacement slope to identify the boundary of the defect.

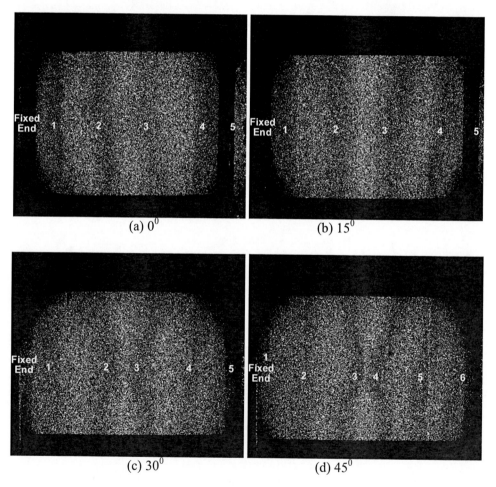

Figure 4.3. The ESPI fringe patterns for different stacking orientations with 3-layer double-sided patching.

To estimate the shape and size of the defect, a new concept is adopted. During testing, whenever the pressure difference is increased, the numbers of the fringes increased and the fringe space between arbitrary two adjacent fringes decreased. Whenever the pressure difference is larger than a threshold pressure, the fringes become too dense and no clear fringe patterns can be identified.

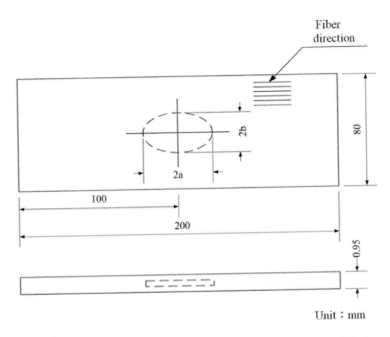

Figure 4.4. The geometrical configuration of an inner defect embedded in a CFRP plate.

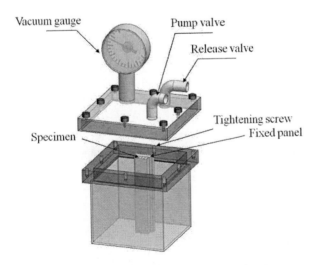

Figure 4.5. Illustration of specimen in a specially designed vacuum chamber.

As shown in Figure 4.6.(b), no clear fringes can be identified from the obtained out-of-plane ESPI image since the pressure difference reached 210.0 mmHg and the produced deformation slightly exceeded the ESPI measurement range.

Inspecting Figure 4.6.(b) carefully, the pixels with higher grey level formed a region which coincides with the defect's location. Generally, the grey level inside the region is about 15 to 50 grey level higher compared to the surrounding areas. Adopting with proper image processing procedure, the defect's location, size and shape can be determined. The detected defect size is about 398.5 mm² which is about 85.5% of the artificial defect.

Despite the fact that the detected shapes do not well match with the embedded shape, this study demonstrated the ESPI method should be a potentially quantitative nondestructive method for detecting the location and size of the defect in composite materials.

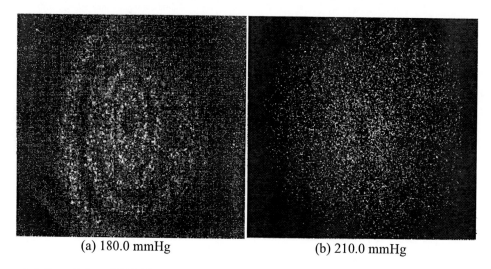

(a) 180.0 mmHg   (b) 210.0 mmHg

Figure 4.6. Out-of-plane ESPI fringe pattern obtained at 180.0 and 210.0 mmHg pressure differences.

4.3. Evaluating Dynamic SIFs of Structure Containing a Crack [14, 31]

When composite structures are in service, occasionally, cracks are introduced due to accidental impact or environmental contamination. Therefore, dynamic properties of composite structures may be changed. And stress intensity factors (SIFs) induced by vibration are produced.

To demonstrate the ESPI method can be used to measure the dynamic behavior of a cracked structure and extracting associated SIF value, as depicted in Figure 4.7, a composite plate of stacking sequence $[0]_{16}$ with 40 mm cracks is located along the clamped edge were used for study.

The AF-ESPI was used to measure the crack opening displacement (COD) of a composite plate with crack located along the clamped edge. Then based on the COD-SIF relationship, the modified K_{III} values were evaluated. The displacement of an edge cracked composite plate w_t can be decomposed into two parts; there are the principal displacement, w_p, and the singular displacement, w_s, i.e.,

$$w_t = w_p + w_s \qquad (4.3.1)$$

Assuming that the existing crack is along the x-axis and located on the clamped end. The displacement w_p and w_s will then possess the following characteristics

$$w_p(x,0^+) - w_p(x,0^-) = 0 \qquad (4.3.2)$$

$$w_s(x,0^+) - w_s(x,0^-) = \Delta \qquad (4.3.3)$$

where $\Delta(x)$ is the crack opening function along the crack and a is the crack length. Since w_s is caused by the singularity of the crack, the singularity can be of the form

$$w_s(x,y) = w_s(r,\theta) = f(\sqrt{r},\theta) \qquad (4.3.4)$$

where r and θ are radial distance and polar angle commonly used in fracture mechanics. Then the displacement difference between crack surfaces and mode III SIFs can be expressed as

$$\Delta w = w_s\big|_{\theta=\pi} - w_s\big|_{\theta=-\pi} = (2\sqrt{\pi}K_{III}/G)\sqrt{r} \qquad (4.3.5)$$

Based on Eqn. (4.3.5), the SIFs are proportional to the slope of $\Delta w - \sqrt{r}$ curve. Assuming the SIF of an orthotropic material has the same form of Eqn. (4.3.5), i.e., $K_{III}^* = \Delta w / \sqrt{r}$, where $K_{III}^* = K_{III} F(D_{ij})$, D_{ij} is the stiffness of an orthogonal material, K_{III}^* is defined as the corrected mode III SIFs.

According to the measurement principle, the displacement field along crack was first evaluated. The typical fringe patterns of the 2nd and 4th mode shapes of the cracked composite plate are shown in Figure 4.8. This is a twist vibration mode, the composite plate can be assumed subjected to a twist moment. Physically, the introduction of the crack generates new free surfaces, the new free surfaces cannot constrain the displacement of the region above the crack.

Therefore, the fringes above the crack are almost parallel to the fiber direction and intersect the crack; moreover, the nodal line was shifted and rotated due to the crack-introduced geometrical non-symmetry.

For the fourth mode, there is a local mode shape which is induced by the crack right above the crack, on the other region, there is a fringe close to the left-upper corner only. Then the displacement between crack introduce surfaces can be extracted from AF ESPI fringe patterns by counting the fringe order from crack tip to the free edge of the specimen. And then the COD can be evaluated by curve fitting and then mode III SIFs are evaluated.

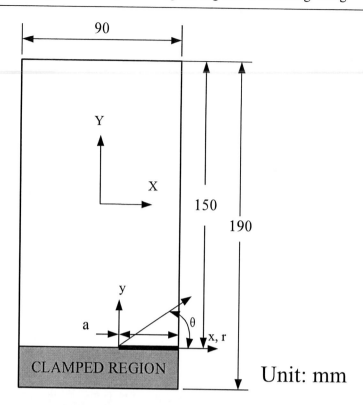

Figure 4.7. The cracked specimen with [0]16 stacking sequence.

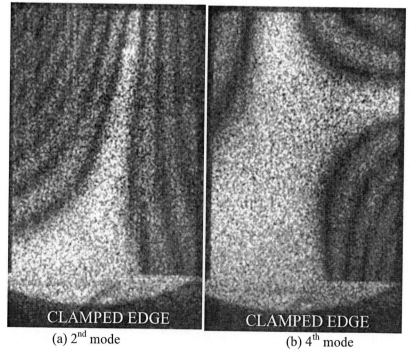

(a) 2nd mode (b) 4th mode

Figure 4.8. The mode shapes of the ESPI fringe patterns of composite plate with 40mm crack length on the clamped edge.

CONCLUSION

In this chapter, the principles and applications of ESPI were introduced. From the above-mentioned descriptions, it can be known that ESPI is a versatile tool to be employed in the measurement or analysis for various problems. In-plane and out-of-plane displacements under static or vibration loading for the objects with diffuse surface can be individually measured. ESPI not only provides the full-field but also be a highly sensitive and non-contact experimental technique. Although there are still great numbers of transparent objects or objects with specular or poor reflectivity, the displacements of them still can be measured by spraying white paint on the surface of object to increase the optical intensity reflected by the object if this manner is allowable.

Both optical arrangements for in-plane and out-of-plane ESPI can also be integrated to obtain the three dimensional displacement. In addition, the contrast of fringe pattern and sensitivity to displacement of AF-ESPI are both significantly improved with respect to the traditional ESPI in vibration measurement.

Therefore, AF-ESPI is an attractive method for the investigation of vibration measurement because all of the optical components used in AF-ESPI are essentially the same as those of the traditional ESPI.

REFERENCES

F. Untersenher, J. Hansen, B. Schlesinger, H*olography Handbook: Marking Holograms the Easy Way*, 3rd ed., Ross Books, 1996.

H. Fan, J. Wang, Y. Tan, "Simultaneous Measurement of Whole In-plane Displacement Using Phase-shifting ESPI", *Optics and Lasers in Engineering*, Vol. 29, pp. 249-257, 1997.

B. Kemper, D. Dirksen, J. Kandulla, G. von Bally, "Quantitative Determination of Out-of-plane Displacement by Endoscopic Electronic-speckle-pattern Interferometry", *Optics Communications*, Vol. 194, pp. 75-82, 2001.

M. J. Huang, Z. C. Liu, J. H. Jhang, "Self-marking Phase-stepping Electronic Speckle Pattern Interferometry (ESPI) for 3D Displacement Measurement on Cathode Ray Tube (CRT)-panels", *Optics and Lasers in Engineering*, Vol. 38, pp. 245-260, 2002.

N. Butters and J. A. Leendertz, "Holographic and Video Techniques Applied to Engineering Measurement", *J. Transactions of the Institute of Measurement and Control*, Vol. 4, pp. 349-354, 1971.

J. Moore and J. R. Tyrer, "Phase-stepped ESPI and Moiré Interferometry for Measuring Stress-intensity Factor and J-integral", *Experimental Mechanics*, Vol. 35, No. 4, pp. 306-314, 1995.

Y. T. Liu and H. M. Shang, *"Single Mode Optical Fiber Electronic Speckle Pattern Interferometry"*, Vol. 25, pp. 103-109, 1996.

E. Bendek, I. Lira, M. Francois, C. Vial, "Uncertainty of Residual Stress Measurement by Layer Removal", *Int. J. of Mechanical Sciences*, Vol. 48, pp. 1429-1428, 2006.

X. Dai, H. Yun, Q. Pu, "Measuring Thickness Change of Transparent Plate by Electronic Speckle Pattern Interferometry and Digital Image Correlation", *Optics Communications*, Vol. 283, pp. 3481-3486, 2010.

J. T. Malmo and E. Vikhagen, "Vibration Analysis of a Car Body by means of TV-holography", *Experimental Techniques*, Vol. 12, pp. 28-230, 1988.

P. Slangen, L. Berwart, C. de Veuster, J. C. Golinval, Y. Lion, "Digital Speckle Pattern Interferometry (DSPI): A Fast Procedure to Detect and Measure Vibration Mode Shapes", *Optics and Lasers in Engineering*, Vol. 25, pp. 311-321, 1996.

G. Jin, N. K. Bao, P. S. Chung, "Application of Nondestructive Testing Methods to Electronic Industry Using Computer-aided Optical Metrology", *Optics and Lasers in Engineering*, Vol. 25, pp. 81-91, 1996.

G. S. Spagnolo, D. Paoletty, P. Zabetta, "Local Speckle Correlation for Vibration Analysis", *Optics Communications*, Vol. 123, pp. 41-48, 1996.

C. H. Hwang, "Investigation of Vibration Characteristics of Composite Plates Containing Defect by Amplitude-Fluctuation ESPI", Ph. D. Dissertation, Department of Power Mechanical Engineering, National Tsing Hua University, Taiwan, Republic of China, 1996.

W. C. Wang, C. H. Hwang, S. Y. Lin, *"Vibration Measurement by the Time-averaged Electronic Speckle Pattern Interferometry"*, Vol. 35, pp. 4502-4508, 1997.

W. C. Wang and J. S. Hsu, "Investigation of the Size Effect of Composite Patching Repaired on Edge-cracked Plates", *Composite Structures*, Vol. 49, pp. 415-423, 2000.

W. C. Wang, C. W. Su, P. W. Liu, "Full-filed Non-destructive Analysis of Composite Plates", *Composites Part A-Applied Science and Manufacturing*, Vol. 39, pp. 1302-1310, 2008.

W. C. Wang and Y. H. Tsai, "Experimental Vibration Analysis of the Shadow Mask", *Optics and Lasers in Engineering*, Vol. 30, pp. 539-550, 1998.

H. Y. Lin and C. C. Ma, "The Influence of Electrode Designs on Resonant Vibrations Square Piezoceramic Plates", *IEEE Transactions on Ultrasonics, Ferroelectrics, and Frequency Control*, Vol. 53, pp. 825-837, 2006.

C. H. Huang and C. C. Ma, "Vibration of Cracked Circular Plates at Resonance Frequencies", *Journal of Sound and Vibration*, Vol. 236, pp.637-656, 2000.

E. Hecht, Optics, 4th ed., Addison Wesley, 2002.

R. Jones and C. Wykes, "Holographic and Speckle Interferometry", Cambridge University Press, Cambridge, 1989.

Z. C. Liu, *The analysis of ESPI technique and system errors*, Master Thesis, National Chung Hsing University, Taiwan, Republic of China, 2001.

T. Y. Wu, Experimental and Numerical Investigation of Thermal Deformation of Thermoelectric Coolers, Master Thesis, Department of Power Mechanical Engineering, National Tsing Hua University, Taiwan, Republic of China, 2010.

T. B. Chiou, On the Evaluation of Repair Efficiency of Composite Patching, Ph. D. Dissertation, Department of Power Mechanical Engineering, National Tsing Hua University, 1996.

W. C. Wang and T. B. Chiou, "Experimental Evaluation of Repair Efficiency of Composite Patching by ESPI", *Journal of Composite Materials*, Vol. 32, pp.1595-1616, 1989.

A. A. Baker and R. Jones, Bonded Repair of Aircraft Structures, Martinus Nijhoff, Dordrecht, Netherlands, 1988.

R. F. Gibson, Principles of Composite Materials, McGraw-Hill Inc., U.S.A., 1994.

C. H. Day, Nondestructive Testing and Repair Efficiency Evaluation of Composite Materials, Ph. D. Dissertation, Department of Power Mechanical Engineering, National Tsing Hua University, Taiwan, Republic of China, 1996.

W. C. Wang, C. H. Day, C. H. Hwang, T. B. Chiou, "Nondestructive Evaluation of Composite Materials by ESPI", *Research in Nondestructive Evaluation*", Vol. 10, pp. 1-15, 1998.

W. C. Wang and C. H. Hwang, "Experimental Analysis of Vibration Characteristics of an Edge-Cracked Composite Plate by ESPI Method", *International Journal of Fracture*, Vol. 91, No.4, pp. 311-321, 1998.

In: Interferometry Principles and Applications
Editor: Mark E. Russo

ISBN 978-1-61209-347-5
© 2012 Nova Science Publishers, Inc.

Chapter 11

PERIODIC ERROR MEASUREMENT FOR HETERODYNE INTERFEROMETRY

Tony L. Schmitz and Hyo Soo Kim

Department of Mechanical and Aerospace Engineering
University of Florida, Gainesville, FL, US

ABSTRACT

Displacement measuring interferometry offers high accuracy, range, and resolution for non-contact displacement measurement applications. One fundamental accuracy limitation for the commonly selected heterodyne (or two frequency) Michelson-type interferometer is periodic error, which is caused by frequency mixing/leakage between the reference (fixed) and measurement (moving) paths. The periodic error level for a given setup can be measured using the discrete Fourier transform of time-based position data. Alternately, it can be determined using "velocity scanning", where the optical interference signal is observed during constant velocity target motion using a spectrum analyzer and the spectral content is used to calculate the periodic error magnitudes. In this chapter, these techniques are described and demonstrated on experimental data. Using this information, the optical setup can either be adjusted to reduce the periodic error magnitudes or compensation can be applied.

1. PERIODIC ERROR DESCRIPTION

The heterodyne displacement measuring interferometer (DMI) is commonly applied in applications where non-contact displacement information is required. While the heterodyne interferometer offers high accuracy in general, there are a number of error sources which may limit the achievable accuracy including: Abbe error, cosine error, thermal expansion/contraction of the optical components, atmospheric changes which affect the measurement medium (typically air) refractive index, laser wavelength stability, electronics error, vibrations, wavefront non-uniformity/variation, and periodic error. Although any of these error sources many dominate in a given application, this chapter is concerned solely with periodic error and its measurement.

Periodic error is an intrinsic error source that prevents traditional heterodyne DMI configurations from achieving sub-nanometer level accuracy. These systems typically use polarization coding to separate the two (heterodyne) optical frequencies. In this approach, the two frequencies are carried on coincident, linearly polarized, mutually orthogonal laser beams and are separated/recombined using polarization dependent optics.

Due to setup alignments and non-ideal optical performance, mixing between the two heterodyne frequencies can occur, which results in first and/or second-order periodic errors[1] with amplitudes that vary cyclically with the target position (see early work by [1-8]). This unwanted leakage of the reference frequency from the fixed path in the DMI into the measurement (moving) path, and vice versa, may occur due to a number of influences, including non-orthogonality between the ideally linear beam polarizations, elliptical polarization of the individual beams, imperfect optical components, parasitic reflections from individual optical surfaces, and/or mechanical misalignment between the interferometer elements (laser, polarizing optics, and targets).

In a perfect system, a single frequency travels to the fixed target, while a second, single frequency travels to the moving target. Interference of the combined signals yields a perfectly sinusoidal trace with phase that varies, relative to a reference phase signal, in response to motion of the moving target.

However, the inherent frequency leakage in actual implementations produces an interference signal which is not purely sinusoidal (i.e., contains unintended spectral content) and leads to periodic error in the measured displacement. Schmitz and Beckwith [9] summarized the potential periodic error contributors using a Frequency-Path model, which identified all possible paths for each light frequency from the source to detector and predicted the number of interference terms that may be expected at the detector output. For the single pass, heterodyne interferometer shown in Figure 1, it was demonstrated that ten distinct interference terms exist in a fully-leaking interferometer (i.e., each frequency is present in both the moving and fixed paths).

These interference terms may be grouped by optical path change dependency into only four categories: 1) *Optical power* which contributes a constant intensity to the photodetector current independent of optical path changes; 2) *AC reference* terms with phase that varies by one full cycle over the synthetic wavelength, or the distance defined by the difference in wave numbers (i.e., the reciprocal of the wavelength) between the source frequencies and occur at the split frequency, or the difference between the heterodyne frequencies; 3) *DC interference*, which are Doppler shifted up from zero frequency during target motion and represent the intended signal in homodyne (single frequency) interferometers; and 4) *AC interference* terms which produce a time-harmonic variation in the detector current at the split frequency and are Doppler shifted up or down during target motion depending on direction. With respect to periodic errors, the leakage-induced *AC interference* term leads to second-order error, while the *AC reference* terms cause first-order error.

---

[1] First and second-order periodic errors exhibit spatial frequencies of one and two cycles per displacement fringe, respectively.

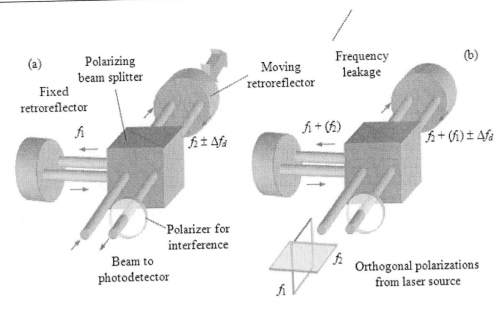

Figure 1. Depictions of a) ideal heterodyne interferometer behavior; and b) frequency leakage (indicated by the frequency terms in parentheses). The two frequencies, f_1 and f_2, are ideally linearly polarized and orthogonal. This enables the polarizing beam splitter to separate them based on their polarization states.

The corresponding frequency content during constant velocity motion is depicted in Figure 2. In addition to the intended *AC interference* signal (Doppler shifted by f_d to a lower frequency relative to the split frequency, $f_1 - f_2 = \Delta\omega/2\pi$, for the selected motion direction), the leakage-induced *AC interference* (up-shifted by f_d) and *AC reference* terms are also observed within the phase locked loop (PLL) modulation bandwidth of the phase measuring electronics. All three terms contribute to the measured displacement and yield the first and second-order periodic errors.

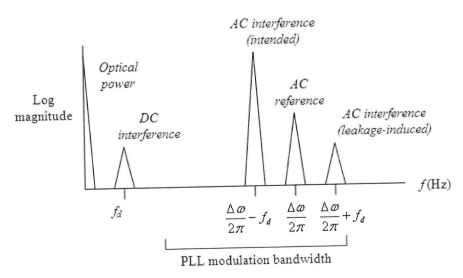

Figure 2. Frequency spectrum for constant velocity motion in polarization coded heterodyne interferometer. The additional frequency content, caused by frequency leakage, leads to periodic error.

2. ANALYSIS OF TIME-DOMAIN POSITION DATA

Periodic error magnitudes may be identified by calculating the discrete Fourier transform of time-domain displacements. Because the Fourier transform expects time-periodic signals, the displacements should be recorded under constant velocity conditions. Considerations should also be made for leakage and aliasing. Because it is generally challenging to exactly capture an integer number of periodic error oscillations in the measurement time interval, leakage can be reduced by ensuring that many full periods are acquired. To prevent aliasing, an appropriate sampling frequency must be selected based on the interferometer configuration, source wavelength, and moving target velocity. The spatial period for first-order error is $\frac{\lambda}{FF}$, where λ is the laser source wavelength and FF is the fold factor (equal to two for a single pass configuration where retroreflector targets are used in the fixed and moving paths). Similarly, the spatial period for second-order error is $\frac{\lambda}{2FF}$. For a He-Ne source with a wavelength of 632.8 nm and a fold factor of two, the corresponding first and second-order periods are 316.4 nm and 158.2 nm, respectively. The associated frequency (in Hz) for the periodic error terms is the moving target velocity divided by the spatial period. The Nyquist-Shannon sampling theorem requires that the (digital) sampling frequency, f_s, is at least two times the highest frequency contained in the measured signal. This enables the maximum velocity, v_{max}, to be calculated which avoids aliasing. For first-order error, the maximum velocity is:

$$v_{max} = \frac{f_s \cdot 316.4}{2 \cdot 1 \times 10^6} \text{ (mm/s)}, \quad (1)$$

where f_s is expressed in Hz. For second-order error, the maximum velocity is:

$$v_{max} = \frac{f_s \cdot 158.2}{2 \cdot 1 \times 10^6} \text{ (mm/s)}. \quad (2)$$

Since both first and second-order periodic error are present in general, Eq. 2 should be used to identify the maximum permissible velocity for the selected sampling frequency. To ensure a high-fidelity representation of the sampled signal, however, a fraction of this maximum velocity should be specified.

After performing the Fourier transform, the frequency axis is normalized to periodic error order by multiplying the frequency vector by $\frac{\lambda}{FF}$. In this way, at a normalized frequency, or error order, axis value of one, the first-order periodic error magnitude is determined. Similarly, the second-order error magnitude is provided at an error order axis value of two.

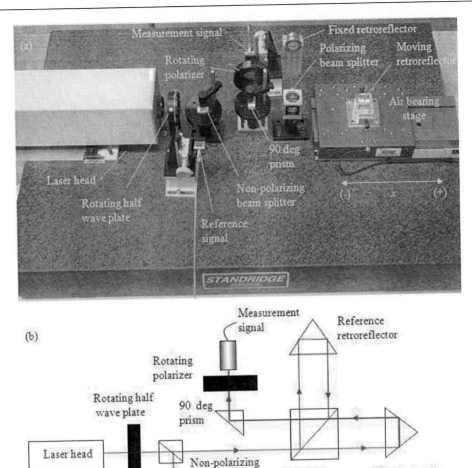

Figure 3. a) Photograph of single pass, heterodyne interferometer experimental setup; and b) schematic of setup.

To demonstrate the variation in periodic error with optical setup, an experimental platform was designed and constructed. A photograph and schematic of the platform are provided in Figure 3. The orthogonal, linearly polarized beams with a frequency difference of approximately 3.65 MHz (Helium-Neon laser source with a Zeeman split) first pass through a half wave plate. Rotation of the half wave plate enables variation in the apparent angular alignment (about the beam axis) between the polarization axes and polarizing beam splitter; deviations in this alignment lead to frequency mixing in the interferometer. The light is then incident on a non-polarizing beam splitter (80% transmission) that directs a portion of the beam to a fiber optic pickup after passing through a fixed angle sheet polarizer (oriented at 45 deg to the nominal laser orthogonal polarizations). The pickup is mounted on a two rotational degree-of-freedom flexure which enables efficient coupling of the light into the multi-mode fiber optic. This signal is used as the phase reference in the measurement electronics.

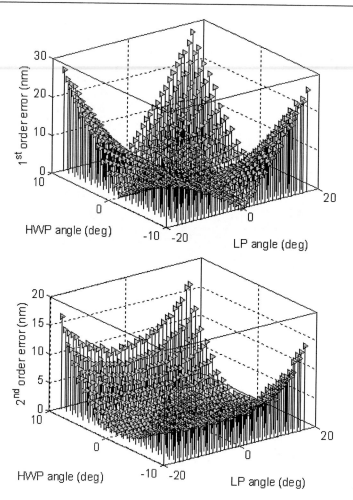

Figure 4. Periodic errors for half wave plate/linear polarizer parameter study. The errors were obtained from the discrete Fourier transform of time-domain position data. First-order periodic error is shown in the top panel and second-order in the bottom.

The remainder of the light continues to the polarizing beam splitter where it is (ideally) separated into its two frequency components that travel separately to the moving and fixed retroreflectors. Motion of the moving retroreflector is achieved using an air bearing stage. After the beams are recombined in the polarizing beam splitter, they are directed by a 90 deg prism through a linear polarizer (or analyzer) with a variable rotation angle. Rotation of this polarizer also leads to variation in the periodic error magnitudes. Finally, the light is launched into a fiber optic pickup. This serves as the measurement signal in the measurement electronics (0.3 nm resolution for the single pass configuration). Displacement measurements were performed using the platform in Figure 3. over a range of half wave plate, HWP, angles (-10 deg to +10 deg from the nominal orientation) and linear polarizer, LP, angles (-20 deg to +20 deg from nominal). For each test at the selected {HWP, LP} setting, the displacement was recorded during constant velocity of the air bearing stage, the discrete Fourier transform of the time-domain displacement was calculated, and the magnitudes of the first and second-order periodic error were determined. The results are provided in Figure 4. It is seen that substantial periodic error levels can be obtained at large misalignments.

3. VELOCITY SCANNING

Another approach to measuring periodic error magnitude is "velocity scanning" where the optical interference signal is recorded during constant velocity target motion using a spectrum analyzer. This was first described by Patterson and Beckwith [10] and explored further by Badami and Patterson [11]. In this method, the magnitudes of the individual periodic error contributors are isolated in frequency during constant velocity motion. Using this information, the optical setup can either be adjusted to reduce their magnitudes or error compensation can be applied. To enable calculation of the periodic error magnitudes from the spectral data, phasor diagrams can be used to develop the required relationships. As described previously, the photodetector current contains not only the desired *AC interference* term, but also the leakage-induced *AC interference* term and two *AC reference* terms. Because the frequency offset is the same (or nearly so) for the two *AC reference* terms, they cannot generally be individually distinguished in the spectrum analyzer display. Therefore, they will be considered as a single term with identical frequency and phase in this analysis. These three terms (intended and leakage-induced *AC interference* and *AC reference* signals), depicted in the Figure 2. spectrum, may be described using three separate phasors in the complex plane.

First, consider the intended *AC interference* term. It can be described as the phasor $\vec{\Gamma}_0 = \Gamma_0 e^{i(\Delta\omega t - FF \cdot kx + \phi_{22} - \phi_{11})} = \Gamma_0 e^{i(\phi + \phi_0)}$, where Γ_0 is the magnitude (photodector current units of Amperes), FF is equal to two for the system shown in Figure 3, $k = 2\pi/\lambda$, x is the displacement of the moving target, $\phi = \Delta\omega t - FF \cdot kx$ (rad) is the nominal phase change due to the measurement target motion, and $\phi_0 = \phi_{22} - \phi_{11}$ (rad) is the (assumed arbitrary) initial phase. This phasor rotates at $\Delta f = f_1 - f_2$ in the complex plane with no motion and $\Delta f \pm f_d$ depending on the direction while the measurement target is in motion. Alternately, the exponential notation may be replaced by the rectangular coordinate representation:

$$\vec{\Gamma}_0 = \Gamma_0 e^{i(\phi + \phi_0)} = \Gamma_0 \cos(\phi + \phi_0)\vec{j} + \Gamma_0 \sin(\phi + \phi_0)\vec{k},$$

which specifically identifies the real ($\vec{j}$ axis) and imaginary ($\vec{k}$ axis) components. The measurement target position is ideally determined from the instantaneous phase of $\vec{\Gamma}_0$. Under constant velocity conditions, for example, the instantaneous phase grows linearly with time, as does the target position. Conceptually, the phase measuring electronics frequency shift this term back to zero for no motion, or near zero during target motion, by subtracting the reference (split) frequency. For the remainder of the analysis, this frequency-shifted condition is considered so that the $\vec{\Gamma}_0$ phasor is rotating at f_d for constant velocity motion; a counter clock-wise rotation for the selected target direction is assumed. Note that after the frequency translation, $\phi = -FF \cdot kx$. See Figure 5.a.

Similarly, the *AC reference* term can be expressed in rectangular coordinates as:

$$\vec{\Gamma}_1 = \Gamma_1 \cos(\phi_1)\vec{j} + \Gamma_1 \sin(\phi_1)\vec{k}.$$

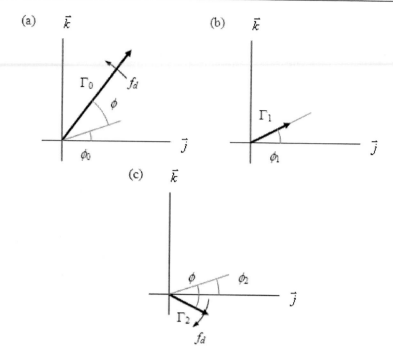

Figure 5. Phasor diagrams for a) intended *AC interference* signal; b) *AC reference* signal; and c) leakage-induced *AC interference* signal.

The orientation of this phasor (see Figure 5b) does not vary with time; its direction is fixed by the (assumed arbitrary) initial phase ϕ_1 which, in general, is assumed to differ from ϕ_0. Finally, the leakage induced *AC interference* term can be expressed in rectangular coordinates as:

$$\vec{\Gamma}_2 = \Gamma_2 \cos(\phi - \phi_2)\vec{j} - \Gamma_2 \sin(\phi - \phi_2)\vec{k} .$$

This phasor (see Figure 5.c) rotates in the clockwise direction (for counter-clockwise $\vec{\Gamma}_0$ rotation) due to the opposite sign of the Doppler shift. It is assumed that its arbitrary initial phase ϕ_2 differs, in general, from both ϕ_0 and ϕ_1.

Prior to determining the periodic error in the general case, consider the presence of only $\vec{\Gamma}_0$ and $\vec{\Gamma}_1$, and then only $\vec{\Gamma}_0$ and $\vec{\Gamma}_2$, individually. The initial phases are selected to be zero for now to enable direct comparison to reference [11]. Figure 6 depicts the superposition of the intended *AC interference* and *AC reference* phasors ($\vec{\Gamma}_0$ and $\vec{\Gamma}_1$, respectively) at progressing times during constant velocity motion. In Figure 6.a, an arbitrary time is selected where the nominal phase (from the intended *AC interference* term) is zero. For zero initial phases, ϕ_0 and ϕ_1, both phasors are directed along the positive real axis. At a later time in Figure 6.b, the nominal phase is $\phi = \frac{\pi}{2}$ rad, but the actual phase, ϕ', is less than the nominal due to the vector addition of $\vec{\Gamma}_0$ and $\vec{\Gamma}_1$. Recall that the orientation of $\vec{\Gamma}_1$ does not change for the frequency translated condition. The phase error, $\Delta\phi = \phi - \phi'$, is therefore positive and depends on the magnitude of $\vec{\Gamma}_1$. Similar to Figure 6.a, the phase error in Figure 6.c. is again zero. In Figure 6.d, the error is negative, but equal in magnitude to the situation depicted in Figure 6.b. Figure 6e demonstrates the corresponding single cycle of phase error variation per 2π rad of nominal phase change for first-order periodic error. By vector addition, the phase error is: $\Delta\phi = \phi - \phi' = \phi - \tan^{-1}\left(\dfrac{\Gamma_0 \sin(\phi)}{\Gamma_0 \cos(\phi) + \Gamma_1}\right)$ (rad) and the corresponding first-order periodic error is:

$$\varepsilon_1 = \frac{1}{FF}\frac{\lambda}{2\pi}\left(\phi - \tan^{-1}\left(\frac{\Gamma_0 \sin(\phi)}{\Gamma_0 \cos(\phi) + \Gamma_1}\right)\right) \text{ (nm)}. \tag{3}$$

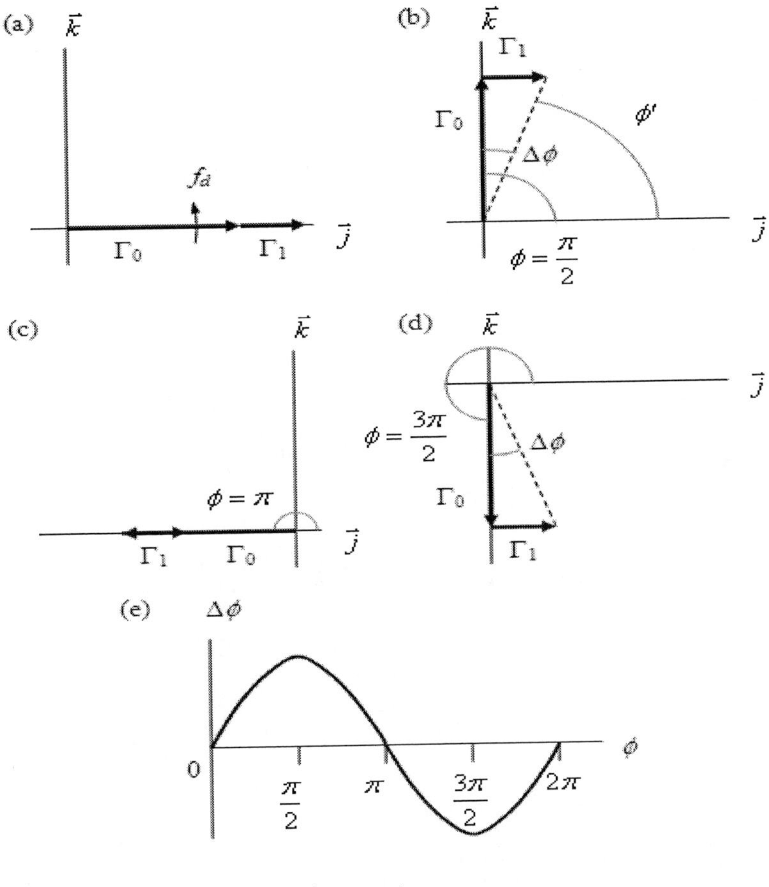

Figure 6. Periodic error in the presence of $\vec{\Gamma}_0$ and $\vec{\Gamma}_1$ only for various nominal phase angles (rad). a) 0; b) $\dfrac{\pi}{2}$; c) π; and d) $\dfrac{3\pi}{2}$. The single cycle of phase error variation per 2π rad of nominal phase change is shown in e).

If the maximum first-order periodic error, $\varepsilon_{max,1}$, is assumed to occur when $\phi = \dfrac{\pi}{2}$ (see Figure 6.b), then:

$$\varepsilon_{max,1} = \frac{1}{FF}\frac{\lambda}{2\pi}\left(\frac{\pi}{2} - \tan^{-1}\left(\frac{\Gamma_0}{\Gamma_1}\right)\right) \text{ (nm)}, \qquad (3.a)$$

which is equivalent to Eq. 5 in reference [11] for the small angle approximation. (Equation 5 in reference [11] identifies the maximum phase error magnitude as $\dfrac{\Gamma_1}{\Gamma_0}$.)

Figure 7 shows the situation when only $\vec{\Gamma}_0$ (intended *AC interference* signal) and $\vec{\Gamma}_2$ (leakage-induced *AC interference* signal) are considered. Again assuming zero initial phases, an arbitrary time may be selected when both phasors are directed along the positive real axis; see Figure 7.a. Because the vectors are counter-rotating, the geometries shown in Figs. 7b through 7h are obtained for nominal phase values of { $\dfrac{\pi}{4}$, $\dfrac{\pi}{2}$, $\dfrac{3\pi}{4}$, π, $\dfrac{5\pi}{4}$, $\dfrac{3\pi}{2}$, and $\dfrac{7\pi}{4}$ } deg. The characteristic two cycle phase error variation per 2π rad of nominal phase change (second-order periodic error) is depicted in Figure 7.i. The phase error is calculated according to: $\Delta\phi = \phi - \phi' = \phi - \tan^{-1}\left(\dfrac{(\Gamma_0 - \Gamma_2)\sin(\phi)}{(\Gamma_0 + \Gamma_2)\cos(\phi)}\right)$ (rad) and the corresponding second-order periodic error is:

$$\varepsilon_2 = \frac{1}{FF}\frac{\lambda}{2\pi}\left(\phi - \tan^{-1}\left(\frac{(\Gamma_0 - \Gamma_2)\sin(\phi)}{(\Gamma_0 + \Gamma_2)\cos(\phi)}\right)\right) \text{ (nm)}. \qquad (4)$$

If the corresponding maximum periodic error is assumed to be obtained when $\phi = \dfrac{\pi}{4}$ rad (see Figure 7.b), so that $\sin(\phi) = \cos(\phi) = \dfrac{\sqrt{2}}{2}$, then the maximum second-order periodic error, $\varepsilon_{max,2}$, is:

$$\varepsilon_{max,2} = \frac{1}{FF}\frac{\lambda}{2\pi}\left(\frac{\pi}{4} - \tan^{-1}\left(\frac{\Gamma_0 - \Gamma_2}{\Gamma_0 + \Gamma_2}\right)\right) \text{ (nm)}, \qquad (4.a)$$

which agrees with Eq. 6 from reference [11] for the small angle approximation. (Equation 6 in reference [11] identifies the maximum phase error magnitude as $\dfrac{\Gamma_2}{\Gamma_0}$.)

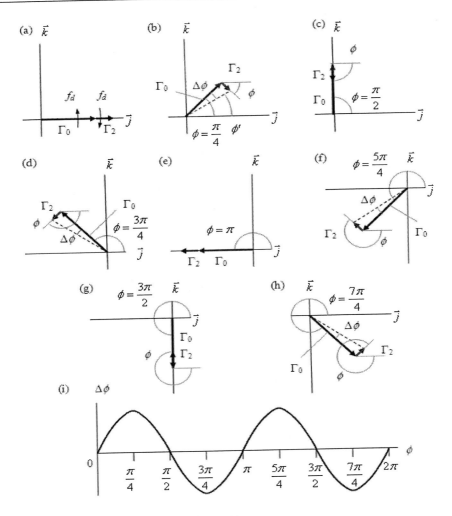

Figure 7. Periodic error in the presence of $\vec{\Gamma}_0$ and $\vec{\Gamma}_2$ only for various nominal phase angles (rad). a) 0; b) $\frac{\pi}{4}$; c) $\frac{\pi}{2}$; d) $\frac{3\pi}{4}$; e) π; f) $\frac{5\pi}{4}$; g) $\frac{3\pi}{2}$; and h) $\frac{7\pi}{4}$. The two cycles of phase error variation per 2π rad of nominal phase change is shown in i).

In general, however, all three phasors, $\vec{\Gamma}_0$, $\vec{\Gamma}_1$, and $\vec{\Gamma}_2$, are present and the initial phases, ϕ_0, ϕ_1, and ϕ_2, may be assumed to be nonzero, unequal, and uncorrelated. In this case, Eqs. 1 and 3 may not accurately describe the first and second-order periodic error magnitudes in the measured phase/position for all combinations of input parameters. To treat the general case, an expression for the phase error must first be determined. Figures 5a, 5b, and 5c show the individual phasors with arbitrary phases. They are superimposed in Figure 8. Based on this geometry, the phase error is:

$$\Delta\phi = \phi + \phi_0 - \phi' = \phi + \phi_0 - \tan^{-1}\left(\frac{\Gamma_0 \sin(\phi+\phi_0) + \Gamma_1 \sin(\phi_1) - \Gamma_2 \sin(\phi-\phi_2)}{\Gamma_0 \cos(\phi+\phi_0) + \Gamma_1 \cos(\phi_1) + \Gamma_2 \cos(\phi-\phi_2)}\right) \text{ (rad)} \quad (5)$$

and the corresponding periodic error is:

$$\varepsilon = \frac{1}{FF}\frac{\lambda}{2\pi}\Delta\phi \text{ (nm)}. \tag{6}$$

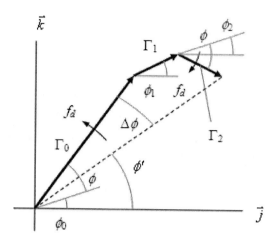

Figure 8.

Note that the error is dependent on the nominal phase (of the intended *AC interference* signal), the three phasor magnitudes, and the initial phases of the three phasors. To evaluate Eq. 6, and identify the periodic error order magnitudes, Monte Carlo simulation is applied [12]. This enables the (assumed) uniformly distributed, unknown, uncorrelated initial phases to be incorporated. The required steps are:

1. define the values for FF, λ, Γ_0, Γ_1, and Γ_2;
2. select random, uniformly distributed values of ϕ_0, ϕ_1, and ϕ_2 from the range $-\pi \leq \phi_i \leq \pi$, where i = 0, 1, 2;
3. compute $\Delta\phi$ from Eq. 5 (note that this is a time-domain, periodic function under constant velocity);
4. compute ε from Eq. 6;
5. calculate the discrete Fourier transform of the result from step 4 and normalize the frequency axis to error order (multiply by $\frac{\lambda}{FF}$) to identify the individual periodic error contributions from each order; and
6. return to step 2.
7. After many iterations, the periodic error magnitude for each order is selected from the resulting distributions.

Because spectrum analyzers typically display power data using a logarithmic (dBm) scale, a measurement unit conversion is required for the preceding analysis. To convert from magnitudes, γ_i, in dBm to the (linear) Ampere units for Γ_i included in the previous descriptions, the conversion shown in Eq. 7 is applied, where $i = 0, 1, 2$.

$$\Gamma_i = 10^{\frac{\gamma_i}{20}} \tag{7}$$

Data were collected using a spectrum analyzer (the γ_0, γ_1, and γ_2 spectral peaks were measured in dBm) for different levels of frequency mixing by varying the linear polarizer and half wave plate angles from their nominal orientations using the setup shown in Figure 3. Equations 3, 4, and 6 were applied to compute the corresponding periodic error. Note that Monte Carlo simulation was used to evaluate Eq. 6, which enabled the uniformly distributed, uncorrelated initial phases to be randomly selected over many iterations. In the following analyses, the maximum values from the simulations are presented. Figure 9. displays the case where the linear polarizer angle was varied about its nominal orientation (indicated as zero), while the half wave plate angle was fixed at 10 deg from its nominal angle. A strong variation for γ_1, the *AC reference* term, is observed while γ_0 and γ_2, the intended *AC interference* and leakage induced *AC interference* terms, respectively, are nearly constant. The first-order errors calculated by Eqs. 3 and 6 increase with larger misalignment angles and agree with the magnitudes calculated from the position data using the discrete Fourier transform. See the top panel of Figure 10. However, the second-order errors computed using Eqs. 4 and 6 do not agree. As shown in the bottom panel of Figure 10, the Eq. 6 results more closely follow the second-order error calculated from the position data.

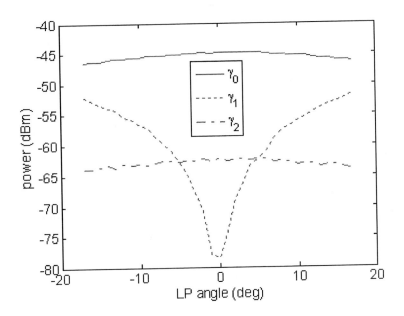

Figure 9. Variation of γ_0, γ_1, and γ_2 with linear polarizer (LP) angle. The half wave plate (HWP) was fixed at 10 deg from its nominal orientation (a large misalignment configuration). Strong variation of γ_1 is observed.

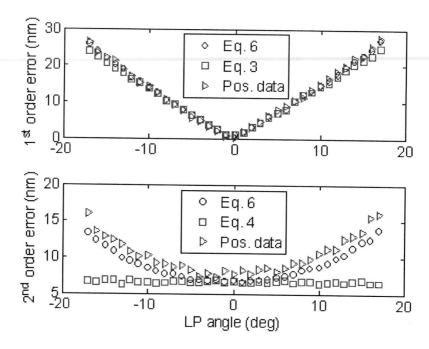

Figure 10. The magnitudes of first (top) and second-order (bottom) periodic errors calculated by Eqs. 3, 4, and 6 are compared to magnitudes computed using the discrete Fourier transform of position data. The agreement is good for first-order error, but only Eq. 6 reproduces the second-order error.

Figure 11. shows the difference between the Eq. 3, 4, and 6 calculations and position data (discrete Fourier transform) magnitudes for first and second error-order errors; the data from Figure 10. was analyzed. It is seen that the Eq. 3, 4, and 6 results agree with the position data for small linear polarizer angular misalignments. For large misalignments, however, Eqs. 3 and 4 provide less accurate estimates (2.6 nm difference for first-order error and 9.5 nm difference for second-order error at the largest misalignment). Equation 6, on the other hand, agrees to within 1.2 nm for first and 3.0 nm for second-order error. These results show that Eq. 6, which considers all three spectral peaks, provides a more accurate estimate of the first and second-order periodic errors than Eqs. 3 and 4, respectively, which consider only two periodic error components - either γ_0 and γ_1 (first-order, Eq. 3) or γ_0 and γ_2 (second-order, Eq. 4) - especially for significant misalignments from nominal.

Results for a medium misalignment case (5 deg half wave plate angular misalignment) are provided in Figs. 12 and 13. Trends in γ_0, γ_1, and γ_2 variation similar to those identified in Figure 9. are observed. This yields the same first and second-order periodic error behavior shown in Figs. 10 and 11. Again, Eq. 6 more closely agrees with the position data periodic error magnitudes.

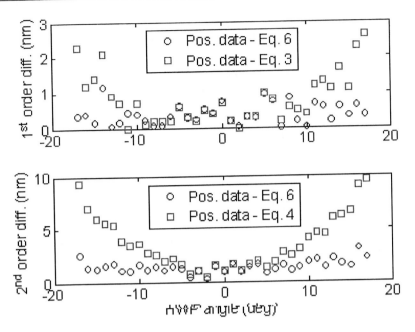

Figure 11. Differences between magnitudes from Eqs. 3, 4, and 6 and discrete Fourier transform of position data; first-order (top) and second-order (bottom). The differences were calculated from the errors displayed in Figure 10.

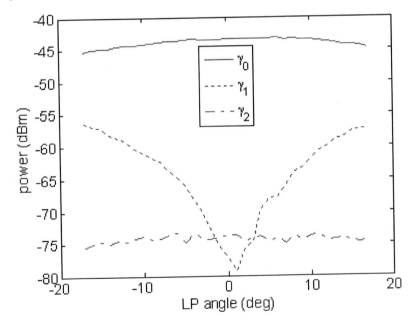

Figure 12. Variation of γ_0, γ_1, and γ_2 with linear polarizer (LP) angle. The half wave plate was fixed at 5 deg from its nominal orientation (a medium misalignment configuration). Strong variation of γ_1 is again observed.

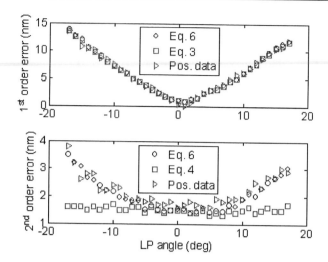

Figure 13. The magnitudes of first (top) and second-order (bottom) periodic errors calculated by Eqs. 3, 4, and 6 are compared to magnitudes computed using the discrete Fourier transform of position data. The agreement is good for first-order error, but only Eq. 6 reproduces the second-order error.

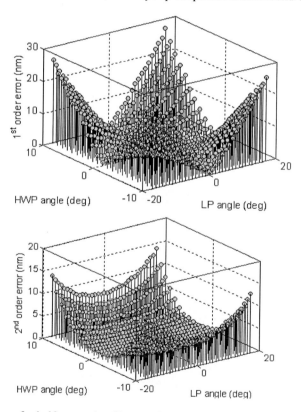

Figure 14. Periodic errors for half wave plate/linear polarizer parameter study obtained from the spectrum analyzer data and Monte Carlo evaluation of Eq. 6 (the maximum values from the simulation are shown). First-order periodic error is displayed in the top panel and second-order in the bottom.

Figure 14. shows first and second-order periodic errors from multiple measurements calculated using a Monte Carlo evaluation of Eq. 6 (populated using spectrum analyzer data), where the maximum values are again shown. These results can be compared to the Figure 4. results obtained using the discrete Fourier transform of time-domain position data. Good agreement is observed.

CONCLUSION

This chapter discussed measurement techniques for periodic error in heterodyne displacement measuring interferometry. First and second-order periodic error measurement using the discrete Fourier transform of time-based position data was discussed first. Limitations imposed by leakage and aliasing were addressed. Second, velocity scanning was described. In this method, the optical interference signal is observed during constant velocity target motion using a spectrum analyzer and the spectral content is used to calculate the periodic error magnitudes. A phasor diagram graphical representation was used to establish the relationship between spectral peak magnitudes and first and second-order periodic error magnitudes. The techniques were demonstrated on experimental data. Using this information, the optical setup can either be adjusted to reduce the periodic error magnitudes or compensation can be applied; see, e.g., [13-15]. As with any measurement technique, to fully describe the measurement result the associated uncertainty should be addressed. Additional information on periodic error measurement uncertainty is provided in [16].

ACKNOWLEDGMENTS

This work was supported by the National Science Foundation (DMI-0555645) and Agilent Technologies, Inc. Any opinions, findings, and conclusions or recommendations expressed in this material are those of the authors and do not necessarily reflect the views of these agencies. The authors would also like to thank V. Badami and J. Beckwith for helpful discussions.

REFERENCES

[1] Fedotova G. Analysis of the measurement error of the parameters of mechanical vibrations. *Measurement Techniques* 1980; 23(7):577-580.

[2] Quenelle R. Nonlinearity in interferometric measurements. *Hewlett-Packard Journal* 1983; 34(4):10.

[3] Barash V, Fedotova G. Heterodyne interferometer to measure vibration parameters. *Measurement Techniques* 1984; 27(7):50-51.

[4] Sutton C. Nonlinearity in length measurements using heterodyne laser Michelson interferometry. *Journal of Physics E: Scientific Instrumentation* 1987; 20:1290-1292.

[5] Estler WT. High-accuracy displacement interferometry in air. *Applied Optics* 1985; 24:808-815.

[6] Bobroff N. Residual errors in laser interferometry from air turbulence and nonlinearity. *Applied Optics* 1987; 26(13):2676-2682.

[7] Steinmetz C. Sub-micron position measurement and control on precision machine tools with laser interferometry. *Precision Engineering* 1990; 12(1):12-24.

[8] Bobroff N. Recent advances in displacement measuring interferometry. *Measurement Science and Technology* 1993; 4:907-926.

[9] Schmitz T, Beckwith J. An investigation of two unexplored periodic error sources in differential-path interferometry. *Precision Engineering* 2002; 27(3):311-322.

[10] Patterson S, Beckwith J. Reduction of systematic errors in heterodyne interferometric displacement measurement, In: Proceedings of the 8th International Precision Engineering Seminar (IPES), Compiegne, France, 1995. pp. 101-104.

[11] Badami V, Patterson S. A frequency domain method for the measurement of nonlinearity in heterodyne interferometry. *Precision Engineering* 2000; 24(1):41-49.

[12] Kim, HS, Schmitz T. Periodic error calculation from spectrum analyzer data. *Precision Engineering* 2010; 34:218-230.

[13] Schmitz, T, Chu, D, Kim, HS. First and second order periodic error measurement for non-constant velocity motions. *Precision Engineering* 2009; 33:353-361.

[14] Schmitz, T, Chu, D, Houck III, L. First order periodic error correction: Validation for constant and non-constant velocities with variable error magnitudes. *Measurement Science and Technology* 2006; 17:3195-3203.

[15] Schmitz, T, Houck III, L, Chu, D, Kalem, L. Bench-top setup for validation of real time, digital periodic error correction. *Precision Engineering* 2006; 30:306-313.

[16] Schmitz, T, Kim, HS. Monte Carlo evaluation of periodic error uncertainty. *Precision Engineering* 2007; 31(3):251-259.

In: Interferometry Principles and Applications
Editor: Mark E. Russo

ISBN 978-1-61209-347-5
© 2012 Nova Science Publishers, Inc.

Chapter 12

MAXIMUM LIKELIHOOD ESTIMATION OF OPTICAL SIGNAL PARAMETERS

V. S. Sobolev

Institute of Automation and Electrometry Siberian Branch of Russian Academy of Sciences, Novosibirsk, Russia

ABSTRACT

Development of algorithms for the optimal estimates of the optical signals parameters in the present time is the task number 1. This is due to the rapid development of optoelectronics and, in particular, the such its directions, as fiber communications, optical disk memory, optical location and interferometry. In the last case it is very imported, as it is not possible to use corrected codes.The subtlety of the problem is that, unlike radio and radar signals, where the noise and the signals are statistically independent, receiving optical signals is accompanied inevitably by so far shot noise, which variance is strictly proportional to the intensity of the signal itself. With this in mind, the entire rich arsenal of existing optimal algorithms for estimating the parameters of signals in noise for receiving optical signals can not be directly applied. This paper is results of the long-term research aimed the solving the problem of obtaining optimal estimates of optical signals parameters with the above especially the accompanying noise. On an example of the Gaussian optical video and radio pulses are deduced and solved the likelihood equation and get the expressions for the boundaries of the Cramer-Rao determining the quality of fetched ratings. The problem is solved for the three main photodetection methods: counting the number of photoelectrons emitted at specified time intervals, fixing the time of emission of each photoelectron and analog detection . The reliability of the algorithms and expressions for the boundaries of the Cramer-Rao confirmed by computer simulation. Its results are given in the article.

INTRODUCTION

The amazing capabilities of products of modern industry is largely determined by the connection technology of optics, microelectronics and computer technology. This is particularly true for areas such as fiber communications, laser location, confocal microscopy, metrology, measurement technology, as well as your CD and DVD storage. Optics, with its high (up to 1015 Hz) carrier frequencies and the ability to create and process two-dimensional and even three- dimensional signals, opened broad prospects developing new areas of computer science. However, at this stage, it can not compete with the possibilities of electronics and computer technology in terms of conversion and signal processing. Therefore, to realize the potential of optics in full, you must convert the optical signals into electrical by a photodetector, and then produce amplification, digitization and the necessary transformations in the electronics and computer technology. It is important to note that the photodetection process differs substantially from the radio frequency signal reception. Electromotive force at the output of the antenna is proportional to the tension of the received electromagnetic field, while the output signal of the photodetector is proportional not to the tension of the optical field, but its intensity. The main difference lies in the fact that the quantum nature of light and the photoelectric effect, determining the photodetection processes, lead to the inevitable until the shot noise. This feature of converting optical signals into electrical signal requires the developer to optoelectronic systems, a detailed review of the features of modern methods of photodetection, knowledge of statistics, the output electrical signal detector and an adequate structure of its models. Unlike the situation in radio communications and radar systems, when accompanying signal noise can be considered independent, the optical signals reception, as already mentioned, is accompanied by shot noise, whose characteristics are closely related to the signal itself: the variance of shot noise is directly proportional to the instantaneous intensity of the optical signal. In this connection well-known algorithms for optimal estimates of radio signals parameters can not be directly used to obtain the precision values of optical signals parameters.

1. Introduction to the Theory of Optimal Estimates of Signal Parameters in Presence of Noise

Development of the majority of radio and radio-physical instruments and devices through improved design and manufacturing technology has a limit defined by the presence of natural and artificial noise. The problems of the signals optimal reception with accounting the accompanying noise statistics were first raised and solved by V.A. Kotelnikov [1]. They can be summarized as follows: the signal at the receiver output is set as an arbitrary combination $\zeta(t)$ of the signal $S(t, g, l, r)$ and noise $n(t)$, where g, l, r - are unknown estimated parameters. The essence of the problem is to obtain the optimal estimates of unknown parameters with the highest accuracy on the base of the reseived signal realisation, accompanied by the noise, to determine the values of these parameters. Because of the random nature of noise, received signal is a random process. Thereby, the most complete knowledge of the interesting signal parameters is contained in posterior probability density of the received realization $\zeta(t)$. With regard to the problem of estimating the single parameter g, posteriori probability density $Pps(g) = P(g/\zeta)$ is the conditional probability density. The condition is the adopted realization $\zeta(t)$.

In accordance with the theorem of conditional densities, we can write

$$P(g/\zeta) \cdot P(\zeta) = P(g) P(\zeta/g) \qquad (1.1)$$

Then

$$Pps = \frac{P(g)P(\zeta/g)}{P(\zeta)} \qquad (1.2)$$

Here P(g) - a priori probability density of appearance of the signal with the parameter g. The probability density P(ζ) is unknown, but its value from the signal parameters does not depend, and therefore it can be replaced by a constant k, whose value is determined from the normalization condition a posteriori probability density

$$\int_G Pps(g)dg = 1 \qquad (1.3)$$

where G - the sphere of all possible values parameter g. The conditional probability density of the realization $\zeta(t)$, considered as a function of the unknown parameter g, P(ζ/g) is called the likelihood function and is usually denoted as L(g). It shows how, if it is received realization of $\zeta(t)$, is one possible value of the estimated parameter g more plausible than others. Thus, the relation (1.2) can now be written in the form

$$Pps(g) = kPpr\, L(g) \qquad (1.4)$$

This means that for a known distribution of the parameter g and the known function of the likelihood expression (1.4) should reach a maximum, under the condition that the estimate of the parameter g will correspond exactly to its true value, i.e. when $\hat{g} = g_*$. Thus, to obtain optimal estimates of the unknown parameter a developer of the optimum receiver must derive, based on a priori information on the distribution of this parameter and the known statistics of the noise, an expression for the posterior probability density and create an algorithm and solver that will find maximum a posteriori probability density as a function of values of the required parameter g. The value of this parameter corresponding to the maximum of likelihood function should be taken as his best estimate. In estimation theory is proved that if a solution to the problem exists, then this estimate will be closer than the other to the true value of an unknown parameter. In most practical cases, the prior distribution of the estimated parameter is not known. Then we can assume that its distribution has equal probability. In this situation, the posterior probability coincides with the likelihood function, and the corresponding estimates are the most exactly. To get the best estimate, as stated above, it is necessary to find the position of the maximum likelihood function (or its logarithm, which in some cases significantly simplifies the search algorithm). In accordance with known rules of the extremum search it is need to take a derivative of this function and equate it to zero. Thus, the likelihood equation solution gives the optimum on accuracy estimate of the desired parameter.

If a few parameters are unknown, then the likelihood function is differentiated by each of them and make the system of likelihood equations. The solutions of this system yields values of joint optimal estimates of all parameters. An important part of the optimal estimation theory is the definition of deviations of the resulting estimates from the true values. This question correspond us to the so-called Rao - Cramer borders [2-4], defining the minimum variance of the resulting estimates.

2. Current State of the Optical Signals Parameters Estimates Problem

The history of the problem of optical signals parameters optimal estimates can be divided into two time phases. The first phase ends of the sixties years of last century, when the photodetectors were so imperfect that their own noise, in most cases were significantly higher than the shot noise signal, and optical signals themselves were received in conditions of strong background illumination. The second phase began when the photodetectors were so perfect that their intrinsic noise can be neglected, and when, instead of optical communication through the atmosphere appeared fiber communication and optical disk drives, where the background illumination can be completely excluded. In this vein, it is clear the contents of the world first beautifully written monograph of V.S. Shestov [5]. It shows the ways of obtaining optimal estimates of optical signals parameters in the presence of independent noise, the methods of optimal reception of optical signals on fluctuating in time two-dimensional field disturbers, and methods of constructing multi-channel detection systems. Now, when the quality of photodetectors increased immeasurably, the main source of noise became inevitable satellites of light converting into an electrical signal - shot noise. At the same time come to the fore the question of finding new appropriate algorithms to obtain the best estimates of optical signals parameters. In this context, it's appeared so far only a small number of publications devoted to solving this problem. of The problems of reception and optimum processing of optical signals in the modern formulation are resolved in a number of monographs on laser communications (A.G.Sheremetev [6-8], V.M. Pratt [9]), optical location (A.A. Kuriksha [10]) and astrophysical measurements (I.Y. Terebyzh [11]).

In recent years a number of publications appears devoted to the optimal reception of optical signals in presence of the shot noise [16-20], but their main focus - is a search of algorithms that reduce the effects of turbulent fluctuations of the atmosphere on the accuracy of estimates, but not the parameters estimation. The main vector of the work is related to the solution of the above problem by using not single, but the matrix of receivers. Problems are solved as the tasks of detecting binary signals with amplitude modulation but not as problem of obtaining estimates of signal parameters. Thus an article [21] M. Cole and K. Kiasaleh is devoted to optimal estimates of optical communication signals intensity in turbulent atmosphere. The problem is solved for the case of modulation «ON – OF». The closest to this paper publication dedicated to the optimal estimates, is an article H. Hagen, [22], which solved the problem of obtaining maximum likelihood estimates of the Gaussian pulse parameters, but in the presence of independent noise. And although it has a section where dependent shot noise is discussed, but the solution of likelihood equation it is obtained under the assumption of independent noise.

3. MODERN METHODS OF PHOTODETECTION AND THE MODELS OF RECEIVED ELECTRICAL SIGNALS

Progress in quantum optics and the theory of photodetection [23-27] allow us to establish a connection between the states of optical fields and photoelectron statistics. Based on these results can be read to classify sources of optical signals:

a) high stability, single-mode, single-frequency lasers,
b) thermal sources,
c) squeezed light sources [28].

Photodetection signals obtained using high stability lasers leads to a Poisson distribution emitted photoelectrons. This means that the probability to count n photoelectrons in the observation interval T is defined as,

$$P(n,T) = \frac{(\lambda_0 T)^n}{n!} \exp(-\lambda_0 T) \qquad (3.1)$$

where λ_0 is an average speed of photoemission. Photoelectron statistics for detecting signals from sources of heat depends on the ratio of measurement time and the characteristics of temporal and spatial coherence of light, as well as the state of polarization and the so-called degeneracy parameter, defined as the average number of photoelectrons emitted in the interval of coherence [26] or as the ratio of variance of the noise caused by fluctuations in the intensity of light to the variance due to shot noise. Statistical distribution of counts at a given time of measurement in general obeys a binomial distribution with a negative exponent. For a small measurement time compared to the coherence time of this distribution is transformed into a Bose-Einstein statistics, in the case of long duration intervals of counting the number of photoelectrons and low light intensity, as well as in the case high stability laser flux distribution of photoelectrons obeys StatisticsPoisson.

Exotic signal source that emits light in the squeezed state, promises to give the sub-Poisson photoelectron flux, ie flow, the dispersion is lower than that of the Poisson distribution. But so far, unfortunately, these sources have not yet been established, although intensive research in this area are already over 10 years. Note that the smallest variance in the number of photoelectron counts at a given time (unless you consider the possibility of squeezed light) gives the radiation resulting to Poisson statistics. The average number of electrons emitted per unit time and dispersion of flow are then

$$\kappa = \lambda_0; \quad \sigma^2 = \lambda_0. \qquad (3.2)$$

where λ_0 - parameter of the Poisson distribution.

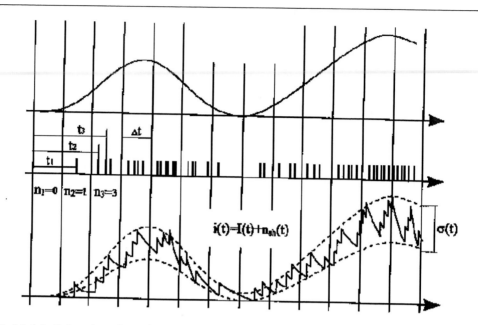

Figure 1. Model of photodetection. Above it is the intensity of the optical signal, below - corresponding to the photoelectron flux, at the bottom - the signal at the output of the analog detector.

Since in most cases for the creation of optical signals one uses lasers, we have to solve the problem of obtaining optimal parameter estimates optical signals, as a model of the photoelectric signal we assume a Poisson photoelectron flux from the emission rate is proportional to the instantaneous value of the intensity of the optical signal at a given time. In favor of this choice is the fact that, as stated above, photodetection signals from thermal sources at low values of extinction coefficient and obeys a Poisson distribution. Then the basic methods of photodetection and mathematical models derived electrical signals. The term «photodetection» we understand the process of converting the optical signal into an electrical and optical signal by the term as some deterministic function of time, representing the change in light intensity, ie square modulus of tension of the optical field.

$$I(t) = |E(t)|^2. \qquad (3.3)$$

At present there are three main ways to obtain electrical signals from photodetection:

a) a count of the number of photoelectrons at a given interval of time,
b) an analog detection, when due to inertia of the photodetector at its output it is forming an analog signal,
c) a fixing time moments of each act of the photoelectron emission.

A graphic representation of these methods gives Figure 1.

4. OPTIMAL ESTIMATION OF OPTICAL SIGNAL PARAMETERS AT DISCRETE PHOTODETECTION WITH A PHOTOELECTRONS COUNTER

4.1. The Derivation of the Likelihood Equation for Discrete Photodetection in the Presence of Background Radiation

Let on the cathode of photoelectrons counter received optical signal with intensity $I_c(t)$ and background radiation with intensity $I_b(t)$. Then, as we know from the theory of the photoelectric effect [23-26] from the detector output will flow photoelectron pulses at a rate

$$\lambda(t) = \lambda_c(t) + \lambda_b(t) = \quad , \qquad (4.1)$$

where χ is quantum efficiency of the photocathode, and $h\nu$ is photon energy. If the photodetector dark current is known, it can also be taken into account by the appropriate correction of the background radiation values. In accordance with the adopted model, the photoelectron flow (4.1) obeys the Poisson distribution. This means that the probability to count n photoelectrons in the intervals of quantization Δt near the point t_i is defined as:

$$P(n, t_i) = \frac{[\lambda(t_i)\Delta t]^{n_i}}{n_i} \exp(-\lambda(t_i)\Delta t) . \qquad (4.2)$$

The joint probability of counting n_i photoelectrons in the interval Δt at the points $t_1 \ldots t_i \ldots t_N$ (likelihood function), due to the fact of statistical independence of the Poisson counts will be determined by following product:

$$P(n_1 \ldots n_N, t_1 \ldots t_N) = \quad , \qquad (4.3)$$

where N – is a number of signal samples in the interval of measurements (parameters evaluation). To obtain the likelihood equations, it is necessary to find the logarithm of the likelihood function, differentiate it on estimated parameter and set equal the resulting expresson to zero. Taking logarithms and differentiating (4.3) and accounting (4.2), we obtain :

$$\ln P(n_1 \ldots n_N) = \sum_{i=1}^{N} [n_i (\ln \lambda(t_i) - \ln \Delta t) - \ln n_i! - \lambda(t_i)\Delta t], \qquad (4.4)$$

$$\frac{d \ln P(n_1 \ldots n_N)}{dx} = \sum_{i=1}^{N} \left[\frac{n_i \lambda'(t_i)}{\lambda(t_i)} - \lambda'(t_i)\Delta t \right] \qquad (4.5)$$

where $\lambda'(t_i)=d\lambda(t_i)/dx$, and x – is estimated parameter. Then the likelihood equation can be written as follows:

$$\sum_{i=1}^{N}\left[\frac{n_i\lambda'(t_i)}{\lambda_c(t_i)+\lambda_b(t_i)}-\lambda'(t_i)\Delta t\right]=0. \tag{4.6}$$

If, as is usually the case, the entire signal can be stored on the observation interval, the second term (2.6) (in square brackets) at small quantization interval Δt in comparison with the duration of the signal may be replaced by an integral with infinite limits. Then the likelihood equations (4.6) is converted to the final form

$$\sum_{i=1}^{N}\frac{n_i\lambda'(t_i)}{\lambda_c(t_i)+\lambda_b(t_i)}-\int_{-\infty}^{\infty}\lambda'(t_i)dt=0. \tag{4.7}$$

Solving this equation for a given input waveform $\lambda_c(t, x)$, background $\lambda_b(t)$ and obtained implementation of counts $n(t_i)$ for the unknown parameter x, we can find his estimate with minimum variance. If the number of estimated parameters is equal to k, composing and solving the system of k equations of the form (4.7), we obtain optimal on accuracy etimates of all k unknown parameters.

4.2. Optimal Estimates of Optical Signal Parameters on the Example of the Gaussian Pulse in the Absence of Background Radiation and Dark Current

The problem of optimal estimates solved on the example of Gaussian signal (Figure 4.1)

$$I(t)=a\exp\left(-\frac{(t-t_0)^2}{\tau^2}\right) \tag{4.8}$$

This form of the signal is chosen because such pulses are widely used in practice, for example, there are the signals of laser ranging systems, fiber communication, optical devices, CD and DVD storage.

The same shape of pulses have the signals of scattered light from small particles crossing the laser beams. In addition, this pulse shape allows to obtain analytical solutions of the likelihood equation and derive explicit expressions for the Cramer-Rao bounds.

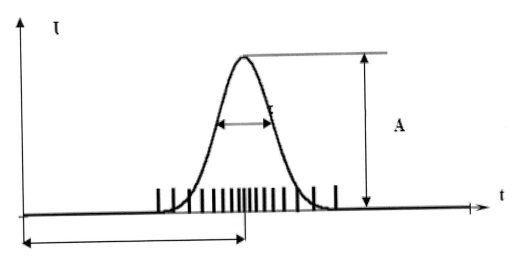

Figure 4.1. A set of sticks on the Figure 4.1. simulates the time points of photoelectrons emission.

As a first step to solve this problem, consider the case when the background radiation and dark current is negligible. Then equation (4.7) takes the form:

$$\sum_{i=1}^{N}\left[\frac{n_i \lambda'(t_i)}{\lambda_c(t_i)}\right] - \int_{-\infty}^{\infty} \lambda'(t_i)dt = 0. \qquad (4.9)$$

Substituting in (4.9) corresponding partial derivatives $\lambda(t, A, \tau, t_0)$ on A, τ, and t_0,, we obtain the following system of likelihood equations for estimating the parameters A, τ, t_0

$$\frac{1}{A}\sum_{i=1}^{N} n_i = \sqrt{\pi}\tau, \quad \sum_{i=1}^{N} n_i(t_i - t_0)^2 = \frac{A\tau\sqrt{\pi}}{2}, \quad \sum_{i=1}^{N} n_i(t_i - t_0) = 0. \qquad (4.10)$$

Solving the system (4.10), we find the following expressions for the joint estimates of all three unknown parameters,

$$A = \sqrt{\frac{\left(\sum_{i=1}^{N} n_i\right)^3}{2\pi \sum_{i=1}^{N} n_i(t_i - t_0)^2}}, \quad \tau = \sqrt{\frac{2\sum_{i=1}^{N} n_i(t_i - t_0)^2}{\sum_{i=1}^{N} n_i}}, \quad t_0 = \frac{\sum_{i=1}^{N} n_i t_i}{\sum_{i=1}^{N} n_i}. \qquad (4.11)$$

Substituting the estimate t_0 from (4.11) in the expressions for the estimates A and τ, we finally obtain:

$$A = \frac{\left(\sum_{i=1}^{N} n_i\right)^2}{\sqrt{2\pi \left[\sum_{i=1}^{N} n_i \sum_{i=1}^{N} n_i t_i^2 - \left(\sum_{i=1}^{N} n_i t_i\right)^2\right]}}, \quad (4.12)$$

$$\tau = \frac{\sqrt{2}}{\sum_{i=1}^{N} n_i} \sqrt{\sum_{i=1}^{N} n_i \sum_{i=1}^{N} n_i t_i^2 - \left(\sum_{i=1}^{N} n_i t_i\right)^2}, \quad (4.13)$$

$$t_0 = \frac{\sum_{i=1}^{N} n_i t_i}{\sum_{i=1}^{N} n_i} \quad (4.14)$$

Thus, having a set of measured values of the photoelectrons number n_i on each interval Δt at times t_i, you can, using the expressions (4.12) - (4.14), find the maximum likelihood estimates of each three unknown parameters of the optical pulse signal. Now we define the quality of these estimates.

4.3. Cramer-Rao Bounds for Joint Estimates of the Gaussian Pulse Parameters in the Absence of the Background and Dark Current

As it's well known [27-29], the quality of the maximum likelihood estimates of the signal parameters defined by a Cramer-Rao bounds. Minimum variance for each of the estimated parameters are the diagonal elements of the correlation matrix of errors, which is the inverse to matrix of Fisher. Elements of the Fisher matrix have the form:

$$J_{mn} = <H_m H_n>, \quad (4.15)$$

where $H = \frac{\partial}{\partial x}[\ln L(x)]$.. Here x is estimated parameter, and $L(x)$ - its likelihood function, the angle brackets denotes statistical averaging. The derivative of the logarithm of the likelihood function for the estimated parameters was derived previously, see (4.5), and if the entire signal, as it is usually the case, confines in the observation interval, it can be represented as:

$$\frac{d\ln L(x)}{dx} = \sum_{i=1}^{N} \frac{n_i \lambda'(t_i)}{\lambda(t_i)} + \int_{-\infty}^{\infty} \lambda'(t)dt = \sum_{i=1}^{N} \frac{n_i \lambda'(t_i)}{\lambda(t_i)} + C_x , \qquad (4.16)$$

where $C_x = \int_{-\infty}^{\infty} \lambda'(t)dt$, $\lambda'(t)$ - derivative of $\lambda(t)$ on the estimated parameter.

In view of (4.5) and (4.16) element J_{mn} of the Fisher matrix takes the form

$$J_{mn} = \langle H_m H_n \rangle = \left\langle \sum_{i=1}^{N}\sum_{j=1}^{N} \frac{n_i n_j \lambda'_m(t_i) \cdot \lambda'_n(t_j)}{\lambda(t_i) \cdot \lambda(t_j)} - C_m \sum_{j=1}^{N} \frac{n_j \lambda'_n(t_j)}{\lambda(t_j)} - C_n \sum_{j=1}^{N} \frac{n_i \lambda'_m(t_i)}{\lambda(t_i)} + C_m \cdot C_n \right\rangle \qquad (4.17)$$

where λ'_m and λ'_n - are the derivatives of $\lambda(t)$ on the corresponding estimated parameters, and C_m, C_n, depending on the position of the element in the Fisher matrix, take the following values:

$$C_A = \int_{-\infty}^{\infty} \lambda'_A(t)dt = \int_{-\infty}^{\infty} \exp\left(-\frac{(t-t_0)^2}{\tau^2}\right)dt = \sqrt{\pi}\tau \qquad (4.18)$$

$$C_\tau = \int_{-\infty}^{\infty} \lambda'_\tau(t)dt = -\frac{2A}{\tau^3}\int_{-\infty}^{\infty}(t-t_0)^2 \exp\left(-\frac{(t-t_0)^2}{\tau^2}\right)dt = -\sqrt{\pi}A \qquad (4.19)$$

$$C_{t0} = \int_{-\infty}^{\infty} \lambda'_{t0}(t)dt = \frac{2A}{\tau^2}\int_{-\infty}^{\infty}(t-t_0)\exp\left(-\frac{(t-t_0)^2}{\tau^2}\right)dt = 0 \qquad (4.20)$$

Statistical averaging of (4.17) given in the Annex to this section, shows that

$$J_{mn} = \sum_{i=1}^{N} \frac{\langle n_i \rangle \lambda'_1(t_i) \cdot \lambda'_2(t_j)}{\lambda^2(t_i)} + \sum_{i=1}^{N}\sum_{j=1}^{N} \frac{\langle n_i \rangle \langle n_j \rangle \lambda'_1(t_i) \cdot \lambda'_2(t_j)}{\lambda(t_i) \cdot \lambda(t_j)} - \\ - C_m \sum_{j=1}^{N} \frac{\langle n_j \rangle \lambda'_n(t_j)}{\lambda(t_j)} - C_n \sum_{j=1}^{N} \frac{\langle n_i \rangle \lambda'_m(t_i)}{\lambda(t_i)} + C_m \cdot C_n. \qquad (4.21)$$

Based on (4.21), we find the matrix Fischer element $J_{A\tau}$.

Taking into account (4.16) and passing in (4.21) from the sums of terms to integrals, we obtain an expression for the sum of the first two terms (4.21) as follows:

$$J_{A\tau 1} + J_{A\tau 2} = -\sqrt{\pi} - \pi A \tau . \tag{4.22}$$

The third, fourth and fifth terms of (4.21) in accordance with (4.21), respectively, are equal:

$$J_{A\tau 3} = C_{x_i} \sum_{i=1}^{N} \frac{\langle n_i \rangle \lambda_2'(t_i)}{\lambda(t_i)} = -\pi A \tau \tag{4.23}$$

$$J_{A\tau 4} = C_{x_i} \sum_{i=1}^{N} \frac{\langle n_i \rangle \lambda_1'(t_i)}{\lambda(t_i)} = -\pi A \tau \tag{4.24}$$

$$J_{A\tau 5} = C_{xi} C_{xj} = -\pi A \tau \tag{4.25}$$

Substituting (4.22) - (4.25) in (4.21), we obtain the final expression for the matrix Fisher element $J_{A\tau}$:

$$J_{A\tau} = -\sqrt{\pi} . \tag{4.26}$$

Further, we will search for the Fisher matrix elements concerning the parameter t_0. At first we find the element J_{At0}. Based on (4.20) and passing in (4.21) to integrals, we obtain the first two terms (4.21) in the form of:

$$J_{At01} + J_{At02} = -\frac{2}{\tau} \int_{-\infty}^{\infty} t \exp\left(-\frac{t^2}{\tau^2}\right) dt + \frac{2A}{\tau^2} \int_{-\infty}^{\infty} \exp\left(-\frac{t^2}{\tau^2}\right) dt \cdot \int_{-\infty}^{\infty} t \exp\left(-\frac{t^2}{\tau^2}\right) dt = 0 \tag{4.27}$$

In accordance with (4.20), the third term (4.21) will have the form:

$$J_{At03} = C_{xm} \sum_{i=1}^{N} \frac{\langle n_i \rangle \lambda_n'(t_i)}{\lambda(t_i)} = \sqrt{\pi}\tau \cdot \frac{2A}{\tau^2} \int_{-\infty}^{\infty} t \exp\left(-\frac{t^2}{\tau^2}\right) dt = 0 \tag{4.28}$$

Similarly, the fourth and fifth terms in (2.21) will equal:

$$J_{At04} = C_{x_n} \sum_{i=1}^{N} \frac{\langle n_i \rangle \lambda_m'(t_i)}{\lambda(t_i)} = 0 \tag{4.29}$$

$$J_{At05} = C_A \cdot C_{t0} = \sqrt{\pi\tau} \cdot 0 = 0 \tag{4.30}$$

From (4.26) - (4.29), we find that.

$$J_{At0} = 0 \tag{4.31}$$

Now we find an element $J_{\tau t0}$. Acting as before, i.e. turning to the integrals in (4.21), we obtain the following values of all five its members:

$$J_{\tau t01} + J_{\tau t02} = -\frac{4A}{\tau} \tag{4.32}$$

$$J_{t03} = \int_{-\infty}^{\infty} \frac{2At}{\tau^2} \exp\left(-\frac{t^2}{\tau^2}\right) dt \int_{-\infty}^{\infty} \frac{A\exp\left(-\frac{t^2}{\tau^2}\right)\left(-\frac{2At^2}{\tau^3}\right)\exp\left(-\frac{t^2}{\tau^2}\right) dt}{A\exp\left(-\frac{t^2}{\tau^2}\right)} = 0 \tag{4.33}$$

$$J_{t04} = \int_{-\infty}^{\infty}\left(-\frac{2At^2}{\tau^3}\right)\exp\left(-\frac{t^2}{\tau^2}\right) dt \int_{-\infty}^{\infty} \frac{A\exp\left(-\frac{t^2}{\tau^2}\right)\left(\frac{2At}{\tau^2}\right)\exp\left(-\frac{t^2}{\tau^2}\right) dt}{A\exp\left(-\frac{t^2}{\tau^2}\right)} = 0 \tag{4.34}$$

$$J_{t05} = \int_{-\infty}^{\infty}\left(\frac{2At}{\tau^2}\right)\exp\left(-\frac{t^2}{\tau^2}\right) dt \int_{-\infty}^{\infty}\left(-\frac{2At^2}{\tau^3}\right)\exp\left(-\frac{t^2}{\tau^2}\right) dt = 0 \tag{4.35}$$

Therefore

$$J_{\tau t0} = \frac{4A}{\tau}. \tag{4.36}$$

Proceeding similarly, given the fact that $J_{mm} = \langle(H_m)^2\rangle$ we get diagonal elements of Fisher matrix J_{AA}, $J_{\tau\tau}$ and J_{t0t0}:

$$J_{AA} = \left(\left\langle\left(\frac{d\ln L(A)}{dA}\right)^{-2}\right\rangle\right)^{-1} = \frac{A}{\sqrt{\pi\tau}} \tag{4.37}$$

$$J_{\tau\tau} = \left(\left\langle\left(\frac{d\ln L(\tau)}{d\tau}\right)^{-2}\right\rangle\right)^{-1} = \frac{\tau}{3\sqrt{\pi}A} \tag{4.38}$$

$$J_{t0t0} = \left(\left\langle \left(\frac{d\ln L(t_0)}{dt_0}\right)^{-2}\right\rangle\right)^{-1} = \frac{\tau}{2A\sqrt{\pi}} \tag{4.39}$$

Thus, we find all elements of the Fisher matrix. Inverting this matrix, we obtain the correlation matrix of errors:

Its diagonal elements represent the optimal estimates variances, respectively, the amplitude A, duration τ, and the signal position t_0 on the time axis, and its off-diagonal elements are the correlation coefficients of the corresponding estimates. On this basis, we can write the following expressions for the minimum relative rms error for estimates of each parameter.

	A	τ	t_0
A	$\dfrac{A(3\pi-2)}{2\sqrt{\pi}\tau(\pi-1)}$	$\dfrac{\sqrt{\pi}}{2(\pi-1)}$	$\dfrac{1}{2(\pi-1)}$
τ	$\dfrac{\sqrt{\pi}}{2(\pi-1)}$	$\dfrac{\tau\sqrt{\pi}}{2A(\pi-1)}$	$\dfrac{\tau}{2A(\pi-1)}$
t_0	$\dfrac{1}{2(\pi-1)}$	$\dfrac{\tau}{2A(\pi-1)}$	$\dfrac{\tau\sqrt{\pi}}{2A(\pi-1)}$

$$\frac{\sigma_A}{A} = \sqrt{\frac{(3\pi-2)}{2A\sqrt{\pi}\tau(\pi-1)}}, \quad \frac{\sigma_\tau}{\tau} = \sqrt{\frac{\sqrt{\pi}}{2A\tau(\pi-1)}}, \quad \frac{\sigma_{t0}}{t_0} = \sqrt{\frac{\tau\sqrt{\pi}}{2At_0^2(\pi-1)}}. \tag{4.40}$$

If, for example, $A=10^9$ e/s, $\tau=10^{-6}$s, then $\dfrac{\sigma_A}{A}$ =3,127%, $\dfrac{\sigma_\tau}{\tau}$ =2,03%, and if $t_0=10^{-2}$s, $\dfrac{\sigma_{t0}}{t_0}$ =0,203%.

CONCLUSION

The urgent for modern technology problem of obtaining optimum on accuracy estimates of the Gaussian optical pulse parameters in the presence of shot noise is solved. The corresponding equations of maximum likelihood are derived and solved. Algorithms of joint and not joint estimations of the pulse amplitude, its width and moment of occurrence are found. Explicit expressions for Kramer- Rao bounds determining the quality of these estimates are derived also.

Thus, there was an opportunity to realize the results of the theory in practice and significantly improve the quality of fiber communication devices, laser location, measurement devices and CD and DVD - memory.

REFERENCES

[1] Kotelnikov, V. A. *The Theory of potential noise immunity.* Moskow, 1956, Gosenergoizdat (in Russian).
[2] Kay, S. M. Fundamentals of statistical signal processing., *Prentise Hall PTR*, 1993, Vol. 1. P. 219.
[3] Levin, B.R., *Theoretical Foundations of Statistical Radio Engineering*, Book Two. Soviet radio, Moscow, 1968, P. 494 (in Russian).
[4] Rao, S.R. *Linear Statistical Methods and Their Application*, Nauka, Moscow, 1968, P. 548 (in Russian).
[5] Shestov, N. S. *Pik out optical signals on a background of random noise*, Sov. Radio, Moscow, 1967, P. 347 (in Russian).
[6] Sheremetev, A.G. *Statistical Theory of Laser Communications*, Svyaz', Moscow,1971, P. 264 (in Russian).
[7] Sheremetev, A. G., Tolparev, R. G. *Lasernaya svyaz', Svyaz'*, Moscow,1974, P. 383 (in Russian).
[8] Sheremetev, A. G. *Coherent optical fiber communication*, Radioand Svyaz', Moscow,1991, P. 188 (in Russian).
[9] Pratt, V. K. *Laser communication systems*, Svyaz', Moscow,1972, P. 232 (in Russian).
[10] Kuriksha, A. A. *Quantum optics and optical radar*, Sov. Radio, Moscow,1973, P. 181 (in Russian).
[11] Terebyzh, V. Y. *Introduction to statistical theory of inverse problems*, Fizmatlit, Moscow, 2005, P. 371 (in Russian).
[12] Terebyzh, V. Y. Limiting resolution for given alternative. *Astrophysics*, 1990,Vol. 33, December, issue.3 (in Russian).
[13] Terebyzh, V. Y. *Image restoration with minimum a priori information*, UFN, 2005, Vol.165, N 2, P. 409 (in Russian).
[14] Terebyzh, V. Y, *Astronomical Journal*, 1999, Vol.76, P. 49 (in Russian).
[15] Terebizh, V. Yu., Cherbunina, O. K. *Astron. And Astroph. Trans.* 1995.Vol. 9, P.159-170.
[16] Saleh, E. A. Estimation of the location optical object with Photodetectors Limited by Quantum Noise, *Appl. Optics*, 1974, Vol. 13, N 8, P. 1824-1827 .
[17] Teih M. C., Rosenberg S. N-Fold Joint Photocounting Distribution for Modulated Laser Radiation: Transmission Through the Turbulent Atmosphere, *Opto-electronics*, 1971, Vol. 3, P. 63-76.
[18] .Davidson, R.S. Estimation of Optical Field Mean Intencities from Photocount Correlations, *Appl. Optics*, 1974, Vol.13, № 9, P. 2171-2176.
[19] Zhu Xiaoming, Kahn, J.M. Free-Spase Optical Communication Thru Turbulence Channels, *IEEE Trans. on Communications*, 2002, Vol.50, № 8. P. 1293-1300.
[20] Vilnrotter, V.A., Srinivasan, M. Adaptive Arrais for Optical Cjmmunications Ressivers, *IEEE Trans. on Communications*, 2002, Vol.50, № 7, P. 1091-1097.
[21] Cole, M., Kiasaleh, K. Signal Intensity Estimators for Free-Spase Optical Communication Thru Turbulent Atmosphere With Array Detectors, *IEEE Trans. on Communications*, 2007,Vol.55, №. 12.

[22] Hagen, N., Kupinsky, M., Dereniak, E. L. Gaussian profile estimation in one dimension, *Appl. Optics*, 2007, Vol.46, No 22, P. 5374-5383.
[23] Glauber, R. Optical coherence and photon statistics (in the book. *Quantum optics and quantum physics*), Mir, 1966, 451 (in Russian).
[24] Loudon, R. *Quantum Theory of Light*. Mir, 1976. 463 (in Russian).
[25] Klauder, Dzh., Sudarshan, E. *Fundamentals of Quantum Optics,* Mir, 1970, 428 (in Russian).
[26] Gudmen Dzh., *Statistical Optics*, Mir, 1988, 527 (in Russian).
[27] Levin, B.R. *Theoretical Foundations of Statistical Radio Engineering*, Moscow, Soviet Radio, 1968, 494 pp. (in Russian).
[28] Sage, E., Mels, J. *Estimation Theory and its application in communication and management*. Moscow, Svyaz', 1976.
[29] Minkoff, J. *Signal Processing*, Boston, London, Arteh House, 2006.

In: Interferometry Principles and Applications
Editor: Mark E. Russo

ISBN 978-1-61209-347-5
©2012 Nova Science Publishers, Inc.

Chapter 13

OPTICAL INTERFEROMETERS: PRINCIPLES AND APPLICATIONS IN TRANSPORT PHENOMENA

Sunil Verma[*,1], *Yogesh M. Joshi*[≠,2] *and K. Muralidhar*[‡,3]

[1] Laser Materials Development and Devices Division,
Raja Ramanna Centre for Advanced Technology, Indore, India
[2] Department of Chemical Engineering,
Indian Institute of Technology Kanpur, Kanpur, India
[3] Department of Mechanical Engineering,
Indian Institute of Technology Kanpur, Kanpur, India

ABSTRACT

Optical techniques are extensively used for high precision diagnostics and process monitoring in physical, biological, and engineering sciences. Interferometry falls in one such class of diagnostics. It relies on changes in the refractive index in the medium arising from variations in the material density. The physical region in which imaging is being carried out is required to be transparent. The light source best suited for an interferometer is a laser. Owing to its features such as greater accuracy, resolution, instantaneous response and non-intrusive nature, interferometry proves to be advantageous and extensively utilized in a broad spectrum of applications. The present chapter deals with the description of laser interferometers in visualization and monitoring of processes involving fluid flow, heat transfer, and mass transfer.

The chapter is divided into two sections. In the first section, we discuss the basic principles of interference and fringe formation. It includes the principles and operations of various interferometer configurations such as Michelson, Mach-Zehnder, holography, phase-shifting, speckle, schlieren and dual-wavelength interferometry. Interferometers

---

[*] [Corresponding author]: Scientist, Laser Materials Development and Devices Division, Raja Ramanna Centre for Advanced Technology, Indore 452013, India, Email: sverma1118@gmail.com, Phone: +91-731-248 8670, Fax: +91-731-248 8650.

[≠] Associate Professor, Department of Chemical Engineering, Indian Institute of Technology Kanpur, Kanpur 208106, India (joshi@iitk.ac.in).

[‡] Professor, Department of Mechanical Engineering, Indian Institute of Technology Kanpur, Kanpur 208016, India (kmurli@iitk.ac.in).

can provide vivid images of temperature and solutal concentration fields. Their real utility is in the quantitative determination of transport properties in addition to heat and mass fluxes. The second section describes the applications of interferometry in studying transient heat conduction, buoyancy-driven convection in a rectangular cavity and superposed fluid layers, and crystal growth from an aqueous solution. These illustrate the utility of interferometry in engineering and research.

1. FUNDAMENTALS

The present section is concerned with image formation when a light beam traverses a transparent medium in which density variation is present.

1.1. Optical Interference

Interferometric measurements are enabled by the following factors:

i. Light waves that are separated by a phase difference will interfere on superposition and produce alternating bands of bright and dark fringes.
ii. Phase difference with respect to a reference is created in a variable refractive index field.
iii. For transparent media, refractive index has a unique relationship with the material density.
iv. Density scales with variables such as temperature and species concentration.

Hence, spatial as well as temporal changes in temperature and concentration will generate corresponding changes in refractive index, phase and ultimately lead to fringe formation. It should be emphasized at this point that interferometry is a *differential* measurement, namely measurements of refractive index in a test section against a reference where refractive index is spatially and temporally a constant. The reference could be vacuum, the ambient, or a liquid bath of uniform chemical composition and temperature.

Since interferometry relies on the phase of the light waves, it is evident that one should employ a light source of unique wavelength (with which the phase is associated). Thus, the light source should be monochromatic. In addition, the phase of the light waves should be stable in time for meaningful fringe patterns to form. These factors require the light source to be *coherent,* a property readily obtainable from lasers, for example, a helium-neon laser. In addition, refractive index fields create an *optical path difference* with the reference wave. Accordingly, it is required that light emerges from a *point source* and the origin for distance measurement is unambiguous. These requirements can be comfortably fulfilled when a laser is used.

Light is electromagnetic radiation with wavelengths λ falling in the visible range (400-800 nm). Light propagates in vacuum at a speed c that is independent of the wavelength. Interference effects are associated with the modulation of the electric field, magnetic field playing only a passive role. The electric field is a vector but for near parallel conditions, it is sufficient to work with its scalar form.

The principle of interference can be illustrated through a simple example. Let a wavefront move through a spatially homogeneous (reference) environment as

$$E_1 = A\sin\{\frac{2\pi}{\lambda}(ct-x)\}$$

The second wavefront moves through the test region where a physical process is in progress. Through changes in refractive index, the wave undergoes a change of phase ϕ and the electric field is obtained as

$$E_2 = A\sin\{\frac{2\pi}{\lambda}(ct-x)+\phi\}$$

Note that the second wave has the same amplitude A and wavelength λ, requiring that a single source be used for passage of light through the test and reference media.

When these two wavefronts are superimposed, the resultant electric field can be calculated as

$$E_1 + E_2 = 2A\cos\frac{\phi}{2}\{\frac{2\pi}{\lambda}(ct-x)-\frac{\phi}{2}\}$$

The amplitude of the resultant wave is

$$= 4A^2\cos^2\frac{\phi}{2}$$

and is perceived as a variation in light intensity. This variation is between zero and $4A^2$ in a sinusoidal form. When seen through a detector (or a camera), the image will look dark below a certain threshold and otherwise, bright. This process is the basis of fringe formation in interferometry. The superposition of two monochromatic coherence wavefronts results in the addition as well as the cancellation of energies and is referred to as the *interference phenomenon*.

Since one wavelength λ corresponds to a phase difference of 2π, the phase difference ϕ can be associated with a path difference δ as per

$$\delta = \frac{\lambda}{2\pi}\phi$$

The relationship between a refractive index field and the phase difference can be established as follows. Refractive index is defined as

$$n = \frac{c_0}{c}$$

where c is the speed of light in the physical region and c_0 is that in vacuum. Since $n > 1$, it is clear that the physical medium serves to slow down the electromagnetic waves. This effect is equivalent to increasing the effective distance to be traversed by the wavefronts. In this context, it is useful to define an optical path length

$$PL_1 = \int n\,dz$$

Here, the integration is in the z-direction along the passage of the light beam. In general, this integration is to be performed along the path of the light ray. Further, in vacuum, $n=1$ and the path length is simply the geometric length of the apparatus in the viewing direction. In other contexts, the effective distance to be covered by the light wave (referred to its speed in vacuum) is greater by a factor that depends on the refractive index.

The reference wave passes through a region of constant refractive index leading to an optical path length

$$PL_2 = \int n_0\,dz$$

On superposition, the path length difference is obtained as

$$\Delta PL = PL_1 - PL_2 = \int (n - n_0)\,dz$$

The equivalent phase difference is calculated from the equation

$$\Delta PL = \frac{\lambda}{2\pi}\phi$$

In a refractive index field, n would vary from one point to another. Accordingly, the path length difference and the phase field will be spatially variable. Lines of constant phase (specifically those that minimize and maximize intensity) will appear as fringes. These would also be lines of constant refractive index of the material in the test section, material density, and ultimately, lines of constant temperature and solute concentration.

Specific optical configurations are described in the present chapter for the generation of fringe patterns and are followed by selected applications.

1.2. Why Use Optical Techniques for Imaging Transport Phenomena?

One could measure fluid properties with external transducers. However, in order to get sufficient data, one has to deploy a large number of such probes, which start interfering with the process under study. On the other hand, only a few probes provide inadequate data. For example, a single moving probe in a liquid medium yields temporally inaccurate data if moved slowly, but stirs the system if moved rapidly. Therefore, external probes are never a satisfactory method for spatially mapping a physical domain. In contrast, optical techniques are *photon probes* that do not affect the process being studied. They map the properties of the process with a spatial resolution of about a micrometer and a temporal resolution of about a millisecond. Their response is practically inertia-free and provides a large volume of data of the process under study. Furthermore, the availability of high speed cameras for image acquisition and powerful computers for data reduction have helped in the revival of optical diagnostics in engineering research in the last decade.

1.3. Optical Imaging Techniques

Various types of optical imaging techniques have been used for studying transport behavior of momentum (fluid mechanics), thermal energy (heat transfer) and transport of molecular species (mass transfer). Overall, these optical techniques can be subdivided into three categories. In the first category, flow marking is carried out by using dyes, bubbles, buoyant seeds, etc. in order to enable direct visualization of the fluid motion. In the second class of techniques, fluid motion is studied by analyzing the frequency of an incident radiation, which is known to undergo Doppler-shift after getting scattered from the particles moving with the flow (e.g. laser Doppler velocimetry). The third category considers all those techniques which analyze behavior of an index of refraction of the medium through which the light beam propagates due to various impetuses such as fluid motion, temperature gradient and/or concentration changes (e.g. shadowgraphy, schlieren, interferometry, etc.). In this chapter we will focus only on the third category of techniques.

The techniques based on the alteration of refractive index n take advantage of unique relation between n and the density of the medium ρ for transparent mediums. Since transported variables such as heat (temperature) or mass of chemical component (concentration) directly affects the density, monitoring of refractive index n directly gives an information about temperature or concentration gradients in the medium. Although various techniques mentioned in the third category require variation in refractive index in order to analyze transport behavior, they have complimentary characteristics. Usually, interferometry is used when the expected changes in the refractive index are small, whereas shadowgraphy and schlieren are used when the changes in the refractive index are large. Non-interferometric techniques such as shadowgraph or schlieren rely on detection of the path of light rays as they traverse through the region of interest. On the other hand, interferometric methods detect optical path-length differences. Principally, the three techniques measure the physical property in an integrated manner, such that the image output is an integration of the physical quantity of interest along the path of the beam through the process chamber. Therefore, these methods show the best results when the transport is two-dimensional and light passes perpendicular to this two dimensional plane. Usually these techniques are employed when the field is nearly two dimensional, i.e. when variation of a physical quantity under investigation is negligible in the third dimension. However, if this assumption is not valid, then variation in the third dimension needs to be obtained by applying analytical techniques such as computerized tomography. The non-interferometric techniques are simpler to implement and difficult to interpret quantitatively, while the interferometric techniques are amenable to quantitative analysis though difficult to set up and maintain.

The application of interferometry are discussed in the context of monitoring the following transport processes: transient heat conduction, solutal transport, buoyancy-driven convection in a rectangular cavity, buoyancy-driven convection in superposed fluid layers, and growth of a crystal from its aqueous solution.

1.4. Imaging Fluid Flow

Fluid motion and convection phenomena can be visualized by optical techniques through either (a) refractive index in the form of optical path difference (or its gradient), (b) displacement of fluid markers as a function of time, or (c) velocities of fluid markers. Flow Visualization by Merzkirch [Merzkirch, 1987] is a classical text for researchers planning to conduct fluid flow and convection imaging. Similarly, text books on Optical Interferometry by Steel [Steel, 1983] and Hariharan [Hariharan, 1985] are a good starting point for familiarizing oneself with the subject. A survey of holographic interferometry is given by Vest in his book [Vest, 1979]. Optical Measurements: Techniques and Applications, edited by Mayinger [Mayinger, 1994] and Schlieren and Shadowgraph Techniques, by Settles [Settles, 2001] present excellent compendia of a variety of optical techniques for fluid flow and convection mapping.

1.4.1. Shadowgraphy

Shadowgraphy employs a collimated beam of light from either a white light source or a laser. The image has uniform contrast if there are no refractive index gradients in the region of interest. On the other hand, due to the curvature of light rays by the refractive index gradients, the portions of the image corresponding to the displaced rays appear bright or dark. Shadowgraph systems are simple and have been extensively used in experimental fluid mechanics research for flow visualization [Goldstein, 1996; Settles, 2001]. Figure 1 shows the schematic drawing of a shadowgraph optical set up and the image acquisition system.

Under the assumption of infinitesimal deviation of light rays inside the inhomogeneous field and also between the exit face of the process chamber and the screen, the linear governing equation for the shadowgraph process is [Schopf, 1996; Verma, 2005]:

$$\frac{I_o(x_i, y_i) - I_s(x_s, y_s)}{I_s(x_s, y_s)} = (L \times D)\left(\frac{\partial^2}{\partial x^2} + \frac{\partial^2}{\partial y^2}\right)\{\ln n(x, y)\}$$

Here I_0 is the intensity at a particular point (x_i, y_i) on the screen in the absence of the field of disturbance; I_s is the intensity at point (x_s, y_s) due to intensities $I_0(x_i, y_i)$ of all the points (x_i, y_i) which are mapped onto (x_s, y_s) in the presence of the field of disturbance; D is the spread of the field of disturbance; L is the distance between the exit surface and the screen; and $n(x, y)$ is refractive index at the location (x, y).

1.4.2. Schlieren

Schlieren technique requires a collimated light system and converging lens with a knife-edge that is introduced halfway into the focal point [Goldstein, 1996; Settles, 2001]. Light rays deflected (by refractive index gradients) toward the knife edge are blocked resulting in dark spots on the image; conversely, rays deflected away from the knife edge show as bright spots on the screen. The knife-edge can be oriented either vertically or horizontally, thus enabling measurement of refractive index gradients in the two directions. The governing equation for the schlieren process under the linear approximation is

$$\frac{I_o - I_s}{I_s} = \frac{2f}{a}\left(\frac{\partial}{\partial y}\{\ln n(x,y)\}\right)$$

where a is the source size and f is the focal length of the lens used for focusing the beam over the knife-edge. As is evident from the above equation, a single integration is needed to extract the refractive index field from the schlieren image as compared to double integration required in the case of solving a Poisson equation for shadowgraph image.

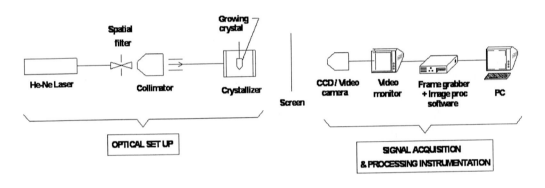

Figure 1. Schematic drawing of shadowgraph optical set up and the image acquisition instrumentation.

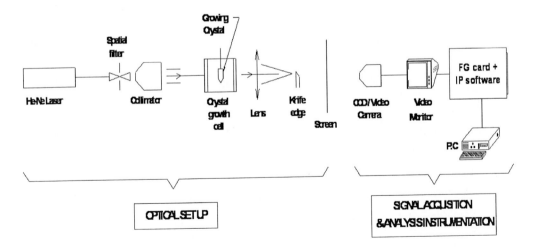

Figure 2. Schematic drawing of a schlieren optical technique.

1.4.3. Interferometric Methods

The physics behind the generation of interferograms is different from shadowgraph and schlieren techniques. In interferometry, the phase change of the test beam with respect to the reference beam is the origin of the formation of interferograms. As a rule, interferometry is used when the expected changes in temperature or concentration are small. If the thermal or solutal changes in the region of interest are large, they limit the usefulness of interferometry in process monitoring for the following reasons: (i) the linearity between density (through temperature or concentration) and refractive index breaks down; and (ii) large gradients in the temperature and concentration fields result in sharp bending of rays. Both of these effects complicate interferometric data analysis.

1.4.3.1. Schlieren-Interferometer

This technique has the characteristics of both non-interferometric and interferometric methods in that it detects the curvature as well as the path length of the light rays. Since the final experimental output is an interference pattern, we have classified it as an interferometric technique. It is also referred to either as a *Wollaston-prism shearing interferometer* or a *differential interferometer* [Mayinger, 1994, pp. 78-83; Settles, 2001, pp. 132-134]. The Wollaston prism uses double refraction to separate an incident light ray into two rays which are polarized orthogonally to each other and diverge from each other at an angle of $\pm\theta/2$ with respect to the original direction of the beam (Figure 3). Collimated monochromatic light is passed through a polarizer, set at an angle of 45° to the direction of separation of the Wollaston prism, so as to produce e- and o- rays of equal intensity on passing through the prism. The parallel-ray polarized beam then passes through the solution in the test cell and is then focused by a lens into the Wollaston prism. As the polarized rays traverse through this prism, they split into e- and o- rays. Incident ray 1 is separated into two rays 1_e and 1_o, representing polarization parallel and perpendicular to the plane of the diagram. On leaving the prism, these two rays have an angle of θ between them. Adjacent incident rays 2, 3… are similarly split, as shown. Since the rays 1 and 2 have a separation of θ between them, ray 1_e coincides with 2_o, and similar pairings occur for all adjacent rays having separation of θ, i.e. the ray 2_e coincides with 3_o, and so on. Since the coinciding rays have orthogonal planes of polarization, there is no interference between them. But when passed through a polarizer oriented at 45° to the separation direction of the Wollaston prism, the emerging paired wavefronts interfere and an interference pattern is obtained on the screen. In short, the collimated image from the experimental test cell is split by the Wollaston prism into two mutually displaced images which then interfere with each other. Therefore, each element of the final image represents the interference between P_{xy} and $P_{x(y+k)}$, where P is the optical path length at image coordinates x and y. If the prism is centered upon the focus of the converging lens, infinite-fringe mode interference results, which resembles a schlieren image. Displacement of the center of the Wollaston prism from the focal point results in wedge fringes to appear in the image. The bends in the wedge fringes are proportional to the refractive index gradients in the test region and can be used for quantitative evaluation. The principal virtue of the schlieren interferometer is the flexibility of being able to record quasi-schlieren images, which facilitate qualitative interpretation, to multiple-fringe images for quantitative measurements.

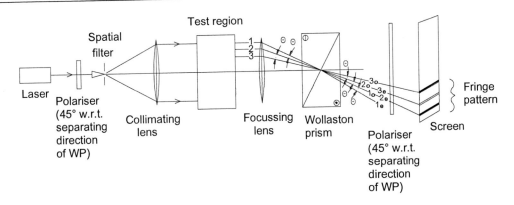

Figure 3. Optical layout of a schlieren (differential) interferometer [Verma, 2008A].

1.4.3.2. Mach-Zehnder Interferometer

The Mach-Zehnder interferometer is a popular configuration for studies in heat and mass transfer in fluids. The optical components, namely beam splitters and mirrors, are inclined exactly at an angle of 45° with respect to the beam direction (Figure 4). The first beam splitter splits the incoming collimated beam into two equal parts, namely the transmitted and reflected beams. The transmitted beam is the *test beam* and the reflected beam is the *reference beam*. The test beam passes through the test region, is reflected by the mirror, and recombines with the reference beam on the plane of the second beam splitter. The reference beam undergoes a reflection at the second mirror and passes through the reference medium and superposes with the test beam at the second beam splitter. On superposition, the two beams produce an interference pattern. This pattern contains information about the variation of refractive index in the test region.

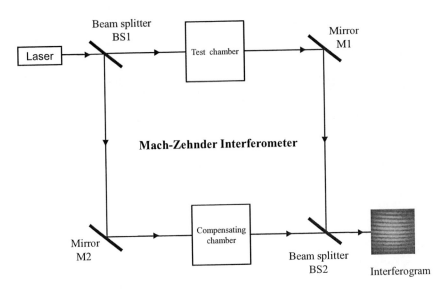

Figure 4. Optical configuration of a Mach-Zehnder interferometer. A compensation chamber may be required in the path of the reference beam to cancel effects arising from an undisturbed medium in the test cell.

The Mach-Zehnder interferometer can be operated in two modes, namely infinite fringe setting and wedge fringe setting (Figures 5(a-d)). In the former, the test and reference beams are set to have identical geometrical path lengths, and the fringes form due to refractive index changes alone. Since each line is a line of constant phase, it is also a line of constant refractive index. The fringe thickness is an inverse measure of the local refractive index gradient, being smaller when the gradients are high. This setting is used for high-accuracy refractive index measurements. In the wedge fringe setting, the mirrors and beam splitters are deliberately misaligned to produce an initial fringe pattern of straight lines. When a refractive-index disturbance is introduced in the path of the test beam, these lines deform and represent refractive index profiles in the fluid. The quantitative methodology of extracting temperature change per fringe shift from the wedge and infinite settings is described below.

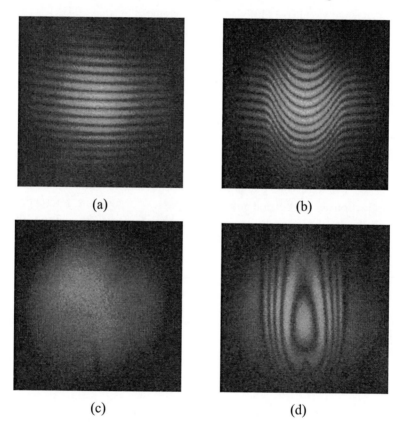

Figure 5. Mach-Zehnder interferograms. (a) Wedge fringes, (b) Candle flame in the wedge fringes, (c) Infinite fringe, and (d) Candle flame in the infinite fringe [Verma, 2009].

Let $n(x,y,z)$ and $C(x,y,z)$ be the refractive index and concentration fields respectively, in the physical domain of interest. With reference to the test chamber, coordinates x and y form the vertical plane, which is perpendicular to the beam propagation direction, and the *z-axis* is the horizontal line along the length of the test section. The light beam propagates parallel to the *z*-axis. Let n_o and C_o be the reference values, as encountered by the reference beam passing through a compensating chamber.

The interferogram is a fringe pattern arising from the optical path difference

$$\Delta PL(x,y,z) = \int_0^S [n(x,y,z) - n_o] \, ds$$

which in terms of concentration is

$$\Delta PL(x,y,z) = \frac{dn}{dC} \int_0^S [C(x,y,z) - C_o] \, ds.$$

The above integral is evaluated along the path of the light ray given by the coordinate *s*. If we neglect higher order optical effects such as refraction and scattering, the path of the light ray will be straight along the length of the test chamber. Since the interference fringes are loci of constant phase or constant phase difference, the optical path difference ΔPL is constant on a given fringe. Therefore,

$$\int_0^L [C(x,y,z) - C_o] \, dz = \frac{\Delta PL(x,y,z)}{dn/dC} = \text{constant}$$

and it follows that

$$\int_0^L C(x,y,z) \, dz - C_o L = \text{constant}$$

for a given fringe.

The integral $\int_0^L C(x,y,z) \, dz$ is equal to $\overline{C}L$, where $\overline{C}$ is the average value of the variable *C(x,y,z)* along the length *L* traversed by the laser beam inside the test chamber. Under the approximation of negligible refraction, the above integral reduces to a line integral of the function *C(x,y,z)* along the *z*-axis. Hence, we get

$$(\overline{C} - C_o) L = \text{constant} = \frac{\Delta PL}{dn/dC} \tag{1}$$

This equation holds for all the fringes in the case of the infinite-fringe setting. Each fringe of the interferogram represents a locus of points over which the average of the concentration field evaluated along the length of the test chamber is constant; i.e. the fringes are isoconcentration contours. In processes governed predominantly by thermal changes, fringes correspond to isotherms. For a change in path length by one wavelength, one fringe shift occurs. Since this change in path length per fringe shift is constant, the change in concentration per fringe shift is also a constant. Defining the function $(\overline{C}-C_o)L$ in Equation (1) as $f(\overline{C},L)$, the concentration at two successive fringes for a given value of L can be obtained as

$$\text{fringe 1:} \quad f_1(\overline{C_1},L) = \frac{\Delta PL}{dn/dC}$$

$$\text{fringe 2:} \quad f_2(\overline{C_2},L) = \frac{(\lambda+\Delta PL)}{dn/dC}$$

where λ is the wavelength of the laser. From these two equations the concentration change per fringe shift can be calculated as

$$\Delta C_E = \frac{1}{L}\left\{f_2(\overline{C_2},L) - f_1(\overline{C_1},L)\right\} = \frac{\lambda/L}{dn/dC} \tag{2}$$

The number of fringes expected in a projection can be estimated from the relation

$$\text{Number of fringes} = \frac{C_2 - C_1}{\Delta C_E}$$

This relation is valid when (a) the refraction effects are negligible and (b) the test beam is traversing through a single fluid.

Now consider the situation wherein concentration gradients are large and the ray traverses a different distance through the cell. Let the two light rays traverse paths L_1 and L_2 respectively inside the test cell, and the corresponding line integrals of the concentration field $C(x,y,z)$ resulting in the average concentrations be $\overline{C_1}$ and $\overline{C_2}$, respectively. Using Equation (1) we can write

$$(\overline{C_2}-C_o)L_2 = (\overline{C_1}-C_o)L_1.$$

Thus, the line integral value $\overline{C_2}$ at a location on the fringe which corresponds to the length L_2 can be expressed in terms of the line integral value $\overline{C_1}$ at some other location corresponding to the ray length L_1 as

$$\overline{C_2} = C_o + \frac{L_1}{L_2}\left(\overline{C_1} - C_o\right).$$

1.4.3.3. Refraction Effects

The presence of strongly refracting fields can modify Equation (2) for the concentration difference per fringe shift. In the present context, a strongly refracting field will arise when there is large concentration or thermal difference in the vertical direction inside the test cell. Hence, the light ray will not travel in a horizontal plane; instead it will bend in the vertical direction, the extent of which depends on the magnitude of the concentration or thermal gradient. Therefore, refraction will introduce an additional path length for the test beam. For the sake of completeness of the discussion on interferometric imaging, we present below the derivation of the formula for calculating the change in concentration per fringe shift under the presence of strongly refracting fields.

Consider the path of the light ray AB through a test cell (Figure 6) when it is affected by refraction effects. Let α be the bending angle at a location P of the test cell. The optical path length from A to B is given by

$$AB = \int_A^B n(x,y,z)\, ds$$
$$= \int_0^L n(x,y,z)\, \frac{dz}{\cos \alpha}.$$

Here, y is a coordinate parallel to the gravity vector and z is parallel to the direction of propagation of light. The length of the test section in the z-direction is indicated as L.

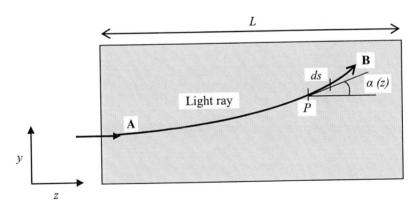

Figure 6. Bending of the light ray due to refraction effects.

Assuming α to be small, $\cos(\alpha)$ can be expressed as

$$\cos \alpha = (1 - \alpha^2)^{1/2}.$$

Using the first two terms of the binomial expansion, one can approximate

$$\cos\alpha \sim 1 - \frac{\alpha^2}{2}$$

Hence, the optical path length is given by

$$AB = \int_0^L n(x,y,z)\left(1-\frac{\alpha^2}{2}\right)^{-1} dz$$

$$= \int_0^L n(x,y,z)\left(1+\frac{\alpha^2}{2}\right) dz \qquad (3)$$

The cumulative bending angle $\alpha(z)$ at any location along the z axis can be easily calculated. It can be shown that the expression for the bending angle $\alpha(z)$ is (Goldstein, 1996)

$$\alpha(z) = \frac{1}{\bar{n}(x,y)}\frac{\partial \bar{n}(x,y)}{\partial y} z \qquad (4)$$

where $\bar{n}(x,y)$ is the average line integral of $n(x,y)$ over the test cell dimension L. Similarly the expression $\partial \bar{n}(x,y)/\partial y$ represents the average line integral of the transverse derivative of $n(x,y)$ over the length L.

Substituting the expression for $\alpha(z)$ from Equation (4) in Equation (3), we get

$$AB = \int_0^L n(x,y,z)\left[1+\frac{1}{2}\frac{1}{\bar{n}^2}\left(\frac{\partial \bar{n}}{\partial y}\right)^2 z^2\right] dz$$

$$= \bar{n}(x,y)L + \frac{1}{6\bar{n}^2}\left(\frac{\partial \bar{n}}{\partial y}\right)^2 L^3$$

The optical path of the reference beam is

$$AB = \int_0^L n_o dz$$

$$= n_o L$$

Hence, the difference in the optical path lengths of the test and reference beams in the presence of refraction effects is

$$\Delta PL = \overline{n}(x,y)L + \frac{1}{6\overline{n}}\left(\frac{\partial \overline{n}}{\partial y}\right)^2 L^3 - n_o L$$

$$= \left(\overline{n}(x,y) - n_o\right)L + \frac{1}{6\overline{n}}\left(\frac{\partial \overline{n}}{\partial y}\right)^2 L^3$$

$$= \left(\overline{C}_1(x,y) - C_o\right)L\frac{dn}{dC} + \frac{1}{6\overline{n}}\left(\frac{\partial \overline{n}}{\partial y}\right)^2 L^3$$

where $\overline{C}_1(x,y)$ represents the average line integral of the temperature field along the direction of the ray at a given point on the fringe. The corresponding ray over the next fringe corresponds to an additional path of λ and can be written as

$$\Delta PL + \lambda = \left(\overline{C}_2(x,y) - C_o\right)L\frac{dn}{dC} + \frac{1}{6\overline{n}}\left(\frac{\partial \overline{n}}{\partial y}\right)^2 L^3$$

where $\overline{C}_2(x,y)$ represents the average line integral of the temperature field along the direction of the ray at a point on the next fringe. The successive temperature difference between the two fringes is

$$\lambda = \left(\overline{C}_2(x,y) - \overline{C}_1(x,y)\right)L\frac{dn}{dC} + \frac{1}{6\overline{n}}\left(\frac{d\overline{n}}{dC}\right)^2\left(\left(\frac{\partial C}{\partial y}\bigg|_2\right)^2 - \left(\frac{\partial C}{\partial y}\bigg|_1\right)^2\right)L^3$$

and the concentration change per fringe shift is

$$\Delta C_E = \frac{\lambda - \left(\frac{1}{6\overline{n}(x,y)}\right)\left(\frac{d\overline{n}}{dC}\right)^2\left(\left(\frac{\partial C}{\partial y}\bigg|_2\right)^2 - \left(\frac{\partial C}{\partial y}\bigg|_1\right)^2\right)L^3}{L\left(\frac{dn}{dC}\right)}$$

Since the gradient in the concentration field is not known before the calculation of the fringe concentration, the factor $\left(\frac{\partial C}{\partial y}\bigg|_2\right)^2 - \left(\frac{\partial C}{\partial y}\bigg|_1\right)^2$ must be calculated from a guessed concentration field. Thus the final calculation of ΔC_E relies on a series of iterative steps with improved estimates of the concentration gradients.

1.4.3.4. Evaluation of Interferograms

The thinned fringes carry the essential information of the process parameter rather than the thick fringe bands. Hence, in order to extract quantitative data from the Mach-Zehnder interferograms, we have to obtain the fringe skeleton. For our process, the thinned fringes contain the information about the path-integrated concentration field. In the sub-sections below, we discuss the methodology for calculating the concentration field from interferograms recorded in the infinite- and wedge-fringe settings.

1.4.3.5. Infinite Fringe Interferograms

Let us consider an infinite fringe interferogram as shown in Figure 7. The presentation is in the context of a concentration field (of salt in water, for example) that varies from C_{sat} to $C_{supersat}$. The reference is to a crystal growth process discussed later in this chapter. The aim of the following discussion is to find the absolute concentration corresponding to each fringe curve. We know that the change in concentration per fringe shift is given by

$$\Delta C_E = \frac{\lambda/L}{dn/dC}$$

and

$$C_2 - C_1 = \frac{\lambda/L}{dn/dC}$$

At the top boundary, the concentration is known to be the saturation concentration C_{sat}, whereas the bottom is at a concentration equal to the supersaturation value $C_{supersat}$. Our aim is to find the concentrations C_1, C_2 and C_3 at the fringes 1, 2 and 3, respectively, in the interferogram (Figure 7) with the given boundary conditions. Although the concentration values at the lower and the upper boundaries are known, assigning concentration to the first fringe appearing near the boundary is not straightforward. This is because the concentration gradients near the boundary result in several fine fringes that are lost because of the finite resolution of the CCD/video camera. Also, the first fringe may be distorted during the image processing operations. Hence, it becomes necessary to assign concentration to the first visible fringe by an appropriate analytical procedure. The methodology of finding the concentration values at the first and the subsequent fringes is presented below.

Step-1: First, identify that region of the interferogram where the fringes are closely packed. The fringes 1, 2, and 3 are three such fringes in the above interferogram (Figure 7).

Step-2: Fitting a second-order polynomial of the type $C(y) = \alpha + \beta y + \gamma y^2$ to the three fringes, we get

$$C_1 = \alpha + \beta y_1 + \gamma y_1^2$$
$$C_2 = \alpha + \beta y_2 + \gamma y_2^2$$
$$C_3 = \alpha + \beta y_3 + \gamma y_3^2$$

where y is the vertical coordinate measured from the lower boundary. An exploded view of the three fringes and the coordinate axis is shown in Figure 8. Using above equations we obtain two simultaneous equations for the two unknowns β and γ:

$$\Delta C_E = C_1 - C_2 = \beta(y_1 - y_2) + \gamma(y_1^2 - y_2^2)$$

$$\Delta C_E = C_2 - C_3 = \beta(y_2 - y_3) + \gamma(y_2^2 - y_3^2)$$

These are solved to obtain the constants β and γ. Here, ΔC_E is the concentration change per fringe shift in the fluid.

Step-3: The next step in the analysis procedure is to compute the concentration gradient near the boundary close to the fringes. In the interferogram shown in Figure 9, we are interested in finding the concentration near the lower boundary.

The concentration gradient at the boundary can be given as

$$\left.\frac{\partial C}{\partial y}\right|_{y=y_o} = \beta + 2\gamma y_o$$

Step-4: Once the concentration and the concentration gradient at the boundary are known, the concentration at fringe 1 near the lower boundary can be obtained as

$$C_1 = C_{boundary} + \left.\frac{\partial C}{\partial y}\right|_{boundary}(y_1 - y_o)$$

and so

$$C_1 = C_{supersat} + (\beta + 2\gamma y_1)(y_1 - y_o)$$

Step-5: ΔC_E is the quantum of concentration change per fringe shift; thus concentration at subsequent fringes 2, 3 and others can be computed by adding it to the concentration of the previous fringe. However, if the concentration gradient is negative in the counting direction, an equivalent amount has to be subtracted.

Step-6: Since the interferogram is digitized and stored in a computer in the form of a matrix of integers, the above procedure of computing concentration can be implemented for any element of the matrix. In our experiment, the concentration is a smoothly varying function, and it is expected that the concentration difference between adjacent rows of the matrix will be small. Hence, the evaluation procedure is applied to the rows coinciding with the fringe locations but not in the gap between them.

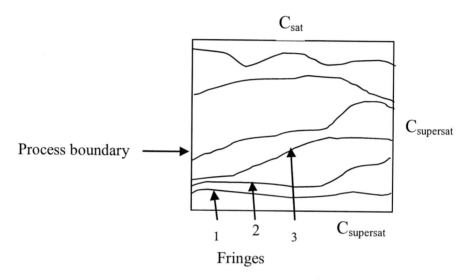

Figure 7. Infinite fringe interferogram of the concentration field in an experiment that develops a concentration distribution.

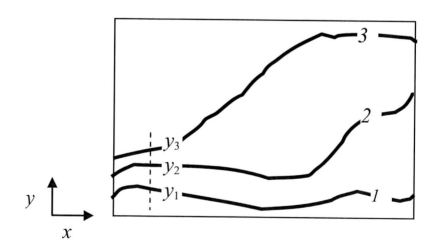

Figure 8. An exploded view of the fringes 1, 2 and 3 along with the coordinate system used.

1.4.3.6. Wedge Fringe Interferograms

The Mach-Zehnder interferometer is not always used with the two interfering wavefronts parallel to each other, as in the infinite-fringe setting discussed above. There is a second mode in which the two interfering wavefronts have a small angle θ between them, introduced deliberately during alignment. Upon interference, they produce an image consisting of bright and dark fringes, representing the loci of constructive and destructive interference, respectively. These parallel and equally spaced fringes are referred to as wedge fringes. The spacing between the wedge fringes is a function of the tilt angle and the wavelength of the laser light used, and is given by

$$d = \frac{\lambda/2}{\sin(\theta/2)}$$

For small tilt angles, the above expression becomes

$$d \sim \frac{\lambda}{\theta}.$$

As θ is decreased to zero, the wedge fringes get farther apart, approaching the infinite-fringe pattern.

When a thermal or concentration field is introduced in the path of the test beam, the phase of the test wavefront gets distorted. Upon interference with the reference wavefront, it manifests itself as a change from straight and parallel fringes to curved fringes. The two interference patterns are shown schematically in Figure 9 and as an exploded view in Figure 10.

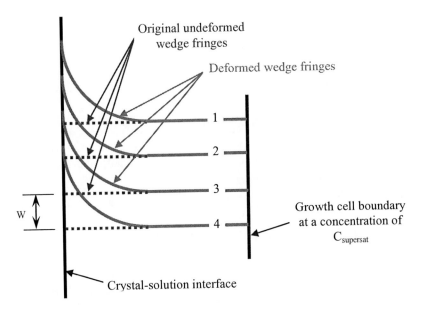

Figure 9. Wedge fringe pattern with and without the field of disturbance.

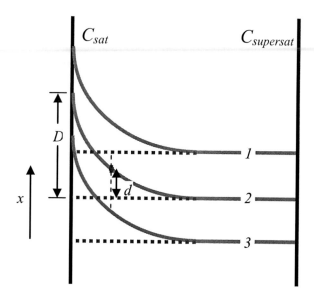

Figure 10. Computing field concentration in the presence of the wedge fringes.

Let us develop a procedure to the solutal concentration at a point x in the interferogram. The deviation of fringe 2 near the crystal-solution interface is D, while at point x is d. Thus it can be written as [Goldstein, 1996]

$$\frac{C(x) - C_{\text{sat}}}{C_{\text{supersat}} - C_{\text{sat}}} = \frac{d}{D}$$

Since all other parameters except $C(x)$ are known, the local concentration can be easily calculated. The concentration gradient dC/dx is obtained from the slope of the fringes at a point x. The above methodology is used for calculating the concentration at points which lie on the wedge fringes. However, if one were to find concentration at a point lying between two wedge fringes, then one has to locate four nearby points on the wedge fringes and compute their respective concentrations. It is then followed by using a suitable interpolation technique to find the concentration at the desired point.

1.5. Imaging Heat Transfer and Mass Transfer

Virtually all of the quantitative refractive index mapping techniques discussed thus far apply to the mapping of both the heat transfer as well as the mass transfer fields. It is assumed that the experiment is essentially two-dimensional along the optical axis, so that the refractive index is constant along each ray of light. Furthermore, an equation that connects refractive index with temperature (via density) is required to convert refractive index into absolute temperature. Interferometry can measure temperature in the following cases: (a) pure liquids having an appreciable dn/dT, (b) systems where changes in concentration are zero or negligible, and (c) applications where a second technique maps the solute concentration.

1.5.1. Mach-Zehnder Interferometry

Mach-Zehnder interferometry can be employed for mapping heat and mass transfer. The quantitative methodology for extracting the field parameter (either temperature or concentration) from wedge- or infinite-fringe mode Mach-Zehnder images has already been discussed in the previous section. Mathematically, let $n(r)$ and $T(r)$ be the refractive index and temperature fields, respectively, in the physical domain being studied. Let n_o and T_o be their reference values, as encountered by the reference beam. The interferogram is a fringe pattern arising from the optical path difference:

$$\delta = \int (n - n_o) dx$$

which, in terms of temperature is

$$\delta = \frac{dn}{dT} \int (T - T_o) dx.$$

Here, the integration is over the path along the ray direction. Other forms of single-pass interferometry and holography can also be used and interpreted in an identical manner. The principal problem with interferometry, for both heat and mass transfer mapping, is calibration. This will not matter for fluid flow (convection) mapping because one is interested in having an observable difference between the moving and static parts of the fluid. Even when experiments require relative temperature or its gradients, the equation relating refractive index to temperature (or density) is required. Such a relationship is unique and obtainable for transparent media via the simplified Lorenz-Lorentz formula

$$\frac{n^2 - 1}{\rho (n^2 + 2)} = \text{constant } (k)$$

The material density can then be related to solute concentration and temperature. In a medium with only temperature variations, typical values of dn/dT obtained are

air, 0.927×10^{-6} K^{-1}

water, 0.88×10^{-4} K^{-1}

It is clear that the number of fringes in water, as a rule, would be considerably larger when compared to air. For a discussion on the generalized Lorenz-Lorentz relationship, please refer to [Kumar, 2008].

1.5.2. Phase Shifting Interferometry

Conventional interferometric mapping is based on obtaining an interferogram, locating the centers of the fringes, measuring the difference between the fringe-center position with that of the reference pattern, and finally interpolating between the fringes to obtain the variable of interest anywhere in the field. This procedure has the shortcomings of being tedious and time consuming and entailing a potential inaccuracy in locating the centers of fringes. Moreover, the data is localized at the fringes, which are often irregular in shape. It necessitates the process of transferring data to a regular grid by the interpolation procedure. The procedure of interpolation also can potentially introduce numerical errors.

Phase-shifting interferometry (PSI) can be used to circumvent the difficulties referred above. First developed by Bruning [Bruning, 1974], PSI involves (a) shifting the phase of one of the two interfering beams so as to get a sequence of phase-shifted interferograms and (b) data reduction to obtain the phase information. Several authoritative reviews exist on phase shifting interferometry [Creath, 1986, 1987 & 1988; Greivenkamp, 1992; Cloud, 1995] and are adopted here in explaining the evaluation steps involved. In general, the phase shifting technique can be incorporated into any type of interferometer systems. The phase of the reference beam is shifted with respect to that of the object beam in steps of $\pi/4$ or $\pi/2$ to get a sequence of phase-shifted interferograms. Various techniques are available to achieve the phase shift [Wyant, 1975 & 1978; Shagam, 1978]. These include mounting the reference mirror on a piezoelectric transducer that moves linearly in proportion to the applied voltage [Wyant, 1982 & 1985].

There are several algorithms reported in the literature for determining the phase of the interferogram using PSI [Creath, 1986 & 1988]. A minimum of three phase-shifted interferograms is required for determining the phase of the object wavefront. This is because the interference equation

$$I = I_{ref} + I_{obj} + 2\sqrt{I_{ref} \times I_{obj}} \cos(\phi_{ref} - \phi_{obj})$$

involves three unknowns, the reference and object beam intensities, and the phase difference between the two interfering beams. The following three-step technique serves as an example:

Let the phase shift between successive interferograms be $\pi/2$ and the absolute phase for the three interferograms be $\theta_1 = \dfrac{\pi}{4}, \theta_2 = \dfrac{3\pi}{4}$ and $\theta_3 = \dfrac{5\pi}{4}$. The set of three equations representing the phase-shifted interferograms are solved simultaneously to get the phase at each point as

$$\phi(x,y) = \tan^{-1}\left\{\frac{I_3(x,y) - I_2(x,y)}{I_1(x,y) - I_2(x,y)}\right\}.$$

Therefore, by detecting the intensity modulation at each pixel and taking the above ratio, the phase at each point of the wavefront is determined.

PSI involves an additional procedure to be performed on the phase calculated through the arctangent equation in order to get the correct final description of the phase. This is necessitated by the nature of the arctangent function, which lies between $-\pi/2$ to $+\pi/2$. As a result, phase values fall within $\pm\pi/2$ regardless of the actual spread of the phase. Therefore, arctangent calculations result in a wrapped up phase. The numerical process of phase unwrapping is carried out in two steps:

[1] Obtain the phase values through the arctangent equation that are confined within $\pm \pi/2$. Noting the discontinuity at every $|\pi/2|$ value, extend these values to fall between $-\pi$ and π. The phase obtained after this step is referred to as modulo 2π phase, and show discontinuity after 2π radians.

[2] Remove the 2π discontinuities by staring from the end of the image in order to obtain a true and continuous map of the wavefront phase.

The phase map obtained by the above procedure is displayed as a gray image wherein the gray value of each pixel corresponds to the phase value at that point. Using 8-bit digitization, the total phase shift (0 to $2n\pi$) is represented quantitatively by gray values in the range of 0 to 255. Using higher bit digitization could further enhance the resolution of the phase measurement, and as a result, the resolution of the physical property being measured.

For PSI based on Mach-Zehnder interferometry, the change in refractive index $\Delta n(x,y)$ corresponding to the phase change of $\Delta\phi(x,y)$ is

$$\Delta n(x,y) = \frac{\lambda}{2\pi L}\Delta\phi(x,y)$$

where L is the length of the chamber through which the beam traverses. In order to get an idea of the resolution of the PSI, a calculation with typical values of parameters in the above equations will show that for a gray value change of unity between adjacent pixels, the minimum detectable Δh and Δn is 1 nm and 10^{-5}, respectively. They are an order of magnitude higher than what is possible with conventional interferometry. It can be further increased by an order of magnitude if 9 or 10 bit digitization is employed. The above equation can be written as

$$\Delta n(x,y) = \frac{N\lambda}{L} \qquad (5)$$

where N is the fringe order.

Finite time is required for translating the reference mirror mechanically in order to obtain the phase shifted interferograms. Thus, conventional phase shifting interferometry has the limitation of being unable to follow a process in real time. Nakadate et al. [Nakadate, 1990 & 1995] overcame this limitation ingeniously by employing polarized light to simultaneously record the required number of phase-shifted interferograms.

1.5.3. Electronic Speckle Pattern Interferometry (ESPI)

When a rough surface is illuminated with a coherent radiation, each facet of the object scatters radiation towards the observer/detector. Here, rough stands for random and microscopic variations of height on the surface of the object with dimension more than the wavelength of light. Since the phase of the light scattered will vary from point to point in proportion to the local surface height, interference occurs between the light scattered from any two facets. Owing to the superposition of individual interference patterns, each arising from a separate facet on the rough surface, a random pattern of interference fringes, called *speckle* is observed. For a detailed discussion on the speckle phenomena, the reader is referred to [Dainty, 1984].

Electronic speckle pattern interferometry (ESPI) has been used for measuring concentration gradients. It can also be used to visualize convection in the form of variations in refractive index in the solution. This technique can be adapted for mapping temperature as well as convection patterns in fluids [Dupont, 1995; Verga, 1997]. In these applications, the speckle pattern is obtained by means of a stationary ground glass plate. When a chamber containing the fluids is interposed between the plate and the CCD camera, the optical path length is altered and a speckle interference pattern of the experiment is obtained. As the optical properties of the fluid in the chamber change, for example, due to changes in temperature, the phase of the speckle pattern changes. It results in changes in the interference pattern between the object and the reference speckles. This is subtracted from the speckle-interference-pattern recorded at the start of the experiment to get the phase variation produced during the time interval. Thereby, the spatial change in the field parameters such as refractive index and temperature are determined.

1.6. Simultaneous Mapping of Heat and Mass Transfer

In single-wavelength interferometry, the interpretation of the fringe pattern becomes difficult when two independent quantities (such as temperature and concentration) affect the phase of the traversing beam. The analysis is then based on the approximation that one of the two parameters has a negligible influence on phase. The approximation may not be not valid in certain processes wherein heat (temperature) and mass transfer (concentration) simultaneously affect the refractive index of the aqueous solution and hence, the phase of the beam traversing the crystallizer. Here we need a diagnostic that can simultaneously map the two parameters. Dual-wavelength interferometry is capable of detecting temperature and concentration and is described below.

1.6.1. Dual-Wavelength Interferometry

Dual-wavelength interferometry can separate the influence of the two competing parameters such as temperature and concentration, provided the sensitivities dn/dC to dn/dT are significantly different for the two wavelengths. This problem is similar to solving two equations for two unknown variables. Using two different wavelengths for mapping the process results in two independent fringe patterns. Dual-wavelength interferometry was first used by El-Wakil and Jaeck to study heat and mass transfer in gaseous boundary layers [El-Wakil, 1964]. Later, Ecker proposed the concept of holography in conjunction with dual-wavelength interferometry for measuring temperature and concentration fields during alloy solidification [Ecker, 1987 & 1988]. Mehta (1990) and Vikram et al. (1990) reported the applicability of this technique for determining the temperature and concentration fields inside liquids. The principle of dual-wavelength interferometry (and holography) as explained in a few reports by Vikram and co-workers [Vikram, 1992A] is described below.

For processes influenced simultaneously by the temperature and the concentration fields, the total change in refractive index can be expressed as

$$\Delta n = \frac{\partial n}{\partial C}\Delta C + \frac{\partial n}{\partial T}\Delta T \qquad (6)$$

where ΔC and ΔT are the changes in the concentration and temperature fields influencing the refractive index field, respectively. Combining the equations (5) and (6) given above, we get

$$\frac{N\lambda}{L} = \frac{\partial n}{\partial C}\Delta C + \frac{\partial n}{\partial T}\Delta T \,.$$

Extending the above discussion to the dual-wavelength interferometry, we get two independent equations for refractive indices n_1 and n_2 corresponding to the two source wavelengths, λ_1 and λ_2, which are then solved for independent estimates of the thermal and the concentration variations [Vikram, 1992A]. Thus,

$$\Delta C = \frac{N_1\lambda_1 \frac{\partial n_2}{\partial T} - N_2\lambda_2 \frac{\partial n_1}{\partial T}}{L\left[\frac{\partial n_1}{\partial C}\frac{\partial n_2}{\partial T} - \frac{\partial n_2}{\partial C}\frac{\partial n_1}{\partial T}\right]}$$

and

$$\Delta T = \frac{N_1\lambda_1 \frac{\partial n_2}{\partial C} - N_2\lambda_2 \frac{\partial n_1}{\partial C}}{L\left[\frac{\partial n_1}{\partial T}\frac{\partial n_2}{\partial C} - \frac{\partial n_2}{\partial T}\frac{\partial n_1}{\partial C}\right]} \,.$$

For accurate determination of ΔC and ΔT, we must know precisely the refractive index and its gradients with respect to temperature and concentration at both the wavelengths and the fringe order. In the absence of experimental values, a suitable analytical methodology has been put forth by Vikram and co-workers [Vikram, 1991, 1992B] for determining the needed data. The refractive index is obtained using the two-constant Cauchy equation. A higher order Cauchy equation involving more constants may be used for greater accuracy. The refractive index gradients with respect to temperature and concentration are determined using Murphy-Alpert [Murphy, 1971] and Lorentz-Lorenz relationships, respectively.

1.7. Three Dimensional Imaging

Since conventional optical techniques can only measure the average refractive index along the optical path (e.g. the thickness of the process chamber), quantitatively meaningful results are possible only in the following situations: Either the experiment is "two-dimensionalized" so that the parameter of interest is constant along the optical path; or truly three-dimensional imaging techniques are adopted for experimentation, such as holography for a direct 3-D imaging or computerized tomography for an indirect 3-D visualization from a set of two-dimensional projection images. The principle of holography and holographic interferometry are explained first followed by the principle of tomography.

1.7.1 Holography and Holographic Interferometry

Holography is a form of interferometry and can be used to achieve new modes of measurement, for example, interference of images of the same object at two different points in time. We explain the conventional holography and how it differs from the holographic interferometry. Various types of holographic interferometry are also discussed.

Holography was discovered by Dennis Gabor in 1948 [Gabor, 1948 & 1949] and is based on the principles of interference and diffraction. Ordinary photography records only the wave intensity while the phase information is lost. In contrast, both the amplitude of the object wave and the phase are recorded in holography [Collier, 1971; Hariharan, 1986 & 2002]. Briefly, the procedure for hologram recording and subsequent reconstruction of the object image is as follows: A coherent beam of light from a laser is split into an illumination beam and a reference beam. The former illuminates the object so that the scattered light from the object strikes a high-resolution photographic plate. The reference beam is also reflected onto the photographic plate, where it interferes with the light from the object. The photographic plate, or *hologram*, is then developed to reveal the interference pattern registered on it.

The object wave can be reconstructed in two different ways, one resulting in a *virtual* image and the other in a *real* image. When the hologram is illuminated by the original reference beam, the interference pattern on it behaves as a diffraction grating, and diffracts the incident light. A part of the diffracted beam is a replica of the object-image wavefront. By looking through the hologram into the direction of the original object, the viewer perceives the waves to be originating from a virtual image of the object located precisely at the original location (Figure 11a). On the other hand, if the back of the hologram is illuminated by a *conjugate* of the reference beam (a beam identical in phase and intensity distribution to the original reference beam but propagating in the opposite direction), the diffracted beam is directed and shaped by the hologram into a real image, corresponding spatially to the original object. Viewing in the direction of the diffracted beam recreates the object through the hologram (Figure 11b).

Holographic interferometry is a refinement of holography [Vest, 1979]. The main difference between classical interferometry and holographic interferometry is that for ordinary interferometry, an object wavefront is continuously compared against a reference wavefront, and the optical path length difference between the two wavefronts is measured to determine changes in the physical parameter of interest. In holographic interferometry, two types of interference phenomena occur - the interference between the reference and the object waves to form a hologram and the interference between two such holograms. This is equivalent to the interference of two coincident images of the same object at two different times. Thus, in classical interferometry, the two waves follow different paths at the same time while in holographic interferometry, they pass through the same path at two different times.

The reader is referred to the edited volume by Mayinger [Mayinger, 1994] and the review by Verma & Shlichta [Verma, 2008A] for details of different types of holographic interferometry. The three forms of holographic interferometry are explained schematically in the Figures 12(a-c). The three holographic interferometric techniques described above have several trade-offs. In the first, one follows the process in real-time, but provides only averaged parameter values along a single direction. The second technique is simple in implementation but provides the parameter values at a particular instant and also provides only averaged values along a single direction. The third can provide a three-dimensional array of parameter values, but again, only at a particular instant of time.

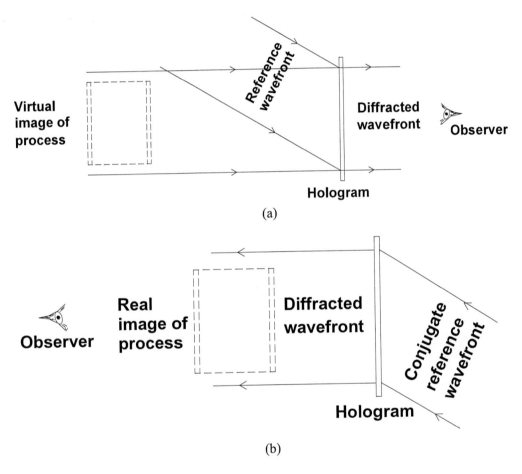

Figure 11. (a) Formation of virtual holographic image with original reference beam used for reconstruction, and formation of a real holographic image with a conjugate reference beam used for reconstruction [Verma, 2008A].

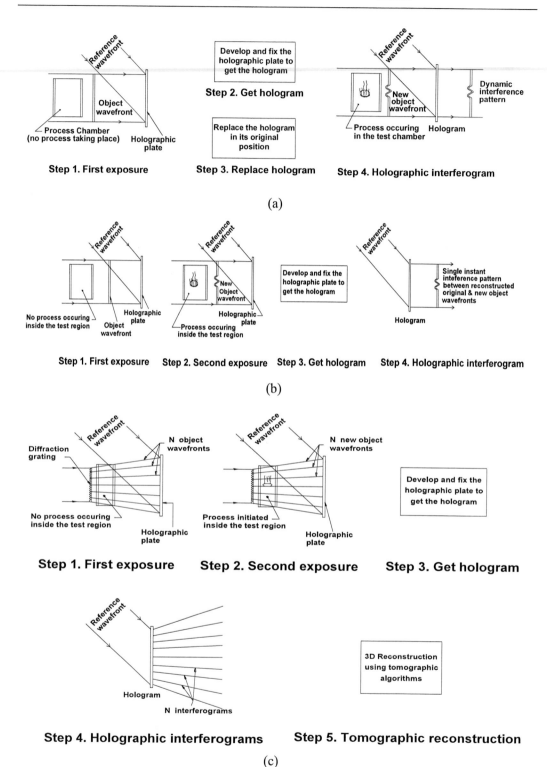

Figure 12. (a) Single-direction, single-exposure, real-time holographic interferometry, (b) single-direction, double-exposure, single-instant holographic interferometry, and (c) multi-directional, double-exposure, single-instant holographic interferometry and tomography [Verma, 2008A].

1.7.2 Computerized Tomography

Computerized tomography is defined mathematically as the process of constructing a continuous two-dimensional function from its one-dimensional line-integrals obtained along a finite number of lines at known locations [Ramachandran, 1971; Herman, 1980; Natterer, 2001]. This can be extended to include reconstruction of a three-dimensional function from its two-dimensional projections taken along a fixed number of known directions. Tomography is essentially a two-step process: first, collecting the projection data and second, reconstructing the three-dimensional function using numerical algorithms.

The tomographic reconstruction methods can be broadly classified as (1) transform methods, (2) series-expansion methods, and (3) optimization methods. The transform methods are computationally efficient, but require a large number of experimental projections to recover the process parameter with some satisfactory accuracy [Lewitt, 1983]. In contrast, the series expansion methods are computationally intensive, but have the advantage of giving satisfactory results even with limited or partial projection data [Censor, 1983]. The optimization approach requires the selection of a suitable functional that has to be extremized under the experimental constraint [Gull, 1986].

Various forms of optical tomography have numerous potential applications in many diverse fields of science and engineering. We shall confine our discussion here to the application of shadowgraphic and interferometric tomography to the study of transport processes. In these cases, the line integrals of refractive index can be converted to point-by-point maps of refractive index at selected planes.

1.7.1. Shadowgraphic Tomography

A three-dimensional pattern of varying refractive index inside the process chamber results in light rays getting refracted out of their original path. This caused a spatial modulation of the intensity distribution with respect to the original intensity and produced a shadowgraph pattern on the screen. Shadowgraph images recorded from several view angles constitute the experimental projection data. The intensity of each pixel in the shadowgraph images is processed numerically to extract the refractive index distribution. This constitutes the numerical input to the reconstruction algorithms. It should be noted that refractive index distribution has one-to-one correspondence with the physical parameter, such as temperature, concentration or fluid flow field in the process chamber. The output of the tomography reconstruction algorithms is a matrix representing the process parameter values over a particular plane in the region of interest. A detailed discussion on the application of shadowgraphic tomography to the process of crystal growth can be found in the work of Verma et al. [Verma, 2006 & 2008B].

1.7.2. Interferometric Tomography

In the case of interferometric tomography, the projection data is in the form of interferograms captured from several angular directions. For example, a Mach-Zehnder interferometer can be used to capture interferograms of the process from several coplanar viewing directions. Such infinite (or wedge) fringe interferograms recorded nearly simultaneously from different view angles constitutes one set of 2D projection data. These interferograms are numerically processed to get the refractive index distribution, which serve as input to the tomography algorithms. It should be noted that data reduction in the case of shadowgraph images is two orders of magnitude difficult as compared to interferograms due to analytical expression relating intensity with the refractive index field in the two processes. Finally, the methodology of implementing the reconstruction algorithms for interferometric tomography is similar to that for shadowgraphic tomography. The reader is referred to Mishra et al. [Mishra, 1998 & 1999] and Muralidhar [2001] for in-depth details of interferometric tomography as applied to heat transfer problems.

1.8. Comparison of Optical Techniques

Each technique described above is accompanied by certain advantages and weaknesses. Shadowgraph and schlieren have the advantages of low cost, ruggedness and simplicity of apparatus, and ease of real-time qualitative interpretation. Gradient interferometers provide an attractive trade-off between simplicity and versatility. Two-beam interferometers are probably the easiest to set up and permit quantitative information. Reconstructive holography has the important advantage of providing rich three-dimensional detail. With regard to sensitivity, an analysis provided in pp. 150-155 of [Merzkirch, 1974] indicates that schlieren, gradient interferometry and two-beam interferometry all have about the same order of magnitude of sensitivity (about $\lambda/100$). The trade-offs between conventional and holographic interferometry are particularly complex. Since holographic interferometry operates in a time-differential, rather than space-differential, manner, it has several advantages over classical interferometry. Unlike the latter, it does not depend critically on the alignment of the two interfering beams. Moreover, the need for high quality optical components is much less critical. Holography also provides capabilities not possible with conventional schlieren and interferometry. A collimated-beam hologram can be made of an experiment and then reconstructed into a real image that can then be processed at leisure to generate schlieren and gradient-interferometric images of variable orientation and fringe width [Smigielski, 1970; Merzkirch, 1987]. By using stored-image holography or by making time-lapse double exposures on the given holographic plate, interferograms which image the change in optical path length with time can be generated [Heffinger, 1966]. All of these advantages, however, must be weighed against the disadvantages of the lack of real-time viewing, labor of reconstruction, and incompatibility with simultaneous mapping of concentration and temperature fields. Electronic-speckle pattern interferometry is different from other interferometric and holographic techniques, in that it can be used under ordinary room lighting conditions. It does not require vibration isolation tables [Lokberg, 1980; Jones, 1989]. Although it is less sensitive to temperature/concentration changes compared to the conventional techniques of interferometry and holography, it can provide useful data under non-ideal laboratory conditions. Since ESPI functions as a time-difference interferometer, it is similar to holographic interferometry. Hence, it does not require accurate adjustments of the interfering beams. Unlike holographic interferometry, it has the advantage that the information is directly recordable onto the CCD sensor without any intermediate recording plate. The primary drawback is the fringe quality, which is quite noisy due to the speckle property itself. As a result, the spatial resolution of the camera limits the resolution of the measurement.

2. APPLICATIONS

The second part of the chapter describes various applications of the above techniques in studying transport phenomena as encountered in engineering applications. These are: (i) crystal growth from solution [Verma, 2007, 2008A, 2008B, 2009 & 2010; Dinakaran, 2010], (ii) heat conduction phenomenon in a differentially heated layer of water [Singh, 2009], (iii) mass transfer from high concentration region to low concentration (dissolution of sugar) [Bhandari, 2009], (iv) natural convection in a differentially heated rectangular cavity of square cross-section (hot lower wall and a cold upper wall) filled with air [Muralidhar, 2005], (v) buoyancy-driven convection in an octagonal cavity half-filled with 50 cSt silicone oil, the rest being air [Punjabi, 2002], (vi) long-time interferograms formed in an octagonal cavity containing silicone oil (50 cSt) floating over water [Punjabi, 2004], (vii) buoyancy-driven convection in a differentially heated cavity that is rectangular in plan (Rayleigh-Benard configuration) [Mishra, 1998], (viii) convection in an eccentric annulus with the gap filled with air [Ranjan, 2005], (ix) steady state heat transfer in an eccentric annulus when the gap is filled with 390 cSt silicone oil [Ranjan, 2005], and (x) convection and heat transfer behind a heated cylinder at low Reynolds numbers [Singh, 2007].

2.1. Transport Phenomena during Crystal Growth from Solution

Crystal growth from solution is the most widely used method for growing large crystals, several centimeters in size [Mullin, 2001]. It is applicable to all classes of materials, including inorganic, organic, metal-organic, macromolecules and complexes. Solution growth offers the convenience of low operating temperatures, less complicated apparatus, and simple instrumentation. More importantly, when the process to be imaged uses optical radiation, the medium under study has to be necessarily transparent at the chosen wavelength of radiation (He-Ne laser wavelength in our experiments). Among the various methods of growth, only solution growth satisfies this requirement. For these reasons, the present work is primarily focused on crystal growth from solution. An ensemble of crystals grown using solution growth are [Paschotta, 2008]: KDP (potassium dihydrogen phosphate, KH_2PO_4) and its deuterated analogue DKDP (KD_2PO_4) for frequency conversion and electro-optic switching (Q-switches and Pockels cell), respectively; TGS (triglycine sulphate) for laser-energy measuring devices; KAP (potassium acid phthalate) for monochromator applications; $LiIO_3$, ZTS [zinc tris (thiourea) sulphate], and many organic and metal-organic crystals for linear and non-linear optical applications.

2.1.1. Importance of Mapping Solution Growth

The ultimate goal of research on crystal growth is to achieve an in-depth understanding of the process with the aim of improving the microscopic and macroscopic homogeneity of the grown crystals, in addition to increasing their size. A crystal growing from solution creates thermal and concentration gradients in the surrounding solution by releasing the heat of crystallization and depleting the solute near the growth surface. The resultant temperature and concentration gradients affect the perfection and stability of the crystal grown [Bunn, 1949]. The change of solution density with temperature ($d\rho/dT$) is negative and the change of solution density with solute concentration ($d\rho/dC$) is positive. Therefore, crystal growth in the Earth's gravitational field is accompanied by a rising buoyant convection current which envelops the crystal, is often oscillatory and unstable, and therefore, drastically modifies the concentration gradient along the growth interface [Chen, 1979A & 1979B]. Thus, the growth history and defect structure of the crystal is a function of the time-dependent spatial distribution of the convection patterns, and of the temperature and concentration profiles in the surrounding solution. Therefore, in order to learn how to grow large defect-free crystals, it is helpful to map the spatial distribution of the thermal and concentration profile around the growing crystal.

2.1.2. Phenomenology of the Growth Process

The rate of crystallization is determined by its slowest stage. Two major stages can be identified: (1) supply of crystallization material from the solution to the crystallization surface; (2) incorporation of this material into the crystal structure, i.e. growth of the crystal. If the first stage is the limiting step, the growth is said to proceed in the *diffusion regime*. If the second stage is the limiting step, growth is said to be operating in the *kinetic regime*. Diffusion regime conditions are prevalent during free convection growth, also referred to as buoyancy-driven growth. In practice, forced flow conditions are adopted. These conditions bring the process into the kinetic regime of growth and are achieved by employing various modes of stirring. Here, the solute is forced to move from the bulk solution towards the crystal surface. There is increased probability of occurrence of defects under irregular or turbulent flow conditions. Under such conditions, the growth rate is high. Therefore, the mass transfer and the fluid flow conditions have to be optimized to avoid deleterious effects such as morphological instability, spurious nucleation, and inclusions. Thus, of particular importance in crystal growth from solution are fluid flow, and heat and mass transfer. Before discussing the results of crystal growth imaging, it is pertinent that a brief description is given of the experimental steps involved in growing a crystal from its aqueous solution.

2.1.3. Supersaturation: Driving Force for Crystal Growth from Solution

The prerequisite for crystallization to take place is the creation of a suitable driving force for the nucleation to occur. In solution growth, this driving force is referred to as *supersaturation*. It is a state of the solution achieved by manipulating its thermal behavior. It can be achieved in the following three ways: (a) *Cooling*: Cooling a saturated solution will result in a supersaturated solution, (b) *Evaporation*: Evaporation of the solvent leads to an increase in concentration of the solution, leading to a supersaturated solution, and (c) at times, a combination of cooling and evaporation is adopted.

The supersaturated state is created by having prior information about the solubility of the material in an appropriate solvent. Solubility of a material in a solvent decides the amount of the material that can be dissolved in it at a particular temperature. If the solubility is too high, it is difficult to grow a large crystal from such a solution. This is because of its sensitivity to thermal fluctuations. At the same time, if the solubility is low, it restricts the growth of large crystals since enough solute is not available for crystallization. The solubility of the material is measured by dissolving it in a continuously stirred solvent; this is done at several temperatures. A relationship so obtained between concentration and temperature fields is referred to as the *solubility diagram*.

A small, good quality crystal is used as seed for initiation of the growth. While immersing the seed crystal, the solution temperature is kept 0.5 degrees above its saturation value so as to dissolve a few surface layers of the crystal. The dissolution helps in the elimination of the physical imperfections on the crystal faces, and also of any surface contamination introduced during crystal processing. After dissolving the crystal for a few minutes, the solution is cooled to its saturation point. A programmed cooling of the solution with an appropriate cooling rate is given from this point onward to continue the growth. The difference in the concentration at saturation temperature and the temperature at which the experiment is performed gives us the extent of supersaturation of the solution. This is the driving force for the growth process and is computed at each stage of imaging experiment.

2.1.4. Optical Imaging of Crystal Growth from Solution

Several research groups working in the field of thermal sciences have also used it for mapping convection during various engineering processes [Rasenat, 1989; Mayinger, 1993; Schopf, 1996; Goldstein, 1996; Settles, 2001]. However, this technique has not been widely used for crystal growth studies. Shiomi et al. used shadowgraphy to measure the temperature profile around a growing Rochelle salt crystal [Shiomi, 1980], while Lenski et al. used a beam deflection technique to measure convection during vapor crystal growth [Lenski, 1991]. Recently, Verma et al. [Verma, S., 2003, 2005C & 2006] have used shadowgraphy for visualizing the buoyancy driven convective patterns around a growing KDP crystal, and studied the gradual build up of this activity with the dimension of the crystal and the time of the growth. A few representative images, shown in Figure 13, demonstrate that buoyant convection increases with the increase of the Grashof number, which is the fluid dynamical dimensionless number representing the strength of the buoyancy driven convection. With the progress of growth, the convection mode changes from laminar (Figure 13a) to chaotic (Figure 13f). The influence of free and forced convection, and the cooling rate on the growth rate and the quality of the grown KDP crystal has also been studied [Verma, S., 2007]. Recently, Dinakaran et al. studied the polar morphology of zinc tris (thiourea) sulphate (ZTS) crystals using this technique [Dinakaran, 2010]. The influence of growth geometry on convection and the morphology of the grown ZTS crystals is shown in time-lapsed shadowgraph images shown in Figure 14.

The schlieren technique was used by Chen et al. measured the magnitude and stability of convection around growing sodium chlorate crystals [Chen, 1979A & 1979B]. Chen also studied the relationship between the convection irregularities observed over a growing sodium chlorate crystal, and formation of defects such as fluid inclusions [Chen, 1977]. Onuma et al. used schlieren techniques to study the buoyancy-driven convection above a growing barium nitrate crystal [Onuma, 1988 & 1989]. Srivastava et al. used schlieren for visualizing convection around a growing KDP crystal [Srivastava, 2004].

Rashkovich and co-workers used a Mach-Zehnder interferometer to measure the diffusion boundary layer thickness under different flow velocities of the solution flowing past the vicinal facets of ADP and KDP crystals [Rashkovich, 1990A & 1990B]. Duan et al. used the high sensitivity of phase-shifting Mach-Zehnder interferometry to map minute convective features around a growing $NaClO_3$ crystal [Duan, 2001]. Recently, Verma et al. have used such an interferometer in both infinite-fringe-width and wedge fringe modes to study the convective field around a KDP crystal growing under free and forced convection conditions in both platform and suspended crystal geometries, and correlated their results with the growth rate and quality of the crystal [Verma, 2005B & 2007].

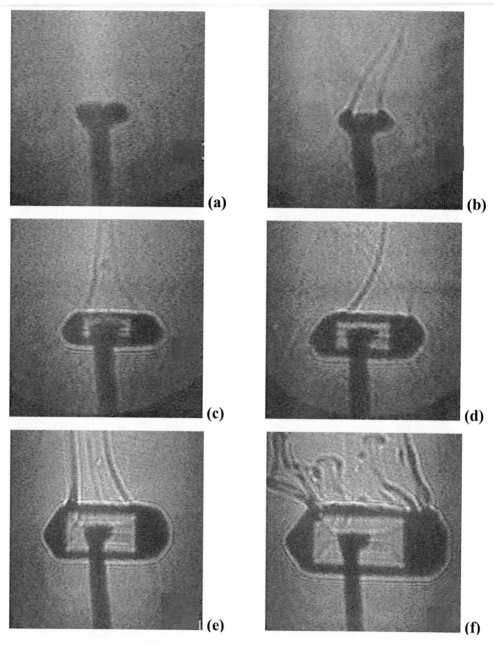

Figure 13. Shadowgraph images show an increase in the strength of buoyancy-driven convection with the growth of the KDP crystal. (a) No buoyant plume is seen at the start; (b) A stable plume starts to appear as the growth begins; (c-e) Buoyant convection intensifies as the crystal grows; (f) The buoyant plume changes from laminar to irregular and finally becoming chaotic. Images adapted from *Imaging techniques for mapping of solution parameters, growth rate, and surface features during the growth of crystals from solution,* Sunil Verma and P.J. Shlichta, Prog. Cryst. Growth & Charact. Materials, 54 (2008) 1-120.

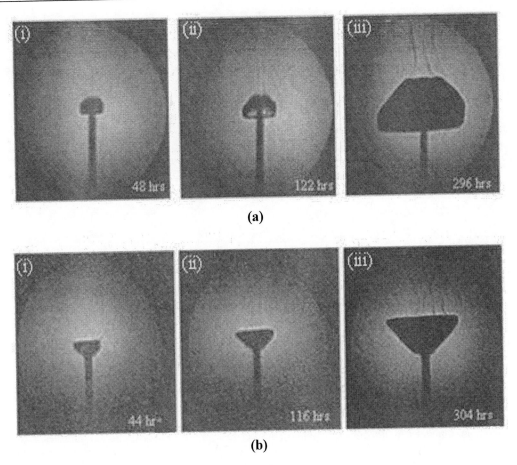

Figure 14. In order to measure the growth rates of {001} and {010} faces of ZTS crystals and their influence on the growth morphology, imaging was performed in two different growth geometries. In the first experiment, the seed was oriented such that (001) face was pointing upwards (a), whereas in the second the (001) face was pointing downwards (b). Shadowgraph images were recorded in a time-lapsed manner for the entire duration of the two experiments. The images were used to calculate precisely the growth rates of {001} and {010} faces. Optical imaging helped in real-time and in-situ observation of the evolution of the growth morphology as a function of the growth geometry and supersaturation. Images adapted from *Optical imaging of the growth kinetics and polar morphology of zinc tris (thiourea) sulphate (ZTS) single crystals*, S. Dinakaran, Sunil Verma, S. Jerome Das, S. Kar and K.S. Bartwal, Cryst. Res. Technol. 45 (2010) 233-238.

Figure 15 shows the concentration gradient inside the crystallizer during KDP growth under buoyancy-driven convection conditions. Due to the growth of crystal, a low density convection plume rises up from the crystal resulting in the accumulation of low density solution in the upper part of the crystallizer as compared to the lower portion of the crystallizer. This results in a concentration gradient inside the crystal, which manifests itself in the form of horizontal fringes in cases of infinite fringe Mach-Zehnder interferometric imaging. This condition is detrimental to the growth of good quality optical crystals and should be avoided.

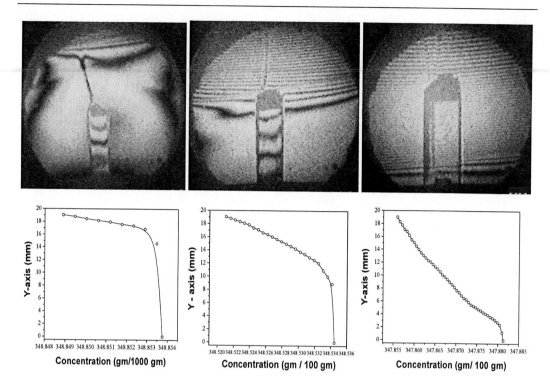

Figure 15. First row - Infinite fringe interferograms of concentration gradient inside the growth chamber at three stages of growth under conditions of free convection. Second row - the computed concentration distribution corresponding to the images in the top row are shown. As the solute-depleted solution accumulates in the upper region of the crystallizer, the concentration diminishes locally, producing a density stratified, gravitationally stable configuration. The Mach-Zehnder interferometer proves to be a very sensitive diagnostic to detect and quantify solution stratification during the crystallization process, a stage that is detrimental for growth of good quality crystals. Images adapted from *Imaging transport phenomena and surface micromorphology in crystal growth using optical techniques*, Sunil Verma and K. Muralidhar, Nat. Acad. Sci. Letts. 33 (2010) 107-121.

Bedarida's group used transmission holographic interferometry as a tool for controlling $NaClO_3$ crystal growth by monitoring the convection around the growing crystal [Bedarida, 1977]. The Chinese group of Yu Xiling adopted real-time, single-exposure holographic interferometry for measuring the diffusivities of KDP and DKDP solutions as a function of temperature and concentration [Xiling, 1992]. The same technique was used to study the convection-driven mass-transport processes at the solid-liquid interface during the growth of DKDP crystals [Xiling, 1996].

Optical techniques have been extensively used for imaging the process of crystal growth aboard space flights. For example, Owen and Shlichta constructed a multi-mode optical observation system (MOOS) for observing crystal growth process using schlieren, shadowgraphy, absorption and shearing interferometry aboard a NASA space shuttle [Owen, 1975; Shlichta, 1985]. Owen also used a Mach-Zehnder interferometer and a holographic unit for studying the growth of NH_4Cl and TGS crystals under microgravity environment [Owen, 1982 & 1986]. Single-exposure holograms were recorded to study the crystal profile microscopically; the interferometric data was used to obtain concentration and temperature profiles near to the crystal; double-exposure holographic interferometry was performed to obtain details of the refractive index changes occurring during growth; and the holographic images were used to observe the development of crystal morphology, growth rate, and the surface structure. Witherow used a multi-purpose holographic system used for investigating the TGS crystal growth experiment aboard Spacelab3 [Witherow, 1987]. The methodology of generating shadowgraph, schlieren, and interferometric images of the growth process in a ground-based laboratory from the holograms recorded in space was also discussed. Solitro, Gatti, Bedarida, and their co-workers developed a fiber-optic based multi-directional holographic interferometer for monitoring convection during crystal growth in space [Solitro, 1989; Gatti, 1989; Bedarida, 1992]. The special volume containing specific contributions dealing with the development of advanced optical diagnostics for crystal growth applications in space was compiled by Trolinger and Lal [Trolinger, 1991]. Experiments on later missions made use of phase-shift interferometry (PSI) and multi-color holographic interferometry (MCHI), developed at Space Sciences Laboratory of the Marshall Space Flight Center (MSFC) for detecting the temperature and concentration field gradients during crystal growth [Witherow, 1994]. Dubois et al. describe the development of an integrated optical set-up for fluid-physics experiments, including crystal growth [Dubois, 1999]. The features of the schlieren technique, electronic speckle pattern interferometry (ESPI), differential interferometry, digital holography, and holographic interferometry were discussed. The performance of different diagnostics was compared for temporal resolution (the recording speed, which is dependent on the hologram refreshing rate), spatial resolution (sensitivity), and dynamic range, using the benchmark experiment of refractive index measurement in Rayleigh-Bénard instability.

The availability of high quality protein crystals is vitally important for determining their structure and subsequently, for drug development [McPherson, 1982 and 2001; Ducruix, 1992]. Shlichta presented a summary of the optical techniques available for imaging crystal growth from solution in general, and for protein crystal growth in particular. The European Space Agency (ESA) has developed a Protein Crystallization Diagnostics Facility (PCDF) equipped with several optical imaging diagnostics for on-line and *in-situ* monitoring of the growth phenomena aboard the International Space Station (ISS) [Pletser, 2001]. Cole *et al.* mapped refractive indices around growing protein crystals by Mach-Zehnder interferometry [Cole, 1995]. McPherson et al. used a phase-shift Mach-Zehnder interferometer to observe the fluid environment around a growing protein crystal in the ground [McPherson, 1999] and in microgravity [McPherson, 2001]. Yin et al. used a Mach-Zehnder interferometer to study the influence of magnetic field on the convection, concentration, solubility and temperature during the growth of lysozyme crystals [Yin, 2001 & 2003].

Several research groups have used Mach-Zehnder interferometer for solutal and thermal field studies of crystal growth process. Onuma et al. measured the surface supersaturation under different flow velocities of the solution flowing over the face of a growing K-alum crystal [Onuma, 1989]. Nakadate and Yamaguchi developed a real-time phase shifting Mach-Zehnder interferometer for concentration measurements [Nakadate, 1990], which was later adopted by Onuma et al. and Kang et al. for concentration measurements around $NaClO_3$ crystal [Onuma, 1993; Kang, 2001]. Kim et al. used a similar interferometer to measure concentration variations at the interface between the solution and the crystal of L-arginine phosphate (LAP) [Kim, Y., 1998A]. Nagashima and Furukawa used this type of interferometer to study the solute distribution during the directional growth of ice crystals, and to understand the origin of morphological instability leading to the dendritic growth

[Nagashima, 2000]. Chen et al. used the Mach-Zehnder interferometer as an aid to monitor and control the growth of large size NaClO$_3$ crystals [Chen, 2002].

Verma et al. used a Mach-Zehnder interferometer in infinite and wedge fringe settings for mapping convection and concentration fields [Verma, 2009]. Figure 16 shows infinite fringe and wedge fringe interferograms of a crystal growing under free convection conditions in top-hanging geometry. The free convection plume and the diffusion boundary layer are clearly visible in infinite fringe interferograms but not in wedge fringe interferograms. Figure 17 shows infinite and wedge fringe interferograms of a crystal growing under forced convection conditions in top-hanging geometry. No buoyancy-driven convection plumes are visible as the crystal is rotating at high RPM and the concentration is uniform inside the crystallizer. Minor fringe distortion can be seen near the crystal-solution interface. Figure 18 shows infinite fringe and wedge fringe interferograms of the crystal growing on a platform under free convection conditions. The concentration gradient is clearly visible in the third interferogram. The free convection plume and the diffusion boundary layers are visible in the infinite as well as wedge fringe interferograms.

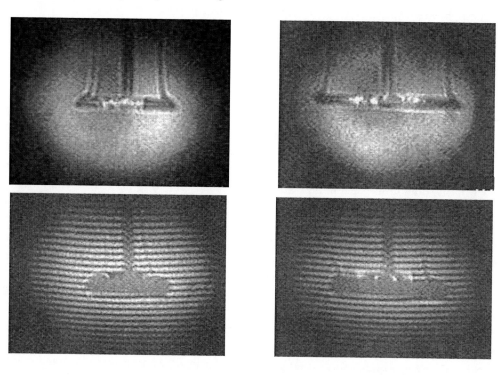

Figure 16. First row - Infinite fringe interferograms of a crystal growing under free convection conditions in the top hanging geometry. . The free convection plume and the diffusion boundary layer are clearly visible. Second row - Wedge fringe interferograms of a crystal growing under free convection conditions in the top hanging geometry. The diffusion boundary layer is not visible as clearly as is in the case of infinite fringe interferogram. Images adapted from *Convection, concentration, and surface features analysis during crystal growth from solution using optical diagnostics,* Sunil Verma and K. Muralidhar, in Recent Research Developments in Crystal Growth, Vol. 5, Transworld Research Network, India, 2009, pp. 141-314.

Onuma et al. used the real-time phase shift interferometer, developed by Nakadate and Yamaguchi, to measure the concentration field around growing and dissolving NaClO$_3$ crystal

[Onuma, 1993; Nakadate, 1990]. Maruyama and co-workers employed a real-time PSI to study the thermal and mass diffusion fields of the aqueous solutions of $NaClO_3$ and $Ba(NO_3)_2$ under microgravity conditions [Maruyama, 1999 & 2002]. Duan et al. used real-time PSI to map convection during $NaClO_3$ crystal growth [Duan, 2001]. McPherson et al. used a phase-shift Mach-Zehnder interferometer to observe the fluid environment around a growing protein crystal on the ground [McPherson, 1999] and in microgravity [McPherson, 2001].

Piano and co-workers have used speckle pattern interferometry for measuring the refractive index variation around a growing KDP crystal [Piano, 2000 & 2001]. The applicability of dual-wavelength holographic interferometry for studying triglycine sulfate crystal growth has been discussed by Vikram and co-workers [Vikram, 1991 & 1992A]. Witherow et al. [Witherow, 1994] describes the efforts in combining the dual-wavelength interferometry with the phase-shifting interferometry for studying crystal growth experiments under microgravity environments.

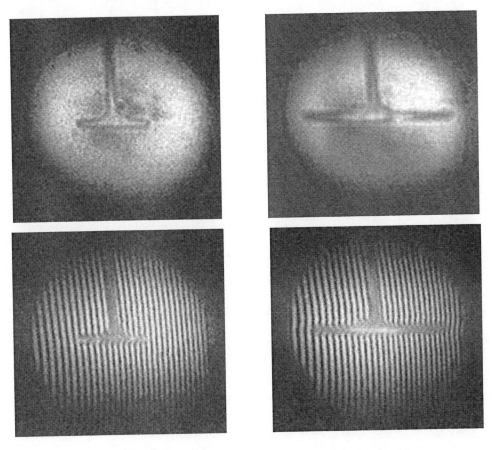

Figure 17. First row - Infinite fringe interferograms of a crystal growing under forced convection conditions in the top hanging geometry. Second row - Wedge fringe interferograms of a crystal growing under forced convection conditions in the top hanging geometry. No buoyancy-driven convection plumes are visible. Minor fringe distortion is visible near the crystal-solution interface. Images adapted from *Convection, concentration, and surface features analysis during crystal growth from solution using optical diagnostics,* Sunil Verma and K. Muralidhar in Recent Research Developments in Crystal Growth, Vol. 5, Transworld Research Network, India, 2009, pp. 141-314.

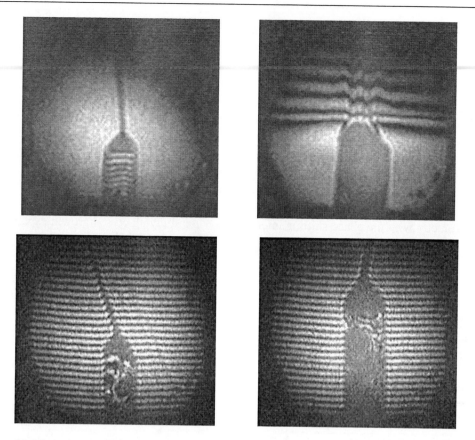

Figure 18. First row - Infinite fringe interferograms of the crystal growing on a platform under free convection conditions. The concentration gradient is clearly visible in the second image. Second row - Wedge fringe interferograms of crystal growing under free convection conditions in a platform geometry. The free convection plume and the diffusion boundary layers are visible in the infinite as well as wedge fringe interferograms. Images adapted from: *Convection, concentration, and surface features analysis during crystal growth from solution using optical diagnostics,* Sunil Verma and K. Muralidhar, Recent Research Developments in Crystal Growth, Vol. 5, Transworld Research Network, India, 2009, pp. 141-314.

2.2. Heat Conduction in a Horizontal Fluid Layer

Figure 19 provides snapshots of interferograms that reveal the heat conduction phenomenon in a differentially heated layer of water [Singh, 2009]. The container is a closed cavity, insulated on the sides, with the top surface maintained at 23°C and the base at 18°C. The room temperature during the experiments was maintained at 20°C. The temperature difference creates a density field and in turn, a refractive index field, thus providing a basis for the formation of interferograms. Since fluids expand with an increase in temperature, the temperature differential creates a density stratified fluid layer, with hot fluid floating over the cold. The thermal field is imaged using a Mach-Zehnder interferometer in the infinite fringe setting. To set the interferometer initially, a compensation chamber filled with water at room temperature is used in the path of the reference beam. Since the surface temperatures are different from the initial value, fringes emerge from the top and the bottom until steady state is reached. Steady state is characterized by straight equi-spaced fringes. The fringe curvature during the initial stages could be the result of imperfect optical alignment that balances the arms of the interferometer. At later times, this error in units of optical path difference is small in comparison to the path difference created by thermal disturbances.

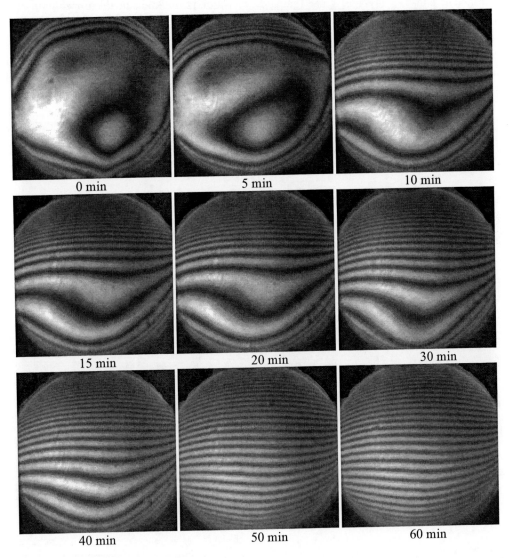

Figure 19. Heat conduction in a differentially heated layer of water with the top surface maintained at 23°C and the lower surface at 18°C. The thermal field is images using a Mach-Zehnder interferometer in the infinite fringe setting. Images adapted from M.Tech. dissertation of Vishnu Singh, Indian Institute of Technology, Kanpur (2009).

2.3. Dissolution of Sugar in Water

In Figure 20, a layer of sugar applied at the base of container is dissolving [Bhandari, 2009]. On evaporation of water, a layer of sugar adheres to the surface. Water is poured in the container until it gets filled. A compensation chamber is used to balance the arms of the interferometer and arrive at an infinite fringe setting. With the passage of time, the sugar slowly dissolves in water. The concentration of sugar in water, thus, changes with time. It is accompanied by the appearance of interferometric fringes. Water adjacent to the lower wall has the highest sugar concentration, diminishing progressively towards the upper wall. During the transient phase, a layer of clear (sugar-free) water is obtained in the upper half of the cavity. Though the physical problem is driven by mass transfer, the overall situation is analogous to heat transfer in a differentially heated cavity. There is a point of difference, however. Sugar at the base creates a region of high concentration, high density and high refractive index while the other wall is one of zero sugar concentration, as in the compensation chamber. Fringes originate here from the lower wall alone.

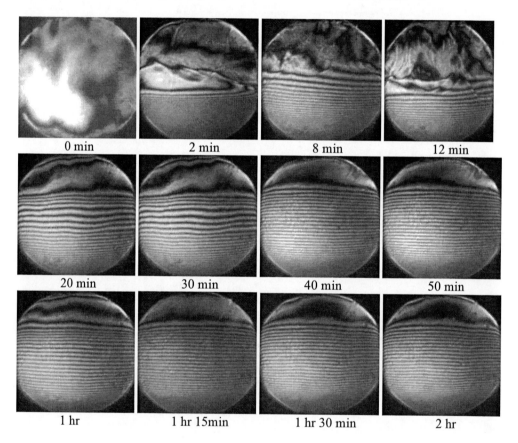

Figure 20. A layer of sugar applied at the base slowly dissolves in water. The concentration of sugar in water changes with time. It is accompanied by the appearance of interferometric fringes. Images adapted from doctoral dissertation of Susheel Bhandari, Indian Institute of Technology, Kanpur (2009).

2.4. Buoyancy-Driven Convection in a Square Cavity

Figure 21 shows natural convection in a rectangular cavity of square cross-section [Muralidhar, 2005]. The fluid medium in the cavity is air and convection is driven by the temperature differential between a hot lower wall and a cold upper wall. The side walls are insulated. Since the lower wall is heated, it is a region of low density when compared to the top surface. Thus, cold fluid is overlaid the hot in a gravitationally unstable configuration. The resulting fluid motion distorts the temperature distribution. When the temperature difference is large, the strength of fluid convection is great and large temperature changes occur close to the boundaries. This phenomenon is akin to the formation of boundary-layers - both hydrodynamic and thermal. The symbol Ra is Rayleigh number and is a measure of the strength of natural convection in the cavity. The first column carries interferometric fringes.

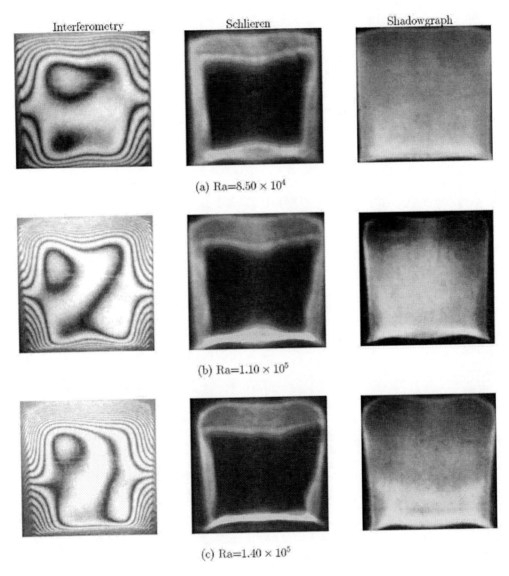

Figure 21. Natural convection in a rectangular cavity of square cross-section. The fluid medium in the cavity is air and convection is driven by the temperature differential between a hot lower wall and a cold upper wall. The side walls are insulated. The symbol Ra is Rayleigh number and is a measure of the strength of natural convection in the cavity. The first column carries interferometric fringes. The second and third columns contain a re-distribution of light intensity as obtained from schlieren and shadowgraph techniques. Images adapted from *Optical imaging and control of convection around a crystal growing from its aqueous solution,* K. Muralidhar, Atul Srivastava and P.K. Panigrahi, in New Developments in Crystal Growth Research, Nova Publishers, 2005, pp. 1-89.

The second and third columns contain a re-distribution of light intensity as obtained from schlieren and shadowgraph techniques. In schlieren, the initial light intensity is rather low but can be suitably adjusted. When the temperature differential across the fluid layer is turned on, regions of high temperature gradients are uniformly illuminated, with low gradient regions remaining dark. In a shadowgraph, the initial intensity distribution is set at a convenient initial value. Temperature gradients can result in either a local increase in intensity or a reduction all the way to zero. Thus, the intensity field may brighten or darken about the initial value. In addition, the resulting intensity of shadowgraph depends on the position of the screen.

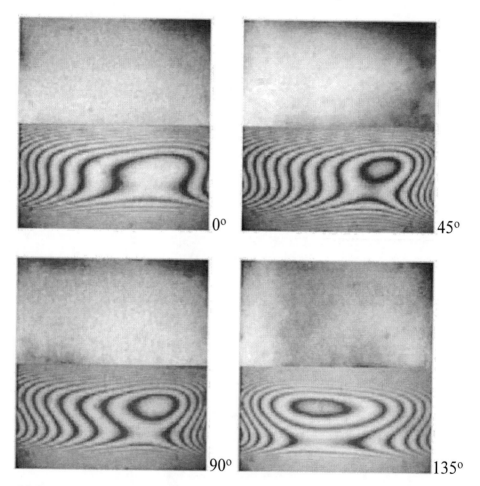

Figure 22. Buoyancy-driven convection in an octagonal cavity half-filled with 50 cSt silicone oil, the rest being air. The lower surface is heated while the top surface is cooled. The figure shows interferometric projections when the thermal field is viewed in various directions – 0, 45, 90, and 135°. Images adapted from *Interferometric study of convection in superposed gas-liquid layers*, Sunil Punjabi, Doctoral dissertation, Indian Institute of Technology, Kanpur (2002).

2.5. Convection in a Fluid Layer with a Free Surface

Figure 22 shows buoyancy-driven convection in an octagonal cavity half-filled with 50 cSt silicone oil, the rest being air [Punjabi, 2002]. The cavity has eight sides and forms an octagon in plan. The lower surface is heated while the top surface is cooled. The thermal arrangement creates cold/heavy fluid at the top floating over hot/light fluid at the bottom, in a gravitationally unstable arrangement. Thus convection patterns are possible in both fluid layers – air and silicone oil. However, the thermal conductivity of air is small when compared to silicone oil. Hence, a considerable fraction of the overall temperature difference occurs in oil and the temperature difference across air is negligibly small. Neither is air set in motion nor do clear fringes form in air. In contrast, convection is set up in silicone oil and fringes that are isotherms are deformed. The omega-shaped fringe (as well as others that are similar but show smaller deviation) is a result of projecting an axisymmetric thermal field along the viewing direction. Figure 22 shows interferometric projections of the refractive index distribution when the thermal field is viewed in various directions – 0, 45, 90, and 135°. Since the figures are practically identical, one may conclude that the thermal field does not depend on the polar angle and may be classified as axisymmetric.

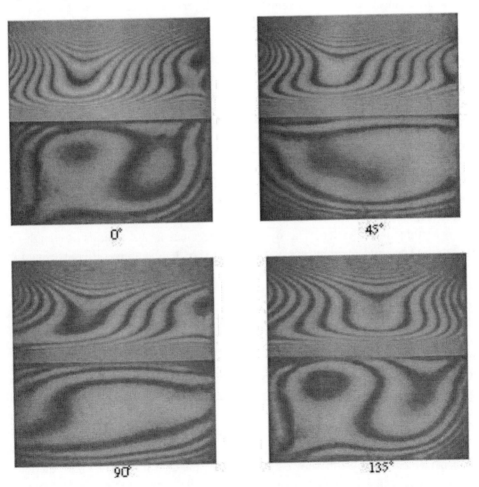

Figure 23. Long-time interferograms formed in an octagonal cavity containing silicone oil (50 cSt) floating over water as recorded from four different angles; Overall temperature difference is 1.8 K, the top plate being cooler than the one at the base. Of special interest are the energy and momentum transfers at the oil-water interface. Images adapted from *Buoyancy-driven convection in two superposed fluid layers in an octagonal cavity,* Sunil Punjabi, K. Muralidhar and P.K. Panigrahi, International Journal of Thermal Sciences, 43 (2004) 849-864.

2.6. Convection in a Two-Fluid Layer System

Figure 23 collects long-time interferograms formed in an octagonal cavity containing silicone oil (50 cSt) floating over water [Punjabi, 2004]. Interferometric projections recorded from four different angles are shown. The flow is driven by an overall temperature difference, the lower surface being warmer than the top surface. Thus, cold/dense fluid set over warm/light fluid generates buoyancy forces and drives fluid convection. Water and oil are immiscible and independent convection patterns are set up in each part of the cavity. Since the fluid conductivities are comparable, a finite temperature difference is available across each fluid layer and detectable fluid convection is to be seen in each half. The overall temperature difference across the cavity is 1.8 K, the top plate being cooler than the one at the base. Of special interest are the energy and momentum transfers at the oil-water interface. When water is set in motion, it can drive the oil layer beneath even if buoyancy is absent. The underlying mechanism is viscosity and the fluid layers are then said to be mechanically coupled. The other possibility is that each fluid layer experiences buoyant convection whose strength scales with the temperature difference appropriate for each of them. The fluid layers are then said to be thermally coupled. In the present experiment, interferometric images show that the nature of coupling is thermal in origin. In addition, the images look different in each view angle, indicating the onset of three dimensional velocity and temperature fields.

2.7. Rayleigh-Benard Convection in a Rectangular Cavity

Figure 24 reports images and analysis related to buoyancy-driven convection in a differentially heated cavity that is rectangular in plan [Mishra, 1998]. The height of the cavity is small in comparison to the linear dimensions of the rectangle (around 10%). The lower surface is warmer than the top. The overall arrangement is the familiar Rayleigh-Benard configuration. The fluid medium considered is air. For temperature differences and cavity height considered, three dimensional fluid motion and temperature distribution are to be expected. The circulatory motion of the fluid particles creates a roll-type of motion. The roll patterns of the flow field are seen as fringe displacement in the interferograms. The corresponding interferograms seen from various viewing directions are shown in the figure. The schematic drawing at the top indicates how a roll is created within the cavity. Here, a parcel of warm fluid rises as a buoyant plume at a certain location, spreads on its way upwards, reaches the cold top, and descends downwards at a second location by gravity. Points where the fluid rises with those where the fluid descends together form a cellular pattern within the cavity. The geometry of the pattern can be understood by recording interferograms from various directions. A sample interferogram is shown in the figure. Here, fringe displacement upwards can be correlated with the rising buoyant plume. Conversely, fringe displacement downwards shows the descent of the cold fluid. The entire thermal field in can be reconstructed by a numerical procedure called tomography. Here, temperatures integrated along the viewing direction are required over the image as input data. An intermediate step that assists conversion of fringe patterns to temperature is fringe thinning. It replaces the fringe bands by a skeleton, using image processing techniques. A sample set of thinned interferograms is also shown in Figure 24, for various view angles.

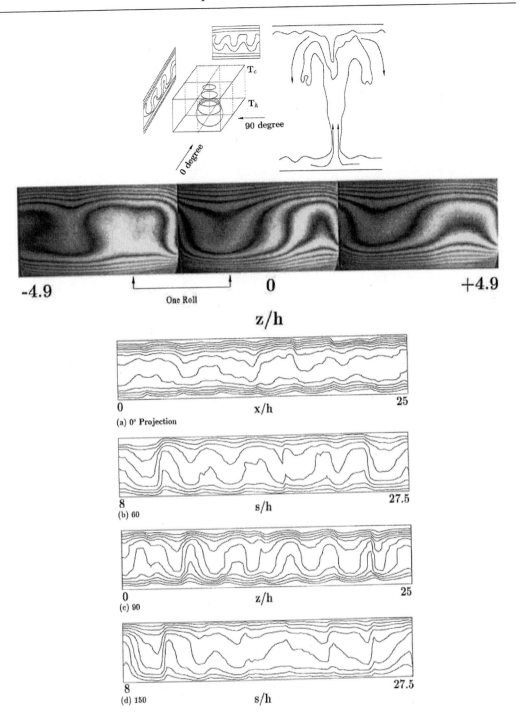

Figure 24. Buoyancy-driven convection in a differentially heated cavity. The roll patterns of the flow field are seen as fringe displacement in the interferograms. The flow pattern is three dimensional and sketched at the top of the figure. The corresponding interferograms seen from various viewing directions are shown below. Images adapted from *Experimental study of Rayleigh-Benard convection using interferometric tomography*, Debasish Mishra, Doctoral dissertation, Indian Institute of Technology, Kanpur (1998).

Figure 25. Interferograms recorded in an eccentric annulus with the gap filled with air. The inner cylinder is heated while the outer cylinder is cooled. The images show a time sequence of interferograms from the initial infinite fringe setting all the way until steady state is reached. Images adapted from *Interferometric study of steady and unsteady convection in cylindrical and eccentric annuli,* Manoj Ranjan, M.Tech. dissertation, Indian Institute of Technology, Kanpur, (2005).

2.8. Convection in an Eccentric Air-Filled Annulus

Figure 25 shows a time sequence of interferograms recorded in an eccentric annulus with the gap filled with air [Ranjan, 2005]. The axes of the two cylinders are horizontal. The cylinder length is roughly ten times the outer cylinder diameter. The inner cylinder is heated while the outer cylinder is kept at the ambient temperature. Heating is by an electric coil wound over a bakelite rod and covered by mica. The heating assembly is placed within the inner cylinder. Thermocouples have to be located over the cylinder length to ensure temperature uniformity along the axis as well as in the angular direction. Initially, the fluid in the annulus is also at the ambient temperature. The fringes are recorded in the infinite fringe setting. Fringes originate from the inner heated boundary. For the temperature difference considered, density differences are large enough for buoyant convection to be set up in the annulus. The fluid is driven upwards near the inner cylinder and flows around the inner periphery of the outer cylinder during the descent. Thus, fringes show an upward displacement closer to the inner cylinder. Owing to the eccentricity of the inner cylinder in its placement with respect to the outer, symmetry in the fringe patterns is broken. The stem attached to the inner cylinder holds it in place and is visible in the recorded images. The fringe patterns yield information on the distribution of heat flux around the inner cylinder. Here, the wall heat flux scales with the derivative of temperature in the radial direction. Since large derivatives will create closely-spaced fringes, fringe spacing itself is indicative of the wall heat flux.

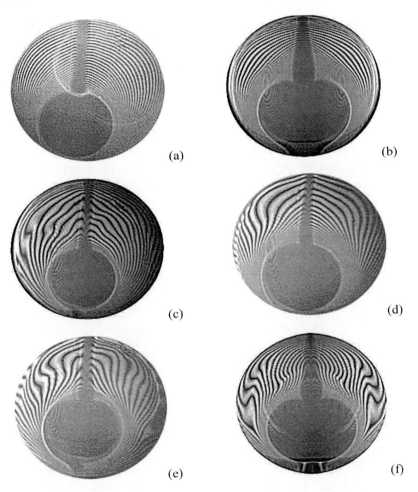

Figure 26. Interferograms recorded during heat transfer in an eccentric annulus when the gap is filled with 50 cSt silicone oil. The applied temperature differences increase from (a) to (e) from 0.2 to 2 K. When compared to air, the principal differences observed in the fringe patterns are an increase in the number of fringes, strong refraction errors near the inner cylinder, symmetric thermal fields about the position of the inner cylinder and weak convection effects. Images adapted from *Interferometric study of steady and unsteady convection in cylindrical and eccentric annuli*, Manoj Ranjan, M.Tech. dissertation, Indian Institute of Technology, Kanpur, (2005).

The interferograms show that fringe spacing is small on the lower side of the cylinder and consequently, the wall fluxes are large. On the top side, fringe spacing is large and the wall fluxes can be expected to be small (in comparison). It is also to be understood that fringe thickness scales inversely as the local temperature gradient. Thick fringes are regions of low gradients. Thus, above the inner cylinder, one obtains a region of warm fluid of nearly uniform temperature.

2.9. Convection in an Eccentric Annulus Filled with Silicone Oil

In Figure 26, interferograms recorded during steady state heat transfer in an eccentric annulus are presented when the gap is filled with 390 cSt silicone oil [Ranjan, 2005]. The applied temperature differences increase from 0.2 to 2 K while going from (a) to (e). When compared to air, several differences are to be observed in the fringe patterns. These include a substantial increase in the number of fringes. In addition, strong refraction errors are seen as errors near the inner cylinder where the temperature gradients are large. However, the thermal fields retain symmetry about the position of the inner cylinder indicative of weak convection effects in the annulus. For small temperature differences, the buoyancy forces cannot overcome the viscous and the fluid is stationary. Heat transfer is then by conduction alone, as seen by near circular fringes for a temperature difference of 0.2 K. In liquids such as silicone oil, experiments with interferometry are to be carefully designed so that refraction errors are small and the fringe density is matched to the camera resolution.

2.10. Vortex Shedding from a Heated Cylinder

Figure 27 shows instantaneous schlieren images for flow past a circular cylinder [Singh, 2007]. Wakes behind heated circular cylinders have been experimentally investigated at a low Reynolds number (100-300). The electrically heated cylinder is mounted in a vertical airflow facility such that the buoyancy aids the inertia of the main flow. The operating parameters, i.e. Reynolds number and Richardson number are varied to examine flow behavior over a range of experimental conditions from the forced to the mixed convection regime. Laser schlieren-interferometry has been used for visualization and analysis of the flow structures. At the Reynolds number studied, the unheated cylinder will periodically shed vortices. The complete vortex shedding sequence has been recorded using a high-speed camera. The results on detailed dynamical characteristics of the vortical structures i.e. their size, shape and phase, Strouhal number, power spectra, convection velocity, phase shift, vortex inception length and fluctuations are reported. On heating, the alteration of organized (coherent) structures with respect to shape, size and their movement is readily perceived from the instantaneous Schlieren images before they reduce to a steady plume. For the circular cylinder, Strouhal number shows a slow increase with an increase in the cylinder temperature, namely the Richardson number. At a critical value, there is a complete disappearance of vortex shedding and a drop in Strouhal number to zero. The corresponding spectra evolve from being highly peaked at the vortex shedding frequency to a broadband appearance when vortex shedding is suppressed. The geometry of the vortex structures transforms to a slender shape before shedding is suppressed. The convection velocity of vortices increases in the streamwise direction to an asymptotic value and its variation is a function of Richardson number. The convection speed abruptly falls to zero at the critical Richardson number. The phase difference of the shed vortices between upstream and downstream location increases with an increase in Richardson number. Velocity profiles show an increase in fluid speed and beyond the critical point, buoyancy forces add enough momentum to cancel the momentum deficit due to the cylinder.

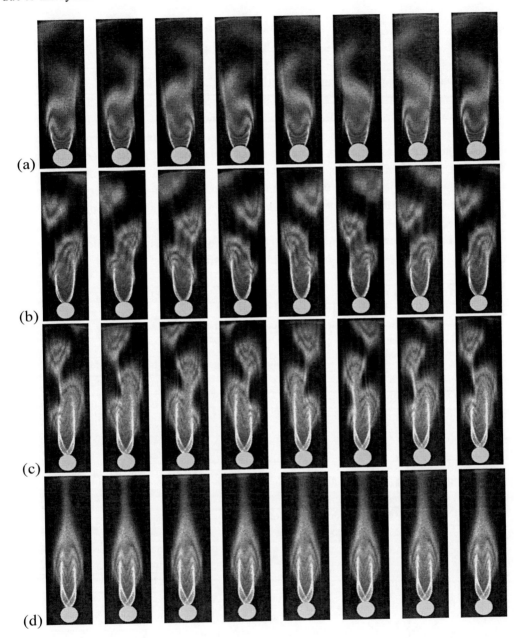

Figure 27. Instantaneous schlieren images (a-e) for flow past a circular cylinder separated by a time interval of one eighth of the time period of vortex shedding at Re=110. (a) Ri=0.052, (b) Ri=0.104, (c) Ri=0.140, and (d) Ri=0.157. For Ri ≥0.157, images show steady patterns. Images adapted from *Effect of buoyancy on the wakes of circular and square cylinders - a schlieren-interferometric study*, S.K. Singh, P.K. Panigrahi and K. Muralidhar, Experiments in Fluids, 43 (2007) 101-123.

The applications described above are representative examples wherein optical interferometers have been deployed for imaging the transport phenomena. There are many other applications which are not discussed here but are equally interesting. For example, the reader can benefit from reading the following references [Hesselink, 1988; Tiziani, 1989; Rao, 1990; Mayinger, 1993; Lehnhoff, 1993; Naylor, 2003; Prasanna, 2011A & 2011B].

CONCLUSION

The present chapter provides an introduction to the subject of interferometry as a method of measurement of temperature in fluids and solute concentration in liquids. Table 1 provides the summary of optical techniques for measuring fluid flow, heat and mass transfer. As a rule, a coherent source of light such as a laser is required and the medium is transparent and non-scattering. The record originates in refractive index variations and is an image of the density field in the form of fringes. The data contains information related to the line-of-sight averaged density and hence temperature as well as concentration. In addition, a variety of alternative optical configurations can be explored by developing composite diagnostic tools.

Table 1. Summary of optical techniques for measuring fluid flow, heat and mass transfer

PARAMETER	TYPE OF MEASUREMENT	TYPE OF TECHNIQUE	TECHNIQUE
FLUID FLOW / CONVECTION	Refractive index as a function of temperature T and concentration C	Non-interferometric	Shadowgraph
			Schlieren
		Interferometric	Schlieren-interferometry
			Mach-Zehnder interferometry
HEAT and MASS TRANSFER	Refractive index	Interferometry	Mach-Zehnder interferometry
			Phase shifting interferometry
			Speckle pattern interferometry
HEAT and MASS TRANSFER	Refractive index and its gradients with respect to T and C	Two-λ interferometry	Differential fringe analysis
3D IMAGING OF FLUID FLOW / HEAT / MASS TRANSFER	Refractive index	Holography	Real-time holographic interferometry
			Double-exposure holographic interferometry
			Holographic tomography
		Tomography	Shadowgraphic tomography
			Schlieren tomography
			Interferometric tomography

Specific advantages of the method of interferometry can be listed as follows:

i. The measurement takes place for the entire cross-section and as opposed to point measurements, it classifies as a field scale technique.

ii. Data is localized at the fringes. Thus, temperature and concentration are available at all points along a fringe. In practice, the resolution is limited by the number of pixels available in the camera.
iii. The measurement is instantaneous without any time lag. In experiments, temporal resolution is limited by that of the camera.
iv. An image, namely the interferogram is a two-dimensional projection of the material density distribution. However, information in three dimensions can be recovered by reconfiguring the interferometer for holography or using the principles of tomography.
v. A variety of optical configurations are available. These can be used to adjust the sensitivity of the measurement. Alternatives such as isotherms, profiles, and gradients can be explored for the image content.
vi. The breadth of applications reported in this chapter shows that interferometry can play an important role in industrial processes and applications.

Interferometry may be limited by the need for frequent alignment and a controlled atmosphere. Ideally, the optics should be computer controlled. A compensation chamber may be required for sensitive measurements. The presence of a solid phase such as particulates will result in scattering and diminish the utility of interferometry. These subjects form the focus of future research.

Acknowledgments

SV is indebted to Dr. P.J. Shlichta for several useful discussions and valuable advice. His approval to use some of the material from our earlier joint work [Verma, 2008A] for the present text is gratefully acknowledged. He is grateful to Dr. Prabhat Munshi for constant encouragement and unflagging support. SV also acknowledges the useful advice received from Dr. P.K. Gupta. KMD gratefully acknowledges close collaboration with his fellow colleagues Dr(s). P.K. Panigrahi and Prabhat Munshi. He also acknowledges the contributions of his past graduate students, namely Debasish Mishra, Sunil Punjabi, Atul Srivastava, S.K. Singh, and Manoj Ranjan. KMD and YMJ thank their current students, Susheel Bhandari and Vishnu Singh for assistance in recording interferograms.

Nomenclature

$T_{supersat}$	Supersaturated (or growth) temperature
T_{sat}	Saturation temperature
$\Delta T = T_{supersat} - T_{sat}$	Supercooling
$C_{supersat}$	Supersaturated concentration
C_{sat}	Saturated concentration
$\Delta C = C_{supersat} - C_{sat}$	Driving force for crystal growth (concentration is expressed in units of grams of solute per 1000 gm of solvent)

Nomenclature (Continued)

$\sigma = \dfrac{C_{supersat} - C_{sat}}{C_{sat}}$	Relative supersaturation
NLO	Non-linear optics
KDP	Potassium Dihydrogen Phosphate
I_s	Intensity on screen in the presence of disturbance
I_o	Intensity on screen in the absence of disturbance
n	Refractive index of the solution
λ	Wavelength of the laser light
t	Thickness of film
θ	Angle of incidence
r	Radius of the cluster
$\theta_{sol}, \theta_{qtz}, \theta_{air}$	Angle of incidence of the light at the quartz window, the angle of refraction of the light ray into the quartz window, the angle of refraction of the light ray in the air, respectively
$n_{sol}, n_{qtz}, n_{air}$	Refractive index of the KDP solution, the refractive index of the quartz window and the refractive index of the quartz window
$Gr = \rho \dfrac{\partial \rho}{\partial C} \dfrac{(C_o - C_i)}{\mu^2} g L^3$	Grashof number
$Re = \dfrac{\omega L^2}{\nu}$	Reynolds number
$Sc = \dfrac{\mu}{\rho D}$	Schmidt number
ρ	Density of the solution at a certain temperature in kg/m^3
$\partial \rho / \partial C$	Gradient of density with concentration
C_o	Bulk solute concentration in mole/m^3
C_i	Interface concentration near to the crystal
g	Acceleration due to gravity in m/s^2
L	Characteristic length in m
D	Diffusion coefficient of solute in the solution in m^2/s
k	Gladstone-Dale constant
ΔC_E	Concentration change per fringe shift
$n(x, y)$	Refractive index at the location (x, y) in the plane
c_o	Velocity of light in vacuum
c	Velocity of light in the physical region of refractive index n
E	Electric field
δ	The path difference
ϕ	The phase difference ϕ can be associated with a path difference δ

REFERENCES

[Bedarida, 1977] F. Bedarida, L. Zefiro and C. Pontiggia, Holographic control of diffusion coefficients in water solutions: Crystal growth from solutions, in: *"Applications of*

Holography and Optical Data Processing", E. Marom and A. A. Friesem (Eds), (Pergamon Press, Oxford, 1977), pp. 259-265.

[Bedarida, 1992] F. Bedarida, G.A. Dall'Aglio, L. Gatti and F. Solitro, The sensitivity of an optical fiber holographic interferometer, *Proc. 8th Euro. Symp. on Materials and Fluid Sciences in Microgravity*, Brussels, Belgium, ESA SP-333 (August 1992), 321-323.

[Bhandari, 2009] Susheel Bhandari, Doctoral dissertation, Indian Institute of Technology, Kanpur, India (2009).

[Bruning, 1974] J.H. Bruning, D.R. Herriott, J.E. Gallagher, D.P. Rosenfeld, A.D. White and D.J. Brangaccio, Digital wavefront measuring interferometer for testing optical surfaces and lenses, *Appl. Opt.*, 13 (11) (1974) 2693-2703.

[Bunn, 1949] C.W. Bunn, Crystal growth from solution. II. Concentration gradients and the rates of growth of crystals, *Discuss. Faraday Society*, 5 (1949) 132-144.

[Censor, 1983] Y. Censor, Finite series-expansion reconstruction methods, *Proc. IEEE*, 71 (3) (1983) 409-419.

[Chen, 1977] P.S. Chen, *Convection irregularities during solution growth and relation to crystal-defect formation*, Ph.D. dissertation, Univ. of Southern California, Los Angeles, CA, USA, (1977), 213 pp.

[Chen, 1979A] P.S. Chen, W.R. Wilcox and P.J. Shlichta, Free convection about a rectangular prismatic crystal growing from a solution, *Int. J. Heat Mass Transfer*, 22 (1979) 1669-1679.

[Chen, 1979B] P.S. Chen, P.J. Shlichta, W.R. Wilcox and R.A. Lefever, Convection phenomena during the growth of sodium chlorate crystals from solution, *J. Cryst. Growth*, 47 (1979) 43-60.

[Chen, 2002] W.C. Chen, D.D. Liu, W.Y. Ma, A.Y. Xie and J. Fang, The determination of solute distribution during growth and dissolution of $NaClO_3$ crystals: the growth of large crystals, *J. Cryst. Growth*, 236 (2002) 413-419.

[Cloud, 1995] Gary Cloud (with contributions from K. Creath), Phase shifting to improve interferometry, Chapter 22, in: *"Optical Methods of Engineering Analysis"*, Gary Cloud, (Cambridge University Press, 1995), pp. 477-491.

[Cole, 1995] T. Cole, A. Kathman, S. Koszelak, and A. McPherson, Determination of local refractive index for protein and virus crystals in solution by Mach-Zehnder interferometry, *Analytical Biochem.*, 231 (1995) 92-98.

[Collier, 1971] R.J. Collier, C.B. Burckhardt and L.H. Lin, *"Optical Holography"*, (Academic Press, New York, 1971), 605 pp.

[Creath, 1986] K. Creath, Comparison of phase-measurement algorithms, *SPIE* 680 (1986) 19-28.

[Creath, 1987] K. Creath, WYKO systems for optical metrology, *SPIE* 816 (1987) 111-126.

[Creath, 1988] K. Creath, Phase-measurement interferometric techniques, Chapter V, in: *"Progress in Optics"*, Vol. XXVI, E. Wolf (Ed.), (Elsevier Science Publishers B.V., Netherlands, 1988), pp. 349-393.

[Dainty, 1984] J.C. Dainty (Ed.), *"Laser Speckle and Related Phenomena"*, 2nd ed., (Springer-Verlag, Berlin, 1984), 342 pp.

[Dinakaran, 2010] S. Dinakaran, Sunil Verma, S. Jerome Das, S. Kar and K.S. Bartwal, Optical imaging of the growth kinetics and polar morphology of zinc tris (thiourea) sulphate (ZTS) single crystals, *Cryst. Res. Technol.* 45 (3) (2010) 233-238

[Duan, 2001] L. Duan and J. Z. Shu, The convection during NaClO$_3$ crystal growth observed by the phase shift interferometer, *J. Cryst. Growth*, 223 (2001) 181-188.

[Dubois, 1999] F. Dubois, L. Joannes, O. Dupont, J.L. Dewndel and J.C. Legros, An integrated optical set-up for fluid-physics experiments under microgravity conditions, *Meas. Sci. Technol.*, 10 (1999) 934-945.

[Ducruix, 1992] A. Ducruix and R. Giege, *Crystallization of Nucleic Acids and Proteins*, (Oxford Univ. Press, 1992), 331 pp.

[Dupont, 1995] O. Dupont, J.L. Dewandel and J.C. Legros, Use of electronic speckle pattern interferometry for temperature distribution measurements through liquids, *Opt. Letts.*, 20 (1995) 1824-1826.

[Ecker, 1987] A. Ecker, *Solidification front dynamics examined by holographic interferometric measurement of temperature and concentration using transparent model systems, Final Report, NASA* (March, 1987).

[Ecker, 1988] A. Ecker, Two-wavelength holographic measurement of temperature and concentration during alloy solidification, *J. Thermophys. Heat Transfer*, 2(3) (1988) 193-196.

[El-Wakil, 1964] M.M. El-Wakil and C.L. Jaeck, A two-wavelength interferometer for the study of heat and mass transfer, *J. Heat Transfer*, 79 (1964) 464-466.

[Gabor, 1948] D. Gabor, A new microscopic principle, *Nature*, 161 (1948) 777.

[Gabor, 1949] D. Gabor, Microscopy by reconstructed wavefronts, *Proc. Soc. Roy.*, A197 (1949) 457-484.

[Gatti, 1989] L. Gatti, F. Solitro, F. Bedarida, P. Boccacci, G.A. Dall'Aglio and L. Zefiro, Three-dimensional measurements of concentration fields in crystal growth by multi-directional holographic interferometry, *SPIE* 1162 (1989) 126-131.

[Goldstein, 1996] R.J. Goldstein and T.H. Kuehn, Optical Systems for Flow Measurement: Shadowgraph, Schlieren, and Intreferometric Techniques, in: *"Fluid Mechanics Measurements"*, R. J. Goldstein (Ed.), (Taylor & Francis, New York, 1996), pp. 451-508.

[Greivenkamp, 1992] J.E. Greivenkamp and J.H. Bruning, Phase shifting interferometry, Chapter 14, in: *"Optical Shop Testing"*, 2nd Ed., Daniel Malacara (Ed.), (John Wiley & Sons, Inc., New York, 1992), pp. 501-598.

[Gull, 1986] S.F. Gull and T.J. Newton, Maximum entropy methods, *Appl. Opt.*, 25 (1986) 156-160.

[Hariharan, 1985] P. Hariharan, *"Optical Interferometry"*, (Academic Press, Sydney, 1985), 303 pp.

[Hariharan, 1986] P. Hariharan, *"Optical Holography: Principles, Techniques and Applications"*, (Cambridge University Press, Cambridge, 1986), 319 pp.

[Hariharan, 2002] P. Hariharan, *"Basics of Holography"*, (Cambridge University Press, Cambridge, 2002), 161 pp.

[Heffinger, 1966] L.O. Heffinger, R.F. Wuerker and R.E. Brooks, Holographic interferometry, *J. Appl. Phys.*, 37 (1966) 642-649.

[Herman, 1980] G.T. Herman, *"Image Reconstruction from Projections: The Fundamentals of Computerized Tomography"*, (Academic Press, New York, 1980), 316 pp.

[Hesselink, 1988] L. Hesselink, Digital image processing in flow visualization, *Ann. Rev. Fluid Mech.*, 20 (1988) 421-485.

[Jones, 1989] R. Jones and C. Wykes, *"Holographic and Speckle Interferometry"*, 2nd Ed., (Cambridge University Press, 1989), p. 368.

[Kang, 2001] Q. Kang, L. Duan and W.R. Hu, Mass transfer process during the NaClO3 crystal growth process, *Int. J. Heat and Mass Transfer*, 44 (2001) 3213-3222.

[Kim, 1998A] Y.K. Kim, B.R. Reddy and R.B. Lal, Laser and Mach-Zehnder interferometer for in-situ monitoring of crystal growth and concentration variation, *SPIE* 3479 (1998) 172-180.

[Kumar, 2008] N.V.N. Ravi Kumar, K. Muralidhar, and Y.M. Joshi, On refractive index of aging suspensions of laponite, *Applied Clay Science*, 42 (2008) 326-330.

[Lehnhoff, 1993] H.-H. Bartels-Lehnhoff, Computer aided evaluation of interferograms, *Expt. in Fluids*, 16 (1993) 46-53.

[Lenski, 1991] H. Lenski and M. Braun, Laser beam deflection: a method to investigate convection in vapor growth experiments, *SPIE* 1557, *"Crystal Growth in Space and Related Optical Diagnostics"*, J.D. Trolinger, Ravindra B. Lal (Eds.), (1991) 124-131.

[Lewitt, 1983] R.M. Lewitt, Reconstruction algorithms: Transform methods, *Proc. IEEE*, 71 (3) (1983) 390-408.

[Lokberg, 1980] O.J. Lokberg, Electronic speckle pattern interferometry, *Phys. Technol.*, 11 (1980) 16-22.

[Maruyama, 1999] S. Maruyama, T. Shibata and K. Tsukamoto, Measurement of diffusion fields of solutions using real-time phase-shift interferometer and rapid heat-transfer control system, *Exp. Therm. Fluid Sci.*, 19 (1999) 34-48.

[Maruyama, 2002] S. Maruyama, K. Ohno, A. Komiya and S. Sakai, Description of the adhesive crystal growth under normal and micro-gravity conditions employing experimental and numerical approaches, *J. Cryst. Growth*, 245 (2002) 278-288.

[Mayinger, 1993] F. Mayinger, Image-forming optical techniques in heat transfer: revival by computer-aided data processing, *Trans. ASME*, 115 (1993) 824-834.

[Mayinger, 1994] F. Mayinger (Ed.), *"Optical Measurements: Techniques and Applications"*, (Springer-Verlag, Berlin, 1994), 463 pp.

[McPherson, 1982] A. McPherson, *"Preparation and Analysis of Protein Crystals"*, (Wiley, New York, 1982), 371 pp.

[McPherson, 1999] A. McPherson, A.J. Malkin, Y.G. Kuznetsov, S. Koszelak, M. Wells, G. Jenkins, J. Howard and G. Lawson, The effects of microgravity on protein crystallization: Evidence for concentration gradients around growing crystals, *J. Crystal Growth*, 196 (1999) 572-586.

[McPherson, 2001] A. McPherson, *"Crystallization of Biological Macromolecules"*, 2nd ed., (Cold Spring Harbor Laboratory Press, 2001) 586 pgs. (1st ed. 1998).

[Mehta, 1990] J.M. Mehta, Dual wavelength interferometric technique for simultaneous temperature and concentration measurement in liquids, *Appl. Opt.*, 29 (13) (1990) 1924-1932.

[Merzkirch, 1974] W.F. Merzkirch, *"Flow Visualization"*, (Academic Press, New York, 1974), 250 pp.

[Merzkirch, 1987] W.F. Merzkirch, *"Flow Visualization"*, 2nd Ed., (Academic Press, New York, 1987), 260 pp.

[Mishra, 1998] Debasish Mishra, *Experimental Study of Rayleigh Benard Convection using Interferometric Tomography*, Ph.D. dissertation, Indian Institute of Technology, Kanpur, India (1998).

[Mishra, 1999] Debasish Mishra, K. Muralidhar and P. Munshi, Interferometric study of Rayleigh-Benard convection using tomography with limited projection data, *Numerical Heat Transfer*, 12 (1999) 117-136.

[Mullin, 2001] J.W. Mullin, *Crystallization*, 4th ed., (Butterworth-Heinemann, Oxford, 2001).

[Muralidhar, 2001] K. Muralidhar, Temperature field measurement in buoyancy-driven flows using interferometric tomography, *Ann. Rev. Heat Transf.*, 12 (2001) 265-375.

[Muralidhar, 2005] K. Muralidhar, Atul Srivastava and P.K. Panigrahi, Optical imaging and control of convection around a crystal growing from its aqueous solution, in *New Developments in Crystal Growth Research* (Nova Publishers, New York, 2005) pp. 1-89.

[Murphy, 1971] C.G. Murphy and S.S. Alpert, Dependence of refractive index temperature coefficients on the thermal expansivity of liquids, *Am. J. Phys.*, 39 (7) (1971) 834-836.

[Nagashima, 2000] K. Nagashima and Y. Furukawa, Time development of a solute diffusion field and morphological instability on a planar interface in the directional growth of ice crystals, *J. Cryst. Growth*, 209 (2000) 167-174.

[Nakadate, 1990] S. Nakadate and I. Yamaguchi, Patent describing the technique of real-time phase shifting interferometry, *Japanese Patent # H02-287107* (1990).

[Nakadate, 1995] S. Nakadate, Real-time fringe pattern processing and its applications, *SPIE* 2544 (1995) 74-86.

[Natterer, 2001] F. Natterer, *"The Mathematics of Computerized Tomography"*, (Philadelphia, PA, SIAM 2001), 222 pp.

[Naylor, 2003] D. Naylor, Recent developments in the measurement of convective heat transfer rates by laser interferometry, *Int. J. Heat Mass Transfer*, 24 (3003) 345-355.

[Onuma, 1988] K. Onuma, K. Tuskamoto and I. Sunagawa, Role of buoyancy driven convection in aqueous soultion; A case study of $Ba(NO_3)_2$ crystal, *J. Cryst. Growth*, 89 (1988) 177-188.

[Onuma, 1989] K. Onuma, K. Tuskamoto and I. Sunagawa, Measurement of surface supersaturations around a gowing K-alum crystal in aqueous solution, *J. Cryst. Growth*, 98 (3) (1989) 377-383.

[Onuma, 1993] K. Onuma, K. Tsukamoto and S. Nakadate, Application of real time phase shift interferometer to the measurement of concentration field, *J. Cryst. Growth*, 129 (1993) 706-718.

[Owen, 1975] W.A. Owen et al., *Advanced Application Flight Experiment: Effect of Residual and Transient Convection on Crystallization in a Spaceflight Environment, Semi-annual Report, NASA Code. No. 638-10-00-01-00* (Jet Propulsion Laboratory, Pasadena, CA, USA, March 1975; see also later reports November 1975 and June 1976.

[Owen, 1982] R.B. Owen, Interferometry and holography in a low-gravity environment, *Appl. Opt.*, 21 (8) (1982) 1349-1355.

[Owen, 1986] R.B. Owen, R.L. Kroes and W.K. Witherow, Results and further experiments using Spacelab holography, *Opt. Lett.*, 11 (7) (1986) 407-409.

[Paschotta, 2008] Rüdiger Paschotta, *Encyclopedia of Laser Physics and Technology* (Wiley-VCH, Berlin, 2008) 844 pp.

[Piano, 2000] E. Piano, G.A. Dall'Aglio, S. Crivello, R. Chittofrati and F. Puppo, New optical techniques for crystal growth from fluids, *Materials Chem. Phys.*, 66 (2000) 266-269.

[Piano, 2001] E. Piano, G.A. Dall'Aglio, R. Chittofrati, S. Crivello and F. Puppo, A non-destructive interferometric technique for analysis of crystal growth and fluid dynamics, *Ann. Chim. Sci. Mat.*, 26 (2001) 23-28.

[Pletser, 2001] V. Pletser, O. Minster, R. Bosch, L. Potthast and J. Stapelmann, The protein crystallization diagnostics facility: status of the ESA program on the fundamentals of protein crystal growth, *J. Cryst. Growth*, 232 (2001) 439-449.

[Prasanna, 2011A] S. Prasanna and S.P. Venkateshan, Heat flux and temperature field estimation using differential interferometer, *ASME Journal of Heat Transfer*, (2011) in press.

[Prasanna, 2011B] S. Prasanna and S.P. Venkateshan, Construction of two dimensional temperature field from first derivative fields, *Exp. Therm. Fluid Science* (2011) in press.

[Punjabi, 2002] Sunil Punjabi, *Interferometric study of convection in superposed gas-liquid layers*, Doctoral dissertation, Indian Institute of Technology, Kanpur, India (2002).

[Punjabi, 2004] Sunil Punjabi, K. Muralidhar and P.K. Panigrahi, Buoyancy-driven convection in two superposed fluid layers in an octagonal cavity, *Int. J. Therm. Sci.*, 43 (2004) 849-864.

[Ramachandran, 1971] G.N. Ramachandran and A.V. Lakshminarayanan, Three-dimensional reconstruction from radiographs and electron micrographs: Application of convolutions instead of Fourier transforms, *Proc. Nat. Acad. Sci. USA*, 68 (9) (1971) 2236-2240.

[Ranjan, 2005] Manoj Ranjan, *Interferometric study of steady and unsteady convection in cylindrical and eccentric annuli*, M.Tech. dissertation, Indian Institute of Technology, Kanpur, India, (2005).

[Rao, 1990] V. Rao, C. Sobhan and S.P. Venkateshan, Differential interferometry in heat transfer, *Sadhana* 15 (2) (1990) 105–128.

[Rasenat, 1989] S. Rasenat, G. Hartung, B.L. Winkler and I. Rehberg, The shadowgraph method in convection experiments, *Exp. Fluids*, 7 (1989) 412-420.

[Rashkovich, 1990A] L.N. Rashkovich and B.Yu. Shekunov, Hydrodynamic effects in growth of ADP and KDP crystals in solution. I. Growth kinetics, *Sov. Phys. Crystallogr.*, 35 (1) (1990) 96-99.

[Rashkovich, 1990B] L.N. Rashkovich and B.Yu. Shekunov, Morphology of growing vicinal surface; Prismatic faces of ADP and KDP crystals in solutions, *J. Cryst. Growth*, 100 (1990) 133-144.

[Schopf, 1996] W. Schopf, J.C. Patterson and A.M.H. Brooker, Evaluation of the shadowgraph method for the convective flow in a side heated cavity, *Exp. Fluids*, 21 (1996) 331-340.

[Settles, 2001] G.S. Settles, *"Schlieren and Shadowgraph Techniques"*, (Springer, Berlin, 2001), 376 pp.

[Shagam, 1978] R.N. Shagam and J.C. Wyant, Optical frequency shifter for heterodyne interferometers using multiple rotating polarization retarders, *Appl. Opt.*, 17 (1978) 3034-3035.

[Shiomi, 1980] Y. Shiomi, T. Kuroda and T. Ogawa, Thermal analysis of a growing crystal in an aqueous solution, *J. Cryst. Growth*, 50 (1980) 397-403.

[Shlichta, 1985] P.J. Shlichta, *Crystal growth in a spaceflight environment: final report, Materials processing in space environment*, Experiment # 770100, (September 1985), pp. 1-123, (Jet Propulsion Laboratory, Pasadena, CA 91109).

[Singh, 2007] S.K. Singh, P.K. Panigrahi and K. Muralidhar Effect of buoyancy on the wakes of circular and square cylinders - a schlieren-interferometric study, *Experiments in Fluids*, 43 (2007) 101-123.

[Singh, 2009] Vishnu Singh, M.Tech. dissertation, Indian Institute of Technology, Kanpur, India (2009).

[Smigielski, 1970] P. Smigielski and A. Hirth, *Proc. 9th Internat. Congress on High-Speed Photography*, W.G. Hyzer and W.G. Chase (Eds.), (SMPTE, New York, 1970), pp. 321-326.

[Solitro, 1989] F. Solitro, L. Gatti, F. Bedarida, G.A. Dall'Aglio and L. Michetti, Multi-directional holographic interferometer (MHOI) with fiber optics for study of crystal growth in microgravity, *SPIE* 1162 (1989) 62-65.

[Srivastava, 2004] A. Srivastava, K. Muralidhar and P.K. Panigrahi, Comparison of interferometry, schlieren and shadowgraph for visualizing convection around a KDP crystal, *J. Cryst. Growth*, 267 (2004) 348-361.

[Steel, 1983] W.H. Steel, *"Interferometry"*, 2nd ed., (Cambridge University Press, Cambridge, 1983), 320 pp.

[Tiziani, 1989] H.J. Tiziani, Optical methods for precision measurements, *Opt. Quant. Electronics*, 21 (1989) 253-282.

[Trolinger, 1991] J.D. Trolinger and R.B. Lal (Eds.), *"Crystal Growth in Space and Related Optical Diagnostics"*, *SPIE* 1557 (1991) 1-296.

[Verga, 1997] A. Verga, P. Baglioni, O. Dupont, J.L. Dewandel, T. Beuselinck and J. Bouwen, Use of electronic speckle pattern interferometers for the analysis of convective states of liquids in weightlessness, *SPIE* Vol. 3172, *Optical Technology in Fluid, Thermal, and Combustion Flow III* (1997) 194-210.

[Verma, 2003] Sunil Verma, S.K. Sharma, K. Muralidhar and V.K. Wadhawan, Optical imaging techniques in crystal growth research, in: *"Crystal Growth of Technologically Important Electronic Materials"*, K. Byrappa, H. Klapper, T. Ohachi & R. Fornari (Eds.), (Allied Publishers, Bangalore, 2003), pp. 616-619.

[Verma, 2005A] Sunil Verma, K. Muralidhar and V.K. Wadhawan, Convection during growth of KDP crystals: Flow visualization and modeling, *Ferroelectrics*, 323 (2005) 25-37.

[Verma, 2005B] Sunil Verma, K. Muralidhar and V.K. Wadhawan, Convection and concentration mapping during crystal growth from solution using Mach-Zehnder interferometry and computerized tomography, *Proc. 3rd Asian Conf. Crystal Growth & Crystal Technology (CGCT-3)*, Oct. 16-19, 2005, Beijing, China.

[Verma, 2005C] Sunil Verma, A. Srivastava, V. Prabhakar, K. Muralidhar and V.K. Wadhawan, Simulation and experimental verification of solutal convection in the initial stages of crystal growth from an aqueous solution, *Ind. J. Pure & Appl. Phys.*, 43 (2005) 24-33.

[Verma, 2006] Sunil Verma, K. Muralidhar and V.K. Wadhawan, Determination of concentration field around a growing crystal using shadowgraphic tomography (Chap. 14), In: *"Computerized Tomography for Scientists and Engineers"*, P. Munshi (Ed.), (CRC Press, New York, 2006) 158-174

[Verma, 2007] Sunil Verma, *Convection, concentration, and surface feature analysis during crystal growth from solution using shadowgraphy, interferometry and tomography*, Ph.D. dissertation, Indian Institute of Technology, Kanpur, India (2007).

[Verma, 2008A] Sunil Verma and P.J. Shlichta, Imaging techniques for mapping of solution parameters, growth rate, and surface features during the growth of crystals from solution, *Prog. Cryst. Growth & Charact. Materials*, 54 (2008) 1-120

[Verma, 2008B] Sunil Verma and K. Muralidhar, Three-dimensional reconstruction of convective features during crystal growth from solution using computerized tomography, In: *"CT2008: Tomography Confluence"*, P. Munshi (Ed.), *American Institute of Physics* CP 1050 (2008) 103-114

[Verma, 2009] Sunil Verma and K. Muralidhar, Convection, concentration, and surface features analysis during crystal growth from solution using optical diagnostics (Chap. 5), in: *Recent Research Developments in Crystal Growth*, Vol. 5 (2009) 141-314 (Transworld Research Network, India)

[Verma, 2010] Sunil Verma and K. Muralidhar, Imaging transport phenomena and surface micromorphology in crystal growth using optical techniques, *Natl. Acad. Sci. Lett.* 33 (5 & 6) (2010) 107-121

[Vest, 1979] C.M. Vest, *"Holographic Interferometry"*, (John Wiley, New York, 1979), 465 pp.

[Vikram, 1990] C.S. Vikram, H.J. Caulfield, G.L. Workman, J.D. Trolinger, C.P. Wood, R.L. Clark, A.D. Kathman and R.M. Ruggiero, *Two-color holographic concept (T-CHI), Final Tech. Rep.*, Contract No. NAS8-38078, NASA George C. Marshall Space Flight Center (April 1990).

[Vikram, 1991] C.S. Vikram, W.K. Witherow and J.D. Trolinger, Refractive properties of TGS aqueous solution for two-color interferometry, SPIE 1557, *"Crystal Growth in Space and Related Optical Diagnostics"*, J.D. Trolinger and R.B. Lal (Eds.), (1991) 197-201.

[Vikram, 1992A] C.S. Vikram and W.K. Witherow, Critical needs of fringe-order accuracies in two-color holographic interferometry, *Exp. Mech.*, (March 1992) 74-77.

[Vikram, 1992B] C.S. Vikram, W.K. Witherow and J.D. Trolinger, Determination of refractive properties of fluids for dual-wavelength interferometry, *Appl. Opt.*, 31 (34) (1992) 7249-7252.

[Witherow, 1987] W.K. Witherow, Reconstruction techniques of holograms from Spacelab-3, *Appl. Opt.*, 26 (12) (1987) 2465-2473.

[Witherow, 1994] W.K. Witherow, J.R. Rogers, B.R. Facemire, S.D. Armstrong, J.D. Trolinger, D. Weber and C.S. Vikram, Methods to detect and measure gradients in fluids and materials processing, *Proc. 6th International Symp. on Experimental Methods for Microgravity Materials Science*, San Francisco (USA), Feb. 27- March 3, 1994, R.A. Schiffman and B. Andrews (Eds.), The Minerals, Metals and Materials Society, 33-37.

[Wyant, 1975] J.C. Wyant, Use of an AC heterodyne lateral shear interferometer with real-time wavefront correction systems, *Appl. Opt.*, 14 (1975) 2622-26.

[Wyant, 1978] J.C. Wyant and R.N. Shagam, Use of electronic phase measurement techniques in optical testing, *Proc. 11th Congr. of the International Commission for Optics*, Madrid, 10-17 September 1978, J. Bescos, A. Hidalgo, L. Plaza and J. Santamaria (eds.) (Sociedad Espanola de Optica, Madrid) pp. 659.

[Wyant, 1982] J.C. Waynt, Interferometric optical metrology: Basic systems and principles, *Laser Focus,* (May 1982) 65-71.

[Wyant, 1985] J.C. Wyant and K. Creath, *Laser Focus*, (November 1985), pp. 118.

[Xiling, 1992] Yu Xiling and Yue Xuefeng, Holographic studies of diffusivities of KDP and DKDP in the solutions, *Cryst. Res. Technol.*, 27 (1992) 825-830.

[Xiling, 1996] Yu Xiling, Sun Yi, Jiang Huizhu and Zhnag Shujun, Growth kinetics of the metastable tetragonal phase of potassium dideutrium phosphate (DKDP) crystals, *J. Cryst. Growth*, 166 (1996) 195-200.

[Yin, 2001] D. Yin, Y. Inatomi and K. Kuribayashi, Study of lysozyme crystal growth under a strong magnetic field using a Mach-Zehnder interferometer, *J. Cryst. Growth*, 226 (2001) 534-542.

[Yin, 2003] D.C. Yin, Y. Inatomi, N.I. Wakayama and W.D. Huang, Measurement of temperature and concentration dependences of refractive index of hen-egg-white lysozyme solution, *Cryst. Res. Technol.*, 38 (9) (2003) 785-792.

Chapter 14

PHASE-STEPPING ALGORITHMS: OVERVIEW AND SIMULATIONS

Jan A.N. Buytaert[*] *and Joris J.J. Dirckx*[†]
University of Antwerp - Laboratory of BioMedical Physics
Groenenborgerlaan 171, B-2020 Antwerpen, Belgium

Abstract

We present a (non-exhaustive) overview of phase-shifting algorithms often used in interferometry. When performing phase-stepping (or -shifting), phase differences in a periodic intensity profile are changed stepwise (or continuously), and the resulting irradiance distributions are recorded at each step (or bucket). The wanted phase can be obtained from the arctangent of the ratio between two combinations of the observed irradiances, according to the phase-shifting algorithm (PSA) used. There are many such combinations and thus different PSAs, each with specific performance and properties.

We briefly discuss some error sources which might influence the performance and quality of interferometry measurements. The robustness against these error sources is strongly dependent on the PSA used.

We intently gathered as many popular PSAs and some of their properties from literature as we could find. Many discrepancies, however, were found between authors' statements, not to mention typos in the published articles and chapters. We meticulously sorted the typos out, listed up the algorithms and ran several computer simulations on all of them to confirm which algorithms perform best in the presence of some straightforward error sources.

PACS 42.87.Bg, 42.30.Rx

Keywords: interferometry, phase-shifting, phase-stepping, algorithms, arctangent, phase evaluation

1. Introduction

Optical methods are widely used in the academic and also industrial world for automatic inspection of object shapes, deformations and defects. The popularity of these approaches

---
[*]E-mail address: jan.buytaert@ua.ac.be
[†]E-mail address: joris.dirckx@ua.ac.be

can be attributed to their sensitivity and accuracy, and the fact that they are non-contacting and non-destructive. Full-field optical measurement techniques, like holographic interferometry, speckle metrology, moiré profilometry and other grid techniques, often provide their measured data in the form of the phase of a periodic intensity profile. This intensity profile is created by projection and sometimes superposition (interference) of light or structured light patterns, which results in sinusoidal alternating light and dark bands or fringes. A recording of such an intensity profile is thus called a fringe image.

A preferred and powerful *phase evaluation method* (PEM) to retrieve the phase encoded in fringes, is the use of *phase-stepping* (or *-shifting) algorithms*. When performing phase-stepping (or -shifting), phase differences in the fringes are induced stepwise (or continuously), and the resulting irradiance distributions are recorded at each step (or bucket). The wanted phase can be obtained from the *arctangent* of the ratio between two combinations of the observed irradiances, according to the *phase-shifting algorithm* (PSA) used. There exist many such combinations and thus different PSAs, each with specific performance and properties. We focus on temporal PSAs, which require at least three fringe patterns to be captured sequentially in time by a camera.

The need for *phase-unwrapping* is inevitable when PSAs are used. An arctangent function normally delivers a phase result φ which lays in the range $]-\frac{\pi}{2},\frac{\pi}{2}]$. However, by using the four-quadrant inverse tangent function available in many software packages, the phase is obtained in the interval $-\pi < \varphi \leq \pi$. This arctangent function, often called *atan2*, requires both a sine and cosine input (respectively numerator and denominator) and uses the signs of these values to figure out in which quadrant the phase φ is situated. The result is called the wrapped phase φ. If multiple fringes are present, there are (multiple) 2π phase discontinuities between all fringes or 2π phase transitions. Phase-unwrapping aims to unwrap or integrate the phase to a continuous surface. Finally, after phase evaluation and phase-unwrapping, scaling from radians to millimeters or inches is needed.

The world of PSAs is a jungle. A big diversity of algorithms is available in literature, accompanied by useful properties and information as well as liberal claims and many typographical errors. We took it upon ourselves to gather all popular PSAs we could find in the literature, to sort them out and correct any typos we could identify, and to list them up for other authors to use. Furthermore, though in the ideal case all algorithms deliver the same result, there are numerous error sources which influence the performance and accuracy of a PSA. We did not just look up the sensitivity of a PSA for a certain source of error from the literature, but performed some straightforward simulations on all algorithms to verify which methods perform best.

The text is structured as follows. First we will give a theoretical description of phase-shifting and discuss some analytical solutions. This is followed by a review of some major PSA families and an overview of all algorithms listed in a table. Next, this chapter contains a synopsis of several typical sources of error and how we simulated these errors. Finally, we discuss the results of the simulations and the performance of the algorithms.

2. Method Families

First we distinguish two main classes in the PSA methods, then subdivide them by the number of phase-shifted fringe images (or *steps*) that are required, and subdivide them again by the size of the required phase-shift step. The first *class* α are the *generic* PSAs which use known phase values ψ_i resulting in fringe images I_i. The second *class* β are the *specific* PSAs for more dedicated applications. And within classes, several families exist.

We will refer to any PSA in the following consistent manner: ' *#-phase-shift-author(s)* *[year]* '
with # the number of fringe images in the algorithm, than the required phase-shift in between consecutive fringe images, followed by the author(s) who (first) published the algorithm in a certain year. We get for instance: the well known 4-$\frac{\pi}{2}$-Bruning et al. [1974] algorithm.

2.1. α Class: Generic Methods

2.1.1. Analytical Solutions

In this text, we will write a fringe image as:

$$I(x,y) = a(x,y) \left[\frac{1}{2} + \frac{1}{2}\cos(\varphi(x,y))\right] + b(x,y) \tag{1}$$

with unknowns: $a(x,y)$ the amplitude modulation, $b(x,y)$ the background offset, and $\varphi(x,y)$ the wanted phase distribution.

2.1.1.1. Three-Step Algorithms

Consequently, when obtaining at least $n = 3$ different fringe images we can solve the equations for their three unknown variables. Hence it is required to record a series of fringe images I_i with a varied reference phase ψ_i:

$$I_i(x,y) = a(x,y) \left[\frac{1}{2} + \frac{1}{2}\cos(\varphi(x,y) + \psi_i)\right] + b(x,y) \tag{2}$$

If $\psi_1 = -\alpha$, $\psi_2 = 0$ and $\psi_3 = \alpha$, the three equations can be solved for the unknown phase φ at each location (x,y):

$$I_1(x,y) = a(x,y) \left[\frac{1}{2} + \frac{1}{2}\cos(\varphi(x,y) - \alpha)\right] + b(x,y) \tag{3}$$

$$I_2(x,y) = a(x,y) \left[\frac{1}{2} + \frac{1}{2}\cos(\varphi(x,y))\right] + b(x,y) \tag{4}$$

$$I_3(x,y) = a(x,y) \left[\frac{1}{2} + \frac{1}{2}\cos(\varphi(x,y) + \alpha)\right] + b(x,y) \tag{5}$$

$$\varphi(x,y) = \arctan\left(\left[\frac{1-\cos(\alpha)}{\sin(\alpha)}\right] \frac{I_1 - I_3}{2I_2 - I_1 - I_3}\right) \tag{6}$$

If we choose $\Delta\psi = \frac{\pi}{2}$ or thus $\alpha = \frac{\pi}{2}$, we obtain the well-known 3-$\frac{\pi}{2}$-Wyant et al. [1984] algorithm:

$$\frac{I_1 - I_3}{2I_2 - I_1 - I_3} \tag{7}$$

Note that in Table 1, which presents an overview of all algorithms, the numerator and denominator of the above formula (7) are reversed and the sign of the denominator is changed. This only generates a phase offset of $\frac{\pi}{2}$, intended to give all algorithms in the table the same offset.

If we choose $\alpha = \frac{2\pi}{3}$ we obtain the 3-$\frac{2\pi}{3}$-Bruning et al. [1974] algorithm:

$$\frac{\sqrt{3}(I_1 - I_3)}{2I_2 - I_1 - I_3} \tag{8}$$

Note that in Table 1 with all the algorithms, a cyclic permutation of I_i was applied compared to the above equation (8). This again only generates a phase offset, similar to the above. Also note that in Table 1 the phase steps are not (all) distributed symmetrically around zero but start at zero and climb from there. This also delivers an offset. All algorithms in the table are written in a form to possess the same offset.

2.1.1.2. Four-Step Algorithm

Using the basic expression for a fringe image, namely (2), with $n = 4$ fringes images and $\psi_1 = 0$, $\psi_2 = \frac{\pi}{2}$, $\psi_3 = \pi$ and $\psi_4 = \frac{3\pi}{2}$, we can again solve to $\varphi(x, y)$ analytically:

$$I_1(x, y) = a(x, y) \left[\frac{1}{2} + \frac{1}{2}\cos(\varphi(x, y))\right] + b(x, y) \tag{9}$$

$$I_2(x, y) = a(x, y) \left[\frac{1}{2} + \frac{1}{2}\cos\left(\varphi(x, y) + \frac{3\pi}{2}\right)\right] + b(x, y) \tag{10}$$

$$I_3(x, y) = a(x, y) \left[\frac{1}{2} + \frac{1}{2}\cos(\varphi(x, y) + \pi)\right] + b(x, y) \tag{11}$$

$$I_4(x, y) = a(x, y) \left[\frac{1}{2} + \frac{1}{2}\cos\left(\varphi(x, y) + \frac{3\pi}{2}\right)\right] + b(x, y) \tag{12}$$

$$\varphi(x, y) = \arctan\left(\frac{I_4 - I_2}{I_1 - I_3}\right) \tag{13}$$

We obtain the well-known 4-$\frac{\pi}{2}$-Bruning et al. [1974] algorithm:

2.1.1.3. Five-Step Algorithm

Again using the basic expression for a fringe image, namely (2), with $n = 5$ fringes images and $\psi_1 = -2\alpha$, $\psi_2 = -\alpha$, $\psi_3 = 0$, $\psi_4 = \alpha$ and $\psi_5 = 2\alpha$, we can again solve for $\varphi(x, y)$

analytically:

$$I_1(x,y) = a(x,y)\left[\frac{1}{2} + \frac{1}{2}\cos(\varphi(x,y) - 2\alpha)\right] + b(x,y) \tag{14}$$

$$I_2(x,y) = a(x,y)\left[\frac{1}{2} + \frac{1}{2}\cos(\varphi(x,y) - \alpha)\right] + b(x,y) \tag{15}$$

$$I_3(x,y) = a(x,y)\left[\frac{1}{2} + \frac{1}{2}\cos(\varphi(x,y))\right] + b(x,y) \tag{16}$$

$$I_4(x,y) = a(x,y)\left[\frac{1}{2} + \frac{1}{2}\cos(\varphi(x,y) + \alpha)\right] + b(x,y) \tag{17}$$

$$I_5(x,y) = a(x,y)\left[\frac{1}{2} + \frac{1}{2}\cos(\varphi(x,y) + 2\alpha)\right] + b(x,y) \tag{18}$$

$$\varphi(x,y) = \arctan\left([2\sin(\alpha)]\frac{I_2 - I_4}{2I_3 - I_1 - I_5}\right) \tag{19}$$

It was shown that a value of $\alpha = \frac{\pi}{2}$ minimizes the sensitivity of the algorithm for phase-shift miscalibration errors [Schreiber and Bruning, 2007]. In this form the algorithm is known as the 5-$\frac{\pi}{2}$-Schwider et al. [1983] or also often called the 5-$\frac{\pi}{2}$-Hariharan and Oreb [1987] algorithm:

$$\varphi(x,y) = \arctan\left(\frac{2(I_2 - I_4)}{2I_3 - I_1 - I_5}\right) \tag{20}$$

2.1.2. Least-Squares Algorithms

Generic formulas exist to derive whole families of algorithms. The N-bucket least-squares algorithms for instance have been derived by Bruning et al. [1974], by fitting the phase data in a least-squares way to a sinusoidal function. The generic expressions for this family of algorithms, with a series of $n = N$ fringes images, goes as follows:

$$I_i(x,y) = a(x,y)\left[\frac{1}{2} + \frac{1}{2}\cos(\varphi(x,y) + \psi_i)\right] + b(x,y) \tag{21}$$

$$= a(x,y)\left[\frac{1}{2} + \frac{1}{2}\cos\left(\varphi(x,y) + \frac{2\pi(i-1)}{N}\right)\right] + b(x,y) \tag{22}$$

$$\varphi(x,y) = \arctan\left(\frac{\sum_{i=1}^{N} I_i \sin\left(\frac{2\pi(i-1)}{N}\right)}{\sum_{i=1}^{N} I_i \cos\left(\frac{2\pi(i-1)}{N}\right)}\right) \tag{23}$$

The analytical solutions (7), (8) and (13), from section 2.1.1. can all be derived from (23) and are thus also members of this family.

2.1.3. Averaging Algorithms

Schwider et al. [1983] developed a method called *the averaging technique* in an attempt to reduce phase errors from phase-shift miscalibration, phase-shift variation and detector non-linearity (discussed later in section 2.3.). To achieve this, two datasets I_i and I'_i each of $n = N$ fringe images are recorded (so double the amount). Instead of calculating the phase map twice and averaging the two results, the data is merged in one calculation:

$$\varphi = \arctan\left(\frac{N_1 + N_2}{D_1 + D_2}\right) \quad (24)$$

For N_1 and D_1 one initially uses the numerator an denominator of either the three- or the four-step phase-shifting algorithm, respectively equation (7) and (13), with images of the first dataset I_i. For N_2 and D_2 one uses the same basic equation but now of course using the images I'_i of the second dataset.

Schmit and Creath [1995] found a clever way to extend this method requiring a smaller amount of data, namely $n = (N + 1)$ instead of $2N$ recordings. They still require two datasets, but these overlap: the first series of images I_i consists of the first N recordings from the $(N + 1)$ in total, and the second dataset I'_i contains the last N of $(N + 1)$ recordings. However, compared to the basic equation the numerator N_2 and denominator D_2 are switched and one has a minus, and then equation (24) is used again.

$$N_2 = -D'_1 \quad (25)$$
$$D_2 = N'_1 \quad (26)$$

This switching of numerator and denominator combined with a minus sign creates a $\frac{\pi}{2}$ backward shift to make I'_i correspond with the first dataset I_i. If the four-step method of equation (13) is used as the basic equation for this averaging technique, one calls the derived equation as being from *class* A. If the three-step method of equation (7) is used, the derivation is of *class* B.

This process can be applied successively on $n = (N + 2)$ fringe images as two sets of $(N + 1)$ recordings, using the newly obtained numerator and denominator from equation (24) as prescription for the new N_i and D_i. In this way, only $(N + 2)$ recordings are needed to achieve the same effect of a dataset of $3N$ images. Depending on the total number of images n used in the extended averaging technique and the basic equation that is used, we refer to a certain algorithm as nA or nB. For instance, the basic equations themselves are called 4A and 3B. As an example, we will now derive the 4B and 5B algorithm iteratively using the extended averaging method from 3B.

The algorithm denoted 4B, means we need $n = 4 = N + 1 = 3 + 1$ recordings using the basic equation of class B, which is the 3B algorithm given in equation (7). The first 3 images are combined as

$$\frac{N_1}{D_1} = \frac{I_1 - I_3}{2I_2 - I_1 - I_3} \quad (27)$$

and the last 3 images as

$$\frac{N_2}{D_2} = \frac{-(2I_3 - I_2 - I_4)}{I_2 - I_4} \quad (28)$$

Through equation (24) we get

$$\varphi(x,y) = \arctan\left(-\frac{I_1 - 3I_3 + I_2 + I_4}{I_1 - 3I_2 + I_3 + I_4}\right) \quad (29)$$

If we continue on from the result in (29) and call it the 4B algorithm, we can calculate 5B as follows:

$$\frac{N_1}{D_1} = -\frac{I_1 - 3I_3 + I_2 + I_4}{I_1 - 3I_2 + I_3 + I_4} \quad (30)$$

$$\frac{N_2}{D_2} = \frac{I_2 - 3I_3 + I_4 + I_5}{I_2 - 3I_4 + I_3 + I_5} \quad (31)$$

Through equation (24) we get

$$\varphi(x,y) = \arctan\left(\frac{I_1 + 2I_2 - 6I_3 + 2I_4 + I_5}{-I_1 + 4I_2 - 4I_4 + I_5}\right) \quad (32)$$

In case of the 5A phase-shifting algorithm, we need $n = 5 = N+1 = 4+1$ recordings using the basic equation of class A: the 4A algorithm in equation (13). The first 4 images are combined in

$$\frac{N_1}{D_1} = \frac{I_4 - I_2}{I_1 - I_3} \quad (33)$$

and the last 4 images in

$$\frac{N_2}{D_2} = \frac{-(I_2 - I_4)}{I_5 - I_3} \quad (34)$$

Through equation (24), one finds the previously discussed 5-$\frac{\pi}{2}$-Schwider et al. [1983] or 5-$\frac{\pi}{2}$-Hariharan and Oreb [1987] method shown in equation (20).

The Schwider et al. [1983] averaging technique, the Schmit and Creath [1995] (A and B) extended averaging technique and the Schmit and Creath [1996] multiple averaging technique generate a big family of algorithms.

2.1.4. $(N+1)$-bucket Algorithms

Larkin and Oreb [1992] developed a family of algorithms with $n = (N+1)$ symmetrically distributed phase-shifts over one full fringe period, called family type A:

$$\varphi(x,y) = \arctan\left(\frac{\sum_{i=2}^{N} I_i \sin\left(\frac{2\pi(i-1)}{N}\right)}{\frac{I_1+I_{N+1}}{2} + \sum_{i=2}^{N} I_i \cos\left(\frac{2\pi(i-1)}{N}\right)}\right) \quad (35)$$

Surrel [1993] published a similar family, i.e. type B:

$$\varphi(x,y) = \arctan\left(\frac{\frac{I_1-I_{N+1}}{2}\cot\left(\frac{2\pi}{N}\right) - \sum_{i=2}^{N} I_i \sin\left(\frac{2\pi(i-1)}{N}\right)}{\frac{I_1+I_{N+1}}{2} + \sum_{i=2}^{N} I_i \cos\left(\frac{2\pi(i-1)}{N}\right)}\right) \quad (36)$$

Especially type B should correct well for phase-shift errors [Schreiber and Bruning, 2007].

2.1.5. $(N+4)$-bucket Algorithms

Hibino et al. [1997] created an improved version of the $(N+1)$-bucket family. His method requires $n = (N+4)$ fringe frames and compensates for quadratic phase-shift errors and harmonics of φ present in the evaluated phase φ up to the order $j = N = n - 4$.

$$\tan \varphi = \frac{\frac{1}{4}(I_1 + I_2 - I_{N+3} - I_{N+4})\frac{\sin\left(\frac{3\pi}{N+2}\right)}{\sin^2\left(\frac{2\pi}{N+2}\right)} + \sum_{i=2}^{N+3} I_i \sin\left(\frac{2\pi}{N+2}\right)\left(i - \frac{N+5}{2}\right)}{\frac{1}{4}(I_1 - I_2 - I_{N+3} + I_{N+4})\frac{\cos\left(\frac{3\pi}{N+2}\right)}{\sin^2\left(\frac{2\pi}{N+2}\right)} + \sum_{i=2}^{N+3} I_i \cos\left(\frac{2\pi}{N+2}\right)\left(i - \frac{N+5}{2}\right)} \quad (37)$$

2.1.6. WDFT Algorithms

Windowed discrete Fourier transform (WDFT) algorithms by Surrel [1996] are designed to be insensitive to harmonics up to the order $j = (N-2)$ (even with linear phase-shift miscalibration), requiring $n = (2N-1)$ images I_i.

$$\varphi(x,y) = \arctan\left(\frac{-\sum_{i=1}^{N-1} i(I_i - I_{2N-i}) \sin\left(\frac{2\pi i}{N}\right)}{NI_{N-1} - \sum_{i=1}^{N-1} i(I_i + I_{2N-i}) \sin\left(\frac{2\pi i}{N}\right)}\right) \quad (38)$$

2.1.7. DFT Algorithms

Recently another approach was brought to our attention from a slightly different field [Pfeiffer et al., 2008], which uses a discrete Fourier transfer (DFT) in the phase-shifting dimension of the dataset. The intensity oscillation of a location (x,y) over several phase-shifted fringe images along i (which should also be sinusoidal) can be written in a Fourier series:

$$I_i(x,y) = \sum_j \gamma_j(x,y) \cos \varphi_i(x,y) + j\psi_i \quad (39)$$

with $\psi_i = i\frac{2\pi}{N}$ and $n = mN$ as the number of fringe images (for $m, N \in \mathbb{N}$). We then take a DFT of equation (39) along the i dimension. By analyzing the resulting Fourier spectrum F and finding the maximal amplitude (which should be located at multiple m and with $j = 1$) we can approximate equation (39) as

$$I_i(x,y) \approx \gamma_0(x,y) + \gamma_1(x,y) \cos(\varphi_1(x,y) + \psi_i) \quad (40)$$

$$\sim a(x,y)\left[\frac{1}{2} + \frac{1}{2}\cos(\varphi(x,y) + \psi_i)\right] + b(x,y) \quad (41)$$

$$\quad (42)$$

with $\varphi(x,y) = \varphi_1(x,y)$. The phase φ is then retrieved by

$$\varphi(x,y) = \arctan\left(\frac{\text{Im}[F(m)]}{\text{Re}[F(m)]}\right) \quad (43)$$

Note that this method is essentially different from the Takeda et al. [1982] method, which also uses (discrete) Fourier transformations but applied on the local spatial frequency variation of fringes in one fringe image. In that case, the Fourier analysis is performed in a direction of the spatial domain (x, y), and not in the phase-shift direction i. We do not discuss the Takeda et al. [1982] method as it is not a phase-shifting method, although it is a powerful phase-evaluation method [Buytaert and Dirckx, 2011].

2.2. β Class: Specific Methods

2.2.1. Carré (Like) Algorithms

In all previous algorithms, it is assumed that the phase-shift $\Delta\psi$ is known. In the method developed by Carré [1966] this is not a stringent requirement.

Using the basic expression for a fringe image, namely (2), with $n = 4$ fringe images and $\psi_1 = -\frac{3}{2}\alpha$, $\psi_2 = -\frac{1}{2}\alpha$, $\psi_3 = \frac{1}{2}\alpha$ and $\psi_4 = \frac{3}{2}\alpha$ we can again solve to phase φ analytically:

$$I_1(x,y) = a(x,y)\left[\frac{1}{2} + \frac{1}{2}\cos\left(\varphi(x,y) - \frac{3}{2}\alpha\right)\right] + b(x,y) \quad (44)$$

$$I_2(x,y) = a(x,y)\left[\frac{1}{2} + \frac{1}{2}\cos\left(\varphi(x,y) - \frac{1}{2}\alpha\right)\right] + b(x,y) \quad (45)$$

$$I_3(x,y) = a(x,y)\left[\frac{1}{2} + \frac{1}{2}\cos\left(\varphi(x,y) + \frac{1}{2}\alpha\right)\right] + b(x,y) \quad (46)$$

$$I_4(x,y) = a(x,y)\left[\frac{1}{2} + \frac{1}{2}\cos\left(\varphi(x,y) + \frac{3}{2}\alpha\right)\right] + b(x,y) \quad (47)$$

$$\varphi(x,y) = \arctan\left(\left[\tan\left(\frac{1}{2}\alpha\right)\right]\frac{(I_2 - I_3) + (I_1 - I_4)}{(I_2 + I_3) - (I_1 + I_4)}\right) \quad (48)$$

with

$$\frac{1}{2}\alpha(x,y) = \sqrt{\left(\frac{3(I_2 - I_3) - (I_1 - I_4)}{(I_2 - I_3) + (I_1 - I_4)}\right)} \quad (49)$$

which allows the (possibly nonuniform) phase-shifts to be calculated and to be known for every location (x, y). Brought together we end up with:

$$\varphi(x,y) = \arctan\left(\frac{\sqrt{[3(I_2 - I_3) - (I_1 - I_4)] \cdot [(I_2 - I_3) + (I_1 - I_4)]}}{(I_2 + I_3) - (I_1 + I_4)}\right) \quad (50)$$

The above equation (50) is determined in the interval $]-\frac{\pi}{2}, \frac{\pi}{2}]$, even when using *atan2*, because of the square root. By determining the correct sign of the square root, *atan2* can again deliver results within $]-\pi, \pi]$. This $\pm$ sign can for instance be obtained by

$$-\text{sign}(I_2 - I_3) \quad (51)$$

Through the years, some similar algorithms emerged. According to Osten [2000] an improved version of the 4-α-Carré [1966] algorithm was made by Juptner et al. [1983], and is defined by:

$$\varphi(x,y) = \arctan\left(\frac{I_1 - 2I_2 + I_3 + (I_1 - I_3)\cos(\alpha) + 2(I_2 - I - 1)(\cos(\alpha))^2}{\sqrt{1 - (\cos(\alpha))^2} \cdot [I_1 - I_3 + 2(I_2 - I_1)\cos(\alpha)]}\right) \quad (52)$$

with

$$\alpha(x,y) = \arccos\left(\frac{I_1 - I_2 + I_3 - I_4}{2(I_2 - I_3)}\right) \quad (53)$$

The above equation (52) will be referred to as the 4-α-Juptner et al. [1983] method.

To end with, another Carré-like algorithm emerged but with $n = 5$ fringe images. The 5-α-Stoilov and Dragostinov [1997] algorithm goes as follows:

$$\varphi(x,y) = \arctan\left(\frac{2(I_4 - I_2)}{(I_1 + I_5 - 2I_3) \cdot \sqrt{1 - \left(\frac{I_1 - I_5}{2(I_2 - I_4)}\right)^2}}\right) \quad (54)$$

However, as the authors themselves already mention in their paper, [Stoilov and Dragostinov, 1997], the method only works well for $\Delta\psi = \alpha = \frac{\pi}{2}$. In other words, this method is not a specific (class B) or Carré-like method with unknown phase-shift, but rather a class A method like the 5-$\frac{\pi}{2}$-Schwider et al. [1983] algorithm in equation (20).

2.2.2. $(2+1)$-bucket Algorithm

This method has been developed to be used in dynamic situations (as is the case in time-sensitive measurements or in the presence of vibrations or perturbations), and allows short acquisition times. It uses only two (subsequent) fringe patterns from an incremental series of $\frac{\pi}{2}$ phase-shifts acquired in rapid sequence with a video camera. As the third required equation, the intensity offset $(a(x,y) + b(x,y))$ is used, obtained by averaging (and sometimes filtering) two previous fringe images which are phase $\Delta\psi = \pi$ apart [Kerr et al., 1990].

Using the basic expression for a fringe image, namely (2), with $n = 2$ fringes images (of a larger continuous series) and $\psi_{k+1} = 0$ and $\psi_{k+2} = \frac{\pi}{2}$, we get:

$$I_{k+1}(x,y) = a(x,y)\left[\frac{1}{2} + \frac{1}{2}\cos(\varphi(x,y))\right] + b(x,y) \quad (55)$$

$$I_{k+2}(x,y) = a(x,y)\left[\frac{1}{2} + \frac{1}{2}\cos\left(\varphi(x,y) + \frac{\pi}{2}\right)\right] + b(x,y) \quad (56)$$

In combination with the average intensity I_a, obtained by averaging I_{k+1} with a previous image I_{k-1} of phase-shift difference $\Delta\psi = \pi$,

$$I_a(x,y) = \frac{1}{2}I_{k+1}(x,y) + \frac{1}{2}I_{k-1}(x,y) \qquad (57)$$

$$= \frac{a(x,y)}{2}\left[\frac{1}{2} + \frac{1}{2}\cos(\varphi(x,y))\right] + \frac{b(x,y)}{2}$$

$$+ \frac{a(x,y)}{2}\left[\frac{1}{2} + \frac{1}{2}\cos(\varphi(x,y) - \pi)\right] + \frac{b(x,y)}{2} \qquad (58)$$

$$= a(x,y) + b(x,y) \qquad (59)$$

we obtain the $(2+1)$-$\frac{\pi}{2}$-Angel and Wizinowich [1988] algorithm:

$$\varphi(x,y) = \arctan\left(\frac{I_2 - I_a}{I_1 - I_a}\right) = \arctan\left(\frac{I_2 - a - b}{I_1 - a - b}\right) \qquad (60)$$

2.2.3. $(2+2)$-bucket Phase-Difference Algorithm

Finally, there are methods which allow immediate calculation of the phase-difference $\Delta\varphi = (\varphi' - \varphi)$ between two sets of fringe patterns I_i and I'_i, without specifically calculating their individual phases φ and φ'. These methods are useful if one is interested in deformations, rather than the shape profile itself.

An exemplary method, requiring the least amount of images, is the $(2+2)$-$\frac{\pi}{2}$-Owner-Petersen [1991] algorithm, using $n = 4$ images I_i with phase-shifts $\Delta\psi = \frac{\pi}{2}$:

$$\Delta\varphi(x,y) = \varphi'(x,y) - \varphi(x,y) = 2\arctan\left(\frac{I'_1 - I_2}{I'_2 - I_1}\right) \qquad (61)$$

with ψ_i: $\psi_i = (i-1)\frac{\pi}{2}$ and $\psi'_i = (i-2)\frac{\pi}{2}$.

2.3. Error Types

PSA error sources are often categorized in the following groups:

1. Phase type errors (f.i. phase-shift miscalibrations)

2. Intensity type errors (f.i. detector non-linearity)

or in these subdivisions:

1. Error sources associated with data acquisition (f.i. detector non-linearity)

2. Environmental effects (f.i. vibration or air turbulence)

3. Errors associated with defects in optical and/or mechanical design (f.i. phase-shift miscalibrations)

2.3.1. Phase-shift Errors

Most PSAs rely on shifting the phase in a known manner as part of the acquisition process. The expected variation of the intensity at any point in the interference pattern should be purely sinusoidal with a linear change in the phase, cf. equation (1).

If the increment in phase-shift of an n phase-step algorithm deviates from $\frac{2\pi}{n}$ with ϵ, then the fringe phase-shift ψ_i becomes ψ_i'

$$\psi_i' = \psi_i + i\epsilon = \psi_i(1 + \epsilon_1) \tag{62}$$

and thus the obtained measured object phase φ' will deviate from the true phase φ by $\Delta\varphi(x,y) = \varphi'(x,y) - \varphi(x,y)$. This error type is called *linear miscalibration*.

In reference to equation (2), a general phase-shift error can be described as

$$\psi_i' = \psi_i(1 + \epsilon_1 + \epsilon_2\psi_i + \epsilon_3\psi_i^2 + \cdots) \tag{63}$$

Using equation (63), a linear phase-shift error has $\epsilon_i = 0$ except for $\epsilon_1 \neq 0$. A quadratic phase-shift error is defined as $\epsilon_i = 0$ except for $\epsilon_2 \neq 0$. We will perform numerical simulations on these topics. The phase-shift can also be different at different field positions, which is hence named a spatially nonuniform phase-shift. We indicate some algorithms that should cope well with this kind of error source, according to Schreiber and Bruning [2007], in our PSA Table 1. In our own simulations we only use spatially uniform shifts, cf. section 3.2..

Phase-shift errors should be minimized by for instance the 4-α-Carré [1966] and the 5-$\frac{\pi}{2}$-Schwider et al. [1983] algorithm. As the 4-α-Carré [1966] algorithm effectively calculates the actual (even non-uniform) phase-shift at each location, cf. expression (49), this method cannot suffer from an error in the phase-shift increment. The 5-$\frac{\pi}{2}$-Schwider et al. [1983] method is tuned to be robust against linear miscalibration, cf. expression (20), and should according to the literature even be insensitive to other errors in addition to the linear error.

2.3.2. Detector Non-linearities

Non-linearity of the detector only becomes important when the dynamic range (and contrast) of the fringes is large. But, a large dynamic range is often desired as it influences the height resolution positively in some techniques, like *liquid crystal moiré profilometry* [Buytaert and Dirckx, 2007]. In the case of low dynamic range (detector bit-depth), detector non-linearity is insignificant.

Gamma correction, gamma non-linearity, gamma encoding or often simply *gamma* is the name for the non-linearity in the coding and decoding of luminance in video or still image systems. Gamma correction can, in the simplest cases, be defined by the following power-law expression

$$I_{out} = I_{in}^{\gamma} \tag{64}$$

with both the input and output intensities normalized and gamma $\gamma \geq 0$.

In general, when an image is recorded with $\gamma \neq 1$, the dependency of the recorded intensity distribution I'_i on the real intensity I_i can be described as follows

$$I'_i(x,y) = \alpha I_i(x,y) + \beta I_i^2(x,y) + \gamma I_i^3(x,y) + \delta I_i^4(x,y) + \cdots \qquad (65)$$

$$= a(x,y)\left[1 + \sum_{k=1}^{j} s_k(x,y)\cos(\varphi_k(x,y) + k\psi_i)\right] + b(x,y) \qquad (66)$$

with j the maximum order of harmonic components, s_k and φ_k amplitude and phase of the k-th component, and $\varphi_1 = \varphi$ the unperturbed object phase. We will perform numerical simulations on this topic, cf. section 3.3.. Expressions (65) and (66) show that harmonics can be present in the original fringe patterns (not to be confused with harmonics of the phase φ).

Note that detector non-linearity of a linear projection system is analogous to a setup with a linear detector but with non-linearity in the structured light pattern projection (and demodulation), again f.i. with *liquid crystal projection moiré profilometry* [Buytaert and Dirckx, 2010].

2.3.3. Source Stability

The stability of the interferometer light source is important, both in light frequency and amplitude.

Frequency instabilities can give rise to corresponding instabilities in the generated fringe patterns.

The effect of projected light intensity fluctuations depends on the algorithm and number of phase-steps n used, as the standard deviation of the measured phase is given by

$$\sigma_\varphi \approx \frac{1}{S\sqrt{n}} \qquad (67)$$

with S the signal-to-noise ratio of the detector. We will perform numerical simulations on this topic, cf. section 3.4..

2.3.4. Vibration Errors

Mechanical stability is paramount in interferometry. Some algorithms cope better than others with vibration. The square root of the number n of phase-steps in the algorithm is roughly in inverse proportion to the phase standard deviation σ_φ:

$$\sigma_\varphi \approx \frac{\sigma_{vib}}{aI_a\sqrt{n}} \qquad (68)$$

with σ_{vib} the standard deviation of the intensity fluctuation, a the fringe modulation factor and I_a the average intensity. We indicate some algorithms that cope well with this kind of error source in our PSA Table 1, based on simulations by Schreiber and Bruning [2007] with 39 (of our 84) PSAs.

2.3.5. Optical Design Errors

Aberrations can be created by the overall quality of the interferometer (components). We will not go into this error source further.

2.3.6. Extraneous Fringes

When using lasers to generate fringe images, reflections from surfaces within the coherence length can interfere with one another and create extraneous fringes. This coherent effect creates an error $\Delta\varphi$ according to Schwider et al. [1983]:

$$\Delta\varphi(x,y) = \arctan\left(\frac{q\sin\left(\eta(x,y) - \varphi(x,y)\right)}{1 + q\cos\left(\eta(x,y) - \varphi(x,y)\right)}\right) \tag{69}$$

The effect of spurious reflections can be removed by several approaches. However, also the algorithm that is used can be important. We will not go into this error source further.

2.3.7. Quantization Errors

The accuracy of A/D conversion also depends upon the bit-depth. The influence of (limited) bit-depth varies with the algorithm used. We will not go into this error source further.

2.3.8. Air Turbulence Errors

Air currents or air turbulence can cause local temporal phase changes in the fringe patterns. Algorithms exist which are more robust against errors caused by these phenomena. In PSA Table 1, we have indicated algorithms from a subset of 39 algorithms that cope well according to Schreiber and Bruning [2007] with this kind of error source.

2.3.9. Fringe Harmonics

Keep in mind that a certain mentioned robustness against an error source in PSA Table 1 only holds when the fringes are purely sinusoidal (without harmonics), or when none of the other errors occur simultaneously.

Harmonics in the fringe patterns might become important in certain moiré profilometers, f.i. in those which use Ronchi rulings as projected structured light pattern [Buytaert and Dirckx, 2010]. A Ronchi ruling is photolithographic plate with a constant-interval opaque and transparent square grid. If such a grid is projected and used for optical demodulation, one then obtains *triangular* (instead of sinusoidal) moiré fringes.

Also, the previously mentioned gamma detector non-linearity is in fact a summation of even and uneven fringe harmonics, cf. equation (66). The triangular fringes consist only of uneven harmonics of the basic sine wave, cf. expressions (72) combined with (66).

We will perform numerical simulations on these topics, cf. section 3.5..

Note that in most papers harmonics in the eventual phase φ are studied. Our interest lies more in harmonics of the fringe pattern. For the phase harmonics, we refer to [Schreiber and Bruning, 2007].

3. Simulation Results

3.1. No Error

We implemented 84 PSA algorithms all with known phase-shifts of either $\frac{2\pi}{3}$, $\frac{\pi}{2}$, $\frac{\pi}{3}$, $\frac{\pi}{4}$ or $\frac{\pi}{6}$, including several n-α-Pfeiffer et al. [2008], 4-α-Carré [1966] and 4-α-Juptner et al. [1983] algorithms and the 5-$\frac{\pi}{2}$-Stoilov and Dragostinov [1997] method. They are all listed in Table 1, except for the n-α-Pfeiffer et al. [2008] methods.

Using MATLAB R2010a (The MathWorks, Inc.) we created 11 perfect concentric sinusoidal fringe images (401×401 pixels) with a specified phase-shift in between, cf. figure 1(a-d). In the case of moiré profilometry, these fringes can be considered as altitude lines on a topographic map. The resulting height distribution, after phase-unwrapping, corresponds to an ideal cone, cf. figure 1(e-g). For all algorithm, the phase measurement should be perfectly free of error for a linear detector and in absence of errors on the fringe images.

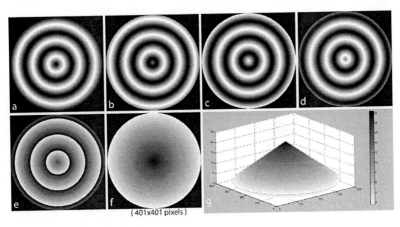

Figure 1. *a-d*) first four (of eleven) fringes images with gradually climbing $\frac{\pi}{2}$ phase-shift offset. *e*) wrapped phase distribution from arctangent calculation. *f*) phase-unwrapped phase/height distribution. *g*) three-dimensional representation of *f*.

Our first goal was to verify if our 84 PSA algorithms all perform correct and identical on a perfect undistorted dataset, as they theoretically should, after correction of all typographical errors from literature and after matching the phase offsets of the different arctangent methods that is. We achieved an identical result for 77 of 84 algorithms. Seven algorithms still deliver a small root-mean-square (RMS) error ranging from RMS% = 0.0036% till 0.22%.

$$\text{RMS} = \sqrt{\frac{\sum_{i=1}^{I}\sum_{j=1}^{J}\left(\mu(i,j)-\varphi(i,j)\right)^{2}}{IJ}} \tag{70}$$

$$\text{RMS\%} = \frac{\text{RMS}}{2\pi}.100 \tag{71}$$

with (I, J) the pixel dimensions of a fringe image, and $\mu(i,j)$ the ideal phase value and $\varphi(i,j)$ the simulated phase value of each image location. These small errors were detected in the case of

- 4-α-Juptner et al. [1983] PSAs with $\alpha = \frac{2\pi}{3}, \frac{\pi}{2}, \frac{\pi}{3}, \frac{\pi}{6}$

but with $\alpha = \frac{\pi}{4}$ the method does work perfectly. And, though Osten [2000] suggested that the 4-α-Juptner et al. [1983] is an improved version of 4-α-Carré [1966], we found the derived algorithms to perform slightly worse and prone to errors (originating in the denominator calculation of the phase-shift α, cf. equation (53)). Other algorithms who did not achieve the theoretical perfect result for some unknown reason were:

- 8-$\frac{2\pi}{3}$-de Groot [1995] PSA

- 8-$\frac{\pi}{2}$-Hibino et al. [1997]-N PSA

- 8-$\frac{\pi}{3}$-de Groot [1995] PSA

3.2. Phase-shift Errors

3.2.1. Linear Miscalibration

The first simulated deviation from a perfect phase-shifting dataset is a linear phase-shift error of 10%, cf. figure 2. Remember from section 2.3.1. and equation (63) that this comes down to $\epsilon_1 = 0.1$ while all other $\epsilon_i = 0$. Though ϵ_1 has the same value for all phase-shift steps, the absolute error-value ϵ depends on and scales with the phase-shift value $\Delta\psi$, as can be deduced from expression (62).

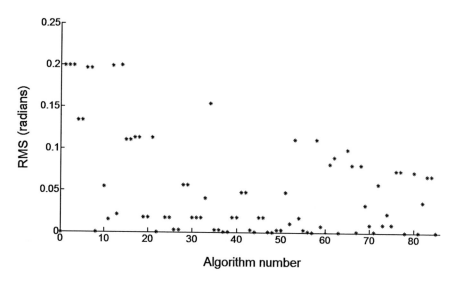

Figure 2. RMS phase errors obtained by 84 phase-shifting algorithms in a simulation with a 10% linear phase-shift error.

Figure 2 shows that the RMS error induced by a 10% linear phase-shift miscalibration simulation differs between algorithms from approximately zero to 0.2 radians. An undistorted perfect result was found for all

- 4-α-Carré [1966] PSAs

which is not surprising as they calculate the actual phase-shift and use this value in the arctangent calculation. As stated before, the 4-α-Juptner et al. [1983] algorithms should do as well but are prone to error because of outliers. Only

- 7-$\frac{\pi}{2}$-Schmit and Creath [1995]-B PSA
- 8-$\frac{\pi}{2}$-Schmit and Creath [1996]-A PSA

also perform extremely well (RMS% $< 10^{-4}$%). Four other methods achieve near-perfection (RMS% < 0.001%):

- 6-$\frac{\pi}{2}$-Schmit and Creath [1995]-B PSA ($\times 2$)
- 7-$\frac{\pi}{2}$-de Groot [1994] PSA (also Schmit and Creath [1995]-A family)
- 11-$\frac{\pi}{2}$-de Groot [1995] PSA

Finally, 13 more algorithms are very robust against this error (RMS% < 0.01%), all positivily indicated in PSA Table 1.

*Best-performing algorithm(s) taking the number of required fringe images ($\sim$ measurement time) into account:

- 4-α-Carré [1966] PSA
- 5-$\frac{\pi}{2}$-Schmit and Creath [1995]-B PSA

The exact same performance for the PSAs was found with $\epsilon_1 = 0.2$.

3.2.2. Quadratic Miscalibration

Quadratic phase-shift error simulations with $\epsilon_2 = 0.05$, $\epsilon_2 = 0.1$ and $\epsilon_2 = 0.2$ (while all other $\epsilon_i = 0$ in equation (63)) were performed. For different ϵ_2 some variation exists in the performance of PSA algorithms. Not surprisingly,

- 4-α-Carré [1966] PSAs especially with $\alpha = \frac{\pi}{3}, \frac{\pi}{4}$ and $\frac{\pi}{6}$

again perform well. However, the algorithm showing extreme robustness against all three quadratic phase-shift errors, was the recent

- 9-$\frac{\pi}{4}$-Estrada et al. [2009] PSA

which in all three case remained below RMS% < 0.1%. The next well-performing method over all cases was

- 6-$\frac{\pi}{3}$-Hibino et al. [1997] PSA

In addition, for $\epsilon_2 = 0.05$ (and $\epsilon_2 = 0.1$) we found

- 7-$\frac{\pi}{3}$-Hibino et al. [1997] PSA
- 7-$\frac{\pi}{3}$-Surrel [1993] PSA

- 10-$\frac{\pi}{3}$-Surrel [1996] PSA

also to be a good choice. For $\epsilon_2 = 0.2$, the following methods deserve mentioning:

- 6-$\frac{\pi}{3}$-Bruning et al. [1974] PSA
- 6-$\frac{\pi}{3}$-Pfeiffer et al. [2008] PSA

*Best-performing algorithm(s) taking the number of required fringe images ($\sim$ measurement time) into account:

- 4-α-Carré [1966] with $\alpha = \frac{\pi}{3}, \frac{\pi}{4}$ and $\frac{\pi}{6}$

3.2.3. Linear + Quadratic Miscalibration

When combining both linear and quadratic phase-shift errors, by choosing $\epsilon_1 = \epsilon_2 = 0.1$ (while all other $\epsilon_i = 0$ in equation (63)), we end up with the same story as in the pure quadratic miscalibration case.

3.3. Detector Non-linearities

A first series of simulations related to detector (or projection and demodulation) gamma non-linearity were performed for $\gamma = 0.5$, $\gamma = 0.75$, $\gamma = 0.85$, $\gamma = 1.15$, $\gamma = 1.25$ and $\gamma = 1.5$ in formula (64). The effect of a gamma deviation on fringes is illustrated in figure 3. The 84 PSAs showed similar performance in all simulated cases of γ. Figure 4 illustrates different behavior of a selection of algorithms for a simulation with $\gamma = 0.5$. We also performed an extra simulation using the alternative equation (65) for gamma non-linearity with $\alpha = 0.8$ and $\beta = 0.2$. Not surprisingly, also in this case the same algorithms as before performed well.

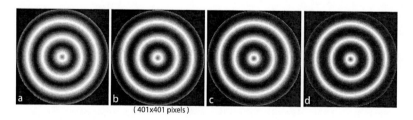

Figure 3. Four fringe images with the same offset ψ_i but with a simulated gamma non-linearity of A) $\gamma = 1$, B) $\gamma = 1.25$, c) $\gamma = 1.5$ and d) $\gamma = 2$.

The best algorithm for non-linearities appeared to be

- 8-$\frac{\pi}{4}$-Pfeiffer et al. [2008] PSA

followed closely by

- 9-$\frac{\pi}{4}$-de Groot [1995] PSA

and seven $\frac{\pi}{3}$-algorithms

- 6-$\frac{\pi}{3}$-Bruning et al. [1974] PSA
- 6-$\frac{\pi}{3}$-Pfeiffer et al. [2008] PSA
- 7-$\frac{\pi}{3}$-Larkin and Oreb [1992] PSA
- 7-$\frac{\pi}{3}$-Surrel [1993] PSA
- 10-$\frac{\pi}{3}$-Surrel [1996] PSA
- 11-$\frac{\pi}{3}$-Hibino et al. [1995] PSA
- 11-$\frac{\pi}{3}$-Surrel [1996] PSA

Something remarkable in these simulations was that the seven (of 15) $\frac{\pi}{3}$-algorithms had exactly the same RMS value. The same statement is valid for most (25 of 44) of the $\frac{\pi}{2}$-algorithms, though with a slightly higher but fairly good RMS value, thus deserving a positive sign in Table 1. And even all fourteen (of 14) $\frac{2\pi}{3}$-algorithms had a (high but) constant RMS value.

*Best-performing algorithm(s) taking the number of required fringe images ($\sim$ measurement time) into account:

- 6-$\frac{\pi}{3}$-Bruning et al. [1974] PSA
- 6-$\frac{\pi}{3}$-Pfeiffer et al. [2008] PSA

but still with a rather high number of fringe images. If fewer images are required

- 4-$\frac{\pi}{2}$-Bruning et al. [1974] PSA
- 4-$\frac{\pi}{2}$-Pfeiffer et al. [2008] PSA

Some more related simulation results can be found in Schreiber and Bruning [2007], where a special case is mentioned for $\gamma = 2$. From our own simulation for $\gamma = 2$, we can conclude that all algorithms which performed good for the previous different values of γ now achieve a perfect result (RMS% $< 10^{-13}$%) in the case of $\gamma = 2$. Even one extra algorithm joined the ranks, namely

- 7-$\frac{\pi}{3}$-Hibino et al. [1997] PSA

3.4. Source Stability

To simulate amplitude variations of the source, we varied the amplitudes $a(x, y)$ in equation (2) for different fringe images i randomly. We were unable to find a general trend in the performance of the PSAs. However, we did notice that the size of the amplitude variations did not influence whether a PSA performed better than another algorithm, while a permutation in the order of the amplitude variations did make a big difference.

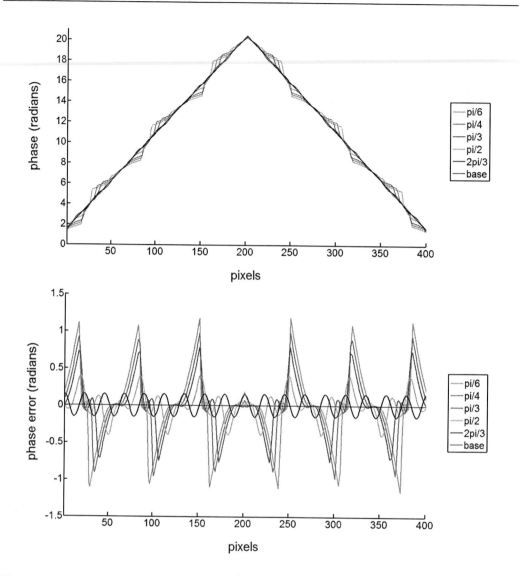

Figure 4. Illustration of the different sensitivity of a selection of PSAs for $\gamma = 0.5$.
Top) phase (height) profile of the theoretical cone,
Bottom) phase difference with the theoretical cone, using 3-$\frac{2\pi}{3}$-Bruning et al. [1974] (blue, RMS% = 1.75%), 3-$\frac{\pi}{2}$-Wyant et al. [1984] (cyan, RMS% = 2.97%), 3-$\frac{\pi}{3}$-Kreis [2005] (red, RMS% = 5.46%), 4-$\frac{\pi}{4}$-Kreis [2005] (magenta, RMS% = 6.96%) and 3-$\frac{\pi}{6}$-Kreis [2005] (green, RMS% = 8.56%).

3.5. Fringe Harmonics

3.5.1. Triangular Fringes

Finally, we simulated fringes consisting of higher order fringe pattern harmonics. The first simulation concerns a triangular (in stead of sinusoidal) fringe pattern, which comes down to uneven harmonics in the fringes. For perfect triangular fringes, we use the following variables in equation (66):

$$i = 1, 3, 5, 7, \ldots$$
$$j = 1000 \qquad (72)$$
$$s_k(x,y) = \frac{8(-1)^{(k-1)/2}}{\pi^2 k^2}$$

In this case, the best performing algorithm was

- 8-$\frac{\pi}{4}$-Pfeiffer et al. [2008] PSA

followed closely by

- 9-$\frac{\pi}{4}$-de Groot [1995] PSA

Furthermore all $\frac{2\pi}{3}$- and seven $\frac{\pi}{3}$-algorithms perform very well and have (the same constant) RMS% = 0.22%. The seven algorithms are the same ones as for detector non-linearities (cf. section 3.3.):

- 6-$\frac{\pi}{3}$-Bruning et al. [1974] PSA
- 6-$\frac{\pi}{3}$-Pfeiffer et al. [2008] PSA
- 7-$\frac{\pi}{3}$-Larkin and Oreb [1992] PSA
- 7-$\frac{\pi}{3}$-Surrel [1993] PSA
- 10-$\frac{\pi}{3}$-Surrel [1996] PSA
- 11-$\frac{\pi}{3}$-Hibino et al. [1995] PSA
- 11-$\frac{\pi}{3}$-Surrel [1996] PSA

All $\frac{\pi}{2}$-algorithms also have a constant RMS value as well but higher.

*Best-performing algorithm(s)** taking the number of required fringe images ($\sim$ measurement time) into account:

- 6-$\frac{\pi}{3}$-Bruning et al. [1974] PSA
- 6-$\frac{\pi}{3}$-Pfeiffer et al. [2008] PSA

but still with a rather high number of fringe images. If fewer images are required:

- 3-$\frac{2\pi}{3}$-Bruning et al. [1974] PSA
- 3-$\frac{2\pi}{3}$-Pfeiffer et al. [2008] PSA

3.5.2. Uneven Harmonics

In the next simulations, only the uneven fringe harmonics were active, starting at a 10% amplitude of the first harmonic and decreasing with the order of the harmonic k.

$$\begin{aligned} i &= 1, 3, 5, 7, \ldots \\ j &= 1000 \\ s_1(x, y) &= 1 \\ s_k(x, y) &= 0.1 \frac{3}{k} \end{aligned} \qquad (73)$$

Not surprisingly, the results are more or less the same as in the case of triangular fringes. However, four extra $\frac{\pi}{3}$-algorithms join the seven others in their good performance, though with a slightly different RMS. So, these also get a positive sign in Table 1:

- 4-$\frac{\pi}{3}$-Kreis [2005] PSA
- 5-$\frac{\pi}{3}$-Kreis [2005] PSA
- 6-$\frac{\pi}{3}$-Hibino et al. [1997] PSA
- 8-$\frac{\pi}{3}$-de Groot [1995] PSA

Again, both groups of $\frac{2\pi}{3}$- and $\frac{\pi}{2}$-algorithms each have a constant RMS value.

*Best-performing algorithm(s) taking the number of required fringe images ($\sim$ measurement time) into account:

- 6-$\frac{\pi}{3}$-Bruning et al. [1974] PSA
- 6-$\frac{\pi}{3}$-Pfeiffer et al. [2008] PSA

but still with a rather high number of fringe images. If fewer images are required:

- 3-$\frac{2\pi}{3}$-Bruning et al. [1974] PSA
- 3-$\frac{2\pi}{3}$-Pfeiffer et al. [2008] PSA

3.5.3. Even Harmonics

We investigate a similar situation, but now with even harmonics, also starting at a 10% amplitude of the first harmonic and decreasing with the order of the harmonic k.

$$\begin{aligned} i &= 1, 2, 4, 6, 8, \ldots \\ j &= 1000 \\ s_1(x, y) &= 1 \\ s_k(x, y) &= 0.1 \frac{2}{k} \end{aligned} \qquad (74)$$

Many algorithms (37 of 84) behave extremely robust against even fringe harmonics (RMS% $< 10^{-13}$%). They are all positively indicated in Table 1, among them

- n-α-Pfeiffer et al. [2008] PSA with $\alpha = \frac{\pi}{2}, \frac{\pi}{3}$ and $\frac{\pi}{4}$

Again, all $\frac{2\pi}{3}$-algorithms have a constant RMS value.

*Best-performing algorithm(s) taking the number of required fringe images ($\sim$ measurement time) into account:

- 4-$\frac{\pi}{2}$-Bruning et al. [1974] PSA

3.5.4. Even and Uneven Harmonics

Finally, we combined both even and uneven fringe harmonics, starting at a 10% amplitude of the first harmonic and decreasing with the order of the harmonic k.

$$
\begin{aligned}
i &= 1, 2, 3, 4, 5, \ldots \\
j &= 1000 \\
s_1(x, y) &= 1 \\
s_k(x, y) &= 0.1 \frac{2}{k}
\end{aligned}
\tag{75}
$$

This simulation closely resembles the case of detector non-linearity errors, given in (66), as it also consists out of a combination of even and uneven harmonics.

The results are similar to section 3.3. and 3.5.1., so the best algorithm again was

- 8-$\frac{\pi}{4}$-Pfeiffer et al. [2008] PSA

followed closely by

- 9-$\frac{\pi}{4}$-de Groot [1995] PSA

One extra PSA, so eight (and not seven) $\frac{\pi}{3}$-algorithms all positively indicated in PSA Table 1, is doing well, namely

- 8-$\frac{\pi}{3}$-de Groot [1995] PSA

It will not come as a surprise that these 10 PSAs can also be identified by the cross section of the robust algorithms for even and the robust algorithms for uneven harmonics.

Again, all $\frac{2\pi}{3}$-algorithms have a constant RMS value.

*Best-performing algorithm(s) taking the number of required fringe images ($\sim$ measurement time) into account:

- 6-$\frac{\pi}{3}$-Bruning et al. [1974] PSA

4. Discussion

It is clear from the simulations that specific algorithms are needed for each specific problem or error source. Nevertheless, as can be seen in Table 1, a few algorithm have an all-round robust behavior for many types of errors. By counting the number of positive signs, the methods with the best overall performance are easily identified. We then end up with four PSAs with four plus signs:

- 6-$\frac{\pi}{2}$-Schmit and Creath [1995]-A PSA
- 7-$\frac{\pi}{3}$-Surrel [1993] PSA
- 8-$\frac{\pi}{2}$-Schmit and Creath [1996]-*Bell6* PSA
- 10-$\frac{\pi}{3}$-Surrel [1996] PSA.

In general, a higher number of recorded fringe images n reduces noise and phase-errors in the resulting phase $\varphi(x,y)$, of course at the cost of more recording time. This is confirmed by the four all-round performing methods: they all have a relatively high number of images (≥ 6).

Although still often mentioned in recent literature, a.o. [Schreiber and Bruning, 2007], and in equation (20) and section 2.3.1., the 5-$\frac{\pi}{2}$-Schwider et al. [1983] or 5-$\frac{\pi}{2}$-Hariharan and Oreb [1987] algorithm should do well in the case of linear and quadratic miscalibration. Its performance, however, was not that impressive (anymore). Apparently, over time several new methods caught up with it and perform better.

At the end of section 2.1.4., it was said that PSAs derived from equation (36) should be robust against linear and even quadratic phase-shift miscalibration. This is certainly found to be true for the

- 7-$\frac{\pi}{3}$-Surrel [1993] PSA

of the $(N+1)$-bucket family type B. However,

- 5-$\frac{\pi}{2}$-Schwider et al. [1983] PSA (also in $(N+1)$-bucket family type B)
- 4-$\frac{2\pi}{3}$-Surrel [1993] PSA

did not show any remarkable performance.

PSAs derived from equation (37) in section 2.1.5. should show robustness for quadratic phase-shift errors (and phase harmonics up to the order $j = n - 4$). This appears to be true. The following methods of this $(N+4)$-bucket family were indeed robust against quadratic phase-shift errors:

- 5-$\frac{2\pi}{3}$-Hibino et al. [1997] PSA with $j = 5 - 4 = 1$
- 6-$\frac{\pi}{2}$-Schmit and Creath [1995]-A PSA (also $(N+4)$-bucket family) with $j = 6-4 = 2$
- 8-$\frac{\pi}{2}$-Hibino et al. [1997]-U PSA with $j = 8 - 4 = 4$

In an article by Hibino et al. [1997], four other algorithms though not belonging to the $(N+4)$-bucket family are mentioned with quadratic phase-shift robustness (and with a defined phase harmonic robustness of order j). These statements were confirmed for:

- 6-$\frac{\pi}{3}$-Hibino et al. [1997] PSA with $j = 1$
- 7-$\frac{\pi}{3}$-Hibino et al. [1997] PSA with $j = 2$

while the following did not show resistance against quadratic errors:

- 8-$\frac{\pi}{2}$-Hibino et al. [1997] PSA with $j = 2$
- 9-$\frac{\pi}{2}$-Hibino et al. [1997] PSA with $j = 2$

Finally, we confirm the statement of Schmit and Creath [1995] that with the same number of frames involved, the class B algorithms (derived from the three-step PSA (7)) cause less phase-error in case of phase-shift miscalibration compared to class A methods. And that algorithms from class A (derived from four-step PSA (13)) are in turn more robust against second-order detector non-linearity than class B. This is easily noticed by looking at the plus-signs in Table 1. Furthermore, it is clear from Table 1 that Schwider et al. [1983] and Schmit and Creath [1995] succeeded in making the (extended) averaging technique algorithms robust against linear phase-shift errors and non-linearity.

The innovative method described by Pfeiffer et al. [2008] and explained in section 2.1.7. is a phase-shifting method which does not make use of a ratio of fringe image combinations. Therefor it is not listed in PSA Table 1. Nonetheless, the approach performs very well for several types of error sources and must be taken under consideration when looking for an algorithm. The 8-$\frac{\pi}{4}$-Pfeiffer et al. [2008] is the best method to correct for detector non-linearities, for triangular fringes and for even and uneven fringe harmonics. The 6-$\frac{\pi}{3}$-Pfeiffer et al. [2008] algorithm also corrects well for detector non-linearities, for triangular fringes, for even, uneven and even+uneven fringe harmonics, and for quadratic phase-errors. The 3-$\frac{2\pi}{3}$-Pfeiffer et al. [2008] method does well in the case of triangular fringes and uneven fringe harmonics. The 4-$\frac{\pi}{2}$-Pfeiffer et al. [2008] method manages in the case of gamma non-linearities. And all n-α-Pfeiffer et al. [2008] PSAs with $\alpha = \frac{\pi}{2}, \frac{\pi}{3}$ and $\frac{\pi}{4}$ cope well with even fringe harmonics. However, on an undistorted dataset, this method deviates from perfection with a small error up till 0.008366 radians.

5. Conclusions

An important and big field in optics is phase-stepping and phase-shifting interferometry. With this publication, we aim to provide an extensive and accurate (but non-exhaustive) list of phase-stepping and phase-shifting algorithms with different phase-steps which will serve as a useful overview and reference for the scientific and industrial community. We meticulously listed, corrected and tested the algorithms on their performance. Nonetheless, we advice to not just blindly follow the literature or even this article. One should always try to identify all error sources in a setup, and test/simulate which PSA performs best on this specific combination of errors, for which this work can be a guide.

steps	algorithm		phase-steps	reference	robustness						generic family					
N	$\tan\varphi =$	$\tan(\varphi-\frac{\pi}{4})=$	$\Psi_i =$	author(s)	(non)uniform $\Delta\Psi$	(un)even harmon.	quad $\Delta\Psi$ error	lin $\Delta\Psi$ error	vibration	turbulence	non-lin detect	N-bucket	Ext Averaging	multi Averaging	$(N+1)$-bucket	WDFT	$(N+4)$-bucket
3	$\frac{\sqrt{3}(I_3-I_2)}{2I_1-I_2-I_3}$	$\frac{-2I_1+I_2-\sqrt{3}(I_2-I_3)+I_3}{2I_1-I_2-\sqrt{3}(I_2-I_3)-I_3}$	$(i-1)\frac{2\pi}{3}$	Bruning et al. [1974]	U							×	B				
3	$\frac{I_1-I_3}{I_1-2I_2+I_3}$		$(i-1)\frac{\pi}{2}$	Wyant et al. [1984]								×					
3	$\frac{2I_1-3I_2+I_3}{\sqrt{3}(I_2-I_3)}$		$(i-1)\frac{\pi}{3}$	Kreis [2005]													
3	$\frac{(2+\sqrt{2})I_1-(2+2\sqrt{2})I_2+\sqrt{2}I_3}{-\sqrt{2}I_1+(2+2\sqrt{2})I_2-(2+\sqrt{2})I_3}$		$(i-1)\frac{\pi}{4}$	Kreis [2005]													
3	$\frac{(2+\sqrt{2})(I_1-2I_2+I_3)}{\sqrt{2}(I_3-I_1)}$		$(i-2)\frac{\pi}{4}$	Kreis [2005]													
3	$\frac{(3\sqrt{3}-5)I_1+(\sqrt{3}-2)I_2+(7-4\sqrt{3})I_3}{(5-3\sqrt{3})I_1+(2\sqrt{3}-3)I_2+(\sqrt{3}-2)I_3}$		$(i-1)\frac{\pi}{6}$	Kreis [2005]													
3	$\frac{(2+\sqrt{3})(-I_1+2I_2-I_3)}{I_1-I_3}$		$(i-2)\frac{\pi}{6}$	Kreis [2005]													
4	$\frac{\pm\sqrt{(3(I_2-I_3)-I_1+I_4)(I_2-I_3+I_1-I_4)}}{I_1+I_4-I_2-I_3}$		$(i-1)\frac{2\pi}{3}$	Carré [1966]	N	U		+									
4	$\frac{\sqrt{3}(I_3-I_2)}{I_1-I_2-I_3+I_4}$	$\frac{-I_1+I_2-\sqrt{3}(I_2-I_3)+I_3-I_4}{I_1-I_2-\sqrt{3}(I_2-I_3)-I_3+I_4}$	$(i-1)\frac{2\pi}{3}$	Larkin and Oreb [1992]	N	U									A		
4	$\frac{I_1-3I_2+3I_3-I_4}{\sqrt{3}(I_1-I_2-I_3+I_4)}$	$\frac{I_1-3I_2+3I_3-I_4+\sqrt{3}(I_1-I_2-I_3+I_4)}{I_1-3I_2+3I_3-I_4-\sqrt{3}(I_1-I_2-I_3+I_4)}$	$(i-1)\frac{2\pi}{3}$	Surrel [1993]	U	U									B		
4	$\frac{I_4-I_2}{I_1-I_3}$	$\frac{2(I_3-I_2)}{I_1-I_2-I_3+I_4}$	$(i-1)\frac{\pi}{2}$	Bruning et al. [1974]	N	E	+				+	×	A				
4	$\frac{\pm\sqrt{(3(I_2-I_3)-I_1+I_4)(I_2-I_3+I_1-I_4)}}{I_1+I_4-I_2-I_3}$		$(i-1)\frac{\pi}{2}$	Schwider et al. [1993]	N	U		+					B				
4	$\frac{\pm\sqrt{(3(I_2-I_3)-I_1+I_4)(I_2-I_3+I_1-I_4)}}{I_1+I_3+I_4-3I_2}$		$(i-1)\frac{\pi}{3}$	Carré [1966]	N	U		+									
4	$\frac{\pm\sqrt{(3(I_2-I_3)-I_1+I_4)(I_2-I_3+I_1-I_4)}}{\frac{1}{5}(I_1-I_2-I_3+I_4)}$		$(i-1)\frac{\pi}{3}$	Kreis [2005]	N	U	+	+									
4	$\frac{\sqrt{3}(2I_1-I_2-I_3)}{I_1+I_4-I_2-I_3}$		$(i-1)\frac{\pi}{4}$	Carré [1966]	N	U	+	+									
4	$\frac{I_1+I_4-I_2-I_3}{I_1+I_4-I_2-I_3}$		$(i-1)\frac{\pi}{6}$	Carré [1966]	N	U	+	+									
5	$\frac{3\sqrt{3}(I_2-I_4)}{-2I_1-I_2+6I_3-I_4-2I_5}$		$(i-3)\frac{2\pi}{3}$	Hibino et al. [1997]	U	U											

continued on next page

steps	algorithm		phase-steps	reference	robustness							generic family				
N	$\tan\varphi =$	$\tan(\varphi-\frac{\pi}{4}) =$	$\Psi_i =$	author(s)	(non)uniform $\Delta\psi$	(un)even harmon.	quad. $\Delta\psi$ error	lin $\Delta\psi$ error	vibration	turbulence	non-lin detect	N-bucket	Ext Averaging	multi Averaging	$(N+1)$-bucket	WDFT	$(N+4)$-bucket
5	$\frac{\sqrt{3}(I_1-2I_2+2I_4-I_5)}{I_1+2I_2-6I_3+2I_4+I_5}$		$i\frac{2\pi}{3}$	Surrel [1996]	U											×	
5	$\frac{3I_1-6I_2+4I_3-2I_4+I_5}{2I_1-2I_2-4I_3+2I_4-2I_5}$	$\frac{I_1-8I_2+8I_3-4I_4+3I_5}{5I_1-4I_2-I_5}$	$(i-1)\frac{\pi}{2}$	Bi et al. [2004]	E						+				AB		
5	$\frac{I_1+I_5-2I_3}{2(I_4-I_2)}$	$\frac{2I_3+2I_4-I_1-2I_2-I_5}{I_1-2I_2-2I_3+2I_4+I_5}$	$(i-1)\frac{\pi}{2}$	de Groot [1995]	N	E			+	+	+		A				
5	$\frac{I_1+I_5-2I_3}{2(I_4-I_2)}$		$(i-3)\frac{\pi}{2}$	Schwider et al. [1983]		E					+						
5	$\frac{4I_1-4I_2-6I_3-I_4+4I_5}{I_1+2I_2-6I_3+2I_4+I_5}$	$\frac{3I_3+I_4-3I_2-I_5}{I_1-I_2-2I_3+3I_4}$	$(i-1)\frac{\pi}{2}$	Kreis [2005]							+						
5	$\frac{2(I_4-I_2)}{I_1+I_5-2I_3}$		$(i-1)\frac{\pi}{2}$	Schmit and Creath [1995]	E		+						B				
5	$(I_1+I_5-2I_3)\sqrt{1-\left(\frac{I_1-I_5}{2(I_2-I_4)}\right)^2}$		$(i-1)\frac{\pi}{2}$	Stoilov and Dragostinov [1997]	E						+						
5	$\frac{2(I_2-2I_3+I_4)}{-2I_2+2I_4-I_5}$	$\frac{I_1-4I_2+4I_3-I_5}{-I_1-4I_3+4I_4-I_5}$	$(i-1)\frac{\pi}{3}$	Surrel [1997]	U												
5	$\frac{\sqrt{3}(2I_1-3I_2-4I_3+5I_5)}{8I_1+3I_2+3I_3+4I_4-6I_4-5I_6}$		$(i-1)\frac{\pi}{3}$	Kreis [2005]		E											
6	$\frac{2I_1-I_2-2(I_3+I_4-2I_5)-I_6}{-3I_2+4I_6}$	$\frac{-I_1-3I_2+4I_3+4I_4-3I_5-I_6}{3I_1-5I_2+5I_5-3I_6}$	$(i-1)\frac{\pi}{2}$	Ishii and Onodera [1995]	U	E		+	+	+	+		A				
6	$\frac{I_1-4I_2+3I_5}{4(I_3-I_2+I_4-I_5)}$	$\frac{-3I_2+4I_3+4I_4+3I_5-I_6}{4(I_3-I_2+I_4-I_5)}$	$(i-1)\frac{\pi}{2}$	Schmit and Creath [1995]	E		+	+	+	+	+		B				
6	$\frac{I_1-5I_2+10I_3-3I_5-I_6}{4I_4+I_5-I_1-3I_2-I_6}$	$\frac{I_1-I_2+6I_3+6I_4-I_5-I_6}{I_1-I_2+2I_3+2I_4-I_5-I_6}$	$(i-1)\frac{\pi}{2}$	Schmit and Creath [1995]	N	E					+						
6	$\frac{-\sqrt{3}(I_2+I_3-I_5-I_6)}{-2(I_2+I_3-I_5-I_6)}$		$(i-1)\frac{\pi}{3}$	Surrel [1996]	EU		+				+						
6	$\frac{-\sqrt{3}(5I_1-6I_2-17I_3+17I_4+6I_5-5I_6)}{I_1-26I_2+25I_3+25I_4-26I_5+I_6}$		$(i-\frac{7}{2})\frac{\pi}{3}$	Bruning et al. [1974]	U	U	+	+			+	×					
7	$\frac{7(I_3-I_5)-(I_1-I_7)}{8I_4-4(I_2+I_6)}$		$(i-4)\frac{\pi}{2}$	Hibino et al. [1997]	U	E		+	+	+	+		A				
7	$\frac{8I_4-4(I_2+I_6)}{4I_4-2I_2-2I_6}$	$\frac{5I_5-5I_2+10I_4-10I_5-I_6+I_7}{I_1-I_2-10I_3+10I_4+5I_5-5I_6}$	$(i-1)\frac{\pi}{2}$	de Groot [1994]	N	E					+						
7	$\frac{I_1-3I_3+3I_5-I_7}{-2I_2+4I_4-2I_6}$		$(i-1)\frac{\pi}{2}$	Hibino et al. [1995]				+			+		B				
7	$\frac{2I_2-4I_4+2I_6}{-4I_2+2I_3+8I_4+2I_5-4I_6-I_7}$	$\frac{-I_2+4I_3+2I_6-I_7}{I_1-I_2-3I_3-3I_5-I_6+I_7}$	$i\frac{\pi}{2}$	Surrel [1996]	N	E		+			+						
7	$\frac{-I_1-4I_2+2I_3+8I_4+2I_5-4I_6-I_7}{I_1-2I_2-7I_3+7I_5+2I_6-I_7}$	$\frac{-3I_2-3I_3+4I_4+I_5-I_6-I_7}{I_1-2I_2-7I_3+7I_5+2I_6-I_7}$	$(i-1)\frac{\pi}{2}$	Zhang et al. [1999]	E			+			+						

continued on next page

steps	algorithm	phase-steps	reference	robustness							generic family				
N	$\tan\varphi =$	$\Psi_i =$	author(s)	(non)uniform $\Delta\Psi$	(un)even harmon.	quadr. $\Delta\Psi$ error	lin $\Delta\Psi$ error	vibration	turbulence	non-lin detect	N-bucket	Ext Averaging	multi Averaging	$(N+1)$-bucket	WDFT	$(N+4)$-bucket
7	$\dfrac{-I_2-I_3+I_5+I_6+\frac{2}{3}(I_1-I_7)}{\sqrt{3}(I_2-I_3-I_5+I_6)}$	$(i-1)\frac{\pi}{3}$	Hibino et al. [1997]	U	E	+	+									
7	$\dfrac{\sqrt{3}(-I_2-I_3-I_5+I_6)}{I_1+I_2-I_4-2I_4-I_5+I_6+I_7}$	$(i-1)\frac{\pi}{3}$	Larkin and Oreb [1992]	N	EU		+	+		+				A		
7	$\dfrac{I_1+I_2-I_3-2I_4-I_5+I_6+I_7}{I_1-3(I_2+I_3-I_5-I_6)-I_7}$	$(i-1)\frac{\pi}{3}$	Surrel [1993]	N	EU		+	+		+				B		
8	$\dfrac{\sqrt{3}(I_1+I_2-I_3-2I_4-I_5+I_6+I_7)}{2(I_1-I_8)-7(I_2-I_7)+17(I_4-I_5)}$	$(i-\frac{9}{2})\frac{2\pi}{3}$	de Groot [1995]	U	U											
8	$\dfrac{(I_1+I_8)+4(I_2+I_7)-15(I_3+I_6)+10(I_4+I_5)}{-2I_1+I_2+6I_3+3I_4-3I_5-6I_6-I_7+2I_8}$	$(i-\frac{9}{2})\frac{\pi}{2}$	Hibino et al. [1997]	N	E		+			+						
8	$\dfrac{4(I_1-I_8)-2(I_2-I_7)+14(I_3-I_6)+20(I_4-I_5)}{-3(I_1+I_8)+I_2+I_7-17(I_3+I_6)+19(I_4+I_5)}$	$(i-1)\frac{\pi}{2}$	Schmit and Creath [1996]	E		+	+			+						
8	$\dfrac{I_1-I_2-I_3+I_5-I_7}{-2I_2+4I_4-3I_6+I_8}$	$(i-1)\frac{\pi}{2}$	Schmit and Creath [1996]	E			+	+		+						
8	$\dfrac{I_1-3I_3+4I_5-2I_7}{-3I_2+7I_4-5I_6+I_8}$	$(i-1)\frac{\pi}{2}$	Schmit and Creath [1996]	E			+	+		+						
8	$\dfrac{I_1-5I_3+7I_5-3I_7}{-4I_2+11I_4-8I_6+I_8}$	$(i-1)\frac{\pi}{2}$	Schmit and Creath [1996]	E			+	+	+	+						
8	$\dfrac{-8I_1+11I_3-4I_7}{-5I_2+13I_4-11I_6+I_8}$	$(i-1)\frac{\pi}{2}$	Schmit and Creath [1996]	E			+	+		+						
8	$\dfrac{I_1-11I_3+15I_5-5I_7}{-(I_1+I_2+I_7+I_8)+2(I_4+I_5)}$	$(i-\frac{9}{2})\frac{2\pi}{3}$	de Groot [1995]	EU							×					
9	$\dfrac{-\frac{1}{2}(I_1-I_9)+(I_2-I_8)+7(I_3-I_7)+9(I_4-I_6)}{(I_1+I_9)+4(I_2+I_8)+4(I_3+I_7)-4(I_4+I_6)-10I_5}$	$(i-1)\frac{\pi}{2}$	Hibino et al. [1997]	N	E		+			+						
9	$\dfrac{-I_2-I_4+I_6+I_8-2I_5}{I_1+I_2-I_4-I_6+I_8+I_9-2I_5}$	$(i-1)\frac{\pi}{4}$	de Groot [1995]		EU	+	+			+						
9	$\dfrac{(\sqrt{2}+1)I_2+(3\sqrt{2}+5)I_4-(4\sqrt{2}+5)I_6-(\sqrt{2}+1)I_8}{-\frac{1}{2}I_1+(2\sqrt{2}+3)I_3-(3\sqrt{2}+5)I_5+(2\sqrt{2}+3)I_7-\frac{1}{2}I_9}$	$(i-1)\frac{\pi}{4}$	Estrada et al. [2009]			+	+			+						
10	$\dfrac{\sqrt{3}(I_3-I_1-3I_2-3I_4+I_5+6I_6+6I_7-3I_8-3I_9-I_{10})}{I_1-I_2-7I_3-11I_4-6I_5+6I_6+11I_7+7I_8+I_9-I_{10}}$	$(i-1)\frac{\pi}{3}$	Surrel [1996]	U	EU		+	+		+						
11	$\dfrac{-4(I_2+I_{10})+12(I_4+I_8)-16I_6}{(I_1-I_{11})-8(I_3-I_9)+15(I_5-I_7)}$	$(i-1)\frac{\pi}{2}$	de Groot [1995]	N	E	+	+	+		+						
11	$\dfrac{\sqrt{3}(I_2+4I_3+7I_4+6I_5-6I_7-7I_8-4I_9-I_{10})}{-2I_1-5I_2-6I_3-I_4+8I_5+12I_6+8I_7-I_8-6I_9-5I_{10}-2I_{11}}$	$(i-6)\frac{\pi}{3}$	Hibino et al. [1995]	N	EU		+			+						
11	$\dfrac{I_1-2I_2-6I_3-4I_4+5I_5+12I_6+5I_7-4I_8-6I_9-2I_{10}+I_{11}}{\sqrt{3}(I_1-2I_2-6I_3-4I_4+5I_5+12I_6+5I_7-4I_8-6I_9-2I_{10}+I_{11})}$	$i\frac{\pi}{3}$	Surrel [1996]	N	EU		+			+						×

Table 1. Phase-shifting/stepping algorithms overview, inspired by Schreiber and Bruning [2007] but corrected and extended.

Acknowledgments

Gathering and correcting the multitude of algorithms in this overview and getting insight in the literature was helped greatly through personal communications with Horst Schreiber, Kieran Larkin, Joanna Schmit and Adam Styk. The work of Schreiber and Bruning [2007] formed the inspiration for this paper. Furthermore, Irene Zanette, Simon Rutishauser and Christian David brought the Pfeiffer et al. [2008] method to our attention. We thank all of them for their assistance.

Financial support was provided by the Research Foundation - Flanders in the form of a post-doctoral fellowship. We also acknowledge the Fondation Belge de la Vocation.

References

JRP Angel and P Wizinowich. *ESO Conference on Very Large Telescopes and their Instrumentation, Vol.1*, chapter A method for phase shifting interferometry in the presence of vibration, pages 561–567. 1988.

H Bi, Y Zhang, KV Ling, and C Wen. Class of 4+1-phase algorithms with error compensation. *Appl Opt*, 42(21):4199–4207, 2004.

JH Bruning, DR Herriott, JE Gallagher, DP Rosenfeld, DA White, and DJ Brangaccio. Digital wavefront measuring interferometer for testing optical surfaces and lenses. *Appl Opt*, 13:2693, 1974.

JAN Buytaert and JJJ Dirckx. Design considerations in projection phase-shift moiré topography based on theoretical analysis of fringe formation. *J Opt Soc Am A*, 24(7): 2003–2013, 2007.

JAN Buytaert and JJJ Dirckx. Phase-shifting moiré topography using optical demodulation on liquid crystal matrices. *Opt Laser Eng*, 48(2):172–181, 2010.

JAN Buytaert and JJJ Dirckx. *Optical Methods for Solid Mechanics*, chapter Fringe projection profilometry. John Wiley & Sons, 2011.

P Carré. Installation et utilisation du comparateur photoélectrique et interférentiel du bureau international des poids et mesures. *Metrologia*, 2:13–23, 1966.

P de Groot. Derivation of algorithms for phase-shifting interferometry using the concept of a data-sampling window. *Appl Opt*, 34(22):4723–4730, 1995.

PJ de Groot. *Proc SPIE vol 2248*, chapter Long-wavelength laser diode interferometer for surface flatness measurements, pages 136–140. SPIE, 1994.

J Estrada, F Mendoza-Santoyo, and M de la Torre. *Fringe 2009: 6th International Workshop On Advanced Optical Metrology*, chapter Digital dynamic-fringe pattern frequency carrier, using wideband phase-shifting algorithms, pages 78–86. Springer, 2009.

P Hariharan and BF Oreb. Digital phase-shifting interferometry: a simple error-compensating phase calculation algorithm. *Appl Opt*, 24:2504–2505, 1987.

K Hibino, BF Oreb, DI Farrant, and KG Larkin. Phase shifting for nonsinusoidal waveforms with phase-shift errors. *J Opt Soc Am A*, 12(4):761–768, 1995.

K Hibino, DI Farrant, BK Ward, and BF Oreb. Phase-shifting algorithms for nonlinear and spatially nonuniform phase shifts. *J Opt Soc Am A*, 14(25):918–930, 1997.

Y. Ishii and R. Onodera. Phase-extraction algorithm in laser-diode phase-shifting interferometry. *Opt Lett*, 20(18):1883–1885, Sep 1995.

W Juptner, TM Kreis, and H Kreitlow. *Proc SPIE vol 398*, chapter Automatic evaluation of holographic interferograms by reference beam phase shifting, pages 22–29. SPIE, 1983.

D Kerr, F Mendoza Santoyo, and JR Tyrer. Extraction of phase data from electronic speckle pattern interferometric fringes using a single-phase-step method: a novel approach. *J Opt Soc Am A*, 7(5):820–826, 1990.

T Kreis. *Handbook of holographic interferometry: optical and digital methods*. John Wiley & Sons, 2005.

KG Larkin and BF Oreb. Design and assessment of symmetrical phase-shifting algorithms. *J Opt Soc Am A*, 9(10):1740–1748, 1992.

W Osten. *Optical methods in experimental solid mechanics*, chapter Digital processing and evaluation of fringe patterns in optical metrology and non-destructive testing, pages 308–363. Springer-Verlag, 2000.

M Owner-Petersen. Digital speckle pattern shearing interferometry: limitations and prospects. *Appl Opt*, 30:2730–2738, 1991.

F Pfeiffer, M Bech, O Bunk, P Kraft, EF Eikenberry, CH Bronnimann, C Grunzweig, and C David. Hard-x-ray dark-field imaging using a grating interferometer. *Nature Materials*, 7:134–137, 2008.

J Schmit and K Creath. Extended averaging technique for derivation of error-compensating algorithms in phase-shifting interferometry. *Appl Opt*, 34:3610–3619, 1995.

J Schmit and K Creath. Window function influence on phase error in phase-shifting algorithms. *Appl Opt*, 35:5642–5649, 1996.

H Schreiber and JH Bruning. *Optical shop testing*, chapter Phase-shifting interferometry, pages 547–655. John Wiley & Sons, 3rd edition, 2007.

J Schwider, R Burrow, KE Elssner, J Grzanna, R Spolaczyk, and K Merkel. Digital wavefront measuring interferometry: some systematic error sources. *Appl Opt*, 22:3421–3432, 1983.

J Schwider, O Falkenstoerfer, H Schreiber, A Zoeller, and N Streibl. New compensating four-phase algorithm for phase-shift interferometry. *Opt Eng*, 32(8):1883–1885, 1993.

G Stoilov and T Dragostinov. Phase-stepping interferometry: Five-frame algorithm with an arbitrary step. *Opt Laser Eng*, 28:61–69, 1997.

Y Surrel. Phase stepping: a new self-calibrating algorithm. *Appl Opt*, 32(19):3598–3600, 1993.

Y Surrel. Design of algorithms for phase measurements by the use of phase stepping. *Appl Opt*, 35(1):51–60, 1996.

Y Surrel. Design of phase-detection algorithms insensitive to bias modulation. *Appl Opt*, 36(4):805–807, 1997.

M Takeda, H Ina, and S Koboyashi. Fourier transform profilometry for the automatic measurement of 3-d object shapes. *Appl Opt*, 22(24):3977–3982, 1982.

JC Wyant, CL Koliopoulos, B Bhushan, and OE George. An optical profilometer for surface characterization of magnetic media. *ASLE trans*, 27(2):101–113, 1984.

H Zhang, MJ Lalor, and DR Burton. Robust, accurate seven-sample phase-shifting algorithm insensitive to nonlinear phase-shift error and second-harmonic distortion: a comparative study. *Opt Eng*, 38:1524, 1999.

In: Interferometry Principles and Applications
Editor: Mark E. Russo

ISBN 978-1-61209-347-5
©2012 Nova Science Publishers, Inc.

Chapter 15

GENERALIZED CARRÉ MULTI-STEP PHASE-SHIFTING ALGORITHMS

Jiří Novák, Pavel Novák and Antonín Mikš
Czech Technical University in Prague, Department of Physics, Prague, Czech Republic

ABSTRACT

Phase shifting is a well-known technique which is used in many areas of science and engineering. Phase-shifting algorithms are used extensively in optical interferometry. This chapter describes a group of multi-step phase-shifting algorithms for phase evaluation of interferometric measurements. Phase shifting algorithms are introduced and analyzed, with a constant but arbitrary phase shift between captured frames of the irradiance of the interference field. The phase-shifting algorithms are similarly derived as so called Carré algorithm, which was firstly described in 1966. The phase evaluation process is not dependent on linear phase shift errors using these algorithms. An advantage of the described algorithms over common phase-shifting algorithms is their ability to determine the phase shift value at each point of the detection plane. Moreover, a complex error analysis of proposed algorithms is performed and the algorithms are compared to several common error compensating phase stepping algorithms.

Keywords: interferometry, phase-shifting algorithms, Carré algorithm, error analysis

1. INTRODUCTION

Optical interferomtery [1,2] is a well-established sort of measuring methods which are successfully applied in many fields of science, engineering, and biomedicine. Techniques for automatic evaluation of interferometric measurements [2] were deeply studied in recent years, because of the rapid development of digital interferometric techniques and their applications in optical measuring methods in industrial practice. Depending on the specific character of the measurement problem to be solved, various phase evaluation methods can be used [1,2]. The trend is to automatize the measurement process and analysis as much as possible using computers in order to obtain more robust measurement systems and higher measurement accuracy in specific measurement condition.

Phase shifting interferometry (PSI) is a well adapted interferometric measurement technique [1-27], which is widely used in the field of optics. Its principle is based on the phase modulation of the recorded measurements of the intensity of the interference field. The phase, which is encoded in the intensity variations of recorded interferograms, is retrieved point-by-point from several recorded intensity frames by special mathematical algorithms. The phase shifting technique was firstly applied for analysis of interferometric measurements in the field of classical two beam interferometry [1,2], and nowadays it is the most widely used technique for evaluating of interference fields in many areas of science and engineering. Phase-shifting technique became a favorite method in recent years because of its possibility to obtain full-field phase data with a high accuracy and in real time. The first reference to this kind of technique dates back to 1966, when P.Carré published his paper [9]. The rapid development of PSI techniques began in the early 1970's [4-8] with main applications to interferometric optical testing and wavefront sensing. Later the main part of research was devoted to development of the theory around PSI algorithms less sensitive to various error sources and thus more robust and high accuracy interferometric measurements.

Since publishing the first PSI algorithm by Carré [9] a number of different types of phase shifting algorithms have been developed [1-3,10-27]. They are based on phase modulation of the detected interference pattern. While many different approaches to phase evaluation exist, various types of evaluation algorithms differs in the number of detected phase modulated interference patterns, values of phase shifts between interferograms, and the susceptibility of different algorithms to errors in the phase shift, noise, vibrations, and nonlinearities of irradiance recorded by the photodetector. In recent years many papers have been reported on PSI algorithms design and compensation of possible errors that may occur during practical measurements [1-3,14-44]. While only three irradiance measurements are required in principle to make an unambiguous and very accurate phase measurement, frequently more than a minimum number of measurements are made to reduce the errors in practice. Various PSI algorithms have been proposed and their properties were analyzed with respect to both systematic and random errors using different approaches. A nice overview of PSI algorithms and their susceptibility to different types of errors is given in Refs. [1,2]. While there are many different approaches to the phase evaluation, the differences between the various algorithms relate to the number of interferograms, the phase shift between these interferograms, and the susceptibility of the algorithm to the systematic and random errors.

In practice the most important type of errors that may affect the resulting accuracy of the described phase measurement technique are errors due to incorrect phase shifts between data frames. The primary sources of these generally nonlinear errors are vibrations and incorrect phase shifter calibration. Almost in all phase shifting evaluation algorithms it is assumed that the phase shift between individual intensity measurements is known in advance. If the actual phase shift differs from the considered value, errors are introduced into the phase evaluation process.

However, there exist PSI algorithms that assume the phase shift to be of an unknown but constant value. The first published PSI algorithm was the so called Carré algorithm, which needs to take four interferograms with a constant phase shift. Errors of this sort could result from an incorrect size of phase steps (i.e. steps are the same but incorrect) and nonuniformities of the phase shift across the pupil (linear phase shift in a converging or diverging beam). Several phase-shifting algorithms have been developed to minimize the effects of these phase shift variations. In the following section, we will review the fundamentals of PSI and show various PSI algorithms that are insensitive to linear phase-shift errors and generalize the so called Carré algorithm [9].

2. PHASE-SHIFTING TECHNIQUE

The phase-shifting technique for analysis of interferometric measurements is based on an evaluation of the phase of the interference signal using phase modulation of this interference signal. The intensity distribution I of the detected phase modulated interference signal at some point (x,y) of the detector plane is given by

$$I_i(x, y) = A(x, y) + B(x, y) \cos[\Delta\varphi(x, y) + \psi_i] \quad i = 1,\ldots, N, \qquad (1)$$

where $A(x, y)$ is the function of the background irradiance, $B(x, y)$ is the function of the amplitude modulation, $\Delta\varphi(x, y)$ is the phase difference of the wave fields that interfere, and ψ_i is the phase shift of the i-th irradiance measurement. To determine phase values $\Delta\varphi$ it is necessary to solve nonlinear equations (1) for each point (x,y) of the detected interference pattern. If the phase shift values ψ_i are properly chosen in advance, there remain only three unknowns A, B and $\Delta\varphi$ that must be determined from the recorded intensity values. We must carry out at least $N = 3$ intensity measurements with known phase shift values ψ_i to calculate phase $\Delta\varphi$. The phase shift can be implemented during measurement by various means [1-3], e.g. by a very precise piezoelectric transducer with a little mirror that is shifted in the path of the reference beam. It is necessary to note that the phase values calculated with any phase shifting algorithm lie in the range $[-\pi,\pi]$. The discontinuous distribution of the calculated phase, i.e. the wrapped phase, is caused by the nonlinear reconstruction function in phase shifting algorithms. The phase values must be correctly unwrapped using suitable mathematical techniques [45-50].

3. GENERALIZED CARRE MULTI-STEP PHASE-SHIFTING ALGORITHMS

Almost all the existing phase-shifting algorithms are based on the assumption that the phase-shift at all pixels of the intensity frame is equal and known. However, it may be very difficult to achieve this case in practice. In case that phase shift values ψ_i between particular intensity frames are not determined in advance, but the phase shift is considered constant between the intensity frames, it is necessary to perform at least $N = 4$ phase shifted measurements in order to determine phase values $\Delta\varphi$, because the interference equation (1) consists of four unknowns A, B, ψ and $\Delta\varphi$. This method, as we will see later, does not depend on miscalibration of the phase shifter, i.e. these algorithms are insensitive to the linear phase shift errors. Such phase shifting algorithms derived for five, six and seven frames are described and analyzed in this section.

We will now describe a group of multi-step phase shifting algorithms that are insensitive to linear phase shift errors. We assume constant but unknown phase shift values ψ between recorded images of the irradiance of the observed interference field. From the above-mentioned assumption we can derive a group of phase calculation algorithms. The very well known Carré algorithm is a phase shifting algorithm of this type for four frames ($N = 4$) [5]. Generally, we can write for the intensity at each point of the recorded interferogram

$$I_k = A + B \cos\left[\Delta\varphi + \frac{(2k - N - 1)}{2}\psi\right], \quad k = 1, \ldots, N, \qquad (2)$$

where ψ is the constant phase shift between captured frames, A is the mean irradiance, and B is the modulation of the interference signal. For a simpler description of the phase evaluation algorithms we make the following denotation

$$a_{jk} = I_j - I_k, \quad b_{jk} = I_j + I_k. \qquad (3)$$

An advantage of the described type of phase calculation algorithms is the possibility to calculate the phase shift at all pixels (x,y) of the detector plane in order to control the distribution of the phase shift over the interferogram. The above-mentioned phase evaluation algorithms are not affected by miscalibration of the phase shift device and can be used if the phase shift is nonuniformly distributed over the area of the photodetector.

The Carré algorithm is a variation of the four-step algorithm, but the reference phase shift between measurements is treated as an unknown and solved for in the analysis. We have four equations (2) for the intensity which contain four unknowns A, B, ψ and $\Delta\varphi$, where the reference phase shift ψ is equal to 2α. We can write for each point of the recorded interferograms ($N = 4$)

$$I_1 = A + B \cos[\Delta\varphi - 3\alpha], \qquad (4a)$$

$$I_2 = A + B \cos[\Delta\varphi - \alpha], \qquad (4b)$$

$$I_3 = A + B \cos[\Delta\varphi + \alpha], \qquad (4c)$$

$$I_4 = A + B \cos[\Delta\varphi + 3\alpha]. \qquad (4d)$$

The solution for the reference phase shift can be found by expanding previous four equations and applying the trigonometric identity for sine or cosine.

We obtain for the constant phase shift between captured intensity frames

$$\cos\psi = \cos 2\alpha = \frac{(I_1 - I_4) - (I_2 - I_3)}{2(I_2 - I_3)} = \frac{a_{14} - a_{23}}{2a_{23}}. \tag{5}$$

This equation can be solved at each measurement point and the reference phase shift ψ can also be determined at each point. The previous equation allows spatial variations of the phase shift to be identified. The solution for the phase at each measurement point can be calculated from the following formula

$$\mathrm{tg}\Delta\varphi = \mathrm{tg}\alpha \frac{(I_1 - I_4) + (I_2 - I_3)}{(I_2 + I_3) - (I_1 + I_4)} = \mathrm{tg}\alpha \frac{a_{14} + a_{23}}{b_{23} - b_{14}}. \tag{6}$$

If we use a well-known trigonometric identity

$$\mathrm{tg}\alpha = \sqrt{\frac{1 - \cos 2\alpha}{1 + \cos 2\alpha}}, \tag{7}$$

then we obtain the resulting algorithm for the phase

$$\mathrm{tg}\,\Delta\varphi = \frac{\sqrt{(I_1 - I_4 + I_2 - I_3)(-I_1 + I_4 + 3I_2 - 3I_3)}}{(I_2 + I_3 - I_1 - I_4)} = \frac{\sqrt{(a_{14} + a_{23})(3a_{23} - a_{14})}}{b_{23} - b_{14}}. \tag{8}$$

The recorded signal modulation B can be expressed from Eq.(4) as

$$B = \sqrt{\frac{(I_2 - I_3)^3 \{(I_1 - I_4 + I_2 - I_3)^2 - 2(I_1 - I_4)^2 - 2(I_2 - I_3)^2 + (I_2 + I_3 - I_1 - I_4)^2\}}{\{3(I_2 - I_3) - (I_1 - I_4)\}\{4(I_2 - I_3)^2 - (I_1 - I_4 - I_2 + I_3)^2\}}} =$$

$$= \sqrt{\frac{(a_{23})^3 \{(a_{14} + a_{23})^2 - 2(a_{14})^2 - 2(a_{23})^2 + (b_{23} - b_{14})^2\}}{\{3(a_{23}) - (a_{14})\}\{4(a_{23})^2 - (a_{14} - a_{23})^2\}}}. \tag{9}$$

The main advantage of this type of PSI algorithm is that it compensates for errors of phase shift as well as for spatial variations of the phase shift. However, the Carré algorithm requires that the phase shift increments at a given point be equal. There is a problem with the conversion of calculated phase values to the phase modulo 2π, because the square root in the numerator (8) produces the absolute value of $\sin\Delta\varphi$. It is evident that we cannot obtain a continuous phase distribution from the described phase calculation algorithms. Because of the character of the evaluation algorithms we calculate only phase values $\Delta\varphi_W^* \in [-\pi/2, \pi/2]$. For unambiguous determination of the wrapped phase values $\Delta\varphi_W$ it is necessary to find for all the algorithms appropriate expressions, which are proportional to functions $\sin\Delta\varphi$ and $\cos\Delta\varphi$. Such expressions are: $\cos\Delta\varphi \sim b_{23} - b_{14}$, $\sin\Delta\varphi \sim a_{23}$. Then according to the sign of these expressions we can determine the wrapped phase values $\Delta\varphi_W \in [-\pi,\pi]$. The discontinuities in phase data can be correctly unwrapped using some of phase unwrapping procedures [45-50]. The formula (8) can be used even in the presence of fairly large variations in the size of the phase step. The Carré algorithm is used as a convenient way to calibrate the phase shift with the solution from Eq.(5). An advantage of the described type of phase calculation algorithms with unknown value of the phase shift is the possibility to calculate the phase shifter at each measurement point (x,y) of the detector plane in order to control the distribution of the phase shift over the interferogram. The above-mentioned phase evaluation algorithm is not affected by miscalibration of the phase shift device and it can be used if the phase shift is nonuniformly distributed over the area of the photodetector.

In the following section we describe multi-frame PSI algorithms which are insensitve to linear phase shift errors. The proposed algorithms are a generalization of well known Carré algorithm technique to a larger number of frames. In the following sections we describe a group of multistep ($N = \{5,6,7\}$) PSI algorithms that can be derived similarly as Carré algorithm. On the basis of Eqs.(1) we derived a large number of algorithms. These algorithms differ in their mathematical properties, especially in their sensitivity to different factors (errors of the phase shifter, noise, external vibration, etc.). For simplification of the text we describe only a principle of derivation of individual multistep algorithms. The complex analysis in Section 4 of PSI algorithms was performed on all combinations of possible algorithms that can be derived using the described formulas. The results are then shown only for the best algorithms, which are the least sensitive to mentioned errors.

Table 1. Expressions for calculation of phase values $\Delta\varphi$ ($N = 5$)

$\tan \Delta\varphi = \sin\alpha \dfrac{2a_{24}}{2I_3 - b_{15}}$	$\tan \Delta\varphi = \dfrac{\cos\alpha + 1}{\sin\alpha} \dfrac{a_{15} - 2a_{24}}{b_{15} - 2I_3}$
$\tan \Delta\varphi = \tan\alpha \dfrac{2a_{15}}{2I_3 - b_{15}}$	$\tan \Delta\varphi = \dfrac{1 - \cos\alpha}{\sin\alpha} \dfrac{a_{24}}{2I_3 - b_{24}}$
$\tan \Delta\varphi = \dfrac{1}{\sin\alpha} \dfrac{a_{15} - 2a_{24}}{2(b_{24} - 2I_3)}$	$\tan \Delta\varphi = \dfrac{1 - \cos\alpha}{\sin 2\alpha} \dfrac{a_{15}}{2I_3 - b_{24}}$
$\tan \Delta\varphi = \dfrac{1}{\tan\alpha} \dfrac{a_{15} - 2a_{24}}{b_{15} - 2b_{24} + 2I_3}$	

3.1. Five-Frame Phase Evaluation Algorithms

Considering $N = 5$ phase shifted intensity measurements, we obtain using Eq.(2) for the constant phase shift between captured intensity frames $\psi = \alpha$ by simple algebra the following formulas

$$\cos\alpha = \frac{a_{15}}{2a_{24}}, \quad \cos\alpha = \frac{b_{15} - 2b_{24} + 2I_3}{2(b_{24} - 2I_3)},$$

and possible mathematical expressions for phase calculation are given by the formulas shown in table 1.

The following five-step algorithms (see table 2) for phase calculation using the phase shifting technique can be derived from the preceding equations (table 1). The phase calculation algorithms were denoted A1-A8.

Table 2. Five-frame phase shifting algorithms (A1-A8)

Algorithm	Expressions for calculation of phase values $\Delta\varphi$
A1	$\tan \Delta\varphi = \dfrac{\sqrt{4a_{24}^2 - a_{15}^2}}{2I_3 - b_{15}}$
A2	$\tan \Delta\varphi = \dfrac{a_{24}\sqrt{4(b_{24} - 2I_3)^2 - (b_{15} - 2b_{24} + 2I_3)^2}}{(b_{24} - 2I_3)(2I_3 - b_{15})}$
A3	$\tan \Delta\varphi = \dfrac{a_{15}\sqrt{4(b_{24} - 2I_3)^2 - (b_{15} - 2b_{24} + 2I_3)^2}}{(b_{15} - 2b_{24} + 2I_3)(2I_3 - b_{15})}$
A4	$\tan \Delta\varphi = \dfrac{a_{24}(a_{15} - 2a_{24})}{\sqrt{4a_{24}^2 - a_{15}^2}\,(b_{24} - 2I_3)}$
A5	$\tan \Delta\varphi = \dfrac{a_{15} - 2a_{24}}{\sqrt{4(b_{24} - 2I_3)^2 - (b_{15} - 2b_{24} + 2I_3)^2}}$
A6	$\tan \Delta\varphi = \dfrac{a_{15}(a_{15} - 2a_{24})}{\sqrt{4a_{24}^2 - a_{15}^2}\,(b_{15} - 2b_{24} + 2I_3)}$
A7	$\tan \Delta\varphi = \dfrac{(4b_{24} - b_{15} - 6I_3)a_{24}}{(2I_3 - b_{24})\sqrt{4(b_{24} - 2I_3)^2 - (b_{15} - 2b_{24} + 2I_3)^2}}$
A8	$\tan \Delta\varphi = \dfrac{(6I_3 - 4b_{24} + b_{15})a_{15}}{(b_{15} - 2b_{24} + 2I_3)\sqrt{4(b_{24} - 2I_3)^2 - (b_{15} - 2b_{24} + 2I_3)^2}}$

One can see that the well-known Schwider-Hariharan algorithm [22,29] can be derived from the algorithm A1 for the value $\psi = \pi/2$. The recorded signal modulation B can be expressed as

$$B = \frac{a_{24}\sqrt{4a_{24}^2 - a_{15}^2 + (2I_3 - b_{15})^2}}{4a_{24}^2 - a_{15}^2}.$$

The mean intensity A of the interference signal can be calculated from the formula

$$A = I_3 - \frac{b_{24} - 2I_3}{2(\cos\alpha - 1)} = \frac{2I_3 \cos\alpha - b_{24}}{2(\cos\alpha - 1)}.$$

For unambiguous determination of the wrapped phase values $\Delta\varphi$ modulo 2π it is necessary to find for all the algorithms appropriate expressions, which are proportional to functions $\sin\Delta\varphi$ and $\cos\Delta\varphi$. Such expressions for five-frame algorithms are: $\cos\Delta\varphi \sim 2I_3 - b_{24}$, $\sin\Delta\varphi \sim a_{24}$. According to the sign of mentioned expressions we can determine the wrapped phase values $\Delta\varphi \in [-\pi, \pi]$.

3.2. Six-Frame Phase Evaluation Algorithms

Considering $N = 6$ phase shifted intensity measurements, we obtain for the constant phase shift between captured intensity frames $\psi = 2\alpha$ using Eq.(2) the following formulas

$$\cos 2\alpha = \frac{a_{16} + a_{25}}{2(a_{25} + a_{34})}, \cos 2\alpha = \frac{a_{34} + a_{16}}{2a_{25}},$$

$$\cos 2\alpha = \frac{b_{16} - b_{25}}{2(b_{25} - b_{34})}, \cos 2\alpha = \frac{a_{25} - a_{34}}{2a_{34}},$$

and several of possible expressions for calculation of phase values $\Delta\varphi$ are given by the formulas shown in table 3.

Table 3. Expressions for calculation of phase values $\Delta\varphi$ ($N = 6$)

$\tan\Delta\varphi = \frac{\sin 2\alpha}{\cos 2\alpha + 1} \frac{a_{34} + 2a_{25} + a_{16}}{b_{34} - b_{16}}$	$\tan\Delta\varphi = \tan\alpha \frac{a_{25} + a_{34}}{b_{34} - b_{25}}$
$\tan\Delta\varphi = \tan\alpha \frac{a_{34} - a_{16}}{b_{16} + b_{34} - 2b_{25}}$	$\tan\Delta\varphi = \sin 2\alpha \frac{2a_{25}}{b_{34} - b_{16}}$
$\tan\Delta\varphi = \tan 2\alpha \frac{a_{25} - a_{34}}{b_{34} - b_{25}}$	$\tan\Delta\varphi = \tan\alpha \frac{a_{25} + a_{16}}{b_{25} - b_{16}}$

Generalized Carré Multi-Step Phase-Shifting Algorithms

For modulation B of the recorded interference signal using six-frame phase shifting algorithms we obtain by solving Eq.(2)

$$B = a_{34} \sqrt{\frac{a_{25}[4a_{25}^2 - (a_{16} + a_{34})^2 + (b_{34} - b_{16})^2]}{[4a_{25}^2 - (a_{16} + a_{34})^2](2a_{25} - a_{16} - a_{34})}}.$$

Table 4a. Six-frame phase shifting algorithms (B1-B10)

Algorithm	Expressions for calculation of phase values $\Delta\varphi$
B1	$\tan \Delta\varphi = \dfrac{\sqrt{4(a_{25} + a_{34})^2 - (a_{16} + a_{25})^2}}{a_{16} + 3a_{25} + 2a_{34}} \dfrac{a_{16} + 2a_{25} + a_{34}}{b_{34} - b_{16}}$
B2	$\tan \Delta\varphi = \dfrac{\sqrt{4a_{25}^2 - (a_{16} + a_{34})^2}}{b_{34} - b_{16}}$
B3	$\tan \Delta\varphi = \dfrac{\sqrt{4(b_{25} - b_{34})^2 - (b_{16} - b_{25})^2}}{b_{16} + b_{25} - 2b_{34}} \dfrac{a_{16} + 2a_{25} + a_{34}}{b_{34} - b_{16}}$
B4	$\tan \Delta\varphi = \dfrac{\sqrt{4a_{34}^2 - (a_{25} - a_{34})^2}}{a_{25} + a_{34}} \dfrac{a_{16} + 2a_{25} + a_{34}}{b_{34} - b_{16}}$
B5	$\tan \Delta\varphi = \sqrt{\dfrac{a_{25} + 2a_{34} - a_{16}}{a_{16} + 3a_{25} + 2a_{34}}} \dfrac{a_{34} - a_{16}}{b_{16} + b_{34} - 2b_{25}}$
B6	$\tan \Delta\varphi = \sqrt{\dfrac{2a_{25} - a_{34} - a_{16}}{a_{16} + 2a_{25} + a_{34}}} \dfrac{a_{34} - a_{16}}{b_{16} + b_{34} - 2b_{25}}$
B7	$\tan \Delta\varphi = \sqrt{\dfrac{3b_{25} - 2b_{34} - b_{16}}{b_{16} + ba_{25} - 2a_{34}}} \dfrac{a_{34} - a_{16}}{b_{16} + b_{34} - 2b_{25}}$
B8	$\tan \Delta\varphi = \sqrt{\dfrac{3a_{34} - a_{25}}{a_{25} + a_{34}}} \dfrac{a_{34} - a_{16}}{b_{16} + b_{34} - 2b_{25}}$
B9	$\tan \Delta\varphi = \dfrac{\sqrt{4(a_{25} + a_{34})^2 - (a_{16} + a_{25})^2}}{a_{16} + a_{25}} \dfrac{a_{25} - a_{34}}{b_{34} - b_{25}}$
B10	$\tan \Delta\varphi = \dfrac{\sqrt{4a_{25}^2 - (a_{16} + a_{34})^2}}{a_{16} + a_{34}} \dfrac{a_{25} - a_{34}}{b_{34} - b_{25}}$

The mean intensity A of the interference signal can be calculated from

$$A = \frac{b_{25}}{2} - \frac{b_{34} - b_{16}}{4\sin 2\alpha \tan 3\alpha} = \frac{b_{25}}{2} - \frac{b_{34} - b_{16}}{4\sin 2\alpha \tan \alpha} \frac{4\cos^2 \alpha - 3}{3 - 4\sin^2 \alpha}.$$

Tables 4a and 4b show selected six-frame algorithms for phase calculation that may be derived from the preceding equations (table 3). The phase calculation algorithms were denoted B1-B17. For unambiguous determination of the wrapped phase values $\Delta\varphi$ modulo 2π we may use expressions, which are proportional to functions $\sin \Delta\varphi$ and $\cos \Delta\varphi$. Such expressions for six-frame algorithms are: $\cos \Delta\varphi \sim b_{34} - b_{25}$, $\sin \Delta\varphi \sim a_{34}$.

Table 4b. Six-frame phase shifting algorithms (B11-B17)

Algorithm	Expressions for calculation of phase values $\Delta\varphi$
B11	$\tan \Delta\varphi = \dfrac{\sqrt{4(b_{25} - b_{34})^2 - (b - b_{25})^2}}{b_{16} - b_{25}} \dfrac{a_{25} - a_{34}}{b_{34} - b_{25}}$
B12	$\tan \Delta\varphi = \sqrt{\dfrac{a_{25} + 2a_{34} - a_{16}}{3a_{25} + 2a_{34} + a_{16}}} \dfrac{a_{25} + a_{34}}{b_{34} - b_{25}}$
B13	$\tan \Delta\varphi = \sqrt{\dfrac{2a_{25} - a_{34} - a_{16}}{2a_{25} + a_{34} + a_{16}}} \dfrac{a_{25} + a_{34}}{b_{34} - b_{25}}$
B14	$\tan \Delta\varphi = \sqrt{\dfrac{3b_{25} - 2b_{34} - b_{16}}{b_{25} - 2b_{34} + b_{16}}} \dfrac{a_{25} + a_{34}}{b_{34} - b_{25}}$
B15	$\tan \Delta\varphi = \dfrac{\sqrt{4(a_{25} + a_{34})^2 - (a_{16} + a_{25})^2}}{a_{34} + a_{25}} \dfrac{a_{25}}{b_{34} - b_{16}}$
B16	$\tan \Delta\varphi = \dfrac{\sqrt{4(b_{25} - b_{34})^2 - (b_{16} - a_{25})^2}}{b_{25} - b_{34}} \dfrac{a_{25}}{b_{34} - b_{16}}$
B17	$\tan \Delta\varphi = \dfrac{\sqrt{4a_{34}^2 - (a_{25} - a_{34})^2}}{a_{34}} \dfrac{a_{25}}{b_{34} - b_{16}}$

3.3. Seven-Frame Phase Evaluation Algorithms

Considering $N = 7$ phase steps, we obtain for the constant phase shift between captured intensity frames $\psi = \alpha$ using Eq.(2) the following formulas

$$\cos 2\alpha = \frac{a_{17} - a_{35}}{2a_{35}}, \quad \cos \alpha = \frac{a_{17} + a_{35}}{2a_{26}}, \quad \cos \alpha = \frac{b_{35} - b_{17}}{2(b_{26} - 2I_4)},$$

$$\cos 2\alpha = \frac{b_{17} + b_{35} - 2b_{26}}{2(b_{35} - 2I_4)}, \quad \cos \alpha = \frac{a_{26}}{2a_{35}},$$

and selected expressions for calculation of phase values $\Delta\varphi$ are given by the formulas shown in table 5.

Table 5. Expressions for calculation of phase values $\Delta\varphi$ ($N = 7$)

$\tan \Delta\varphi = \tan\alpha \dfrac{2a_{26} - a_{17} - a_{35}}{b_{17} - b_{35} - 2(b_{26} - 2I_4)}$	$\tan \Delta\varphi = \dfrac{1}{\sin\alpha} \dfrac{a_{17} - 3a_{35}}{2(b_{26} - 2I_4)}$
$\tan \Delta\varphi = \dfrac{\sin\alpha}{\cos\alpha + 1} \dfrac{2a_{26} + a_{17} + a_{35}}{b_{35} - b_{17}}$	$\tan \Delta\varphi = \sin\alpha \dfrac{2a_{35}}{2I_4 - b_{26}}$
$\tan \Delta\varphi = \dfrac{1}{\sin 2\alpha} \dfrac{a_{17} + a_{35} - 2a_{26}}{2(b_{35} - 2I_4)}$	$\tan \Delta\varphi = \tan\alpha \dfrac{a_{17} + a_{35}}{b_{35} - b_{17}}$

The signal modulation B and the mean intensity A of the interference signal in case of seven-frame phase shifting algorithms can be expressed as

$$B = a_{26}^2 \frac{\sqrt{4a_{35}^2 - a_{26}^2 + (2I_4 - b_{26})^2}}{4a_{26}^2 - (a_{17} + a_{35})^2}, \quad A = I_4 - \frac{b_{35} - 2I_4}{2(\cos\alpha - 1)} = \frac{2I_4 \cos\alpha - b_{35}}{2(\cos\alpha - 1)}.$$

Tables 6a and 6b show selected seven-frame algorithms for phase calculation that may be derived from the described equations (table 5). The phase calculation algorithms were denoted C1-C20. For unambiguous determination of the wrapped phase values $\Delta\varphi$ modulo 2π we may use expressions, which are proportional to functions $\sin \Delta\varphi$ and $\cos \Delta\varphi$. Such expressions for seven-frame algorithms are: $\cos \Delta\varphi \sim 2I_4 - b_{34}$, $\sin \Delta\varphi \sim a_{35}$.

Table 6a. Seven-frame phase shifting algorithms (C1-C9)

Algorithm	Expressions for calculation of phase values $\Delta\varphi$
C1	$\tan \Delta\varphi = \sqrt{\dfrac{3a_{35} - a_{17}}{a_{17} + a_{35}}} \dfrac{2a_{26} - a_{35} - a_{17}}{b_{17} - b_{35} - 2(b_{26} - 2I_4)}$
C2	$\tan \Delta\varphi = \dfrac{\sqrt{4a_{26}^2 - (a_{17} + a_{35})^2}}{a_{17} + a_{35}} \dfrac{2a_{26} - a_{35} - a_{17}}{b_{17} - b_{35} - 2(b_{26} - 2I_4)}$
C3	$\tan \Delta\varphi = \sqrt{\dfrac{2(b_{26} - 2I_4) + (b_{35} - b_{17})}{2(b_{26} - 2I_4) - (b_{35} - b_{17})}} \dfrac{2a_{26} - a_{35} - a_{17}}{b_{17} - b_{35}}$
C4	$\tan \Delta\varphi = \sqrt{\dfrac{b_{35} - b_{17} + 2(b_{26} - 2I_4)}{3b_{35} + b_{17} - 2b_{26} - 4I_4}} \dfrac{2a_{26} - a_{35} - a_{17}}{b_{17} - b_{35} - 2(b_{26} - 2I_4)}$
C5	$\tan \Delta\varphi = \dfrac{\sqrt{4a_{35}^2 - a_{26}^2}}{a_{26}} \dfrac{2a_{26} - a_{35} - a_{17}}{b_{17} - b_{35} - 2(b_{26} - 2I_4)}$
C6	$\tan \Delta\varphi = \dfrac{\sqrt{3a_{35} - a_{17}}}{2\sqrt{a_{35}} + \sqrt{a_{17} + a_{35}}} \dfrac{2a_{26} + a_{35} + a_{17}}{b_{35} - b_{17}}$
C7	$\tan \Delta\varphi = \dfrac{\sqrt{4a_{26}^2 - (a_{17} + a_{35})^2}}{b_{35} - b_{17}}$
C8	$\tan \Delta\varphi = \dfrac{\sqrt{4(b_{26} - 2I_4)^2 - (b_{35} - b_{17})^2}}{2(b_{26} - 2I_4) + (b_{35} - b_{17})} \dfrac{2a_{26} + a_{35} + a_{17}}{b_{35} - b_{17}}$
C9	$\tan \Delta\varphi = \dfrac{\sqrt{b_{35} - b_{17} + 2(b_{26} - 2I_4)}}{2\sqrt{b_{35} - 2I_4} + \sqrt{3b_{35} + b_{17} - 2b_{26} - 4I_4}} \dfrac{2a_{26} + a_{35} + a_{17}}{b_{35} - b_{17}}$

Table 6b. Seven-frame phase shifting algorithms (C10-C23)

Algorithm	Expressions for calculation of phase values $\Delta\varphi$
C10	$\tan \Delta\varphi = \dfrac{\sqrt{4a_{35}^2 - a_{26}^2}}{a_{26} + 2a_{35}} \dfrac{2a_{26} + a_{35} + a_{17}}{b_{35} - b_{17}}$
C11	$\tan \Delta\varphi = \dfrac{a_{35}}{\sqrt{4a_{35}^2 - (a_{17} - a_{35})^2}} \dfrac{a_{17} + a_{35} - 2a_{26}}{b_{35} - 2I_4}$
C12	$\tan \Delta\varphi = \dfrac{a_{26}^2}{(a_{17} + a_{35})\sqrt{4a_{26}^2 - (a_{17} + a_{35})^2}} \dfrac{a_{17} + a_{35} - 2a_{26}}{b_{35} - 2I_4}$
C13	$\tan \Delta\varphi = \dfrac{(b_{26} - 2I_4)^2}{(b_{17} - b_{35})\sqrt{4(b_{26} - 2I_4)^2 - (b_{35} - b_{17})^2}} \dfrac{a_{17} + a_{35} - 2a_{26}}{b_{35} - 2I_4}$

Algorithm	Expressions for calculation of phase values $\Delta\varphi$
C14	$\tan\Delta\varphi = \dfrac{a_{17}+a_{35}-2a_{26}}{\sqrt{4(b_{35}-2I_4)^2-(b_{35}+b_{17}-2b_{26})^2}}$
C15	$\tan\Delta\varphi = \dfrac{2a_{35}^2}{a_{26}\sqrt{4a_{35}^2-a_{26}^2}}\dfrac{a_{17}+a_{35}-2a_{26}}{b_{35}-2I_4}$
C16	$\tan\Delta\varphi = \dfrac{\sqrt{a_{35}(3a_{35}-a_{17})}}{2I_4-b_{26}}$
C17	$\tan\Delta\varphi = \dfrac{a_{26}}{\sqrt{4a_{26}^2-(a_{17}+a_{35})^2}}\dfrac{a_{17}-3a_{35}}{b_{26}-2I_4}$
C18	$\tan\Delta\varphi = \dfrac{a_{17}-3a_{35}}{\sqrt{4(b_{26}-2I_4)^2-(b_{35}-b_{17})^2}}$
C19	$\tan\Delta\varphi = \sqrt{\dfrac{b_{35}-2I_4}{b_{35}-b_{17}+2b_{26}-4I_4}\dfrac{a_{17}-3a_{35}}{b_{26}-2I_4}}$
C20	$\tan\Delta\varphi = \dfrac{a_{35}}{\sqrt{4a_{35}^2-a_{26}^2}}\dfrac{a_{17}-3a_{35}}{b_{26}-2I_4}$
C21	$\tan\Delta\varphi = \dfrac{\sqrt{4a_{26}^2-(a_{17}+a_{35})^2}}{2I_4-b_{26}}\dfrac{a_{35}}{a_{26}}$
C22	$\tan\Delta\varphi = \dfrac{-a_{35}\sqrt{4(b_{26}-2I_4)-(b_{35}-b_{17})^2}}{(b_{26}-2I_4)^2}\dfrac{a_{35}}{a_{26}}$
C23	$\tan\Delta\varphi = \sqrt{\dfrac{b_{35}-b_{17}+2b_{26}-4I_4}{b_{35}-2I_4}\dfrac{a_{35}}{2I_4-b_{26}}}$

4. ANALYSIS OF PHASE-SHIFTING ALGORITHMS

The overall accuracy of given phase measuring techniques depends on systematic and random errors that may affect the measurement process. These error sources are located both in the measurement setup and in the algorithms to calculate the phase. There are many potential factors that may influence the measurement accuracy (incorrect phase shift value, vibrations during phase measurement, noise and nonlinearities in the detection process, etc.). Different approaches can be chosen to analyze the susceptibility of PSI algorithms to errors. In our work we have chosen a Monte Carlo simulation technique of error propagation in particular PSI algorithms that allowed us to consider both individual and simultaneous contributions of different types of errors to the resulting phase error. The similar technique for uncertainty analysis of PSI algorithms was performed by Cordero et al. [44]. However, it was applied only to two popular algorithms, the conventional N-bucket [2] and the self-calibrating $(N+1)$-bucket algorithm [14].

In this section it is similarly studied the impact on the overall accuracy and stability of derived phase evaluation algorithms with respect to both random and systematic errors caused by the phase shifting device and the photodetector. Above-mentioned factors were implemented into a numerical model that can simulate the impact of individual parameters, describing these factors, on the phase measurement accuracy using a Monte Carlo approach as in Ref.[43]. It was supposed that random errors behave as normally distributed quantities $N(\mu, \sigma^2)$ with the mean value $\mu = 0$. The error of the phase shifter can be modeled by the following expression [37]

$$\delta\psi_i = \psi_i \sum_{k=0}^{\infty} c_{k+1} \left(\frac{\psi_i}{2\pi} \right)^k + \sigma_\psi, \qquad (10)$$

where ψ_i is the phase shift, c_k are coefficients of systematic errors of the phase shifting device, and σ_ψ is the random error of the phase shifting device. Coefficients c_k describe the real (nonlinear) behaviour of the chosen phase shifter. However, first two coefficients c_1 and c_2 are probably most significant for the measurement and evaluation process in practice. The standard deviation of the random error distribution can be determined for our model from the accuracy of the phase shifter. Assume now that we use a very precise piezoelectric translator for phase shifting. The non-linearity is then in the range 0.01-0.2% and the repeatability of the shifting is ±(1-10) nm. From the repeatability of the phase shifting device it can be calculated the corresponding phase change from the relation $\delta\varphi = (2\pi/\lambda)\delta W$, where δW is the change of the optical path difference caused by shifting a little mirror in the path of the reference beam, and λ is the wavelength of light. In case of He-Ne laser with the wavelength $\lambda = 632.8$ nm the phase error will be approximately in the range 0.02-0.2 radians. The error of detection of the irradiance of the observed interference field can be modeled due to the following expression

$$\delta I = I \sum_{k=1}^{\infty} d_k I^k + \sigma_I, \qquad (11)$$

where I is the irradiance, d_k are coefficients that characterize systematic errors of irradiance detection, and σ_I is the random error of detection. Coefficients d_k describe the real (nonlinear) behavior of the given photodetector. The most important factor for a real description of the detector response on the incident light is coefficient d_1, which describes the second order non-linear response of the detector. The standard deviation that characterizes the distribution of random errors during the detection process of the interference signal can be simulated as a fraction of the intensity incident onto the detector, i.e. $\sigma_I = pI$, where values of p can be considered in the range 0.1-1% with respect to the properties of the currently produced photodetectors.

Above-mentioned factors were implemented into a numerical model that can simulate the impact of individual parameters, describing these factors, on the phase measurement accuracy. The model of the irradiance distribution for the *i*-th measurement can be expressed as

$$I_i = A + B\cos[\Delta\varphi + \psi_i + \delta\psi_i] + \delta I_i, \qquad (12)$$

where A is the mean intensity of the interference signal, B is the modulation of the interference signal, $\Delta\varphi$ is the phase, ψ_i is the phase shift in *i*-th intensity measurement, $\delta\psi_i$ is the phase shift error, and δI_i is the detection error. The resulting error of phase values $\Delta\varphi$ is then given by $\delta(\Delta\varphi) = \Delta\varphi' - \Delta\varphi$, where $\Delta\varphi'$ are the calculated phase values and $\Delta\varphi$ are the original phase values. The analysis was performed for phase values in the range (-π,π). The root-mean-square $\sigma_{\Delta\varphi}$ of calculated phase errors was chosen as an error characteristic of the accuracy of particular algorithms

$$\sigma_{\Delta\varphi} = \sqrt{\frac{\sum_{k=1}^{M}\delta(\Delta\varphi)_k^2}{M}}, \qquad (13)$$

where M is the number of computer simulations of a phase evaluation. More than 1000 simulation cycles were performed to guarantee the reliability of the results. The analysis of algorithms derived in Section 3 was performed on all combinations of possible algorithms. The results of an error analysis are then shown only for the best PSI algorithms. It is known that the stability and the accuracy of described phase evaluation algorithms with unknown but constant phase shifts also depend on the phase shift value itself. Performing the computer analysis, the optimal phase shift values can be obtained for all PSI algorithms. In case that the phase shift value is near the optimum the phase measurement errors are low. Better algorithms offer a wider range around the optimum phase shift value, where the phase error is comparable. Outside this range errors increase considerably. Firstly, we optimized the algorithms with respect to the phase shift value ψ and from the analysis we obtained the optimum values ψ_{opt} of the phase shift, where the resulting phase error reaches its minimum. The accuracy and sensitivity to the change of the phase shift value has been evaluated on the basis of the root-mean-square phase error.

The optimum phase shift value ψ_{opt} is found, where the phase error function reaches its minimum. These values were obtained from the simulated relationship using a polynomial approximation near the minima of these functions. We can see that the phase error is practically insensitive to the change of the phase shift value with respect to the optimum value $\psi_{opt} = \pi/2$ in a relatively wide range. Furthermore, the algorithm A1 for $\psi = \pi/2$ converts to the well-known Schwider-Hariharan five-frame algorithm [22,29] and thus is one variant of the described PSI algorithms with the optimum phase shift. Similarly, Carré four-frame algorithm [3,9] has its optimum phase shift value $\psi_{opt} = 110°$.

A similar analysis was performed for all above-mentioned PSI algorithms. Different phase evaluation algorithms differ in the value of the optimum phase shift and in the error sensitivity to its change. Some of them are robust with respect to a relatively large change of the optimum phase shift. The results of the analysis can be found in tables 7-9 (five, six and seven-frame algorithms).

Table 7. Optimum phase shift value for five-frame algorithms

Algorithm	Phase shift ψ_{opt} [rad]	Phase shift ψ_{opt} [°]	Algorithm	Phase shift ψ_{opt} [rad]	Phase shift ψ_{opt} [°]
A1	$\pi/2$	90	A5	0.67π	121
A2	0.55π	99	A6	0.63π	113
A3	0.7π	126	A7	0.53π	96
A4	0.6π	108	A8	0.70π	126

Table 8. Optimum phase shift value for six-frame algorithms

Algorithm	Phase shift ψ_{opt} [rad]	Phase shift ψ_{opt} [°]	Algorithm	Phase shift ψ_{opt} [rad]	Phase shift ψ_{opt} [°]
B1	0.4π	72	B10	0.32π	58
B2	0.35π	63	B11	0.7π	127
B3	0.47π	85	B12	0.52π	93
B4	0.43π	78	B13	0.46π	83
B5	$\pi/2$	90	B14	0.46π	83
B6	0.53π	96	B15	0.36π	64
B7	0.53π	96	B16	0.4π	73
B8	0.73π	131	B17	0.39π	71
B9	0.33π	60			

A complex error analysis was carried out with the calculated optimal phase shift values for given algorithms. The influence of errors of the phase shifter and the detector were simulated using the mentioned Monte Carlo technique.

Table 9. Optimum phase shift value for seven-frame algorithms

Algorithm	Phase shift ψ_{opt} [rad]	Phase shift ψ_{opt} [°]	Algorithm	Phase shift ψ_{opt} [rad]	Phase shift ψ_{opt} [°]
C1	0.62π	111	C13	0.70π	127
C2	0.72π	129	C14	0.76π	138
C3	0.69π	124	C15	0.74π	133
C4	0.62π	111	C16	$\pi/2$	90
C5	0.59π	107	C17	0.56π	100
C6	0.30π	54	C18	$\pi/2$	90
C7	0.72π	129	C19	0.59π	107
C8	0.39π	71	C20	0.52π	93
C9	0.35π	63	C21	0.45π	80
C10	0.36π	64	C22	$\pi/2$	90
C11	0.74π	133	C23	0.53π	96
C12	0.75π	136			

The results of the analysis will be presented on a group of selected algorithms and for chosen values of important error parameters (linear systematic error of the phase shifting device 0.5 %, second-order nonlinear systematic error of the phase shifting device 0.1 %, RMS random error of the phase shifting device 0.05 rad, second-order nonlinear systematic error of irradiance measurement 0.5 %, and noise in irradiance measurement 0.5 %). The analysis can be generally applied for different types of error factors and error sources can be considered simultaneously.

The influence of errors of the phase shifter and the detector were simulated using the described model of these errors. The parameters considered in the error analysis of the phase calculation algorithms are shown in table 10.

Table 10. Parameters of error analysis

Coefficient	Value	Units
c_1	0.5	%
c_2	0.1	%
σ_ψ	0.05	rad
d_1	0.5	%
p	0.5	%

On a basis of the performed analysis several algorithms for a different number of steps N were chosen with minimum phase errors and robustness of phase calculation. These algorithms (A1, A2, A5, B7, B12, B16, C6, C18, C21) are compared together to selected error compensating PSI algorithms with the same number of steps (Surrel (1997) [21], Schmit and Creath (1995) [17], Hibino et al. (1995) [18]) and the results are presented in the following text.

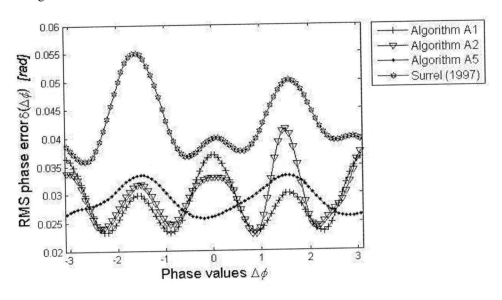

Figure 1. Dependency of phase error $\delta(\Delta\varphi)$ on phase values $\Delta\varphi$ ($N = 5$).

We must again note that proposed algorithms are not sensitive to linear phase shift errors in case that the actual phase shift value is constant and unknown in advance. This fact, which arises from the derivation scheme of algorithms, has been also proved numerically by the performed simulation. Figures 1-3 present the relationship between the calculated RMS phase error $\delta(\Delta\varphi)$ and the phase values $\Delta\varphi$ from the range $(-\pi, \pi)$. This relationship expresses the accuracy and robustness of the compared phase shifting algorithms.

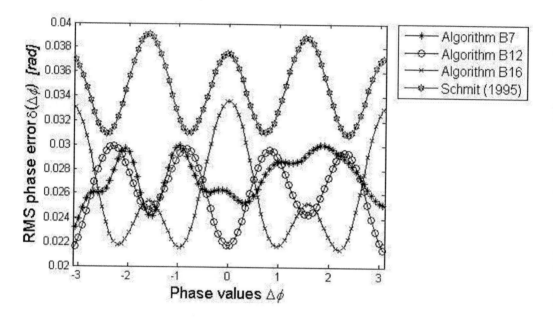

Figure 2. Dependency of phase error $\delta(\Delta\varphi)$ on phase values $\Delta\varphi$ ($N = 6$).

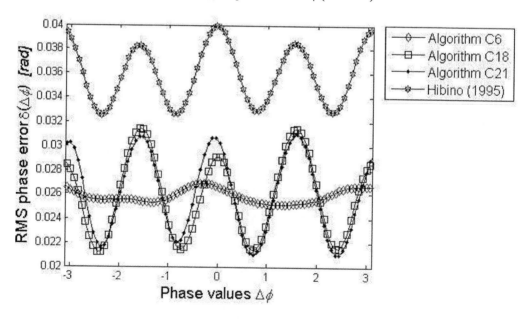

Figure 3. Dependency of phase error $\delta(\Delta\varphi)$ on phase values $\Delta\varphi$ ($N = 7$).

We can see that in comparison to mentioned classical PSI algorithms the derived algorithms perform very well. It is caused especially by their inherent property – insensitivity to linear errors of the phase shifter. However, even without considering these errors, proposed algorithms are comparable to standard phase calculation algorithms, especially the PSI algorithm C6 shows excellent properties (small errors, robustness of phase calculation).

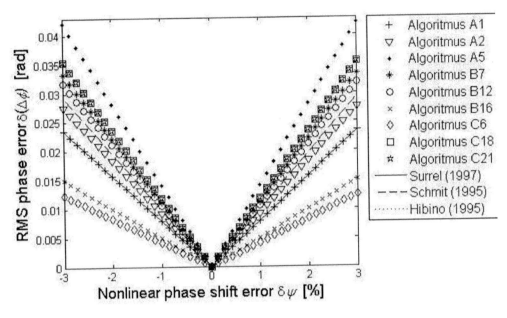

Figure 4. Dependency of phase errors on nonlinear phase shift errors.

Furthermore, we studied the properties of the algorithms with respect to the change of parameters simulating the nonlinearities of the phase shifter device and photodetector. The algorithms described in this work are not sensitive to linear phase shift errors. The dependency of the phase error $\delta(\Delta\varphi)$ on the second order nonlinear behaviour of the phase shifting device is shown in figure 4 for all the compared algorithms.

The dependency of the phase error on the second order nonlinear response of the photodetector is then demonstrated in figure 5. From presented figures we can see that the analyzed PSI algorithms are comparable to classical algorithms and some of them are insensitive to the second order nonlinearity in irradiance detection, namely algorithms C18 and C21.

From the performed error analysis and the comparison to classical PSI algorithms we may conclude that the described multiframe PSI algorithms with an unknown but constant phase shift are comparable to other phase evaluation algorithms. Especially the seven frame algorithm C6 performs very well. Moreover, these algorithms are not sensitive to the phase shifter linear errors in principle and the derived formulas can be used for the calculation of the phase shift value during measurements. A certain disadvantage of these PSI algorithms can be their higher complexity, which corresponds to 40-150% increase of the computation time with respect to conventional algorithms with the same number of steps.

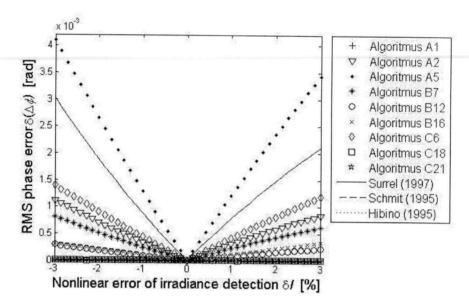

Figure 5. Dependency of phase errors on nonlinear response of detector.

CONCLUSION

The chapter presented a group of formulas for phase calculation in interferometric measurement methods using the phase shifting technique. Multi-frame phase-shifting algorithms, which use five, six and seven intensity frames to determine phase values in interferometric measurements, were described theoretically. The phase shifting algorithms (for $N = 5-7$ frames) were derived similarly as the Carré algorithm. It is assumed that the phase shift between particular captured frames is constant but unknown in advance, i.e. we suppose that the phase shifter reproduces practically the same phase shifts during measurements. In principle the described phase evaluation algorithms are not sensitive to phase shift miscalibration and enable calculation of the actual phase shift value at any point of the detector plane during the measurement process. Several formulas were derived for calculation of the phase shift value. With these algorithms, the phase shifting device can be calibrated and measurement can be carried out if the phase shift is not uniformly distributed over the detector area. Phase calculation algorithms with unknown phase shifts can be applied for any measurement that uses the phase shifting technique for evaluation. The performance (accuracy and robustness) of mentioned algorithms depends on the value of the phase shift. The optimum phase shift value corresponds to each algorithm where the phase error has the lowest value. Furthermore, the robustness of algorithms differs with respect to the phase error susceptibility to the change of the phase shift value. The robust algorithms have small phase errors in a wide interval around the optimum phase shift value. The accuracy and robustness with respect to the initial phase shift values of all described algorithms were studied, and optimum phase shift values for particular algorithms were calculated in order to ensure the lowest possible phase errors of the phase calculation algorithms. On the basis of the proposed model, the Monte Carlo analysis was performed of the influence of the most important parameters that may negatively affect the overall accuracy and robustness of phase calculation algorithms. The algorithms were compared with respect to their accuracy, robustness and computation time. A comparison to common error compensating phase stepping algorithms was performed. From the performed error analysis we may conclude that the described multiframe PSI algorithms with an unknown but constant phase shift are comparable to other phase evaluation algorithms and their advantage of insensitivity to the phase shift miscalibration and a possibility of pointwise phase-shift calculation can be used in practice.

Acknowledgments

This work has been supported by grant MSM6840770022 Ministry of Education of Czech Republic.

References

[1] Malacara, D. (Ed.), *Optical Shop Testing*; Wiley Interscience: New York, 2007.
[2] Malacara, D.; Servín, M.; Malacara, Z., *Interferogram Analysis for Optical Testing*, Taylor & Francis: New York, 2005.
[3] Creath, K., Phase-Measurement Interferometry Techniques, *Progress in Optics*. Vol. XXVI, E. Wolf, (Ed.), Elsevier Science Publishers, Amsterdam, 1988, 349- 393.
[4] Bruning J. H.; Herriott, D. R.; Gallagher, J. E.; Rosenfeld, D. P.; White, A. D.; and D. Brangaccio, J., Digital Wavefront Measuring Interferometer for Testing Optical Surfaces, Lenses, *Appl. Opt.*, 13, 2693-2703 (1974).
[5] Crane R., Interference Phase Measurement, *Appl. Opt.*, 8, 538 (1969).
[6] Wyant J.C., Use of an ac Heterodyne Lateral Shear Interferometer with Real-TimeWavefront Correction Systems, *Appl. Opt.*, 14, 2622-2626 (1975).
[7] Bruning, J. H., Fringe Scanning Interferometers, in Optical Shop Testing, D. Malacara Ed.,1st ed., Wiley: New York, 1978.
[8] Massie N. A.; Nelson, R. D.; and Holly,S., High-Performance Real-Time Heterodyne Interferometry," *Appl. Opt.*, 18, 1797 (1979).
[9] Carré, P., Installation et utilisation du comparateur photoe´lectrigue et Interferentiel du Bureau International des Poids ek Measures, *Metrologia*, 1, 13-23 (1966).
[10] Morgan C. J., Least-Squares Estimation in Phase-Measurement Interferometry, *Opt. Lett.*, 7, 368-370 (1982).
[11] Freischlad, K. R.; Koliopoulos, C. L., Fourier Description of Digital Phase-Measuring Interferometry, *J. Opt. Soc. Am. A*, 7, 542-551 (1990).
[12] Lai, G.; Yatagai, T., Generalized phase-shifting interferometry, *J. Opt. Soc. Am. A*, 8, 822-827 (1991)
[13] Phillion, D. W., General methods for generating phase-shifting interferometry algorithms, Appl. Opt. 36, 8098-8115 (1997)
[14] Surrel, Y., Phase stepping: a new self-calibrating algorithm, Appl. Opt. 32, 3598-3600 (1993)

[15] Malacara-Doblado D.; Dorrío, B. V., Family of detuning-insensitive phase-shifting algorithms, *J. Opt. Soc. Am. A* 17, 1857-1863 (2000)
[16] Surrel, Y., Design of phase-detection algorithms insensitive to bias modulation, *Appl. Opt.* 36, 805-807 (1997)
[17] Schmit, J.; Creath, K., Extended averaging technique for derivation of error-compensating algorithms in phase-shifting interferometry, *Appl. Opt.* 34, 3610-3619 (1995)
[18] Hibino,K; Oreb,B.F.; Farrant, D.I.; Larkin,K.G., Phase-shifting algorithms for nonlinear and spatially nonuniform phase shifts, *J. Opt. Soc. Am. A* 14, 918-930 (1997)
[19] Larkin, K. G.; Oreb, B.F., Design and assessment of symmetrical phase-shifting algorithms, *J. Opt. Soc. Am. A* 9, 1740-1748 (1992)
[20] de Groot, P., Derivation of algorithms for phase-shifting interferometry using the concept of a data-sampling window, *Appl. Opt.* 34, 4723-4730 (1995)
[21] Surrel, Y., Design of algorithms for phase measurements by the use of phase stepping, *Appl. Opt.* 35, 51-60 (1996)
[22] Hariharan, P.; Oreb, B. F.; Eiju,T, Digital phase-shifting interferometry: a simple error-compensating phase calculation algorithm, *Appl. Opt.* 26, 2504-2506 (1987)
[23] Zhang, H.; Lalor, M.J; Buton, D.R, Robust, Accurate Seven-Sample Phase-Shifting Algorithm Insensitive to Nonlinear Phase-Shift Error and Second-Harmonic Distortion: A Comparative Study, *Opt. Eng*, 38, 1544-1533 (1999).
[24] Bi, H.; Zhang, Y.; Link, K. V.; Wen,C., Class of 4+1 Phase Algorithum with Error Compensation, *Appl. Opt.,* 43, 4199-4207 (2004).
[25] Malacara-Doblado D.; Dorrío,B.V.; Family of Detuning-Insensitive Phase-Shifting Algorithms, *J. Opt. Soc. Am. A* 17, 1857–1863 (2000).
[26] Zhu Y.; Gemma, T., Method for Designing Error-Compensating Phase-Calculation Algorithms for Phase-Shifting Interferometry, *Appl. Opt.* 40, 4540–4546 (2001).
[27] Afifi, M.; K. Nassim, K.; Rachafi, S., *Five-frame phase-shifting algorithm insensitive to diode laser power variation, Opt. Commun.* **197** (**2001**), pp. *37*–42.
[28] Schmit J., Creath, K. Window function influence on phase error in phase-shifting algorithms, *Appl. Opt.* 35, 5642-5649 (1996)
[29] Schwider, J.; Burow, R.; Elssner, K.-E.; Grzanna, J.; Spolaczyk,R.; Merkel, K., Digital wave-front measuring interferometry: some systematic error sources, *Appl. Opt.* 22, 3421-3432 (1983)
[30] Schwider J., Phase Shifting Interferometry: Reference Phase Error Reduction, *Appl. Opt.,* 28, 3889-3892 (1989).
[31] Zhao, B.; Surrel, Y. Effect of Quantization Error on the Computed Phase of Phase-Shifting Measurements, *Appl. Opt.,* 36, 2070–2075 (1997).
[32] Hariharan, P., Phase-shifting interferometry: minimization of systematic errors, *Opt. Eng.* 39, 967 –969 (2000).
[33] Liu,Q.; Cai, L. Z.; He, M. Z., Digital correction of wave-front errors caused by detector nonlinearity of second order in phase-shifting interferometry, *Opt.Commun.* 239, 223-228 (2004).
[34] Chen, X.; Gramaglia,M.; Yeazell,J.A., Phase-Shifting Interferometry With Uncalibrated Phase Shifts,'' *Appl. Opt.*, 39, 585–591 (2000).
[35] Hibino,K., Susceptibility of systematic error-compensating algorithms to random noise in phase-shifting interferometry, *Appl. Opt.* 36, 2084-2093 (1997)

[36] Rathjen, C., Statistical properties of phase-shift algorithms, *J. Opt. Soc. Am. A* 12, 1997-2008 (1995)
[37] Zhang, H.; Lalor, M.J.; Burton, D.R., Error-Compensating Algorithms in Phase-Shifting Interferometry: A Comparison by Error Analysis, *Opt. Las. Eng.* 31, (1999) 381.
[38] Kinnstaetter, K.; Lohmann, A.W.; Schwider,J.; Streibl, N. Accuracy of phase shifting interferometry, *Appl. Opt.* 27, 5082-5089 (1988)
[39] Novak,J., Five-step phase-shifting algorithms with unknown values of phase shift, *Optik* 114, (2003) 63-68.
[40] Novák, J.; Novák, P.; Mikš, A., Multi-step Phase-shifting Algorithms Insensitive to Linear Phase Shift Errors, *Opt.Comm.* 281(21) 5302-5309 (2008).
[41] de Groot, P.J.; Deck,L.L., Numerical simulations of vibration in phase-shifting interferometry, *Appl. Opt.* 35, 2172-2178 (1996)
[42] Wizinowich, P.L., Phase shifting interferometry in the presence of vibration: a new algorithm and system, *Appl. Opt.* 29, 3271-3279 (1990)
[43] de Groot, P.J., Vibration in phase-shifting interferometry, *J. Opt. Soc. Am. A* 12, 354-365 (1995)
[44] Cordero, R.R. et al., Uncertainty analysis of temporal phase-stepping algorithms for interferometry, *Opt.Commun.* 275, 144-155 (2007).
[45] Ghiglia, D.C.; Pritt, M.D., *Two-dimensional Phase Unwrapping: Theory, Algorithms and Software.* John Wiley & Sons: New York, 1998.
[46] Herráez, M. A.; Burton, D. R.; Lalor, M. J.; Gdeisat,M.A., Fast Two-dimensional Phaseunwrapping Algorithm Based on Sorting by Reliability Following a Noncontinuous Path, *Appl. Opt.,* 41, 7437–7444 (2002).
[47] Huntley, J. M.; Saldner, H., Temporal Phase Unwrapping Algorithm for Automated Interferogram Analysis, *Appl. Opt.* 32, 3047–3052 (1993).
[48] Herráez, M. A.; Burton, D. R.; Lalor, M. J.; Clegg,D.B., Robust Unwrapper for Twodimensional Images, *Proc. SPIE*, 2784, 106 (1996).
[49] Bioucas-Dias, J.; Valadão,G., Phase Unwrapping via Graph Cuts., *IEEE Trans. on Image Processing* , 16 , 698 - 709 (2007).
[50] Kemao, Q.; Gao,W.; Wang,H., Windowed Fourier-filtered and quality-guided phase-unwrapping algorithm, *Appl. Opt.* 47, 5420-5428 (2008)

In: Interferometry Principles and Applications
Editor: Mark E. Russo

ISBN 978-1-61209-347-5
© 2012 Nova Science Publishers, Inc.

Chapter 16

INTERFEROMETRIC METHODS APPLIED TO POLYMERIC ANALYSIS

Gustavo F. Arenas[1,*], *Nélida A. Russo*[2] *and Ricardo Duchowicz*[2,3]

[1] Laboratorio Laser, Facultad de Ingeniería, Universidad Nacional de Mar del Plata
Mar del Plata, Argentina

[2] Centro de Investigaciones Ópticas (CIOp), CCT La Plata CONICET-CIC
(1900) La Plata, Argentina

[3] Facultad de Ingeniería, Universidad Nacional de La Plata (UNLP), Argentina

ABSTRACT

Several epoxies and photo- or thermal- cured polymers found their way in almost every field of structures manufacturing. These materials can compete with metallic ones and even substitute them in several applications. However, it is well known that their mechanical properties are highly dependent upon the curing process of the matrix. Curing evolution is connected directly to the contraction process occurring in the material. Relevant information include the evolution and final conversion (degree of cure) related to the amount of the chemical cross-linking occurring during cure, the gel point where a phase change from liquid-like to solid-like occurs, the glass transition temperature and the induced residual strain produced during the curing process causing structure distortion and intrinsic strain accumulation. Several cure monitoring techniques have been proposed and applied in the past. Among these techniques, optical approaches seem to be the best candidate in polymer based materials manufacturing monitoring. On this way, different methods based on interferometric techniques by using fiber optics technologies have been developed. In this work, we discuss the feasibility of different approaches. First, we analyze optical sensors based on a Fizeau fiber optic interferometer to measure polymer contraction that occurs during cure with a measure resolution better than 100 nm; second, we discuss the simultaneous application of a cantilever and the Fizeau interferometer, and, finally, the use of a pair of fiber Bragg grating based sensors to uncouple strain and temperature (assuming a thermal related cure process). Properties and relevant information extracted from of the different techniques are discussed.

---

[*] E-mail: garenas@fi.mdp.edu.ar

Keywords: Fiber optic sensor, Fizeau interferometer, shrinkage, Fiber Bragg grating, photocured polymer.

1. INTRODUCTION

Polymer-based composites provide several advantages over traditional materials and are extensively used in many applications, such as aerospace and automotive industries, civil structures, etc. Shrinkage and mechanical changes occurred during the polymerization process are very important issues that should be taken into account for improving the performance of the final composite materials. The properties of the built structures are strongly dependent on processing parameters, resin characteristics, photosensitivity rate, degree of cure and the post-cure method.

Visible light activated composite materials are widely used in dentistry. Typically, they are a blend of fillers dispersed in a photo-polymerizable organic resin matrix with the camphorquinone photoinitiator. When exposed to blue light (absorption peak at 468 nm), the camphorquinone stimulates production of free radicals and thus initiate the polymerization process. During photo-polymerization the polymer shrinks at the same time that it changes from liquid to solid material. As a result, cure-induced strains are generated, which cause internal stresses, in the cured material. Volume shrinkage of the resin is one of the major causes leading to poor accuracy of the fabricated parts [1]. Therefore, its knowledge is of vital importance in the design and optimization of different formulations. The physical properties such as hardness and bulk volume of the polymerized composite are related to the conversion rate (quantity of reacted versus unreacted material) which depends on composition parameters as well as on curing light parameters. Higher conversion rate ensures better hardness, however it generally causes higher shrinkage of the polymerized composite. While hardness is important for the longevity of a restoration, shrinkage manifests as stress which may lead to microgap formation and leakage. Thus, polymerization shrinkage is recognized as one of the main problems in the resin based restorative materials applications. Most of commercial resin composites were found to undergo volume shrinkage in the range of 2–3% [2].

A number of works have been reported on the UV based solidification phenomena of micro-fabrication suitable photopolymers. In recent years, the use of experimental techniques capable of monitoring the curing shrinkage and mechanical changes during the polymerization process is increasingly attracting attention [3-8]. The bonded-disk method allows to obtain a linear shrinkage approximated volumetric shrinkage [9,10].

Several optical techniques has been considered relevant for in-process monitoring, as well as post-process determination of the micro curing strain state induced in the material. Among these, laser interferometric techniques such as the Michelson and the Mach-Zender, that employ a single-frequency laser, show high sensibility and a very good reproducibility [11,12]. However, in these systems, random fluctuations in the polarization state in the guided beam caused by the environmental perturbation of the fiber eigenmodes, should be eliminated.

The most common interferometer is the Fabry-Pèrot (F-P). Its transfer function is the same to the bulk F-P formed by two parallel mirrors with high reflectance. Usually white-light interferometry is used to interrogate the absolute cavity length, in which a broadband source is injected into the device. The reflected light is in periodic distribution along the wavelength in the white-light optical spectrum. The absolute cavity length can be obtained by measuring the wavelength spacing between two adjacent apexes, where there is a 2π phase shift. An optical spectrum analyzer (OSA), a fiber Fabry–Pèrot tunable filter (FFP-TF) or a FFP-TF-based scanning fiber laser can be employed for that purpose [13-17]. One particular version of the F-P is the fiber optic Fizeau interferometer, which employs a very simple set-up and, by its design, is very stable. Different photocured materials have been recently analyzed using this technique [18,19]. On the other hand, another very useful alternative to monitor the polymerization process are Fiber Bragg Grating (FBG) sensors. Apart from being inexpensive and not sensitive to electromagnetic interference, fiber optic sensors are extremely small when compared to thermocouples or strain gauges, have a high accuracy and sensitivity, and offer unique embedding characteristics in relation to a variety of polymeric and composite materials. The mechanical behavior of the glass fiber that contains the grating is not affected when it is embedded into a host material. Therefore, this technique is well suited for the determination of non-uniform axial strains development due to the shrinkage of the resin during polymerization, as well as to identify gelation, vitrification and other postcuring effects [20,21]. Besides, another advantage of FBG sensors is their possibility to be multiplexed in order to simultaneously measure temperature or strain at many points of an object. Reports of strain and cure temperature measurements in epoxy resins and fiber reinforced composites by using Fiber Bragg grating (FBG) sensors have been published [22-26]. In this work, we analyze the use of the last mentioned techniques, i.e. Fizeau interferometer (FI), the simultaneous application of the Fizeau interferometer and a cantilever (FC), and Fiber Bragg Grating (FBG) sensors, in order to obtain relevant information of photocured materials, such as cure-induced shrinkage strains, thermal and vitrification processes, etc. These techniques provide a useful data collection at high acquisition rates and at very low cost. At Section 2, we introduce the analyzed materials as well as the polymer initiator molecules and the photocuring system used to excite the samples. We also define the shrinkage contraction. Next, at Section 3, we analyze the Fizeau interferometer starting with a basic theory, the experimental Fizeau set-up to measure shrinkage and examples of the results obtained with this technique. At Section 4 we introduce vibration measurements by the Fizeau employing a small cantilever with one extreme joined to the sample and a shaker near to the other. Fiber Bragg Grating sensor applied to photocured polymer is discussed in Section 5, starting with the grating operation principle. Next we show the experimental set-up employed to the polymer measurements and by last, results are presented and discussed. Finally, we mention the conclusions.

2. Materials and Photocuring System

2.1. Sample Preparation

The resins were formulated from blends of {2,2-bis[4-(2-hydroxy-3-methacryloxyprop-1-oxy)phenyl]propane} (BIS-GMA) and triethylene glycol dimethacrylate (TEGDMA) at mass fractions 70:30 bis-GMA/TEGDMA. bis-GMA (Esstech, Essington, PA, USA) and TEGDMA (Aldrich) were used as received.

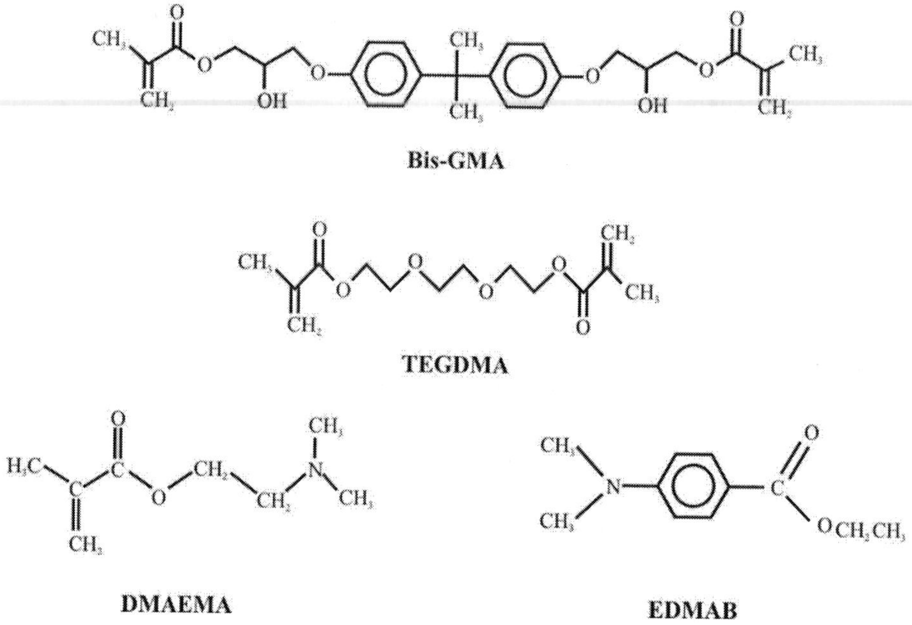

Figure 1. First and second line: common molecular structures of the base monomers (BISGMA) and diluents (TEGDMA), presents in dental composites. Third line: Photopolymerization co-initiators reducing agents used.

The resins were activated for visible light polymerization by the addition of camphorquinone (CQ) and amine reducing agents. The amines were dimethylamino-ethylmethacrylate (DMAEMA) (Aldrich) and ethyl-4-dimethylaminobenzoate (EDMAB) (Aldrich). The structure of the monomers and photoinitiator systems are depicted in Figure 1.

A third resin was prepared by adding nanosilica particles (Aerosil) as filler to blends made within EDMAB type. In this way, it forms a composite with different properties providing complementary insight aspects of photocuring process and better understanding of final material properties.

2.2. Pumping Scheme

The light source employed to cure the resins was assembled from a Light Emitting Diode (LXHL-PB09, Philips LUMILED Luxeon III), with its emittance centered at 470 nm. The LED was selected taking into account that the CQ photoactivator is activated in the wavelength range 400–500nm with an absorption peak at 470 nm. The emission spectrum of the LED source was measured with a calibrated CVI-monochromator (Digikrom 480) and a Si-photodetector.

The power of the photocuring source was set by adjusting the current of a Laser Diode Controller (Thorlabs LDC-500) to the maximum value and half when it was needed (referred as 100% and 50% respectively). In all cases described here, the curing time of the samples was 600 seconds long with constant irradiation.

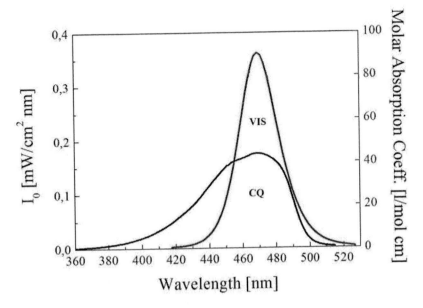

Figure 2. Irradiance of curing LED and molar absorption coefficient for CQ versus wavelength. The value of ε_{CQ} is 42 l mol-1 cm-1. At the top left of the graph: CQ photo activator molecule.

2.3. Shrinkage Analysis

The percent of the linear contraction is calculated by using the expression:

$$Shrinkage\ (\%) = \frac{\Delta d}{d}100 \approx \frac{\Delta V}{V}100 \qquad (1)$$

where d is the final thickness of the sample. A plot of %Shrinkage as a function of the curing time provides a kinetic profile for the curing process. These measurements were carried out under the following conditions: initial separation ($d_0 \approx d$) was kept to (approximately) 100 μm in order to have good visibility in the d-range corresponding to the contraction process; after sample preparation and having waited enough time for proper stabilization were made the photocuring process by turning on the LED source and registering the interferogram signal for further processing and shrinkage quantification.

3. FIZEAU INTERFEROMETER

3.1. FIZEAU INTERFEROMETER BASIC THEORY

Figure 3. presents a schematic view of generic cavity of length d resulting between two interfaces with reflection coefficients r_1 y r_2, and transmission coefficients t_1 for waves travelling forward to second media and t'_1, for the backward situation. In this case the refraction index n_1 is greater than $n_0 \approx 1$, supposing that air as the filling media in the cavity.

As this device is exploited in reflection mode and due to the great value of reflective surface, it is no necessary to have in mind any other further transmissions.

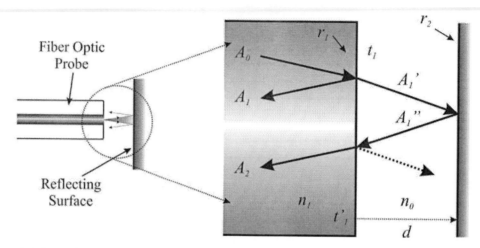

Figure 3. Scheme for the theoretical deduction of a multi-beam interference cavity between two generic surfaces with reflection and transmission coefficients r_1, r_2 and t_1, respectively.

Consider a roughly normal incident wave A_0 into the cavity generating a succession of reflected and transmitted beams, which are considered only A_1 and A_2, as shown in figure x. Assuming no absorption, the incident beam A_0 is partly reflected (A_1) by the end of the fiber optic, and partly transmitted to the air interface (A_1') according to

$$A_1 = r_1 A_0$$
$$A_1' = t_1 A_0 \qquad (2)$$

The diffracted beam then propagates through cavity and is reflected by the second mirror and coupled back to the source fiber giving rise to the following expression

$$A_2 = r_2 t_1 t_1' A_0 \sqrt{\beta} \cdot e^{i\varphi} \qquad (3)$$

The argument φ of this modulation takes into account the phase shift caused by the path difference of the beams and is defined as

$$\varphi = \frac{4\pi n_0 d}{\lambda} \qquad (4)$$

where n_0 is the refraction index of cavity medium ($n_0 \approx 1$ considering air), λ is the working wavelength and d as the distance between surfaces cavity.

The β parameter, commonly known as optical coupling efficiency, equals [27]:

$$\beta = \frac{1}{1+\left(\dfrac{d}{k\omega^2}\right)^2} \quad \text{con} \quad \begin{cases} \omega = \dfrac{a}{ln(\upsilon)} \\ k = \dfrac{2\pi n}{\lambda} \end{cases} \tag{5}$$

depending on the distance of the cavity d, the radius of the optical fiber core a, the wavelength λ, and normalized frequency υ.

In this case, where single-mode fiber is used, can be accepted as valid the assumption of Gaussian propagation within the nucleus and, in this way w is the beam waist and coincides with the radius of the fiber optic [28]. Based on the circumstances of the case, it suffices to consider only the contribution of the first two terms, i.e.

$$A_R = A_1 + A_2 + \cdots \approx A_1 + A_2 \approx \sqrt{R_1}A_0 + \sqrt{R_2}(1-R_1)\sqrt{\beta}A_0 \cdot e^{i\varphi} \tag{6}$$

with $r_1^2 = R_1$, $r_2^2 = R_2$, $t_1 t'_1 = 1 - r_1^2 = 1 - R_1$ and regarding the intensity I_R as the real part of complex conjugate product of A_R, is possible to obtain

$$I_R = I_0\left[R_1 + R_2(1-R_1)^2\beta + 2\sqrt{R_1 R_2 \beta}(1-R_1)cos\varphi\right] \tag{7}$$

In order to achieve a normalized expression of the transfer we may pose the following relationship when both surfaces are parallel, the transference function of this device is given by [27]

$$\frac{I(d)}{I_0(d)} = 1 + \frac{2\sqrt{R_1 R_2 \beta}(1-R_1)}{R_1 + (1-R_1)^2 R_2 \beta}cos\varphi \tag{8}$$

where $I_0(d)$ is the mean intensity that varies exponentially with d, R_1 and R_2 are the reflective surfaces and d is the air-gap distance. As stated above, β value is dependent on the cavity [29,30], and phase change φ due to the optic path in one cavity round trip can be understood as follows: the separation between two intensity maxima corresponding to two consecutive interference fringes occurs at $\Delta d = \lambda/2$. Due to this, the technique precision can reach interesting values lower to a micron. A more complete expression is given in [31], which considers possible misaligning between both reflective surfaces. In our case, we neglect this fact.

3.2. Experimental Set-up for Shrinkage

A fiber optic sensing method based on a Fizeau-type Interferometric scheme was employed for monitoring the evolution of the shrinkage during photopolymerization. Details of the technique were reported elsewhere [18]. Figure 4. depicts the experimental arrangement. A laser diode operating at 1310 nm was used as the coherent source of the interferometer (Mitsubishi 725B8F mounted on Thorlabs KT112 collimation and focusing system, powered with LD2000 Thorlabs Laser Driver). Laser light was coupled to a 2×2 single mode (1.31 μm) fiber directional coupler (Thorlabs 10202A-50). Two optical waves are coupled back into the fiber from the Fizeau interferometer. The optical signal from one of the outputs (labeled as 2) was collected by an InGaAs detector (Thorlabs DET410), amplified, converted to a digital signal and acquired on a personal computer at 500 samples/s. Light reflections from the other output of the coupler (labeled as 3) were eliminated by the employment of an index adapter liquid. An air-gap was developed between the cleaved end (corresponding to the output labeled as 4) of the single mode directional coupler and a reflective surface consisting of a very small piece (1mm×1mm) of thin aluminum foil was placed onto the resin sample. The glass plate was fixed to a positioning system which was aligned in order to ensure that the two reflective planes were parallel (i.e. fiber optic end and aluminum target). As explained with detail above, beams coming from cavity interfere and, if the distance between both reflective surfaces changes, the coherent process generates a temporal modulation (intensity distribution of maxima and minima).

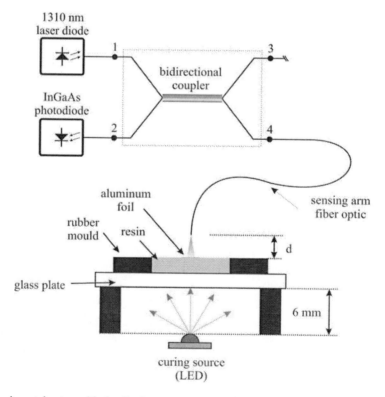

Figure 4. Experimental setup with details for measurement of photocuring resins.

Thus, the separation between two intensity maxima corresponding to two consecutive interference fringes occurs at L = λ/2 or 0.655 μm. Figure 5. shows this with high detail when polymerization took place on a typical commercial composite. Mechanical vibrations give rise to noise that complicate contraction measurements, so it is important to assemble the apparatus in a vibration free environment. The apparatus used in the present experiment was assembled on a vibration free optical table. A rubber ring of inner diameter equal to 10mm was glued to the microscope glass plate. The sample resin was placed onto the glass plate in the centre of the rubber mould and the small piece (1mm×1mm) of reflective thin aluminum paper was placed on top of the sample. The fiber end was aligned parallel to the reflective foil with an initial separation of less than 100 μm. When the system was stabilized and a nearly continuous optical signal of 1.31μm reflected from the interferometer was observed on the oscilloscope, data acquisition commenced five seconds before the LED was turned on, and thus, the optical signal was acquired and processed. For each signal point plotted, three experimental data were averaged.

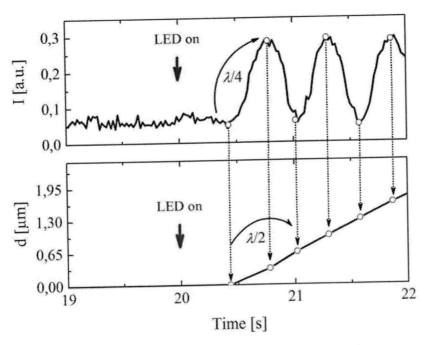

Figure 5. Processing details for a typical resin for the beginning of a curing reaction.

Samples of 1 and 2mm thickness were measured and removed from the microscope slide at the end of each experimental run by using a razor blade. Sample thickness was then measured to within ±2% with a precision micrometer. The initial sample thickness, L_0, was calculated as the sum of the thickness measured immediately after the test and the total ΔL measured from the interferogram register. The samples were irradiated for 320 s and data acquisition continued with the LED unit turned off to complete the 600 s experimental period.

The temperature during polymerization was monitored in specimens of the same thickness to those used in the shrinkage measurements (1 and 2 mm), with fine K-type thermocouples (Omega Engineering Inc., USA) embedded in the resin. The thermocouples were connected to a data acquisition system that registered values of temperature every 1 s.

The samples were irradiated for 320 s and the data acquisition continued with the LED unit turned off to complete the 600 s experimental period. Three replicates were conducted for each experiment. The experimental conditions, i.e. glass plate support, rubber ring and irradiation method, were the same as those for the shrinkage measurements. This ensured that the heat transfer of the different tests was the same.

3.3. Measurement of Contraction/Expansion during the Photocuring Process

The different samples were cured on glass subtract limited by either an aluminum ring with a diameter of 6 mm or a rubber spacer with a diameter of 10 mm [32]. The thickness of the specimens ranged from 0.2 to 5 mm and the diameter of the polymerization light from the LED was large enough to completely cover the sample surface. The surface of the specimens was exposed to uniform light flux. After the polymerization, the final thickness was measured with a micrometer.

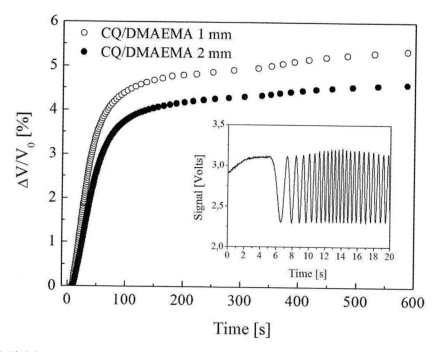

Figure 6. Shrinkage vs. time during photopolymerization for formulations containing 1wt.% CQ/DMAEMA and different sample thickness. LED unit was turned off at 320 s. Inset: Typical interferogram collected from the detector. Cure light was turned on at 5 s.

4. CANTILEVER METHOD

One way to determine viscosity or the degree of solidification of a material can be done by embedding a cantilever that is subjected to oscillation of constant amplitude and frequency. By using this approach, the profile of the cantilever behaves as shown in Figure 7, which has the lower end embedded in the resin and the other end subjected to excitation. As the polymerization process progresses, the embedded end of the beam ceases to be free. As a consequence, the amplitude of the excitation, that remains constant at the upper end, is measured by the Fizeau fiber optic interferometer. The variation in the amplitude at the measured point can be interpreted as a proportional indicator or solidification in the resin being photocured.

It is assumed that this behavior depends mainly on the excitation amplitude and frequency as well as the elastic properties of material used in the beam. Figure 8. shows the response of the beam to different positions prior to being immersed in the resin (i.e. the lower end completely free), which exhibits a very nearly parabolic profile as predicted theoretically for small vibrations. Nonetheless, taking measurements on the oscillation amplitude at a given point as a function of curing time, would indicate the strength of the material under study and thus allowing, for example to know the viscosity and its evolution through time.

A Fizeau fiber optic interferometer, due to its unique characteristics can detect precisely within a high resolution such small vibration amplitude (only a few microns). In this manner, the method provides a minimally invasive high-resolution technique in order to achieve a better knowledge about the vitrification of BisGMA curing resins. This feature provides a vital support in the early stages of photo-polymerization of embedded gratings. Since there is a period in which the structure does not interact with the fiber so it is assumed that changes in the response will be due primarily to changes in temperature. As the resin adheres to the fiber it generates a certain contraction with a consequent effect that will compete with the thermal expansion to finally be the dominant effect. For the case shown, we used an aluminum cantilever of 0.5 mm thick, 2 mm width and 500 mm length, subjected to sinusoidal signal with amplitude of 20 microns and a frequency of 10 Hz by an electromagnetic actuator fed by a signal generator (Hewlett Packard HP33120A). In order to obtain valid conclusion, the resins used were the same presented prior in this work, with the same mould geometry and irradiation profile.

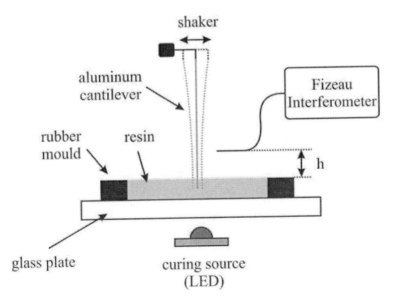

Figure 7. Experimental setup for solidification analysis.

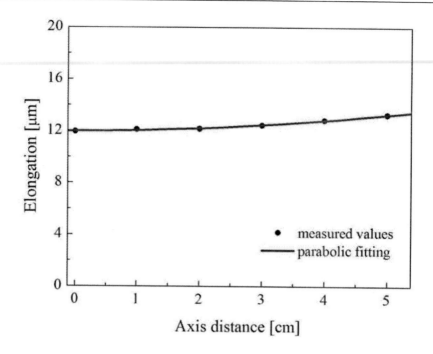

Figure 8. Nearly parabolic profile for a thin aluminum cantilever prior to be imbibed.

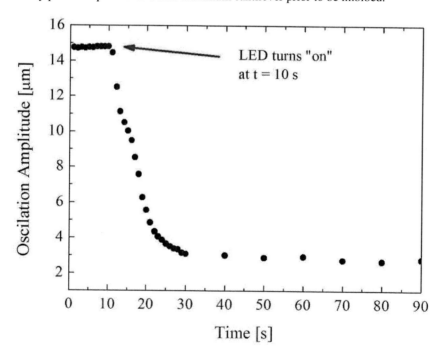

Figure 9. Cantilever amplitude variation for a typical curing process. Note the relative short solidification time compared with the entire process.

The measurement technique involves the acquisition of few interferograms from cantilever oscillation every second throughout the polymerization process. This complex signal must be post-processed to determine the amplitude of oscillation relative to the height "h" measured by the FI. A plot of a typical sequence of values is given in Figure 9.

These points represent a measure of the dynamics of solidification for a normal curing. Alternatively they can be represented in a more understandable way by subtracting the initial value of oscillation amplitude. Figure 10. shows the result, which suggests a clearly first-order exponential behavior with a time constant easy to determine with a general form

$$A_0\left(1-e^{-t/\tau}\right) \tag{9}$$

In other words, it seems to be a dynamic with a time constant that allows us to know the overall time for the resin to solidify. In this manner, the system including the cantilever and the F-I offer us extra information: changes of the system resonance induced by changes in the phase of the polymeric material show us the temporal evolution of the liquid to solid process (Figure 10).

Is important to remark that once equilibrium is reached, not changes no further changes are recorded in the signal amplitude even though the photo-curing continues until 600 seconds for this case. Although the irradiation is suspended early, or even long time after the polymerization completion (i.e. more than 48 hours), there are no changes of any kind.

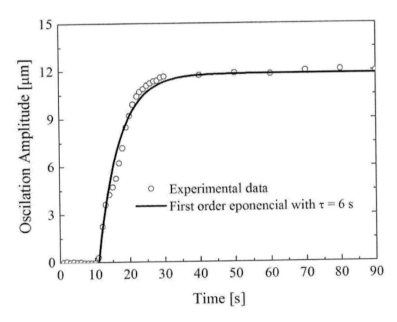

Figure 10. A more comprehensively plot that shows a very simple exponential dynamic of solidification state.

In conclusion, due to this technique we can infer that resin passes rapidly from liquid to gel and then to a vitreous state. After this short transition, polymerization should take place mainly through diffusive processes. The remaining oscillation can be regarded as the characteristic of a cantilever beam with a fixed end (the end is embedded in the resin) and the excitation at the upper end.

5. FIBER BRAGG GRATING APPLIED TO PHOTOCURED POLYMER

5.1. Principle of Operation

On an optical fiber light is confined to the fiber core by total internal reflection due the differences between the refractive index of the core and the cladding layer. On the other hand, a Fiber Bragg Grating (FBG) is a region along the length of an optical fiber's core where a permanent periodic (or quasi-periodic) modulation of the refractive index has been generated by using an intense interference pattern coming from an ultraviolet laser source [33-34]. Light propagating axially within the fiber core experiences high reflectance when the wavelength of the light matches the Bragg wavelength of the grating (λ_B), which is determined by the period of the refractive index variation generated within the core (typically about 500 nanometers):

$$\lambda_B = 2n_{eff}\Lambda \tag{10}$$

where n_{eff} is the effective refractive index of the core over the length of the grating, and Λ is the grating period.

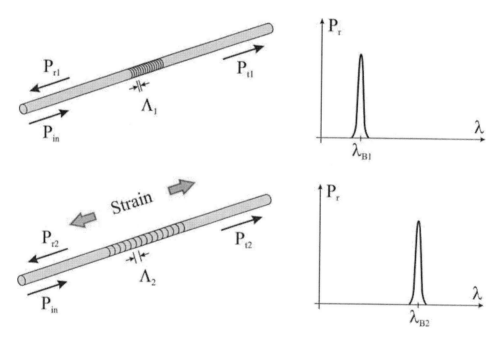

Figure 11. Principle of the Bragg grating sensor.

According to this Bragg condition, a change occurred over n_{eff} or Λ, generates the Bragg wavelength modification, and this is the basic working principle of a FBG sensor [27,35-37]. These changes can be induced by strain or temperature variations. By monitoring changes in the reflected wavelength, the temperature or strain applied to the fiber can be determined. The principle of the Bragg grating sensor is shown in Figure 11, where the input, transmitted and reflected powers, are indicated. As is shown, the FBG can be interpreted as multiple low reflective F-P interferometers.

In principle, the spectral response of a FBG is a function of temperature and strain variations due to effective refractive index and physical dimensions of the optical fiber (then, the grating spacing) are affected by the mentioned parameters [38]. The Bragg wavelength shift ($\Delta\lambda$) caused by a change of strain (ε) and a change of temperature (ΔT) can be expressed as:

$$\Delta\lambda = \lambda(1-p)\varepsilon + \lambda(\alpha_s + \zeta)\Delta T \tag{11}$$

where λ is the Bragg wavelength at the actual state, $\Delta\lambda$ the corresponding Bragg wavelength change occurring in Δt, p the strain–optic coefficient of the optical fiber, α_s and ζ are the thermal expansion coefficient and the thermo-optic coefficient of the fiber, respectively. A value of p = 0.22 is reported in the literature [38]. When the FBG sensor is embedded into a host material and both experience temperature variations, the Bragg equation is modified as follows, to account for the thermally induced axial strains in the fiber due to the mismatch between the thermal expansion coefficients of the optical fiber (α_s) and the host material (α_m):

$$\Delta\lambda = \lambda(1-p)\varepsilon + \lambda(1-p)f(t)(\alpha_m - \alpha_s)\Delta T + \lambda(\alpha_s + \zeta)\Delta T \tag{12}$$

In the generalized expression described by Eq. (12), a dependence with time ($f(t)$) is assumed because at the beginning of the photocuring process, the FBG is inside a fluid material (monomer), and after some irradiation time, more and more material solidifies. Due to results in previous section, this temporal process is assumed as defined in equation (9), where τ is a time constant related to the fluid to solid conversion process. The temporal dependence of the accumulated strain is then:

$$\varepsilon = \frac{\Delta\lambda}{\lambda(1-p)} - \{(f(t)\alpha_m - \alpha_s) + \frac{(\alpha_s + \zeta)}{(1-p)}\}\Delta T \tag{13}$$

5.2. Experimental Set-up

As it was described in Section 3, changes in the vertical dimension of the polymer sample can be monitored by a Fizeau interferometer (Figure 6). The information obtained included the strain and temperature contributions which cannot be easily separated requiring additional measurements. This situation is similar to the strain measurements performed under a horizontal scheme with Bragg gratings (FBGs).

Figure 12. shows the experimental set-up employed with the FBGs. A rectangular rubber mould is filled with the monomer under study. Two FBGs are immersed in the liquid. One of them is in direct contact with the polymeric material and is subject to strain and temperature variations, while the other is placed inside a stainless steel capillary tube in order to monitor the temperature, free from the strain induced during the photocuring process. Alternatively, a K-type thermocouple embedded in the material is used to supply temperature measurements instead of employ a second FBG as temperature sensor.

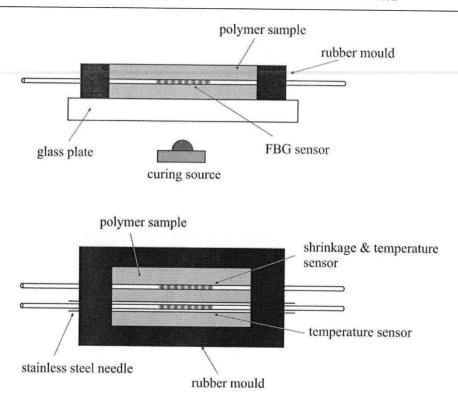

Figure 12. Detailed view of the Bragg gratins sensors incorporated into resin.

5.3. Results and Discussion

Figure 13. (a) and (b) shows the photocuring process progress by means of typical temporal evolutions of the Bragg wavelength and of the temperature of the polymeric sample. Temperature values were obtained with an additional FBG free from material contact or with a thermocouple embedded in the material. The moments when the curing source (LED) is turned on and off, can be clearly identified.

By using equation (13), the strain generated during the photocuring process can be calculated taking into account that the maximum value is obtained when the temperature variation returns to zero. Figure 14. shows the strain temporal evolution for a sample made by adding Aerosil as filler to blends made with bis-GMA/EDMAB.

As was mentioned before, diverse resins were tested in different conditions to determine their behavior throughout polymerization. From the measurements performed we can deduce that the addition of nanosilica particles substantially reduces the strain generated during photocuring and consequently the final contraction of the material. Besides, the presence of the filler encourages the diffusion of heat. Therefore, the increase of temperature experienced by the sample at the initial moments of the curing process is also reduced.

On the other hand, resins based on bis-GMA-TEGDMA activated by the addition of CQ in combination with DMAEMA take longer time for curing and experience much lower shrinkage than that which was activated by CQ/EDMAB.

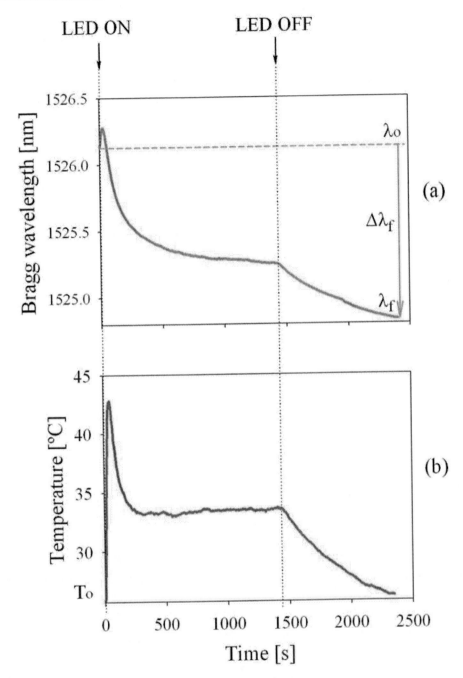

Figure 13. a) Bragg wavelength evolution with time b)Temporal evolution of temperature of the polymeric sample.

Finally, by reducing the power of the photocuring source to 50 % of its maximum value, the contraction of the material is remarkably slower, registering a reduction of the strain and accordingly of the maximum contraction greater than 50 % of the value obtained with the maximum irradiation power.

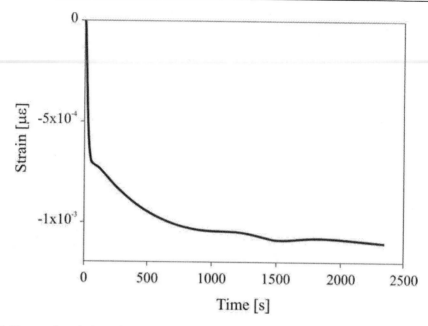

Figure 14. Temporal evolution of strain during the photocuring process.

CONCLUSION

Three techniques were employed in the work: contraction measurements by using a Fizeau interferometer (FI), analysis of the liquid to solid material evolution through vibration studies follow by the FI, and an alternative arrangement that employs Fiber Bragg Gratings (FBG) to determinate residual strain generated by the contraction. Results from these very sensible techniques are complementary.

ACKNOWLEDGMENTS

This work has been financially supported by the Consejo Nacional de Investigaciones Científicas y Técnicas (CONICET, Argentina) – PIP CONICET N° 112-200801-01769; Project I-128 of Facultad de Ingeniería, Universidad Nacional de La Plata, PICT 2005 N° N° 38289 of Agencia Nacional de Promoción Científica y Tecnológica (ANPCYT, Argentina); N.R. thanks to Comisión de Investigaciones Científicas (CIC), Provincia de Buenos Aires, Argentina, for its support (Grant Acta 1313/09).

REFERENCES

[1] Huang Y.M., Jiang C.P. *Int. J. Adv. Manuf. Technol.* 2003; 21: 586–595.
[2] Peutzfeldt A., *Eur. J. Oral Sci.* 1997; 105: 97-116.
[3] Rosin M., Urban A. D., Gärtner C., Bernhardt O., Splieth C., Meyer G. *Dent. Mater.* 2002; 18(7):521-528.

[4]   Lai J. H., Johnson A. E. *Dent. Mater.* 1993; 9: 139-143.
[5]   Cook W. D., Forrest M., Goodwin A. A. *Dent. Mater.* 1999; 15: 447-449.
[6]   Puckett A. D., Smith R. *J. Prosth. Dent.* 1992; 68: 56-58.
[7]   Tarle Z., Meniga A., Ristic M., Šutalo J., Pichler G., Davidson C. L., *J. Oral Rehabil.* 1998; 25: 436-42.
[8]   Sharp L. J., Choi I. B., Lee T. E., Sy A., Suh B. I., *J. Dent.* 2003; 31: 97-103.
[9]   Watts D. C., Marouf A. S., *Dent. Mater.* 2000; 16: 447-451.
[10] Hofmann N., Denner W., Hugo B., Klaiber B., *J. Dent.* 2003; 31(6): 383-393.
[11] Fogelman E. A., Kelly M. T., Grubbs W. T., *Dent. Mater.* 2002; 18: 324-330.
[12] Demoli N., Knežević A., Tarle Z., Meniga A., Šutalo J., Pichler G. *Opt. Commun.* 2004; 231: 45–51.
[13] Wang A. C. L., Childs P. A., Peng G. D. *Opt. Lett.* 2006; 31: 23–25.
[14] Rao, Y. J., Wang X. J., Zhu T., Zhou X. *Opt. Lett.* 2006; 31: 700–702.
[15] Yu B., Wang A. B., Pickrell G. R., Xu J. *Opt. Lett.* 2005; 30: 1452–1454.
[16] Boulet C., Hathaway M., Jackson D. A. *Opt. Lett.* 2004; 29: 1602–1604.
[17] Han M., Zhang Y., Shen F. B., Pickrell G. R., Wang A. B. *Opt. Lett.* 2004; 29: 1736–1738.
[18] Arenas G. F., Noriega S., Vallo C. I., Duchowicz R. *Opt. Commun.* 2007; 271: 581-586.
[19] Mucci V., Arenas G. F., Duchowicz R., Cook W., Vallo C. I. *Dent. Matter.* 2009; 25: 103-114.
[20] Parlevliet P. P., Bersee H. E. N., Beukers A. *Polym. Test.* 2010; 29: 291-301.
[21] Karalekas D., Schizas C. *Materials and Design* 2009; 30: 3705-3712.
[22] Antonucci V., Giordano M., Cusano A., Nasser J., Nicolais L. *J. Comp. Sci. Technol.* 2006; 66: 3273–3280.
[23] Colpo F., Humbert L., Botsis J. *J. Comp. Sci. Technol.* 2007; 67: 1830–1841.
[24] Montanini R., D'Acquisto L. *J. Smart Mater. Struct.* 2007; 16: 1718–1726.
[25] Anttila E. J., Krintilä O. H., Laurila T. K., Lassila L. V. J., Vallittu P. K, Hernberg R. G. R. *Dent. Mater.* 2008; 24: 1720-1727.
[26] Harsch M., Karger-Kocsis J., Herzog, F. *J. Appl. Pol. Sci.* 2008; 107: 719-725.
[27] Grattan K. T. V. and M. B. T. *Optical Fiber Sensor Technology Fundamentals.* Kluwer Academic Publishers, Boston, 2000.
[28] Cibula E. and Donlagic D. *Optics Express* 2007; 15(14): 8719-8730.
[29] Marcuse D. *Bell Syst. Tech. J.* 1977; 56: 703-718.
[30] Rao Y. J., Huang S. L. *Proc. SPIE* 1988; 988: 196-200.
[31] Andres M. V., Tudor M. J., Foulds K. W. H. *Electron. Lett.* 1987; 23: 774-775.
[32] Watts D. C., Cash A. J. *Dent. Mater.* 1991; 7: 281-287.
[33] Othonos A., Kalli K. *Fiber Bragg Gratings.* Artech House: Boston, London, 1999.
[34] Hill K. O., Meltz G. *J. Lightwave Tech.* 1997; 15(8): 1263–1276.
[35] Udd, E., *Fiber Optics Sensors – An introduction for Engineers and Scientists,* Udd E. ed. John Wiley and Sons, 1991.
[36] Berthold III J. W., *Industrial Applications of Fiber Optic Sensors.* 1991: p. 414.
[37] Rao Y. J. *Meas. Sci. Technol.* 1997; 8: 355-375.
[38] Lo Y. L., Chuang H. S. *Meas. Sci. Technol.* 1998; 9: 1543-1547.

In: Interferometry Principles and Applications
Editor: Mark E. Russo
ISBN 978-1-61209-347-5
© 2012 Nova Science Publishers, Inc.

Chapter 17

APPLICATION OF OPTICAL INTERFEROMETRY FOR MEASUREMENT OF (THERMO) DIFFUSION COEFFICIENTS

A. Mialdun[*] and V. Shevtsova[†]

MRC, CP165/62, Université Libre de Bruxelles,
Av. F.D. Roosevelt, 50, B-1050, Bruxelles, Belgium

Abstract

We report on the successful application of the digital interferometry for measuring diffusion and Soret coefficients in transparent organic fluids. The unique feature of this method is that it traces the *transient* path of the system in the *entire* diffusion cell. In this way it is applicable not only for studying thermodiffusive and diffusive transport mechanisms, but also for exploring convective motion. Presently, this method is not widely used for above purpose and, in our view, not because of fundamental limitations but rather due to a lack of properly developed experimental procedures and raw data post-processing. Thus, in this paper our attention is focused on the successive analysis of different steps: the fringe analysis, the choice of reference images, the thermal design of the cell and multi-parameter fitting procedure. Using the interferometry we have measured the diffusion and the Soret coefficients for three binary mixtures composed of dodecane ($C_{12}H_{26}$), isobutylbenzene (IBB), and 1,2,3,4 tetrahydronaphtalene (THN) at a mean temperature of 25°C and 50 wt% in each component. These measurements were compared with their benchmark values and show an agreement within less than 3%.

Keywords: digital interferometry; Fourier transform method; diffusion; thermal diffusion; Soret effect

1. Introduction

Diffusion is a molecular transport of mass in mixtures, which occurs in the presence of a concentration gradient and tends to reduce concentration variations. Mass transport of

---
[*]E-mail address: amialdun@ulb.ac.be
[†]E-mail address: vshev@ulb.ac.be

species caused by the thermal gradient is known as thermal diffusion or the Soret effect [1]. Here, the term "species" may refer to molecules, polymers, or small particles (colloids). There are many important processes in nature and technology, where these phenomena play a crucial role. The composition of hydrocarbon reservoirs is significantly affected by diffusion as well as the Soret effect (due to the presence of geo-thermal gradient) [2], [3]. The effect of thermal diffusion is employed for isotope separation in liquid and gaseous mixtures as well as other separation process [4] that involve colloids, macromolecules or nano-fluides. Another potential applications include high-pressure combustion[5], solidification processes, oceanic convection[6], biological systems[7] and CO_2 geological storage.

The rising interest in the reliable measurement causes appearance of new approaches [8], [9], revisiting early-established techniques [10] and tendency of cross-checking the results obtained by different methods [11],[12]. For this reason the Benchmark of Fontainebleau was performed [11], where several laboratories carried out a quantitative comparison between different experimental methods actively used in 1999-2003 to determine the diffusion, thermal diffusion, and Soret coefficients of three binary liquid mixtures. The Benchmark is still in progress as recently the results from Beam Deflection technique have been introduced to the benchmark [10] and some actively used methods have not yet contributed into Benchmark data base.

The existing methods can be divided into two groups. The first group employs convection arising from compositional and thermal variations in gravity field. In the Rayleigh-Bénard setup, the data are extracted from critical parameters for the oscillatory onset of convection in binary fluid [13]. Thermogravitational column technique [14], [15] relies on coupling between convection and horizontal thermal diffusion in a side-heated vertical slot. However, liquid sampling required in this method may disturb the diffusive process [13].

For the second group convectionless state is crucial for the measurements. The sampling problem exists in the Standard Soret Cell [13], where liquid is placed between two differentially heated copper plates. Modern techniques based on optical methods of observation do not perturb the diffusive process. In the Beam Deflection technique, the Soret cell has transparent walls and evolution of composition is observed via deflection of a laser beam passing through the medium [16], [17]. In the Thermal Lens technique [18], [19] the temperature gradient is created by heat generation resulting from the absorption of light into the fluid. The sample behaves like an optical lens due to the change of refractive index resulting from thermal and compositional gradients. Another important method is the Thermal Diffusion Forced Rayleigh Scattering [20, 21, 22, 23] where a grating created by the interference of two laser beams is converted into temperature grating by a chemically inert dye. A periodic temperature field induces periodic field of concentration due to the Soret effect. Note that the relaxation time is short but a very large number of experimental runs in the same configuration are required for obtaining reliable results [13].

Interferometry is recognized as one of the most precise methods of measuring diffusion coefficients in liquids both in past and nowadays (see e.g. [25]). Strong points of this technique are high sensitivity and ability to monitor concentration field in the whole cell but not in the distinctive points. For these reasons the method is extremely promising for applying in thermodiffusion studies as well, which had been demonstrated in [24].

But at the same time it has to be noted that application of interferometry to measurement of Soret effect is very challenging task. First, very small concentration differences suppose to be measured over experiment cell, hardly exceeding one weight percent in total. Second, very high long-term stability of interferometer is needed, as typical experiment can last few days. And third, in experiments of this type both temperature and concentration of liquid sample are inevitably entering into overall refractive index variation. Moreover, refractive index variation due to concentration change (that is quantity of interest) in most cases is order of magnitude less in comparison with refractivity variation due to temperature gradient. So, careful separation of both factors is vital.

Early attempts to use interferometry for measuring the Soret coefficients date back as early as 1957 (e.g. see [26]), but since then the method did not find a wide use either for reasons listed above or due to lack of properly developed interferogram processing techniques.

The purpose of the current paper is two-fold. First, to report development and recent improvements of the interferometry as applied to thermodiffusion studies. Second, to report results of measurements of the Soret coefficients for the benchmark mixtures: THN-$C_{12}H_{26}$, THN-IBB, IBB-$C_{12}H_{26}$. By comparison of data given by interferometry with accurate and strongly cross-checked data coming from other techniques, we would like to draw a conclusion about perspectives of interferometry in this particular field.

2. Experimental

2.1. Set-up

For the measurements of Soret effect by means of optical diagnostics, we have adopted classical thermodiffusion cell with transparent lateral walls clamped between two thermostabilized copper blocks of $60 \times 50 \times 10$ mm^3 size. A close-up view of the cell is shown in Fig. 1a. Laterally, liquid volume is enclosed in a rectangular cell, which is made of optical quality fused silica (custom made by Hellma) with external dimensions of $22 \times 22 \times 6.3$ mm^3, where $H = 6.3$ mm is the cell height; wall thickness of 2.0 mm leaves $L = 18.0$ mm for the optical path in the fluid in both horizontal directions. The glass frame is sealed between the copper blocks with two gaskets of thermal conductive rubber of 0.2 mm thick (Chomerics, Cho-Therm 1674, thermal conductivity α=1.0 Wm^{-1}K^{-1}). The choice of the cell geometry and sealing material will be discussed in more details in the results section. To prevent excess stress in glass part caused by non-uniform clamping, four equal height spacers were additionally placed between copper blocks and the glass frame.

Each copper block has a hole for sensor (a calibrated NTC thermistor Epcos, B57861S861) and a channel for cell filling of 1.0 mm diameter. External part of the channel has a widening to accommodate standard connectors for fluid management. Inlets of the channels into the cell are located at opposite corners. The outgassed liquid is injected through the channel in the bottom plate. To outgas the liquid, a flow-through degasser (Systec, OEM Mini Vacuum Degasser 0001-6274) was interposed between injecting syringe and cell. The degasser designed for outgassing solvents for HPLC applications does not affect composition of liquid mixtures. Because the deep degassing requires a low flow rate, injection of liquid is done by a syringe pump (KD Scientific, model 210) with flow rate of 0.1 ml/min.

Each copper block is thermostabilized by Peltier element (Altec, 127-1.0 $\times$ 1.0-1.15,

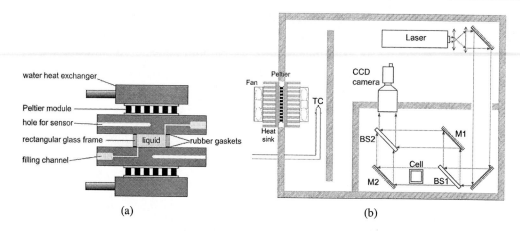

Figure 1. (Color online). (a) Cross-section of the cell. (b) Scheme of the setup.

$P_{max} = 34$ W, $I_{max} = 3.9$ A). The temperature of Peltier elements back sides is kept constant by water heat exchangers connected to the circulation water bath (Haake F3). The thermal contact of Peltier elements with the copper blocks on working side and with the water blocks on back sides was improved by insertion thin sheets of thermal conductive filler. The whole structure was gently fixed by a set of screws with equally applied torque.

Temperature of each copper block is regulated independently by two computer-driven PID controllers (Supercool, PR-59), enabling temperature stability of $\pm(0.01$-$0.02)$ K. Temperature logging is also provided by the controller with the resolution of 10^{-3} K. The data are recorded by computer with the desired sampling rate (up to 20 Hz).

To observe the concentration variation inside liquid mixtures, a digital interferometry was chosen. From a variety of different interferometer schemes (Rayleigh, Gouy, Fizeau, Michelson, Mach-Zehnder), we have selected the last one because it allows simultaneous observation of fringe pattern and the object itself and can easily treat the beam deflection problem.

Schematic drawing of the setup is shown in Fig. 1b. The light source is He-Ne laser (Thorlabs, HRR020) with power of 2 mW and wavelength $\lambda = 632.8$ nm. Laser beam was expanded to cover full area of the cell and then collimated. The collimated beam is splitted by 50R/50T plate beam splitter (50×50 mm^2 in size, Edmund Optics, NT45-854) into the reference and objective arms with cell assembly placed within the last one. Then both beams are redirected by 50 mm diameter mirrors of $\lambda/10$ flatness (Thorlabs, PF20-03-G01) at the second beam splitter to interfere. The resulting interferogram is recorded by CCD camera (JAI, CV-M4+CL) with 1392×1040 effective pixels on a 2/3" sensor and frame rate up to 24 fps. The camera is equipped with objective lens of 70 mm focal length and F2.2-32 aperture (Schneider, TXR 2.2/70-0902). The field of view covered by the imagine system is around 25 mm, so the resolution is 50 pixels/mm.

All optical components are mounted on an optical bench plate of 600×450 mm^2, which is placed on top of the optical table but mechanically decoupled and thermally isolated from the latter. To improve the mechanical stability of interferometer, all elements were mounted on low rigid 25 mm diameter posts. To ensure the thermal stability of interferometer during the experiment (up to 2-3 days), the whole setup including bearing bench plate is placed inside a box made of 3 cm thick foam thermal insulation material (shaded part in Fig. 1b). The box is equipped with air-to-air cooling/heating assembly based on Peltier element (Supercool, AA060-24220000, P = 58W, I = 3.1A) and driven by the dedicated PID controller of the same type as for the cell thermo-regulation. A set of shields made of the same insulation material is inserted in the box to prevent air perturbation over optical paths. The temperature inside the box was kept equal to the mean temperature of liquid with residual fluctuations of ± 0.1 K.

2.2. Digital Interferometry

Interferometry is a trusted and widely used optical technique for measurements of the refractivity of objects, from which related quantities like temperature or concentration can be determined [27]. For a given wavelength λ, the variation of the refractive index $n(x,z)$ includes temperature and concentration contributions

$$\Delta n(x,z) = \left(\frac{\partial n}{\partial T}\right)_{T_0,C_0,\lambda} \Delta T(x,z) + \left(\frac{\partial n}{\partial C}\right)_{T_0,C_0,\lambda} \Delta C(x,z) \qquad (1)$$

where $\Delta T(x,z)$ and $\Delta C(x,z)$ are temperature and concentration changes at point (x,z), here z-axis is in the direction of temperature gradient. At the same time Δn may be obtained from the phase change $\Delta\varphi$, which is measured by interferometry:

$$\Delta n(x,z) = n(x,z) - n(x_0,z_0) = \frac{\lambda}{2\pi L}\Delta\varphi(x,z) \qquad (2)$$

Here L is the optical path in liquid. As it follows from Eq. 2 the variation of refractive index Δn is equivalent to the change of optical phase $\Delta\varphi$. Hereafter we will work with $\Delta\varphi$ as it is measured value and is additive quantity contrary to Δn.

In order to increase the accuracy of phase evaluation of interference fringe patterns beyond the early fringe scanning technique, several procedures have been developed, such as the Fourier transform technique [28] or the temporal phase stepping technique [29]. These procedures are well established now.

2.2.1. Fringe Analysis for Phase-Measuring Interferometry

In optical interferometry, the change in phase between two coherent light waves (reference and objective ones) is the reason of spatial intensity variation. The results of optical interference of two light waves, an interferogram or fringe pattern, is taken by sensor as digital picture. We will shortly address the strategy of extracting phase information from fringe pattern following [30] and [31] since this is fundamental for the discussed experimental technique. The intensity of two superposed waves of the same frequency is

$$i = |A_r + A_o|^2$$

where $A_r = a_r \cdot exp(-i\varphi_r(x,z))$ and $A_o = a_o \cdot exp(-i\varphi_o(x,z))$ are amplitudes of reference and objective beams and φ_r, φ_o are the initial phase angles. Accordingly, the intensity in the fringe pattern at $t = t_1$ is

$$i(x,z,t_1) = i_0(x,z) + m(x,z) \cdot cos[\Delta\varphi(x,z,t_1)] \qquad (3)$$

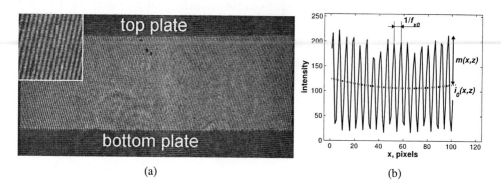

Figure 2. (a) Typical interference pattern of 512 × 1024 pixels size covering full cell width with insert of magnified fringe pattern in top left corner (b) Intensity profile on a horizontal line crossing an interferogram.

where $\Delta\varphi(x, z, t_1) = \varphi_o(x, z) - \varphi_r(x, z)$ is the initial phase shift, $i_0(x, z)$ is the background intensity distribution, and $m(x, z) = 2|a_r| \cdot |a_o|$ is the modulation function. A typical interference pattern recorded by camera is presented in Fig. 2a. Typical intensity profile over the horizontal line $z = H/2$ is shown in Fig. 2b. It is possible to modify the spatial frequency of the fringes, $f_0 = \{f_{x0}, f_{z0}\}$, by changing the incident angle of the object and reference beams through adjusting the inclination angle of one the mirrors.

At time t_2, when the object beam goes through the diffusing media the intensity of the fringe pattern is given by an equation similar to Eq. 3

$$i(x, z, t_2) = i_0(x, z) + m(x, z) \cdot cos[\Delta\varphi'(x, z, t_2)] \qquad (4)$$

$$(5)$$

Thus, $\Delta\varphi'(x, z, t_2) = \Delta\varphi(x, z, t_1) + \Delta\varphi$ where the phase shift $\Delta\varphi$ is caused by refractive index variation of the diffusing media since time $t = t_1$ to time $t = t_2$.

Without violating generality we may suppose that the spatial frequency has x-component only ($f_0 = \{f_{x0}, 0\}$) then Eq. 4 can be rewritten as

$$i(x, z) = i_0(x, z) + c(x, z) \cdot exp(2\pi f_{x0}) + c^*(x, z) \cdot exp(-2\pi f_{x0}) \qquad (6)$$

where $c(x, z) = 0.5 \cdot m(x, z) \cdot exp[i\Delta\varphi'(x, z)]$ is a complex amplitude. The asterisk superscript (∗) denotes a complex conjugate.

To calculate the phase shift $\Delta\varphi'$, we have adopted a powerful two-dimensional Fourier transform technique. Originally developed for treating 1D phase distributions [28], this method was later adapted for 2D phase maps [32]. Performing the discrete Fourier transform of the array $i(x, z)$ yields

$$I(f_x, f_z) = I_0(f_x, f_z) + C(f_x - f_{x0}, f_z) + C^*(f_x + f_{x0}, f_z) \qquad (7)$$

The location of the terms I_0, C and C^* in Fourier domain and its intensity is shown in Fig. 3 a,b. If the introduced mirror tilt (correspondingly f_{x0}, f_{z0}) is appropriate, then these contributions are well separated in Fourier domain. Both terms $C(f_x - f_{x0}, f_z)$ and

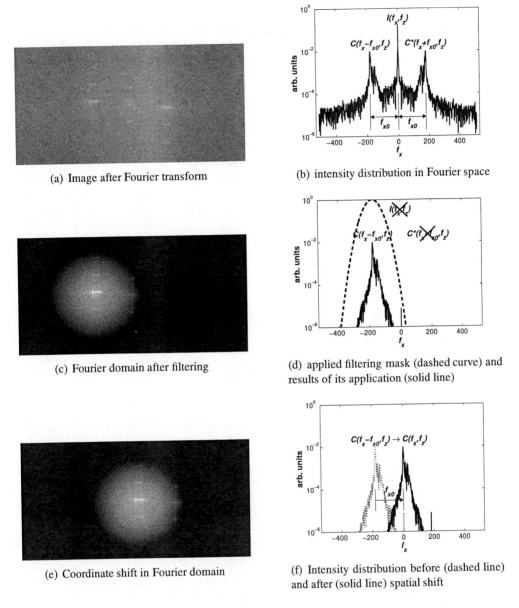

Figure 3. Steps of interferogram digital evaluation by 2D Fourier transform.

$C*(f_x + f_{x0}, f_z)$ contain equal information about the phase shift and a filter is applied in the Fourier space to keep only one of them. Filtering sets to zero all frequencies except those belonging, for example, to the term, $C(f_x - f_{x0}, f_z)$. The resulting image is shown in Fig. 3 c,d while the dashed curve shows the intensity distribution of the applied mask. During this procedure all filtered out terms must be suppressed, but at the same time useful information within the remaining term $C(f_x - f_{x0}, f_z)$ should be preserved.

After that 2D Fourier transform output is rearranged by moving the spectrum on f_{x0} towards origin of Fourier domain. It eliminates the carrier frequency f_{x0} and mathematically gives the term $C(f_x, f_z)$, see Fig. 3 e,f.

In performing the inverse Fourier transform one can reconstruct both amplitude and a spatial distribution of total phase change of the fringe pattern produced by object beam. Accordingly

$$C(f_x, f_z) \to c(x, z) \quad \text{and} \quad \Delta\varphi'(x, z) = arctan\left\{\frac{Im[c(x, z)]}{Re[c(x, z)]}\right\} \tag{8}$$

A similar procedure is applied to the reference image for the evaluation of the phase $\Delta\varphi(x, z, t_1)$, see Eq. 3. The reference interferogram also determines the vector f_0. Subsequently, the phase distribution $\Delta\varphi(x, z, t_1)$ must be subtracted from $\Delta\varphi'$ that corresponds to the fringe pattern recorded after the change of an object. Correspondingly, the required phase shift in Eq. 2 is $\Delta\varphi = \Delta\varphi'(x, z, t_2) - \Delta\varphi(x, z, t_1)$. In this way, the method applies the holography principle and tracks the only posterior optical phase variation in the set of images following the reference image (taken at $t = t_1$).

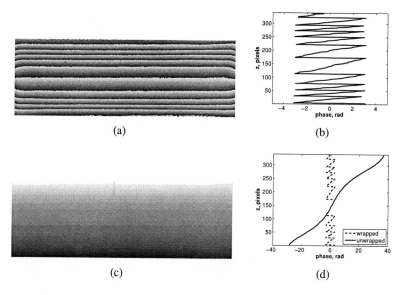

Figure 4. Phase map resulting from Fourier processing: (a) wrapped 2D distribution and (b) wrapped vertical profile; (c) unwrapped 2D map and (d) corresponding vertical profile.

The phase difference calculated by arctangent function is wrapped, which means that it belongs to the range $(-\pi, \pi)$, see Fig. 4a,b. It should be unwrapped to construct the continuous natural phase. It is a simple task for one-dimensional case, while for two-dimensional case and noisy fringes, sophisticated unwrapping techniques are required. As in our case phase maps typically have good quality, we adopt the following approach for phase unwrapping. The procedure starts at a pixel with well defined neighbourhood assuming error-free phase there. Following a spiral path the phase unwrapping is performed by comparing wrapped phase with previously validated neighbours. If the difference is less than π the phase remains unchanged. If the difference between two pixels is more than π, the phase equals its wrapped phase minus 2π. If the difference is less than $-\pi$, the phase equals its wrapped phase plus 2π. By the end the relative phase change between two pixels is placed in the range $-\pi$ and π and a smooth 2D phase map is obtained (see 4c,d).

It should be noted that the processing based on Fourier transform introduces two ambiguities to the extracted phase value. The sign ambiguity is caused by the fact that the term for filtering in Fourier domain, $C(f_x - f_{x0}, f_z)$ or $C^*(f_x + f_{x0}, f_z)$, is usually chosen arbitrarily. The absolute value ambiguity is naturally arising from wrapped character of the calculated phase. These ambiguities can be eliminated by using a priori information about the system (e.g. known temperature of copper blocks and mean concentration of liquid).

2.2.2. Subtraction of Reference Image

The two-dimensional phase distribution (map) contains information about many things. The wave front is distorted by: all optical elements along the beam path $\Delta\varphi_{optics}$, non-uniform air temperature $\Delta\varphi_{air}$, temperature distribution in glass walls $\Delta\varphi_{glass}$, temperature distribution in liquid bulk $\Delta\varphi_{th}$, and finally by concentration distribution in the liquid $\Delta\varphi_C$.

$$\Delta\varphi = \Delta\varphi_{optics} + \Delta\varphi_{air} + \Delta\varphi_{glass} + \Delta\varphi_{th} + \Delta\varphi_C \qquad (9)$$

Each particular interference pattern is formed either by all above factors or by some of them, depending on the state of instrument. An important part of overall processing routine is the separation of contributions. The advantage, which gives vast flexibility to this technique, is that any fringe pattern stored in the computer memory can be taken as a reference depending on the aim of the current processing step.

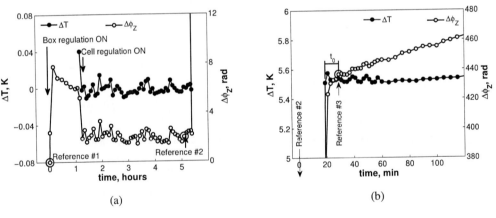

Figure 5. Readings of temperature on sensors (left vertical axis) and phase variation measured optically (right vertical axis) with time. (a) Thermalization process at $T_0 = 25^oC$; (b) Non-isothermal step.

Let us briefly review this approach by considering an example of typical data processing steps. The first step traces the phase distribution during the instrument thermalization process. The step lasts 4-6 hours and aims at stabilizing the temperature of the cell and interferometer at mean temperature of $T_0 = 25^oC$. The very first image, Ref.1 in Fig. 5a, is taken as a reference and processing continues during entire isothermal step. In this case, the reference image keeps information about the initial state of the optics only ($\Delta\varphi_{ref_1} = \Delta\varphi_{optics}$)

and the following phase variation is caused by possible optical path perturbations due to different mean temperatures of air and cell. In Fig. 5a the closed circles (see left vertical axis) show alteration of temperature on sensors while the open circles (see right vertical axes) show the corresponding variation of the phase $\Delta\varphi_z$. Importance of this step is to make sure of complete interferometer stability before experiment is started. At the end of this step, a new reference image ($\Delta\varphi_{ref_2} = \Delta\varphi_{optics} + \Delta\varphi_{air}$) must be chosen to process images after applying the thermal gradient over the cell (see Fig. 5b).

At the second step with typical duration of a few minutes, the phase variation is caused by the temperature differences in the glass wall and liquid bulk; concentration contribution is negligibly small. Because the diffusion characteristic time, τ_D, is much larger than the thermal one, τ_{th}, Eq. 1 can be decomposed. For benchmark liquids $\tau_D = H^2/D \approx 12$h while $\tau_{th} = H^2/\chi \approx 200$ s, here χ is the thermal diffusivity of liquid. Thus the total phase variation at the observed point, affected by temperature gradient, can be written as

$$\Delta\varphi = \Delta\varphi_{ref_2} + \Delta\varphi_{glass} + \Delta\varphi_{th}. \tag{10}$$

Assuming that temperature distribution in the glass is linear ($T = T_0 + \Delta T[z/H - 0.5]$), glass thickness and temperature contrast factor for glass are well known, one can determine contribution of glass walls $\Delta\varphi_{glass}(z)$. Here H is the cell height.

$$\Delta\varphi_{glass}(z) = \left[T_0 + \Delta T\left(\frac{z}{H} - 0.5\right)\right]\left(\frac{\partial n}{\partial T}\right)_{glass} \frac{2\pi L_{glass}}{\lambda}$$

where $L_{glass} = 4mm$ is the optical path in glass. Substituting Eqs. 2 and 10 into Eq. 1 the optically measured temperature distribution inside liquid is

$$T(x,z) = T_0 + [\Delta\varphi(x,z) - \Delta\varphi_{ref_2}(x,z) - \Delta\varphi_{glass}(z)]\frac{\lambda}{2\pi L}\left(\frac{\partial n}{\partial T}\right)^{-1}_{C_0,T_0} \tag{11}$$

This optical measurements of temperature field should be done as soon as the temperature field is established, i.e. within a few thermal times, $t \approx 3-9$ min. Later on optical measurements will start deviating from sensors readings due to concentration contribution, see Fig. 5b beyond point Ref.3. This deviation point, which appears after the temperature difference is established, is the reference point for the third image processing step - thermodiffusion separation, $\Delta\varphi_{ref_3} = \Delta\varphi_{ref_2} + \Delta\varphi_{glass} + \Delta\varphi_{th}$. This new reference interferogram holds information about all inputs into optical phase except concentration variation. Processing the next images with respect to this one provides the phase change, from which a full 2D map of concentration field is extracted

$$C(x,z) = C_0 + [\Delta\varphi(x,z) - \Delta\varphi_{ref_3}(x,z)]\frac{\lambda}{2\pi L}\left(\frac{\partial n}{\partial C}\right)^{-1}_{C_0,T_0} \tag{12}$$

2.2.3. Experimental Limitations and Precision of the Method

The suggested method is based on Fourier processing, and its accuracy strongly depends upon the number of periods in the analyzed signal (i.e. the number of fringes in the interferogram). It means that the carrier fringe system has to be sufficiently dense. Small fringe spacing will also provide a distant peak in the Fourier domain, which is favourable for accurate filtering. At the same time, fringes have to be distinguished in the image. Fringe spacing of $4-6$ pixels seems to be optimal for the method.

The other experimental conditions that strongly influence phase accuracy are the stability of light intensity (good quality laser is needed) and stability of interferometer. Last point is especially important. Perturbations of interferometer by environmental disturbances was analysed in detail in [24] for a similar problem and it was shown that they can affect the measurement. For this particular reason, we took the extra precautions for stabilizing the interferometer as mentioned in section 2.1..

We tested the interferometer stability in two cases. First test is when only the box thermoregulation is switched on; another one is when the box and cell are thermostabilized at the mean temperature. All tests last at least 24 hours. The variations of phase differences with time between the opposite extremes of the field of view (top and bottom of the cell, for example) were analyzed. Typical value of phase variation for the first test of interferometer is $\pm(0.1\text{-}0.2)$ rad. This value agrees well with the common accuracy of interferometry $(2\pi/50)$, which can be found in the literature [31]. Phase stability in the second test is $\pm(0.5\text{-}1.5)$ rad. This deterioration of phase stability is not related to the interferometer itself, but to the stability of cell thermal regulation, which was also confirmed by sensors readings (e.g. see Fig 5a).

The unique feature of this method is that it traces the *transient* path of the system in the *entire* cell. In this way it fits not only for measurements of Soret coefficients but also for studying diffusive transport mechanism.

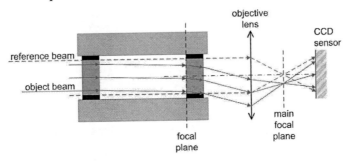

Figure 6. (Color online). Beam deflection problem.

2.2.4. Beam Deflection Problem

It is well known that light beam refracts when passing through medium with refractive index gradient. Both, a constant temperature gradient applied over the cell and gradually increasing concentration gradient definitely cause refractive index variation, which in turn causes deflection of the object beam. This principle is used for measurements of transport coefficients in beam deflection technique; for detail description one can refer to [10], [17], [33]. In our technique it may play negative role as the reference beam, which bypasses the cell, does not suffer from beam deflection. It means that the object beam may interfere with various regions of the reference beam at different time instants. For an ideally collimated beam with perfect plane wave-front this point is not important. However, as the wave-front can be slightly disturbed in reality (e.g. by imperfect optical elements), it is better to exclude the problem *ab origin*. There is a natural way to make a correction for the beam deflection. Namely, an objective lens can return the beam refracted by an object into its initial position being properly focused (see Fig. 6). The problem is reduced to seeking the ideal working focal plane of the objective lens. To sum up, a simple practical rule can be applied for minimizing beam deflection effect on interferometry: imaging system has to be focused at the inner plane of the glass wall nearest to the camera, as it is drawn in Fig. 6. This agrees with conclusions of [34].

3. Results and Discussion

3.1. Cell Optimization

Another important aspect of the experiment design is the overall cell geometry. The cell geometry is a matter of serious optimization and we outline the most influential features. For example, the height of the cell should be large enough to allow easier manipulations with the probing beam. At the same time, it should not be too large since the time of experiment is proportional to the cell height squared ($\tau_D = H^2/D$). The length of the cell should be also large enough since the signal measured by the instrument is directly proportional to the optical path (length). However, a long cell, which is advantageous for the beam deflection technique, is not the best choice for interferometry.

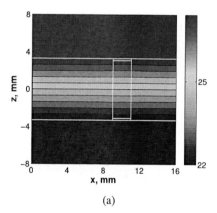
(a)

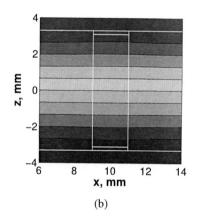
(b)

Figure 7. (Color online). (a) Computed temperature field in the half of the cell; white lines shows the contact lines liquid/copper and liquid/glass, see entire geometry in Fig. 6; (b) The temperature distribution in magnified region near the contact line liquid/glass.

The cell geometry is important, but here we focus our attention on another factor that is really crucial for the measurement: the thermal design of the cell. Non-accurate thermal design can completely discard all advantages of interferometry, since it determines presence and intensity of residual convection inside the cell. Due to drastically different characteristic times for transport of momentum and mass, the presence of convective flows in diffusion cell will significantly alter the concentration field. For example, it was shown that local convective vortex with flow speed of $10 - 20 \, \mu m/s$ can completely homogenize concentration distribution in the region where this vortex exists [24]. So, residual convection has to be reduced as much as possible.

When the thermal gradient is directed against gravity, the mechanical equilibrium in the cell is generally stable. But convection can easily appear at lateral walls of the cell and in the corners where thermal perturbations are inevitable. In our first cell design [24], [35] we faced this problem of horizontal temperature gradient at the corners. Using the advantage of the optical interferometry to monitor all motions in the entire cell during transient process, the zones affected by convection were determined and excluded from consideration.

Unlike to the previous work, in current study we have started with stationary heat transfer calculation of the supposed cell geometry with realistic boundary conditions. We used Comsol Femlab package as it allows easy generation and treatment of complex geometries. The thermal properties of copper, quartz glass, water, rubber and air are taken from a physics handbook. Working on a design of a new cell we have considered different feasible cell geometries, and drew a conclusion that almost complete elimination of lateral temperature gradients is possible, but in very few cases. One of them finally chosen by us is the following: *flat* copper plates (without protrusions and grooves) in the area of contact with liquid and glass walls. Sealing between glass and copper is made of material with thermal conductivity close to that of glass. Nowadays such materials are commercially available as thermal conductive composite rubber. One can see the sketch of geometry in Fig. 1a and Fig. 6. The results of numerical simulation, shown in Figs. 7a, b, demonstrate absence of horizontal temperature gradients near the walls. Experimental observations of temperature field made by interferometer confirm the absence of lateral heat fluxes as well (see Fig. 8a). The concentration field resulting from thermodiffusion separation in this cell (Fig. 8b) also

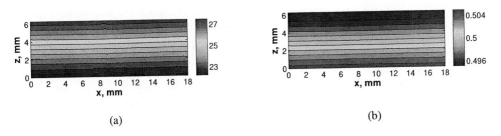

Figure 8. (Color online) Stationary fields in $THN - C_{12}H_{26}$ measured by interferometer. (a)Temperature field; (b) Concentration field.

demonstrates the absence of convective motion since the concentration distribution is practically non-disturbed and varies only along the temperature gradient.

3.2. Extraction of Transport Coefficients

3.2.1. Fitting Equation for Full Path

The present technique gives a unique possibility of increasing measurement accuracy by providing information about concentration distribution along the whole thermodiffusion path. In fact, the method gives two-dimensional concentration field, although the distribution itself is almost one-dimensional. This extensive information can be used for evaluation of convection, but in the case when convection is really negligible, the 1D approach can be used for mathematical description of measurements. Each time a measurement is taken, a full concentration distribution over the thermodiffusion path is determined. The solution of thermodiffusion equations for convection-free case can be written as follows [24]:

$$C(z,t) = C_0 + C_0(1-C_0) S_T \Delta T \left[\frac{1}{2} - \frac{z}{H} - \frac{4}{\pi^2} \sum_{n,odd}^{\infty} \frac{1}{n^2} \cos\left(\frac{n\pi z}{H}\right) exp\left(-n^2 \frac{t}{\tau_r}\right) \right] \quad (13)$$

Here $\tau_r = H^2/\pi^2 D$ is the relaxation time, D and S_T are the diffusion and the Soret coefficients, H is the height of the cell. Sign between two first terms in the Eq.13 depends on the choice of the component which we follow: lighter or heavier. Sign is negative (+) in the case of heavier component (as it is written) and positive (-) in the case of lighter component.

In Eq.(13) there are two unknown parameters, S_T and τ_r. The simplest procedure to find them is to compare the concentration measured at point z_i at time t_j with its value from Eq.13 using initial guess for S_T and τ_r. Then fitting is done iteratively by using the Nelder-Mead algorithm (Matlab), which minimizes the misfit function

$$\delta C(z,t) = \sum_{i,j}^{k,m} [C_{exper}(z_i,t_j) - C_{theor}(z_i,t_j)]^2 \quad (14)$$

by varying the fitting parameters S_T and τ_r. The number of spatial pixel points in the experimental dataset is around $k = 340$ and the number of acquired images is varied in limits $200 < m < 800$. So, the fitting is done with the matrix of at least $k \times m = 340 \times 200$ size. Thanks to such big amount of data this approach gives satisfactory results. However there exists a less obvious fitting parameter which allows essential improvement of the fitting results. This fitting variable, t_0, can be called *initial time*.

The point is that the reference image for extracting concentration distribution is not necessarily located at the very beginning of the separation step (see Fig. 5b). Note that this image has to be taken only after the temperature profile is completely established, see section 2.2.2.. A reliable reference image can often be found $3-9$ minutes after the visible separation starts, although in the above fitting the time of reference image is $t = 0$ and $C_{exper}(z,0) = C_0$, which is not precise. This can be resolved by introducing third fitting variable t_0. Then the theoretical anticipation for experimental concentration profile is given by

$$C_{theor}(z,t) = C(z,t_0+t) - C(z,t_0) + C_0 \quad (15)$$

where both terms $C(z,t_0+t)$ and $C(z,t_0)$ are calculated according to Eq.13. Such correction provides much better fit to the experimental data as it is shown in Fig. 9a. Note that the snapshot of data is presented at the time $t = 0.16\tau_r$, which is far from steady state. The open circles present experimental results; the dashed curve shows $C_{theor}(z,t)$ when two fitting parameters were used; the solid line shows $C_{theor}(z,t)$ when three fitting parameters were used. We should emphasize that margins of the new fitting parameter t_0 is known from the experiment (see Fig. 5b) with rather good accuracy.

Finally, we will discuss the source of errors, which is less evident and more difficult to treat. It arises from the fact that the temperature of copper plates is not strictly constant; it has some minor fluctuations as the regulation is active. In this case experimentally observed profile, considered as pure concentration one, has some input of the thermal nature. This input appears due to optical coupling of both values (i.e. $\Delta \varphi_{th} = const[1 + \delta(t)]$) and can

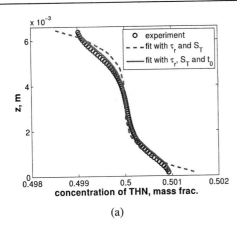 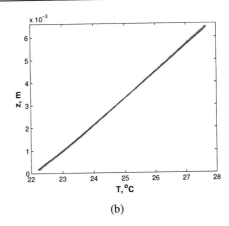

Figure 9. (Color online). (a) Fitting optimization; Experimentally measured $C(z)$ (open circles) and its fitting curves with two (dashed curve) and three (solid curve) fitting parameters for THN-$C_{12}H_{26}$ at $t = 0.16\tau_r$, (b) Ten snapshots of temperature profile for THN-$C_{12}H_{26}$ display small oscillations $T(z)$ near horizontal walls due to temperature instability of copper plates.

be either negligibly small $(max|\delta(t)| \to 0)$ or noticeable depending on ratio of respective contrast factors and temperature fluctuation value. This effect can be seen in Fig. 9b where ten snapshots of temperature profile are shown for THN-$C_{12}H_{26}$. Temperature fluctuations can be seen by eye near horizontal walls where their amplitude achieve 0.5% of signal at the worse case.

Our dedicated study showed that the amplitude of temperature disturbances rapidly decay and do not penetrate deeply into the liquid bulk. However, in some cases (as shown in Fig. 9b) when other uncertainties are added to the signal (e.g. due to low contrast factor) we have to consider near-wall regions as corrupted and crop out them from consideration. For different reason similar approach was applied in [24]. Comparison of S_T obtained on the basis of such transient approach and from steady state justifies the procedure.

3.3. Benchmark Mixtures

Chemicals used for experiments are dodecane (Acros Organics, 99%), isobutylbenzene (Acros Organics, 99.5%) and 1,2,3,4-tetrahydronaphthalene (Acros Organics, 98+%). The above substances were used without further purification. Then three binary solutions were mixed according to the required weight fraction 50wt%.

Temperature difference applied over the cell was constant in all cases and equal to $\Delta T = 5.53 \pm 0.02$ K, while the mean temperature was $T_0 = 298$ K. The Soret coefficients for all three mixtures are positive. So, no hydrodynamic instabilities are expected for these mixtures.

Values of refractive index variation with temperature and concentration, $(\partial n/\partial T)_{T_0,C_0}$ and $(\partial n/\partial C)_{T_0,C_0}$, (the so-called contrast factors), are required for the measurement of species separation. We rely on the values provided in [21], where the measurements were performed by precise and reliable technique using the same quality chemicals provided by the same manufacturer.

Table 1. Evaluation of signal to noise ratio (SNR). Comparison of refractive index variation due to concentration and thermal impacts for investigated mixtures.

Mixture	$\left(\frac{\partial n}{\partial T}\right)^1$ $\times 10^{-4}, 1/K$	δn_T $\times 10^{-6}$	$\left(\frac{\partial n}{\partial C}\right)^1$ $\times 10^{-2}, 1/K$	ΔC $\times 10^{-3}$	Δn_C $\times 10^{-4}$	SNR
THN-$C_{12}H_{26}$	−4.41	8.82	11.7	8.54	9.99	113
IBB-$C_{12}H_{26}$	-4.54	9.08	6.28	3.64	2.29	25
THN-IBB	-4.76	9.52	5.44	2.88	1.57	16

Each experiment consists of two steps and at each of them the images are acquired at specified rate. First, Soret separation step starts when the temperature gradient is established. This step lasts for about 10-12 hours. Second, relaxation step starts when temperature gradient is removed with purpose of measurement the diffusion coefficients in isothermal regime. This step also takes about 10-12 hours. To summarize, during one experiment each coefficient is measured twice: at the *Soret* step we measure diffusion D and Soret S_T coefficients in transient regime. An alternative value of S_T is determined from achieved steady state

$$S_T = -\frac{1}{C_0(1-C_0)}\frac{\Delta C_{st}}{\Delta T}, \qquad (16)$$

and an alternative value of D is obtained from *relaxation* step in isothermal regime.

Majority of the methods used for measurements of transport coefficients works with point measurements at given time or with a gradient. Optical digital interferometry enabled measurements of ΔC between arbitrary points and, correspondingly, reproduce concentration field in the whole two-dimensional cross-section.

The time-dependencies $\Delta C(t)$ for three benchmark mixtures are shown in Fig. 10 on left side while spatial concentration distributions for the same liquid are shown on right side. The concentration profiles are presented at four different times which are very far from the steady state: t/τ_r= 0.11, 0.33, 1.0, 3.0. Note, that stationary values ΔC_{st} in Fig. 10a,c,e should be divided by $\Delta T < 5.53K$ in Eq. 16 as some small regions near horizontal walls are cropped out; see right sides of Fig. 10. Succession of the profiles $C(z)$ demonstrates the development of the diffusion process with time. The small fluctuations of the fitting curve near the horizontal walls, shown via δC in Fig.10f, results in to the scattering of the points on separation curve $\Delta C(t)$ which is more than satisfactory. Nowadays the presented technique is the only one, which can display evolution of the concentration field with time, as shown on the right pictures in Fig. 10.

At the first glance the separation curves $\Delta C(t)$ for various mixtures exhibit different scattering of data points. Reason of that becomes clear if we consider optical properties of media. Let us calculate variation of refractive index of liquid due to temperature inconstancy, $\delta n_T = \delta T(\partial n/\partial T)$ where $\delta T = \pm 0.02K$, and total refractive index difference due

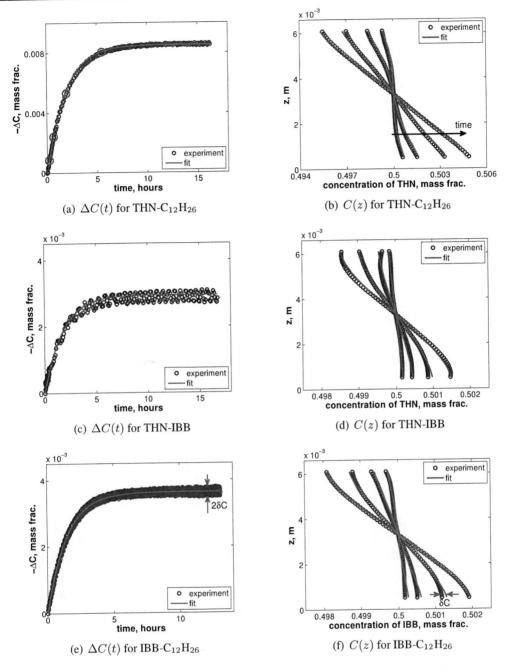

Figure 10. (Color online). Experimental results for benchmark liquids: open circles indicate experimental points and solid curves outline their fitting. Pictures on left side (a),(c),(e) show separation of component with time. Pictures on the right side (b),(d),(f) show the concentration profiles over the height of the cell at non-steady regime with progress in time, $t/\tau_r = 0.11, 0.33, 1.0, 3.0$. The curve for the smallest time is close to be vertical; τ_r is the relaxation time.

Table 2. Diffusion and Soret coefficients for three benchmark mixtures.

	quantity	THN-C$_{12}$H$_{26}$	THN-IBB	IBB-C$_{12}$H$_{26}$
This work	$D^1{}_S, 10^{-10} m^2/s$	6.16 ± 0.05	8.6 ± 0.28	9.4 ± 0.22
	$D^2, 10^{-10} m^2/s$	6.37 ± 0.05	8.43 ± 0.28	9.23 ± 0.22
	$D^3, 10^{-10} m^2/s$	**6.27 ± 0.29**	**8.52 ± 0.12**	**9.32 ± 0.12**
	$S_T, 10^{-3}$ $1/K$	9.24 ± 0.01	3.29 ± 0.11	3.98 ± 0.08
Benchmark values ref.[11]	$D, 10^{-10} m^2/s$	6.21 ± 0.06	8.5 ± 0.6	9.5 ± 0.4
	$S_T, 10^{-3}$ $1/K$	9.5 ± 0.3	3.3 ± 0.3	3.9 ± 0.1

to species separation, $\Delta n_C = \Delta C(\partial n/\partial C)$. Then one can deduce a kind of signal-to-noise ratio (SNR), which is ratio of above values, $\Delta n_C/\delta n_T$. These data are summarized in Table 1. The contrast factors $(\partial n/\partial C)_{T_0,C_0}$ for the last two mixtures are small and it results in poor SNR. Ratio signal/noise for THN-C$_{12}$H$_{26}$ is largest because of largest contrast factor and largest separation ΔC.

Although the SNR value is somehow overestimated (because this specific noise corrupts a part of data only), it can provide a clear idea about method applicability range. For example, in regions where contrast factor $\partial n/\partial C$ vanishes, or separation decreases due to vanishing S_T, the SNR drops down along with accuracy of the technique. There is an evident way to increase the SNR by increasing temperature stability, but capability of this way is limited. As soon as temperature stability will rise from, say ±0.01 K to ±0.001 K (it will extend SNR by order of magnitude), an optical noise will enter into play, which is much more difficult to suppress. To conclude, the transport coefficients can be measured with a good accuracy for the values of SNR as low as 10-15.

The measured values of diffusion and Soret coefficients for the three mixtures are given in Table 2 together with the original benchmark references. The thermodiffusion coefficient $D_T = D \cdot S_T$ can be easily calculated. The results were obtained using 3-parameters fitting procedure (D, S_T, t_0) with value of parameter t_0 close to the measured one, see Fig. 5b. Two values for diffusion coefficient are given in Table 2: from isothermal and non-isothermal measurements. For comparison with benchmark data an average value of D is used.

For all mixtures we have got an excellent agreement for both diffusion and Soret coefficients, our measurements differs from previous benchmark values by 1-3 %.

4. Conclusions

We have discussed in depth the use of digital interferometry for the measurement of diffusion and Soret coefficients. Although the first impression may give the idea that the technique is a combination of known elements (Soret cell and interferometer), the proper design of the novel instrument reveals the unique feature of this method - the observation of the temperature and concentration fields in *transient* regime in the *whole* cross section, and not just in distinctive points. Nowadays the presented technique is the only one, which can display evolution of the concentration field with time, as shown on the right pictures in Fig. 10.

From the first time when we presented this technique [24] the cell design as well as the image processing and interferometer performance have been considerably improved. One of the targets of this paper was to describe this technique in such a way that it becomes practical and widely employable. With this goal in mind, the successive steps of the entire procedure of the interferometric technique were presented with a focus on measuring the diffusion and Soret coefficients. Among all steps four have been selected for in-depth consideration: the fringe analysis, the choice of reference images, the thermal design of the cell and the multi-parameter fitting procedure. In addition, the sources of possible uncertainties were identified and guidelines were drawn for an improvement of the accuracy of the results.

With the renewed technique we have measured the diffusion and the Soret coefficients for three benchmark systems composed of dodecane ($C_{12}H_{26}$), isobutylbenzene (IBB), and 1,2,3,4 tetrahydronaphtalene (THN) at a mean temperature of 25°C and 50 wt% in each component. The results showed an excellent overall agreement with the proposed benchmark references. The data measured by optical interferometry were missing from the benchmark databases. As a conclusion, our results are in the general trend of other ones, and prove that the suggested technique is competitive, while having some attractive advantages.

Acknowledgements

This work is supported by the PRODEX programme of the Belgian Federal Science Policy Office and ESA. The author is indebted to Prof. F. Dubois (ULB, Brussels) for valuable discussions.

References

[1] C.Soret, *Arch. Sci. Phys. Nat.* **2**, (1879) 48.

[2] K. Ghorayeb, A. Firoozabadi, and T. Anraku, *SPE journal* **8**(2) (2003) p.114.

[3] F. Montel, *Entropie* 184–185, (1994) p. 86.

[4] F.M. Richter, N. Dauphas and F.-Z. Tenga, *Chemical Geology* **258** (2009) p. 92.

[5] K. Harstad and J. Bellan, *Int. J. Multiphase Flow* **26** (2000) p.1675

[6] R.M. Schmitt, *Annual Rev. Fluid Mech.* **26** (1994) p.255.

[7] F.J. Bonner and L.O. Sundelöf, *Z. Naturforsch. C* **39** (1984) p.656.

[8] S.A. Putnam and D.G. Cahill, *Rev. Sci. Instrum.* **75** (2004) p.2368.

[9] M. Hartung and W. Köhler, *Rev. Sci. Instrum.* **78** (2007) 084901.

[10] A. Königer, B. Meier and W. Köhler, *Phil. Mag.* **89** (2009) p.907.

[11] J.K. Platten, M.M. Bou-Ali, P. Costeseque et al., *Phil. Mag.* **83** (2003) p.1965.

[12] T. Pollak and W. Köhler, *J. Chem. Phys.* **130** (2009) 124905.

[13] J.K. Platten, *ASME J. of Applied Mechanics*, **73** (2006) p.5.

[14] J.K. Platten, M.M. Bou-Ali, J.F. Dutrieux, *Philosophical Magazine* **83** (2003) p.2001.

[15] P. Blanco P., M.M. Bou-Ali, J.K. Platten, P. Urteaga, J.A. Madariaga, C. Santamara, *J. Chem. Phys.*, **129** (2008) 174504.

[16] P. Kolodner, H. Williams and C. Moe, *J. Chem. Phys.* **88** (1988) p.6512.

[17] R. Piazza, *Phil. Mag.* **83** (2003) p.2067.

[18] M. Giglio and A. Vendramini, *Appl. Phys. Lett.* **25** (1974) p.555.

[19] H. Cabrera, L. Martin-Lopez, E. Sira, K. Rahn, M. Garcia-Sucre, *J. Chem. Phys*, **131** (2009) 031106.

[20] K. Thyagarajan and P. Lallemand, *Opt. Commun.* **26** (1978) p.54.

[21] G. Wittko, W. Köhler, *Phil. Mag.* **83** (2003) p.1973.

[22] C. Leppla and S. Wiegand, *Phil. Mag.* **83** (2003) p.1989.

[23] P. Polyakov and S. Wiegand, *J. Chem. Phys.* **128**, 034505 (2008)

[24] A. Mialdun, V. Shevtsova, *Int. J. Heat Mass Transfer* **51** (2008) p.3164.

[25] S. Maruyama S. and A. Komiya, *J. Flow Visualization and Image Processing*, **13** (2006) p.243.

[26] L.G. Longsworth, *J. Phys. Chem.* **61** (1957) p.1557.

[27] W. Merzkirch, *Flow Visualization*, Academic Press, Orlando, 1987.

[28] M. Takeda, H. Ina and K. Kobayashi, *J. Opt. Soc. Am.* **72** (1982) p.156.

[29] K. Creath, Phase-measurement interferometry techniques, in *Progress in Optics XXVI*, E. Wolf, ed., Elsevier Science Publishers, 1988, p.350.

[30] H.J. Tiziani, *Opt. Quant. Electron.* **21** (1989) p.253.

[31] M. Hipp, J. Woisetschlger, P. Reiterer and T. Neger, *Measurement* **36** (2004) p.53.

[32] W.W. Macy Jr., *Appl. Opt.* **22** (1983) p. 3898.

[33] K.J. Zhang, M.E. Briggs, R.W. Gammon, and J.V. Sengers, *J. Chem. Phys.* **104** (1996) p.6881.

[34] C. Mattisson, D. Karlsson, S.G. Pettersson et al., *J. Phys. D : Appl. Phys.* **34** (2001) p.3088.

[35] A. Mialdun, V. Shevtsova, *Microgravity Sci. Technol.* **21** (2009) p.31.

In: Interferometry Principles and Applications
Editor: Mark E. Russo

ISBN 978-1-61209-347-5
©2012 Nova Science Publishers, Inc.

Chapter 18

MODERN ARTWORK DOCUMENTATION QUALITATIVE EVALUATION OF SECONDARY INTERFERENCE FRINGES: A STANDALONE STRUCTURAL DIAGNOSTIC TOOL IN ARTWORK DOCUMENTATION

Vivi Tornari

Institute of Electronic Structure and Laser (IESL) Foundation
for Research and Technology, Hellas, Greece

EXECUTABLE ABSTRACT

The direct result of visible interference processes is seen as the macroscopic effect of formation of interference fringes or interference fringe patterns. These are generated due to coherence phenomena concerned in wave physics in general and in coherent physics in particular. In regards to the nowadays most common sources to generate coherence phenomena one should reasonably think of a laser light source. This chapter considers the fringe patterns which are generated after interference of coherent light beams and their visual effect of alternate dark and bright fringes as a direct basis for qualitative analysis in Cultural Heritage documentation.

Interferometry is a well known technique for quantitative measurement of shape deformation due to field wise observation of object point's displacement. In optical interferometry a single light beam is divided in two beams travelling separate paths in space and recombining by an optical element to create the phenomenon of interference captured in a screen or detector. In holographic interferometry the process is repeated twice with an initial single beam primarily coherent light divided to an object and a reference beam paths from which the object carries the information of illuminated surface recombined to reference beam in detector plane without use of an optical element. Overlapping of the interference field produces the visual effect of secondary fringe patterns. The holographic interference is possible under strict experimental settings and principles. However demanding the process it is has become possible to use holographic interference in a number of different industrial and medical applications and more recently to be included as a competitive technique in the structural documentation and diagnosis of art objects.

An artwork consist a unique piece of human kind and preservation to the next generations has always been a demand. Since antiquity history-witnessing objects were subjected to the effort of preservation for the next generations to come. In this context structural inspection and diagnosis of Cultural Heritage items requires highly sensitive and accurate techniques which can retrieve inborn and upcoming deterioration well before it becomes visible to the eye. Phase information is the coded quantity in holography interference fringe patterns and offers a unique sensitive detector of structural displacements due to any externally induced factor. Phase changes provide high information content allowing tracing invisible defects under the surface with unbeatable quality and clarity compared to any other known method including x-rays and tomographic techniques. In fact the result of phase encoding in the holographic interference process through the fringe pattern secondary distribution of intensity turns to visual evidence the underneath surface activity of defects. Each defect produce own set of localized secondary interference fringes among the general fringe pattern such that each one uniquely indicate its subsurface effect on the surface. Exact location, size, shape and value of hidden defect and the profile of its deformation can be extracted directly by naked eye within a reasonable error appraisal. The generated fringes have to be separated by the general interference field, isolated, identified and allocated too an internal cause. The basis required understanding the fringes of holographic interferometry and assigning them to underneath effects and shape deformation are given in this chapter. Automatic processes are also considered and difficulties for their implementation specifically in the field of art conservation are described.

Holographic recording can be performed in a variety of optical geometries depending on the investigation aim without interacting with the precious surface or requiring any intervention prior to illumination. It is consider a fully non destructive and not interventive method as ethics of treatment in Cultural Heritage field presuppose. The protocols of investigation involved in this application are presented and explained.

Examples of results are given in characteristic case studies in sufficient range of art objects variety.

The objective of this chapter is to familiarize the reader with the complexity of secondary interference fringes and produce the evidence of the unique source of information that is unfolded in their formation prior to or without the electronic post-processing routines that are usually implemented in the analysis of interferograms.

It should also be highlighted the fact that the artwork application due to the uniqueness of artworks, the strict requirements for their preservation, and last but not least the complexity of the results which cannot feed the known automated routines, is considered through a different approach than other known applications of the same techniques aiming in other fields.

INTRODUCTION

Human kind has been proved restless in creating evidence of its existence throughout history. Preservation of Cultural Heritage (CH) was a concern since antiquity and conservation methods have been developed thereafter and continue to develop till our days. Modern techniques and methods are developed or implemented following the state of the art in technology in any current time. At these very dates such a progress is seen in implementation of laser techniques and sources for analysis, assessment and documentation of materials and CH objects. It is thus not surprising that the ultimate tool for coherent light emission has been introduced in the timeless field of CH preservation as early as the first commercial laser devices came available in the market [1-3].

True it is that the very first application of lasers in CH aimed to "cleaning" of polluted surface brought out the utility as a structural diagnostic tool. The enthusiasm raised by conception of holographic image generation was due to the laser invention one of the most outstanding applications of laser light. It is the holographic principle that allowed for very first time to use interferometry for rough surface investigation and works of art belong to the rough surface category [4-6].

As the branch of science utilising laser sources was expanding so did the experimentation on artwork flourishing new branches of applications for artwork conservation. Soon after the competency of laser coherent interferometry as a hologhraphic interferometry non destructive testing analysis (HINDT) was worldwide accepted [7-12]. HINDT remains most competent in analysis of artworks' rough surfaces despite emerging of many other applications. Contrary to nowadays automation requirements HINDT main advantage remains the direct qualitative performance provided by the appearance of secondary formation of macroscopic fringe patterns [13, 14].

The direct result of any visible light-interference process is seen as the macroscopic effect of a formation of interference fringes or interference fringe patterns. These are generated due to coherence phenomena among same wavelengths concerned in wave physics in general and in coherent physics in particular. In regards to the most common sources of nowadays coherence phenomena the laser light is the chosen source.

This chapter considers the fringe patterns generated after interference of coherent laser light emitted beams of visible wavelengths resulting in the visible by naked-eye effect of alternate dark and bright fringe patterns. These are treated as feed in data of microscopic physical aging processes dominating the structure of mechanical bodies long before any visible aging sign appears as defect on surface [15-20]. Treated as data for quantitative analysis [21] for many years this chapter aims to treat equally important the information of the data to serve for direct qualitative analysis primarily in the field of structural documentation of artworks of Historic and Cultural Heritage.

1. BRIEF ON OPTICS FUNDAMENTALS

Visible light represents only a small portion of the entire electromagnetic spectrum of radiation that extends from high-frequency gamma rays through X-rays, ultraviolet light, infrared radiation and microwaves to very low frequency long-wavelength radio waves. The complex phenomenon of visible light is classically discussed in terms of rays and wavefronts.

Optics is the physical science that studies the origin and propagation of light, how it changes, what effects it produces, and other phenomena associated with it. There are two branches of optics. Physical optics is concerned with the nature and properties of light itself. Geometrical optics deals with the principles governing image-forming properties of lenses, mirrors, and other devices, such as optical data processors.

1.1. Nature of Light

Separate models according to the field of study have been used to describe the nature of light. Since light behaves as particle and wave with specific paths, optical scientists have modelled their observations according approaches that could better describe the revealed phenomena. Christian Huygens[1] championed the view that light is a wave motion, spreading out from a light source in all directions and propagating through an all-pervasive elastic medium called "ether". The laws of reflection and refraction and double refraction in calcite by wave theory were derived. Isaac Newton[2] clearly regarded rays of light as streams of very small particles emitted from a source of light and travelling in straight lines. Although Newton was aware of the phenomenon of light patterns (named after him, the Newton rings) he maintained his basic particle hypothesis dominating the century that followed. Thomas Young double-slit experiment[3] with the generation of the complex interference pattern of shadows demanded a wave interpretation. Augustin Fresnel[4] and his results with double refraction in calcite required light be a transverse wave. For each of the two components of polarised light he developed the Fresnel equations, which give the amplitude of light reflected and transmitted at a plane interface separating two optical media. James Clerk Maxwell synthesized known principles in his set of four equations[5] yielded the prediction for the speed of light as the speed of an electromagnetic wave in ether. From then on, light was viewed as a particular region on the electromagnetic spectrum of radiation. Although the nineteenth century served to place the wave theory on a firm foundation the wave-particle controversy was resumed with vigor by the new difficulties that showed up in situations involving interactions of light with matter. However important quanta and quantum mechanics introduction exceed the purpose of this chapter and the reader should only remember that Max Planck derived the blackbody radiation spectrum assuming that atoms emitted light in discrete energy chunks rather than in a continuous manner[6] whereas Albert Einstein offered the explanation of the photoelectric effect, the emission of electrons from a metal surface when irradiated with light, based on the conception of light as a stream of photons whose energy is related to frequency by Planck's equation[7], and Niels Bohr incorporated the quantum of radiation in his explanation of the emission and absorption processes of the hydrogen atom[8], while Arthur Compton explained the scattering of X-rays from electrons as particle-like collisions between photons and electrons[9]. Louis de Broglie published his hypothesis[10] that subatomic particles are endowed with wave properties using again Planck's constant [Annex I].

---

[1] Treatise on light, 1678
[2] Optics, 1730.
[3] 1801.
[4] Wave Optics, 1821.
[5] 1873.
[6] 1900.
[7] 1905.
[8] 1913.
[9] 1922.
[10] 1924.
[11] 1962.

The Dual Nature of Light

- **Ray Model of Light**
  Describing optical phenomena with the path of light

- **Wave Model of Light**
  -Color is described naturally in terms of wavelength.
  -*Required* in order to explain the interaction of light with material objects with sizes comparable to a wavelength of light or smaller

- **Particle model of light**
  - *Required* in order to explain the interaction of light with individual atoms. At the atomic level, it becomes apparent that a beam of light has a certain graininess to it.

Figure1. Fundamental light division scheme.

The wave-particle duality came into force. Light behaved like waves in its propagation and its fundamental wave phenomena (interference-diffraction), behaves though as particle in its interactions with matter. However electrons as well behave like photons but with the crucial difference that have mass. *Principle of complementarity* expressed by Niels Bohr clarifies that are neither waves nor particles and generally the full intelligibility of a photon or an electron is not exhausted by either model.

Another important distinction between electrons and photons is that electrons obey Fermi statistics, whereas photons Bose statistics. The restriction that derives is that no two electron in the same interacting system be in the same state whereas identical photons with same energy and momentum can and it is this principles that forms the basis for Light Amplification by Stimulated Emission of Radiation or LASER[11]. A profound consequence of the wave nature of particles is embodied in the Heisenberg *principle of indeterminacy* in which particles do not obey deterministic laws of motion. Rather the theory predicts only probabilities. Wave functions are associated with the particles through the fundamental wave equations or quantum mechanics. The wave amplitudes or better the square of wave amplitudes assigned to these particles provide a means of expressing the probability that a particle will be found within a region of space during an interval of time. Thus the *irradiance* (power/area) of these waves at some intercepting surface also proportional to the square of the wave amplitude provides a measure of this probability.

In the theory called quantum electrodynamics, which combines the principles of quantum mechanics with those of special relativity, photons are assumed to interact only with charges. An electron is capable of both absorbing and emitting a photon with a probability that is proportional to the square of the charge. There is no *conservation law* for photons as there is for charge associated with particles. In this unification of this theory light is just another form of matter.

Nevertheless the complementary aspects of particle and wave descriptions of light remains justifying separate use accordingly when appropriate (note summary in figure 1).

The beam path of travelling light is adequately described with rays while wavelike phenomena as interference and diffraction require a wave model and in order to study the microscopic level phenomena light is described in terms of particles and photons. The duality of light in nature is a very interesting and puzzling phenomenon itself with a lot of investigation going on continuously in our days to better define it.

Optical Science

- **Geometrical Optics:** Reflection, Refraction, Imaging

- **Wave Optics:** Coherence, Interference, Diffraction

- **Visual Optics:** Eye seeing, Optical devices

Figure 2. Fundamental optical science division scheme.

Accordingly, Optical science is divided in three broad categories in respect to the study of interest. The geometrical, wave and visual optics embrace the knowledge regarding the modern optics (note summary in figure 2). In the phenomena to follow in this presentation the wave optics are implied.

1.2. Wave Optics

The wave optics treats light distinctively as a wave motion. Coherent optics is a division of wave optics to describe effects of light beams represented mainly by laser beams. Therefore the term coherent optics is directly related to the involved optical phenomena produced by laser beams of light. These phenomena classified as interference and diffraction are observed in nature but in the context of this presentation the interest primarily lies in the manifestations produced by laser light. Interference and diffraction are important interrelated phenomena which are used in many laser applications presented in this book and chapter. Laser light is a narrow beam that by aid of lenses can be expanded to form extended spherical and plane waves. Therefore these waves can be used to interfere in order to study wave phenomena.

1.2a. Interference

According to the description of coherence interference is described as superposition of cosine time dependent waves. The optical frequencies of interest in which the electric and magnetic fields not being directly measured is the intensity of radiation expressed as the square of amplitudes as the major quantity of interest. The two waves are capable of generating various phenomena such as none of them could alone generate. A manifestation of such phenomenon is the interference in which the phase difference between the two waves is expressed as intensity distributions governed by the laws of constructive and destructive wave interference. When the total phase difference δ equals to multiples of odd integers of π

minimum intensity is obtained while with integral multiples the intensity obtained is maximum. Thus an irradiance pattern stable in time and spatial domain is generated for as long as spatial phase difference remains unchanged.

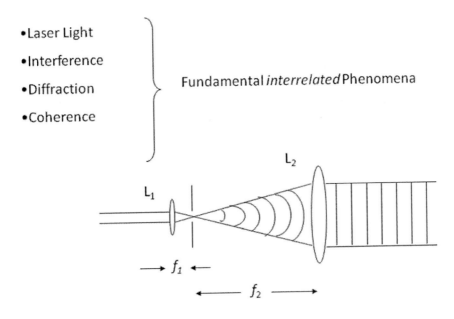

Figure 3. Coherent light can be manipulated by optical elements from which fundamental interrelated phenomena are studied.

Interference is illustrated by waves superposition in fig 4. Two coherent waves travelling in phase they produce a double intensity wave and their interference is termed as constructive while two coherent waves travelling out of phase they produce a minimum or moderate intensity wave and their interference is termed as destructive. It is apparent in this illustration the interrelation of the coherence and interference phenomena.

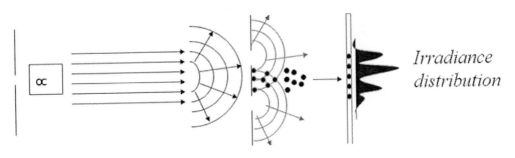

Figure 4. Light is diffracted and coherent wavefronts interfere to produce intensity distributions of phase changes.

1.2b. Diffraction

Diffraction and interference are interrelated phenomena. The fringe pattern produced by the interference experiment can be used as the main optical component provided a photosensitive medium captures it. This optical component carrying the product of interference is termed diffraction grating because if a beam of laser light illuminates it from one side beams will emerge from the other side of the component. The separation of the beams depends from the wavelength of the laser light used and the angle between the recording beams. The diffracted waves can be expressed at the plane of the grating as a wave with real and imagery parts with an attenuation t depended on photosensitive medium absorption. The principles of diffraction derived from the famous in wave optics Huygens experiment with the two slits and the bending of the light "rays". Any obstacle placed in a propagating beam of light will bend the "rays" and act as a secondary source of light with new wavefronts emerging from the obstacle looking as deriving from the obstacle. If some conditions are satisfied such as slit separation, size and observation distance then an irradiance pattern is generated that is the product of diffraction and interference.

The interrelated phenomena of interference and diffraction produce special optical components such as diffraction gratings that can in turn act to an incident light beam on demand. The diffraction order that is the amount of diffracted beams as well as their separation is governed by the Bragg law and is depended on the recording angle and the wavelength.

ANNEX I

1-1.:

$E=h\nu$

Where the constanct of proportionality, *Planck's constant*, has the value of 6.63×10^{-34} J-s

1-2. : A particle with momentum p has an associated wavelength of

$\lambda = h/p$

with *h* planck's constant.

1-3. : Quantum mechanics, or wave mechanics, deals with all particles localised in space and so describes both light and matter. Combined with special relativity, the momentum *p*, wavelength λ, and speed for both material particles and photons are given by the same general equations:

$$p = \frac{\sqrt{E^2 - m^2 c^4}}{c}$$

$$\lambda = \frac{h}{p} = \frac{hc}{\sqrt{E^2 - m^2 c^4}}$$

$$\upsilon = \frac{pc^2}{E} = c\sqrt{1 - \frac{m^2 c^4}{E^2}}$$

m is the rest mass and E is the total energy –sum of kinetic energy $m\upsilon^2/2$ and rest-mass energy mc^2. A crucial difference between particles like electrons and neutrons and particles like photons is that the latter have zero rest mass, so the above equations take the simpler form for photons

$$p = \frac{E}{c}$$

$$\lambda = \frac{h}{p} = \frac{hc}{E}$$

$$\upsilon = \frac{pc^2}{E} = c$$

1-4. : When large numbers of particles are involved, probabilities approach certainties, so that the irradiance E_e of light at a location is proportional to the number of photons passing through the location per second

n (photons/m^2-s)= $E_e/h\nu$

1-5.: Scalar diffraction theory describes the e/m wave in linear, isotropic and homogeneous medium

$$\nabla^2 u(\mathbf{x},t) - \frac{n^2}{c^2}\frac{\partial^2 u(\mathbf{x},t)}{\partial t^2} = 0$$

$$u(\mathbf{x},t)$$

any coordinate at field positions e, b **and** for light waves with $A(\mathbf{x})$, $\varphi(\mathbf{x})$ amplitude and phase of wave, ν optical frequency

$$u(\mathbf{x},t) = A(\mathbf{x})\cos(2\pi\nu t + \varphi(\mathbf{x}))$$

2. SECONDARY INTERFERENCE FRINGES: DEFINITION

The term "secondary" indicates the interference fringes formed as a secondary product of interference between two primary products of interference. The spatial superposition of cosine time dependent waves of same frequency E_1, E_2 with a wave amplitude, ω angular frequency, k_1 and k_2 the wave vectors and φ total phase difference of the superimposed waves define "primary' products of interference:

$$E_1 = a_1 \cos(\omega t - k_1 r)$$

$$E_2 = a_2 \cos(\omega t - k_2 r + \phi)$$

Total phase difference φ give rise to the primary interference and the primary interference fringes are formed if $\varphi \neq 0$.

In coherence phenomena two spatially closed wave fields can be written as total addition of E:

$$E = E_1 + E_2$$

The electric and magnetic frequencies of E are not directly measured in optical fields instead the intensity of radiation I is the quantity of interest:

$$I = |E|^2 = |E_1^2| + |E_2^2| + 2|E_1 E_2|$$

Combining and averaging gets:

$$I = I_1 + I_2 + 2\sqrt{I_1 I_2} \cos \delta$$

With

$$I_1 = a_1^2, \quad I_2 = a_2^2 \quad \delta = k_2 r - k_1 r - \phi$$

The two waves are capable of generating various phenomena such as none of them could alone generate. A manifestation of such phenomenon is the interference in which the phase difference between the two waves is expressed as intensity distributions governed by the laws of constructive and destructive wave interference.

$$I_{min} = I_1 + I_2 - 2(I_1 I_2)^{1/2}$$

at points $\delta = 2N + 1$

$$I_{max} = I_1 + I_2 + 2(I_1 I_2)^{1/2}$$

at points $\delta = 2N\pi$,

with N= 1, 2, 3, ...

We have just written the irradiance pattern of phase differences expressed among two coherent spatially superimposed fields. The optical frequencies (630≥λ≥420 nm) on which we are interested are patterned in a field of cosine distribution as equidistant light and dark regions. It is clear from this discussion that interferometry can be used to convert a phase difference to an irradiance pattern. Gabor [Gabor 1949] invented holography by adding a coherent reference wave to the object wave the interference pattern of which can remain stable and can be captured in space by a screen or recording medium. The produced interference pattern is microscopic consisting a primary light interference field and is irresolvable by naked eye or microscopes and can be revealed on sight by SEM imaging methods. However if this pattern is illuminated by same reference wave light diffraction occurs reproducing the amplitude of object wave. This reproduction is an identical of the object wave as it was at the instant of recording and it is termed a Hologram.

It is the stable interference among two holograms which can be used to profoundly study the secondary fringes for art work documentation sake. The secondary fringes are produced with a variety of optical geometries but are obliged to strict boundary conditions in geometry to satisfy the diffraction grating producing interference along the chosen angle θ, (d·sinθ=m·λ with d: primary fringe distance and m:1,2,3..).

2.1. Secondary Fringe Generation

Secondary fringes are interferometric comparisons of two coherent waves scattered by the same object at another time. The composition of these two waves will be referred as "secondary interference pattern". The term secondary denotes a visible patterning formation of interference fringes projected on a screen or viewed through a photosensitive medium by the naked eye as in the case of optical holographic interferometry. Secondary fringes have been demonstrated with lasers emitting optical, infrared, ultraviolet, microwave and ultrasonic radiation. In consideration of artwork documentation in visible light only results with the optical wavelengths are being illustrated.

2.2. Basic Optical Geometry

The secondary interference patterns are results of linear processes in the sense that two or more optical waves can be recorded at different times but reconstructed at the same time. Therefore one can record the light wave of an object of interest in different times and reconstructed sequentially having created a "history" of the object waves through time. In reconstruction can use two or more of these history events as witnessed by the object waves to create different secondary patterns as a result of sum, difference, time average of the recorded events. It is expected a different interference pattern for each different pairs of waves at least if the original waves were even slightly different so that the resulted $\varphi \neq 0$. A typical optical setup to produce series of separated records is shown in figure 5.

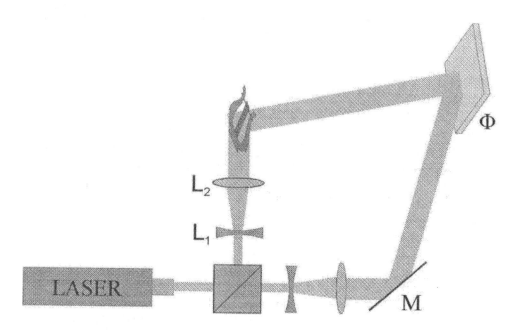

Fiure.5. A schematic of a typical geometry setup to generate phase modulated interference fringes. The setup can be used to produce primary and secondary fringes by single or double exposing the photosensitive medium at position Φ

The irradiance of each reconstructed pair recorded sequentially in time and reconstructed simultaneously is expressed as the irradiance of the object modulated by the constructive and destructive fringes of interferometry, thus a pattern of bright and dark zones. So that the irradiance of the reconstructed wave becomes:

$$I(x,y) = |U_1(x,y) + U_2(x,y)|^2$$
$$= |\alpha(x,y)\exp[-i\varphi(x,y)] + \alpha(x,y)\exp\{-i[\varphi(x,y) + \Delta\varphi(x,y)]\}|^2$$
$$= |2\alpha(x,y)|^2 \{1 + \cos[\Delta\varphi(x,y)]\}$$

(4)

where $2\alpha^2(x,y)$ the amplitude of the image of the object and $\Delta\varphi(x,y)$ the phase difference due to the displacement of the object between exposures and the cosine intensity distribution term $\{1+\cos[\Delta\varphi(x,y)]\}$ describes the fringes. Fringes are curves with constant values of $\Delta\varphi$ with dark fringes being odd and bright fringes even multiples of π, αν example of visible reconstruction of an antique artwork copy is shown at figure 6a, b.

 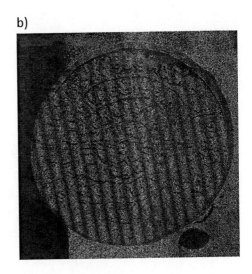

Figure 6a, b. In a) Photo of Bronze copy of Phaistos Disc, and b) photo of example of secondary interference fringes due to mechanical force, the reconstruction shows both the visible B&W image and the equidistant fringe pattern representative of isotropic displacement.

A fringe pattern as the example demonstrated at figure 6b is a visible demonstration of a surface change related to an optical path displacement ($\varphi \neq 0$) due to a physically or artificially induced change which corresponds to a physical quantity such as temperature, pressure, mechanical and other forces, induced either intentionally or not. Thus fringes represent volume reaction translated in terms of displacement in the optical path producing a directly measurable quantity to quantify the reaction of any distinct object to the induced force. Using the laser wavelength to illuminate the investigated surface and define the optical path one holds a powerful ruler to measure microdisplacements from the initial position. The microdisplacements are expressed as multiples (N) of laser wavelengths (λ) in μm units.

Fringe patterns are subject to change, any change in fringe density or fringe formation is related to a change of surface points of the object that represents. The reasoning of its generation is searched to material properties and environment. This is valid for all different types of objects but is critical to the investigation of artworks due to the rarity and preciousness of the worth they represent for which environment and material deterioration are carefully controlled.

2.3. Basic Portable Geometry

The basic optical geometry was used to create the suitable geometry for portable device with digital instead of optical recording medium. In the portable device optical elements are stable or computer controlled operating under a strict investigation procedure with the protocol of basic investigation steps being similar both for laboratory and field expeditions. The portable system presented in figure 7 is developed for artwork examination and is using holographic interferometry optical geometry with a lens collecting the interference speckle-fields to a digital camera. It is termed Digital Holographic Speckle Pattern Interferometry (DHSPI) and it is constructed with the aim to assist documentation of artworks independently of laboratory conditions. Speckle is called the inherent property of coherent laser wave to interfere with neighbouring photons producing speckle fields as bright and dark dot-like or granule-like visual effects. Can be easily observed if a laser beam scatters on a rough diffuse surface [1].

Figure 7. A DHSPI system for artwork examination consists of the laser-optics head, the PC-monitor and peripheral devices (DHSPI is a development of Holography laboratory at IESL/FORTH).

Table I. DHSPI Technical Parameters

Parameters	Values	Affects	Depends on
Ratio RB/OB	1/5	Contrast of Reconstructed image	Surface Reflectance, Surface - CCD Distance
Surface- CCD Distance (DST)	15-70cm	OB Diameter and Image Resolution	Size of object
F.O.V.	Adjustable	OB Diameter and Image Resolution	Collecting Lens, Surface-CCD Distance
Excitation Duration	1-60 sec or ΔT +8°C max	Defect Detection	Artwork Material/construction
Interval time	>0,25sec	Defect Detection	Artwork Material/construction

Fixed Parameters	Value
RB-OB Angle	<2,7°
Spatial Frequency (f=sinθ/λ) Lines/mm	*88,5 max
Wavelength, nm	532
Pixel Size, μm	6,45
CCD	H1392xV1040
Illumination Angle	<10o
Resolvable fringe-pairs on Image Surface	<24 fringe pairs/cm

*88,5 primary fringes/mm are equivalent to 0.57 fringes/pixel, while 1 primary fringe-pair correlates to 1.75 pixels, eg. a CCD of 1040 pixels can record 520 primary fringes and produce 24 secondary fringes/cm of FOV.

The technical parameters of DHSPI system are subjected to system upgrading however obey to strict boundary conditions of holographically produced speckle interferometric recording, shown in TABLE I. Remote investigation presupposes tolerably stable environmental parameters in respect to vibrations.

2.4. Investigation Procedure

In order to generate the secondary fringe patterns which characterise the state of deterioration one should thus be able to generate an induced displacement in surface points with a controlled external excitation. After placing an artwork on the optical table or placing the portable device facing the artwork a controlled thermal excitation is induced on artwork surface while a thermometer records the surface temperature and a chronometer the duration in seconds.

In table II it is tabulated the experimental stepwise procedure. The working procedure starts with the setup alignment and for the case of double exposure recording starts with the first capture of the initial object state as shown in table II. Then a transient excitation is applied which for artwork inspection it is suggested to be a thermal IR-source that is well suited in provoking variety of thermal coefficients and thus material displacement to allow detailed defect visualization. An indicative table of excitation duration is shown in table III. After the induced excitation has been applied then a second exposure with equal time duration as the initial is recorded on the same photosensitive medium. The holographic interferogram is thus recorded and the processing and post-processing routines follow.[10]

Table II. Recording Procedure

Steps	Procedure
1st step	Prepare the Artwork: Align the artwork to diffusely reflect the laser light towards the photosensitive medium with specified FOV
2nd step	Prepare the CCD and take FOV image: Place a white card or a photometer to balance the intensity, adjust aperture, take a FOV photo.
3rd step	Settling time: Shut off the beams and allow the artwork to settle from vibrations
4th step	Exposure time: Turn on the beams for the 1st exposure for duration relevant to the total brightness estimated (the lower the brightness the higher the exposure time required, depends on technical characteristics)
5th step	Excitation time: Turn on the excitation source (IR- lamp) or any device to induce transient displacement. Turn on the chronometer of excitation.
6th step	Exposure time: Turn on the beams for the 2nd exposure for equal time as in 1st exposure Steps 3, 4 and 6 can be repeated as long as displacement is able to produce interference fringes. Steps 3-6 can be repeated at will if 1-2 remain same.

In table III an example of induced excitation and provoked alteration of surface temperature measured from the surface of an icon is indicatively given. In digital recordings several interferograms can be recorded in sequence and be studied as a function of time. In this respect the defect can be observed through its pattern over time from the first appearance till its extinction due to the excitation cease. The artwork surface is getting excess of energy diffusing it to the rest of its volume and the equilibrium is temporarily lost for as long as each material it takes to transmit the received energy excess. During the physical process and before a new state of equilibrium is reached records of surface microdisplacement are captured.

Table III. Indicative Excitation Duration And Provoked Δt From An Icon Sample

Initial surface Temperature °C	Duration of thermal load (sec)	Direction of thermal load in respect to artwork surface	Surface temperature after thermal load °C	Temperature Difference ΔT°C
23	5	front	26	3
23	5	back	26	3
23	10	front	29	6
23	10	back	29	6
23	15	front	30	7
23	15	back	30	7
23	20	front	31	8
23	20	back	31	8

The number of visual fringes (secondary fringe patterns) covering the field of view (FOV) of the investigated surface appears different during the process that material takes to reach equilibrium and it is expected to reach its maximum at two crucial investigation points: a) at the start of the project as soon as the surface gets the stream of excitation and/or b) at the time that all volume contributes to displacement of the surface. The rest of the times the displacement is expected to be equal or increased or decreased relatively to the previous measurement.

2.5. Localised Fringe Patterns: A Structural Defect Detection Key to Artwork Documentation

Holographic interferometry non destructive testing (HINDT) has been successfully transferred from research and industrial applications to the investigation of artworks and antiquities aiming as main priority to assess the state of artwork conservation in terms of accurate qualitative visual detection of hidden defects. The property allowing for holographic interferograms to serve for this ultimate use in artwork objective documentation of condition is the unique ability of generation of the secondary or localised fringe systems which are defined by the distinct limits from the general or whole-body generated displacement. There are several causes for the generation of localised fringe abnormalities in the generally smooth interference field especially within the inhomogeneous structure of artworks. In any fresh constructed artwork, the different materials age variably and react differently to the slow deterioration effects of environment and handling getting a diverse aging effect which results in re-structuring within the material molecules. As an end-effect materials tend to separate and disintegrate producing plethora of de-cohesion phenomena, such as cracks and detachments within the layers, material separation and partition in areas and/or material concentration and weak areas stressed under concentrationed forces. In table IV are summarised most common defect causes and the localised patterns they tend to generate.

Table IV. DEFECT Cause to FRINGE Characteristic:
detachments between layers = concentric fringes

- material voids in bulk = concentric fringes
- inclusions in surface contaminated layers = concentric fringes
- enclosures between layers = concentric fringes
- cracks subsurface = disrupted fringes
- cracks in the support = disrupted fringes
- materials' separation = disrupted fringes
- surface cracks = disrupted fringes
- weak zones = disrupted/alternate fringe density
- crack/detachment propagation = curved fringes

The localised characteristic due to each defect case allow to distinguish separate causes evidenced through distinct "shape" which typify a visual fringe pattern provoked by a corresponding defect. As a reader may note this is not a division according to kind of artwork or material since the visual appearance of fringes due to subsurface defect depends on the type of the defect that is possible to be generated according to the artwork type. To simplify this concept further one could think of "concentric fringes" visualised on the examination of a statue and a painting with the former presenting a circular intensity distribution exactly of the same type as the latter but in the first case is due to material void while in the second to layer detachment.

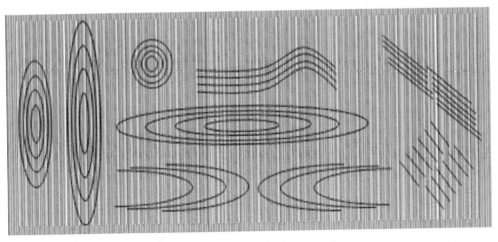

Figure 8. Example of several schematics of concentric and disrupted fringes representing characteristic fringe pattern formation. When appearing restricted locally among a general field of interference fringes indicate a defect cause underneath the surface. May appear in any combination or in repeat order and/or in variety signifying advanced structural deterioration. May appear and extent in any scale from micrometers, millimeters to centimetres depending on the FOV and field spatial frequency.

Most common fringe pattern visual form representing a defect cause is shown in schematics, figure 8. The patterns in artworks indicate defect when they present restricted size and dimensions among a general interference field. Artworks include several numbers of defects and their influence on the surface provokes appearance of several number of fringe patterns in a variety of combinations and scales. The low or high number of fringes measured as fringe density in µm indicates the state of deterioration.

None of the observed patterns can be copied or frauded in order to appear in a new construction of a copied artwork whereas a comprehensive archiving of documented patterns serves as a secret code for tracing originality and impact. Documentation through secondary fringe patterns coding can be correlated to any later documentation and the history of impact can be traced.

2.6. Example 1: Application of Defect Detection on Panel Paintings

The panel painting shown in figure 9 is titled "Saint-Sebastian" belongs to the collection of National Gallery of Athens and is attributed to Rafael. The conservation of a painting starts with documentation of the state-of-conservation to assess the condition prior and after the conservation, using conventional diagnostic methods in conservation such as ranking light, stereo-microscopy, axial tomography and/or x-ray radiography, followed with careful finger knocking since detachments are not visualized with x-ray or other methods and need to be differentiated by sound. In x-ray investigation is uncovered the existence of nails, cracks and holes but detailing and other defects primarily due to detachments between the layers, voids, loss of material, or propagation of cracks and worms, are all remain hidden from x-ray imaging.

The secondary fringes of holographic interferometry methods can respond to the above documentation demands providing visual maps of the defected regions including detachments and if further processed and measured to result in a risk priority map.

Figure 9. In a) the photograph of Saint Sebastian painting, in b) the defected locations as traced from secondary fringes and in c) the risk priority map resulted by fringe density.

In the example shown in figure 9 the painting was examined by the double exposure technique in stable laboratory environment with controlled thermal excitation and surface monitoring. The random high reflectivity of the varnish layer has caused random loss of contrast obstructing the capturing of a normalized intensity photographic image without affecting the fringe formation. Secondary fringes are results of volume changes and it is expected not to be affected by surface inhomogeneous height or reflection. Thus the overall uniform fringe formation due to an overall uniform response of the panel painting volume to the transient temperature alteration is unobstructedly exhibited while is locally interrupted by smaller fringe patterns generated by the various subsurface and bulk defects and material discontinuities. The next step is to extract from the overall distributed intensity field the localized fringe patterns indicating the abnormalities. The procedure requires to locate and isolate, or/and for smaller scale defects to zoom-in, in order to extract those defect-indicating fringe patterns. At the final step the defect-map is seen as in fig9b. Exact location, size and morphology can be studied in great detail. Not any other conventional or not method except holographic recording can provide invisible information in such great detail and fidelity.

The pseudo-colors represent mean fringe value calculated by multiplying the total number of fringe-pairs by half wavelength of the laser used and averaging over various viewpoints starting from frame edge. Red color is the maximum displacement measured and green minimum. The procedure produces a defect-topography and the priority risk map.

The procedure is a two-step process, at first is the production of secondary fringe patterns and second is the post-processing elaboration.

2.7. Example 2: Application of Restoration Assessment on Statuette

Another important range of application in art conservation is the evaluation of interventive conservation and restoration actions. In this respect a methodology was developed to compare the secondary fringes at different time regimes of the artwork under consideration. The methodology implies accurate repositioning and reproduction of environmental conditioning while it is assumed that experimental parameters are kept absolutely same.

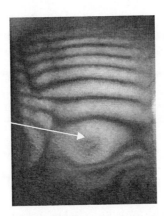

Figure :10 :. An unevenly applied restoration layer was discovered by holography (circle zone at b) and removed by laser cleaning. The statuette re-examined and the uniform fringe distribution seen at c confirmed successful restoration. (photos focused on fringes)

Recently the methodology was extended to the portable system with additional development of algorithms and peripherals. In the example of figure 10 it is shown the feasibility of application to compare a statue before and after cleaning on a statuette of Laboratoire des Researche des Monument Historique- Paris.

Prior to laser cleaning the examination revealed a main discontinuos fringe distribution confirmed as void due to unevenly applied earlier restoration layer. The lasser cleaning removed the contaminated layers and the layer of prior restoration was not able to generate the "void" that cause the fringe pattern abnormality witnessed by the secondary fringe formation before restoration. The successful results of cleaning is confirmed in the normal fringe pattern distribution captured and is shown in figure 10c.

Conclusion

The secondary generated interference fringe patterns offer a powerful tool in the examination and documentation of artworks by providing a non-destructive, non-contacting and remote access method to visually investigate precious objects of artistic, historic and aesthetic value. The method allows application on diverse objects and structures since do not have dependence on surface texture, size, complexity and inhomogeneity of materials. Recent advances allow the method to be applied on-field although in the past the aim encountered considerable difficulties to the vibration isolation requirements of coherent metrology methods. However demanding the experimental boundary conditions for the interference phenomena to take place are equally advanced are the recent developments in software and hardware allowing application in diverse environments and on different conservation aims. The method allows flexibility in experimental methodologies which can be developed according to the problem under consideration and procedures to be adjustable to the artwork in concern. Portable configuration described here as the DHSPI system can be transported to Museum laboratories, galleries, on-field campaigns, and wherever there is demand for artwork examination. It is fast and accurate while its robust, small dimensions and lightweight design minimises prerequisites in handling and transportation. It is quite and can be adjusted to any light-condition. Main consideration remains the average low vibration control and precautions are taken to minimise extraneous motions other than the information provided by the artwork itself. The method has been applied by the author and the co-workers team in major European UNESCO sites, eg. Malta fortifications, St John Cathedral in Valetta, St Savin Cathedral in France, Byzantine wallpaintings in Crete, etc as well as on important paintings eg El Greco, Rafael, Byzantine icons, etc.

Acknowledgments

From this position I wish to acknowledge the support of my affiliated institute (IESL/FORTH) to the Cultural Heritage research and to director Prof. Costas Fotakis and my co-workers Irini Bernikola, Elsa Tsiranidou, Kostas Hatzigiannaki. Special mention to the EC FP5 project LASERACT (EVK4-CT-2002-00096), and FW6 project MULTIENCODE (SSP-0006427) for support on the variety of subjects involved in the presented study. Last but not least a reference to co-workers over the years from art conservation field, National Gallery of Athens, Benaki Museum of Athens, Laboratoire des Researche des Monuments Historique-Paris, the TATE – London, etc.

NOTE FROM THE AUTHOR

The presentation of documentation of works of art through evaluation of *secondary fringes* is confined within the terms of coherent optics and *structural diagnosis* in art conservation as outlined in a number of exemplary studies and applications performed basically at applied holography research laboratories and concerned mainly within Holographic Interferometry Non Destructive Testing (HINDT) and its digital holographic counterparts broadly termed as Electronic Speckle Pattern Interferometry (ESPI) [19-28]. This brief presentation does not review onto other laser techniques that have emerged or been tested and may have probable application on CH conservation [29-33]. The unique properties of laser light, the produced coherent phenomena of interference and diffraction, and the holography principles allowing one to record and reconstruct optical wave fields are dominating the basic needed information for the use of technique and applications in the field of art conservation.

This chapter is facilitated by the context of the present book in interferometry to avoid long theoretical introduction so to have instead a straightforward focus on the context of the chapter. Thus is directed on aiming to reassess the remarkable properties of macroscopic whole field interference fringes that can be especially utilized into the sensitive field of application on precious artworks rather than to explicitly report on the partial subjects consisting it.

However a brief intro to optics and several references are listed if a more detailed study in the partial subjects is desired by the reader.

REFERENCES

[1] Collier R. J., *Optical Holography*, Academic press NY, 1971
[2] Vest C. M., *Holographic interferometry*. 1979, USA: John Wiley & Sons.
[3] Gabor D., *Microscopy by reconstructed wavefronts*. Proc. Roy. Soc., 1949. **A197**: p. 454-487.
[4] Gabor D., *Microscopy by reconstructed wavefronts*. Proc. Roy. Soc., 1951. **64**(II): p. 449-469.
[5] Leith E. N. and Upatnieks J., *Wavefront reconstruction with diffused illumination and three-dimensional objects*. J. Opt. Soc. Am., 1964. **54**: p. 1295-1301.
[6] Okoshi, T., *Three-dimensional imaging techniques*. 1976, London, U.K.: Academic Press.
[7] Asmus, J. F., Guattari, G., Lazzarini L., and Wuerker, R. F., "Holography in the conservation of statuary", Studies in Conservation, 18, (1973)
[8] Wuerker, R. F., and Heflinger, L. O., "Pulsed Laser Holography", Soc. Photo-Optical Instr. Engineers, 9, (1970)

[9] Amadesi S., Gori F., Grella R. and Guattari G., 'Holographic methods for painting diagnostics', *Applied Optics*, vol.13, pp.2009-13, (1974)
[10] V. Tornari, V. Zafiropulos, N. A. Vainos, C. Fotakis, FORTH/IESL, W. Osten, F. Elandaloussi, BIAS, " A holographic systematic approach to alleviate major dillemas in museum operation", EVA 98 Conference on Electronic Imaging and the Visual Arts, Berlin 13-16 November (1998)
[11] V. Tornari, V. Zafiropulos, A. Bonarou, N. A. Vainos and C. Fotakis, "Modern technology in artwork conservation: a laser-based approach for process control and evaluation", Optics and Lasers in Engineering, 34 (October 2000) 309-326
[12] K.D. Hinsch, T. Fricke-Begemann, G. GuKlker, K. Wolff, "Speckle correlation for the analysis of random processes at rough surfaces", Optics and Lasers in Engineering 33 (2000) 87-105
[13] Vivi Tornari, Antonia Bonarou, Vasilis Zafiropulos, Costas Fotakis and Michaelis Doulgeridis, "Holographic applications in evaluation of defect and cleaning procedures", Journal of Cultural Heritage, Volume 1, Supplement 1, 1 August 2000, Pages S325-S329
[14] V. Tornari, D. Fantidou, V. Zafiropulos, N. A. Vainos, and C. Fotakis "Photomechanical effects of laser cleaning: a long-term non-destructive holographic interferometric investigation on painted artworks", SPIE vol. 3411, pp.420-430, (1998)
[15] V. Tornari, A. Bonarou, V. Zafiropulos, L. Antonucci, S. Georgiou, C. Fotakis, "Progressive holographic interferometry recording: A key to monitor long-term structural effects of laser surface treated art objects", ROMOPTO Conference, Bucharest, SPIE vol. 4430, (2000)
[16] V. Tornari, A. Bonarou, V. Zafiropulos, C. Fotakis, N. Smyrnakis, S. Stassinopoulos, "Structural evaluation of restoration processes with holographic diagnostic inspection", *Journal of Cultural Heritage* **4,** pp S347-S354 (2003).
[17] Athanassiou, K. Lakiotaki, V. Tornari, S. Georgiou, C. Fotakis, "Photocontrolled mechanical phenomena in photochromic doped polymeric systems", Appl. Phys. A 76, 97–100 (2003).
[18] Amadesi S., Gori F., Grella R., and Guattari G., *Holographic methods for painting diagnostics.* Applied Optics, 1974. **13**: p. 2009-13.
[19] Asmus J. F., Guattari G., Lazzarini L., and Wuerker R. F., *Holography in the conservation of statuary.* Studies in Conservation, 1973. **18**.
[20] G. Saxby, *Practical Holography*, Prentice Hall Int. UK last edition 2003
[21] Osten, W., F. Elandaloussi, and U. Mieth. *The Bias Fringe Processor- A useful tool for the automatic processing of fringe patterns in optical metrology.* in 3rd International Workshop in Optical Metrology- Series in Optical Metrology. 1998: Akademie Verlag.
[22] Athanassiou, E. Andreou, A. Bonarou, V. Tornari, D. Anglos, S. Georgiou, C. Fotakis, "Examination of chemical and structural modifications in the UV ablation of polymers", Applied Surface Science 197–198, 757–763 (2002)
[23] Bonarou, V. Tornari, L. Antonucci, S. Georgiou, C. Fotakis, "Holographic interferometry for the structural diagnostics of UV laser ablation of polymer substrates", Appl. Physics A, vol. 73, pp 647-651 (2001)
[24] V. Tornari, A. Bonarou, V. Zafiropulos, C. Fotakis, M. Doulgeridis, "Holographic applications in evaluation of defect and cleaning procedures", J. Cult. Heritage **1**, S325-S329 (2000).

[25] Hinch, K.D., et al. *Artwork monitoring by digital image correlation.* in *LACONA V.* 2003: Springer proceedings in Physics.

[26] E. Bernikola, A. Nevin, V. Tornari, "Rapid initial dimensional changes in wooden panel paintings due to simulated climate-induced alterations monitored by digital coherent out-of-plane interferometry", Applied Physics A **95,** pp. 387-399 (2009).

[27] V. Tornari, E. Bernikola, A. Nevin, E. Kouloumpi, M. Doulgeridis, C. Fotakis, "Fully non contact holography-based inspection on dimensionally responsive artwork materials". Sensors 2008, 8, DOI 10.3390/sensors

[28] V. Tornari, "Laser Interference-Based Techniques and Applications in Structural Inspection of Works of Art", *Analytical and Bioanalytical Chemistry*; **387,** 761-80 (2007).

[29] W. Osten, F. Elandaloussi, and U. Mieth, "The BIAS FRINGE PROCESSOR™ - A useful tool for the automatic processing of fringe patterns in optical metrology," *Proc. Fringe'97*, Akademie Verlag Berlin, Berlin 1997, pp. 98-106.

[30] Y.Y. Hung, "Image-shearing camera for direct measurement of surface strains," *Applied Optics,* Vol. 18, (1979) pp. 1046-1051.

[31] D. E. Oliver, "Scanning Laser Vibrometers as Tools for Vibration Measurement and Analysis", *Test Engineering & Management*, pp. 18-21, 1991.

[32] P. Castellini, G.M. Revel, E. P Tomasini, "Laser Doppler vibrometry: a review of advances and applications", *The Shock and Vibration Digest*, **30,** No. 6, pp. 443-456, 1998.

[33] Y.Y. Hung, "Shearography for non-destructive evaluation of composite structures", *Optics and Lasers in Engineering*, Vol. 24, (1996) pp. 161-182.

In: Interferometry Principles and Applications
Editor: Mark E. Russo

ISBN 978-1-61209-347-5
©2012 Nova Science Publishers, Inc.

Chapter 19

PRINCIPLE AND APPLICATION OF OPTICAL INTERFEROMETERS FOR INVESTIGATING THE REFRACTIVE INDEX HOMOGENEITY, BIREFRINGENCE, OPTICAL INDICATRIX, AND SURFACE FEATURES OF CRYSTALS

Sunil Verma[*], *S. Kar, and K.S. Bartwal*

Laser Materials Development & Devices Division, Raja Ramanna Centre for Advanced Technology, Indore, India

ABSTRACT

High quality crystals are required for research and development in the field of optical frequency conversion, optoelectronics, electro-optics and acousto-optics. The crystals are grown by various techniques such as growth from solution, melt, flux and vapor. The as-grown crystal is cut and polished along specific directions to obtain an element for use in the desired device. However, before deploying a crystal element for a particular application it is necessary to assess its optical quality and measure its important optical parameters. In this respect optical interferometric techniques assume great significance. The present chapter deals with the application of the conoscopic interferometry for investigating the optical quality and the optical indicatrix of the crystal, the Mach-Zehnder interferometer and the birefringence interferometer for non-destructive assessment of the optical homogeneity of the crystal, for measurement of the variation in the refractive index along the spatial extension of the crystal sample. Additionally the potential of Michelson interferometry is presented for surface characterization. The first half of the chapter presents the principle and optical instrumentation of the interferometers. The second half presents the application of these interferometric techniques for investigating the above mentioned properties of various classes of crystals such as inorganic, semi-organic and organic.

---

[*] Corresponding author: Sunil Verma, E-mail: sverma1118@gmail.com, Phone: +91-731-248 8670, Fax: +91-731-248 8650

1. CRYSTALLIZATION TECHNIQUES AND OPTICAL CRYSTALS

Crystals are the pillars of modern technology and constitute an essential input for the current photonics revolution. There are various methods of growing crystals. These are, low temperature solution growth, melt growth, flux growth and vapor growth [Hurle, 1994; Mullin, 2001; Byrappa, 2003]. These methods are briefly described below.

Solution growth is the most widely used method for growing large crystals, several centimeters in size [Mullin, 2001]. It is applicable to all classes of materials, including inorganic, organic and metal-organics. It offers the convenience of low operating temperatures and simple instrumentation. However, this method demands optimum growth conditions to achieve the desired size and the crystal quality. Some of the representative optical crystals grown by solution growth and their applications are as follows: KDP (potassium dihydrogen phosphate) and DKDP (deuterated KDP) for frequency conversion and electro-optic switching respectively; TGS (triglycine sulphate) for laser-energy measuring devices; KAP (potassium acid phthalate) for monochromator applications; ZTS (zinc tris thiourea sulphate), and many organic and metal-organic crystals for frequency conversion and other optical applications.

Melt growth is used for materials that melt congruently or near congruently [Hurle, 1994]. The most popular method of growing crystals from melt is the Czochralski method. Here the material to be grown (referred to as charge) is melted in a platinum or iridium crucible by induction or resistive heating. A seed crystal is then lowered into the molten charge at a temperature just above its melting point such that it just touches the top surface of the melt. By minor adjustment of the temperature, a proper growth interface is achieved between the melt surface and the seed crystal. It is followed by slow pulling and rotation. In order to achieve a constant-diameter crystal, several parameters have to be optimized, such as the pulling rate, rate of melt level drop in crucible, heat transfer rates in and out of the system, and the crystal and crucible rotations. The main advantage of this technique is the high growth rate achievable, in the range of several millimeters per hour. This is several orders of magnitude greater than solution growth. Since no foreign molecules are present in the melt, the compositional purity of crystals is high in the case of melt growth when compared with other techniques. Representative crystals grown by melt growth technique and their applications are as follows: $LiNbO_3$ (lithium niobate) and its doped variants for frequency-conversion, electro-optic modulation for switching applications, acousto-optic modulation in SAW devices, photo-refractive applications, waveguide material in integrated optic applications, and as holographic data storage media; Ba:PBO (barium doped lead tetraborate) as a high-laser-damage-threshold material; LBO (lithium triborate) for higher order frequency conversion of Nd:YAG fundamental laser line and as an optical parametric oscillator for radiation source in the visible range; LTB (lithium tetra borate) for dosimetry applications, Nd:YAG (Nd doped yttrium aluminum garnet) as laser host crystal; $Nd:YVO_4$ (Nd doped yttrium vanadate) for diode pumped solid-state laser applications; $LiTaO_3$ (lithium tantalate) for electro-optic switching and high damage threshold applications, and $Ce:BaTiO_3$ (cerium doped barium titanate) and $Bi_{12}SiO_{20}$ (bismuth silicate) for photorefractive and holographic applications.

Flux growth is used when the material to be grown melts incongruently, i.e. the material decomposes before melting, so that crystallization by the conventional melt growth technique is not possible. In order to overcome this difficulty, a secondary component, called flux, is added to the nutrient. It helps in reducing the crystallization temperature so that the desired phase can be crystallized. The nutrient and the flux are placed in a crucible, heated above the liquidus temperature, and allowed to cool. The cooling rate is adjusted so that one (or may be a few) spontaneously nucleated crystals grow to a fairly large dimension. Once the growth is complete, the flux is dissolved away to get the grown crystal. Flux growth has been used to grow KTP (potassium titanyl phosphate) crystal for frequency conversion of Nd:YAG laser line, for electro-optic switching applications, as an optical parametric source of tunable output from visible to mid-IR, and in integrated optics as waveguide based on ion-exchange on KTP substrate; BBO (β-barium borate) for higher-order frequency conversion of Nd:YAG laser line, in electro-optic switching, and for optical parametric oscillator and amplifier applications; CLBO (cesium-doped lithium borate) for UV applications and for generating the 4th and 5th harmonics of the Nd:YAG fundamental laser line.

In order to grow crystals from the vapor phase, the solid nutrient phase is heated above its sublimation temperature. This is followed by transportation of the nutrient and the carrier gas to a region which is at a slightly lower temperature, resulting in the supersaturation of the nutrient phase. Supersaturation, when controlled properly, results in crystalline growth of the nutrient material. Compared to the other three methods of growth discussed above, the growth rate and the size of the crystal achievable in vapor growth are small. This limits its usability for growing bulk crystals.

2. CHARACTERIZING DEFECTS IN CRYSTALS

The crystal growth process is a non-equilibrium process and therefore several defects get incorporated inside the crystals during growth. These can be classified as point defects such as voids, interstitials and substitutions; line defects such as dislocations; planar defects such as stacking faults, striations, impurity banding and growth sector boundaries; and volume defects such as inclusions, bubbles and precipitates. All the above defects influence the optical homogeneity of the crystal. The quality of crystals required for optical applications is very high, and therefore assessing the optical homogeneity, in general, and the refractive index homogeneity, in particular, is an important post growth exercise. There are several techniques reported in literature for characterizing the defects state of the crystal. For example, etching is used for estimating the dislocation density in the crystal; X-ray diffraction topography is used for imaging dislocations; laser scattering technique is used for imaging volume defects; optical transmittance is used to get an estimate of the transparency of the bulk of the crystal; and high resolution X-ray diffraction is used for assessing the crystalline perfection of the crystal by recording the rocking curve. Some of the above techniques are destructive in nature while others require use of X-ray radiation, which involves potential safety hazard and expensive instrumentation. Therefore, development of techniques which are non-destructive and cost effective are required for assessing the optical quality. Optical diagnostics fulfill these criteria and have many other advantages. For example, optical techniques correspond to the use of *photon probes* that do not affect the sample being investigated. They can map the crystal sample with a spatial resolution of less than a micrometer and the response is practically inertia-free. In addition, the optical techniques provide full information about the crystal sample in a single image. Further, the availability of fast image acquisition hardware and advanced computers for data reduction have made the optical diagnostics a preferred choice. Realizing the advantages of the optical techniques, we have developed optical interferometric techniques for assessing the optical homogeneity of the grown crystals and also for measuring a few important optical parameter such as variation of the birefringence (or refractive index) and the position of the optic axis.

The chapter is organized as follows: Section 1 is an introduction to crystallization techniques and optical crystals. Section 2 discusses the various types of defects in crystals and lists various techniques available to detect them. The advantages of optical diagnostics for imaging such defects are also presented. Section 3 discusses various optical interferometers for characterizing the refractive index homogeneity, birefringence, optical indicatrix and surface quality of the crystals. In particular, the principle and instrumentation of conoscopy, Mach-Zehnder interferometer, birefringence interferometer and Michelson interferometer are discussed. Section 4 describes the applications of these interferometers for the purpose described above. The salient conclusions are summarized in Section 5.

3. Optical Interferometers: Principle and Instrumentation

The principles of various optical interferometers used for investigating the optical quality of the crystal are discussed below. In addition, the optical instrumentation developed for the interferometric investigations is described. Crystal sample in the form of a plate of typical size ~ 10×10×1 mm^3 is cut from the as-grown crystal. The surfaces of the crystal plate are smoothened and polished for interferometric studies.

3.1. Conoscopic Interferometer

The conoscopic interferometry is performed to get qualitative information about the optical homogeneity of the grown crystal and the position of the optic axis. The principle of conoscopic interferometry is explained below [Davidson, 2005].

When a converging beam of white light polarized along a specific direction is passed through the crystal plate, the beam splits into e- and o-rays. The phase difference between the two rays is [Born, 1989]

$$\delta = \frac{2\pi d}{\lambda \cos\theta}(n_e - n_o)$$

where d is the thickness of the sample, n_e and n_o are the refractive indices of e and o rays, respectively, and θ is the angle of refraction. Figure 1a shows the schematic representation of the formation of conoscopic fringes and the polarizing microscope (*OLYMPUS*) used for the purpose is shown in Figure 1b. Upon emerging from the crystal plate, the two rays are made to recombine along the axis of an analyzer. This results in concentric fringes of interference colors (isochromes). These are superimposed by a black cross (isogyres), the centre of which is the location of the optic axis (melatope). The interference pattern, commonly called the conoscopic pattern, is recorded by a camera attached to the microscope.

In order to determine the sign of the optic axis, a wave plate (530 nm) is introduced at 45° to the polarizer and between the crystal plate and the analyzer. This results in change in the interference colors of the four quadrants of the conoscopic image. This is compared with the interference color chart (Figure 1c) which gives the information about the optical sign of the crystal. For example, if the order of interference color in the 1st and 3rd quadrants decreases (i.e. moves to the left on the interference color chart), whereas it increases in the 2nd and 4th quadrants, this specific pattern of color change in the four quadrants is characteristic of optically negative crystals.

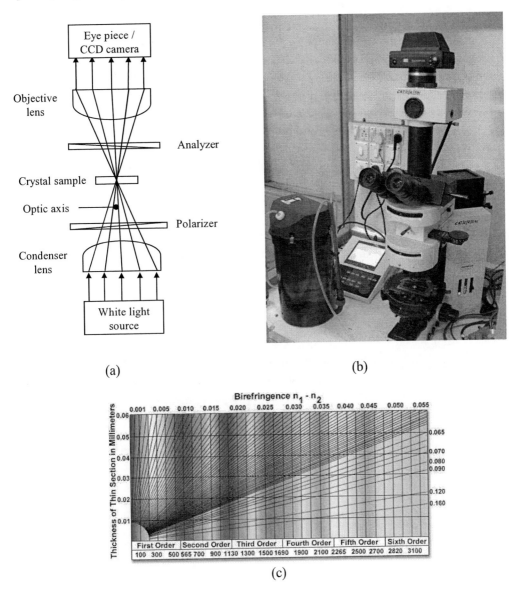

Figure 1. Optical schematic of conoscopy (a), photograph of the polarizing optical microscope (b), and the interference color chart (c)

3.2. Mach-Zehnder Interferometer

A Mach-Zehnder interferometer comprises of four components, two beam-splitters and two mirrors, arranged diagonally on the four corners of a square (or rectangle). The four components have to be exactly at 45° to the input beam direction for proper functioning of the interferometer. An expanded laser beam falls on the first beam splitter (BS1) which splits it in to two beams, one transmitted and the other reflected, in the intensity ratio 50:50. The two beams traverse in orthogonal directions and are incident on two mirrors (M1 and M2) at an angle of 45°. The beams reflected from these mirrors recombine on the plane of a second beam splitter (BS2) resulting in an interference pattern [Hariharan, 1985; Malacara, 1992]. The schematic of the optical set up and the photograph of the Mach-Zehnder interferometer are shown in Figure 2(a, b). Such a four-component Mach-Zehnder interferometer is susceptible to frequent misalignments, owing either to the mechanical vibrations in the surroundings or the environmental fluctuations. Both factors are detrimental to the quality of the experimental data. This was taken care of by design and development of a novel Mach-Zehnder interferometer which was immune to frequent misalignments [Verma, 2009]. The entire experimental set up was placed on a vibration isolation optical table, which further improved the stability of the interferometer. A pair of wedge compensators were used to correct for any residual wedge between the interfering wavefront.

The crystal sample whose refractive index homogeneity is to be tested is placed in one of the beam paths. The wavefront passing through the sample carries the information of the refractive index variations of the sample in the form of spatial variations of the phase of the wavefront. This is called the test wavefront. The other wavefront passes through a uniform refractive index region and called the reference wavefront. The two wavefronts interfere to produce a fringe pattern which carries information of the refractive index homogeneity of the crystal sample under test.

It may be noted that the requirement of the surface flatness and the parallelism of the opposite faces of the crystal plate is very stringent for obtaining the desired Mach-Zehnder interferogram. This is because the Mach-Zehnder interferometer is extremely sensitive to the flatness variations on the sample surface (fraction of a wavelength) and also any residual wedge in parallelism (a few arc seconds) of the crystal plate. Both these factors results in phase change of the test wavefront and hence result in fringe artifacts which are not characteristic of the refractive index homogeneity of the crystal sample under test. For these reasons, application of the Mach-Zehnder interferometer for investigating the bulk refractive index homogeneity of the crystal sample requires special sample preparation for desired surface flatness and parallelism.

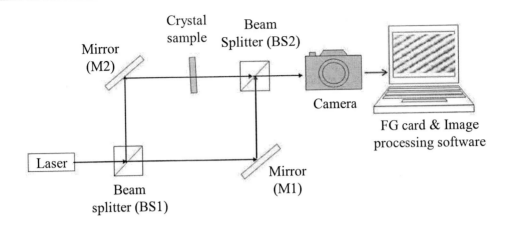

(a)

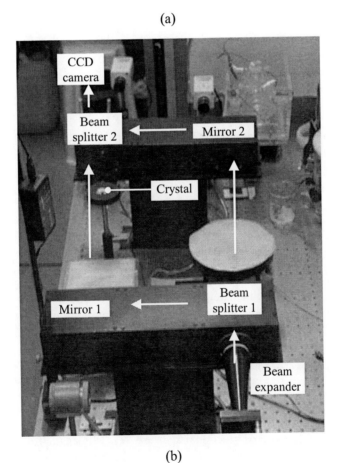

(b)

Figure 2. The optical schematic of the Mach-Zehnder interferometer (a), and the photograph of the interferometer (b) [Dinakaran, 2010B].

Mathematically, the origin of fringe formation is understood as follows [Mayinger, 1994]. The optical path length (OPL) traversed by the test wave is

$$OPL_1 = \int n\, dz$$

where the integration is along the passage of the light beam (z-direction). On the other hand the reference wave passes through a region of constant refractive index resulting in an optical path length

$$OPL_2 = \int n_o\, dz$$

The fringe pattern is the result of optical path difference (OPD) given by

$$OPD = OPL_1 - OPL_2 = \int (n - n_o)\, dz$$

The phase difference corresponding to the above optical path difference is

$$\phi = \frac{2\pi}{\lambda}(OPD)$$

In a spatially varying refractive index field, the path length difference and hence the phase field will also be spatially variable. Experimentally, the loci of constant phase manifest itself as bright and dark fringes in the interferogram. These fringes are processed to get information of the refractive index field of the material or process under investigation.

3.3. Birefringence Interferometer

This interferometer is used for quantitative determination of the spatial variation of the birefringence of the crystalline sample and for qualitative assessment of the defects state of the crystal. The birefringence interferogram gives a cumulative information about all those defects which influence the birefringence, such as stress, cracks, inclusions, major spatial variation in the dislocation density inside the crystal, and lattice misorientations due to these defects [Rao, 1991]. The principle of this interferometer is described below [Henningsen & Singh, 1989].

An expanded and collimated laser beam of wavelength λ is polarized such that the plane of polarisation of the incident beam makes an angle of 45° with the optic axis of the crystal element (thickness L and cross-sectional aperture W). This angular orientation is obtained by relative alignment of the polariser direction and the crystal element. As a result of above orientation, each ray splits in to e- and o- components of equal amplitudes at each point on the crystal sample. These components traverse through the crystal element with their characteristic velocities, and at the exit face of the crystal emerge with a definite optical path difference. Since the two components are orthogonally polarized, they cannot interfere. However, when an analyzer is placed after the crystal, the components of e- and o- rays along the analyzer axis interfere to produce an interference pattern referred to as the birefringence interferogram [Bloss, 1961]. Figure 3(a, b) show the schematic of the optical set up and the photograph of the birefringence interferometer.

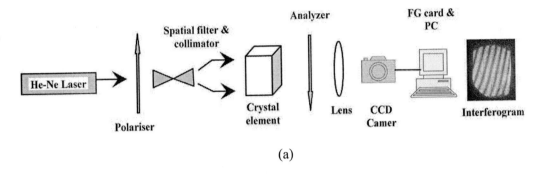

(a)

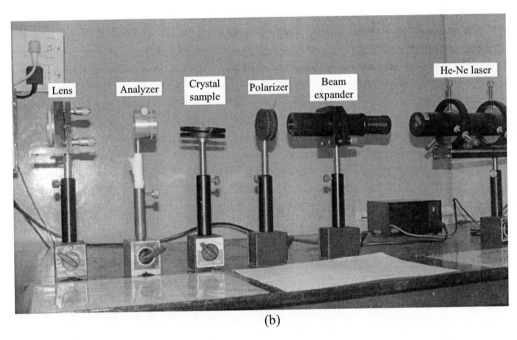

(b)

Figure 3. Optical schematic of the birefringence interferometer (a), and the photograph of the interferometer (b) [Kar, 2008A].

The variation in birefringence is computed from the interferogram in the following manner. The optical path difference between the interfering e- and o- components is

$$OPD = \left[L \times (n_e - n_o)\right] + \left[L \times \Delta(n_e - n_o)\right] + \left[\Delta L \times (n_e - n_o)\right]$$

The phase difference corresponding to the above optical path difference is

$$\phi = \frac{2\pi}{\lambda} \times \left\{\left[L \times (n_e - n_o)\right] + \left[L \times \Delta(n_e - n_o)\right] + \left[\Delta L \times (n_e - n_o)\right]\right\}$$

The first term on the right is the phase difference arising due to the birefringence of the crystal, the second term is the phase difference arising due to the variation of birefringence across the sample cross-section and the third term is the phase difference arising due to the variation in the sample thickness across its cross-section.

We consider three different experimental scenarios to understand the influence of each of the three terms in the above equation:

(i) Perfect crystal element with no defects (hence Δ(n$_e$-n$_o$)=0), and no variation in the sample thickness (i.e. ΔL=0)

For a perfect crystal element (zero defects), there will not be any variation in the values of the n_e and n_o components of the refractive indices across the cross-sectional aperture (W) of the crystal element of thickness (L). In this situation, the second and third term in the above equation vanish, respectively. Therefore, the fringes in the birefringence interferogram will be only due to the first term in the above equation. Since birefringence, $\Delta n=(n_e-n_o)$, is constant along a given direction of the crystal, therefore the interferogram will have only one fringe (bright or dark fringe) across the entire cross-section of the sample, depending upon whether the phase difference satisfies the condition for constructive or destructive interference.

(ii) Crystal element having defects in it (hence Δ(n$_e$-n$_o$) ≠ 0) and no variation in the sample thickness (i.e. ΔL=0)

Usually a crystal element has several different types of defects, which invariably results in change in the value of the components of the refractive indices across its cross-section, and hence variation in the sample birefringence across its cross-section. As a result the birefringence interferogram has several fringes depending upon the extent of variation in the birefringence across the sample cross-section. The resulting birefringence interferogram can be used to obtain a quantitative assessment of the variation in birefringence, $\Delta(n_e - n_o)$ across the cross-section of the crystal element. A variation in birefringence across any two points on the crystal sample will result in corresponding phase diference. A phase difference of 2π (or an optical path difference of λ) between the two points of the sample results in one fringe change in the interferometer. By counting the number of fringes in the interferogram, the total variation in the birefringence across the sample thickness can be quantified. Further, the shape of the fringe provides us the information that whether the birefringence is periodically varying or random across the sample cross-section.

(iii) Crystal element having no defects (hence Δ(n$_e$-n$_o$)=0) but minor variation in the sample thickness (i.e. ΔL ≠ 0)

Another factor that could result in additional optical path difference between the *e*- and *o*- rays inside the crystal element is the minor variations in the crystal thickness, ΔL, along its the cross-section.

Sensitivity of a birefringence interferometer vis-a-vis a Mach-Zehnder interferometer to the deviation from perfect flatness and paralleism of the opposite faces of the crystal element:

As described in the previous section, use of a Mach-Zehnder interferometer for assessing the optical homogeneity of the crystal requires very stringent conditions on the surface flatness and parallelism of the opposite faces of the crystalline element. These requirements are considerably relaxed when a birefringence interferometer is used for assessing the optical homogeneity of the crystal. This is explained below:

Let the variation in the thickness of the crystal element due to either surface imperfections or the paralleism of the opposite faces at two be ΔL. A birefringence interferometer will result in a phase difference

$$\Delta \varphi_{BI} = (n_o - n_e) \times \Delta L$$

The same imperfections when measured with a Mach-Zehnder interferometer will result in a phase difference

$$\Delta \varphi_{MZ} = n \times \Delta L$$

Since $(n_o - n_e) \ll n$ (less by two orders of magnitude), therefore comparining the above equations for the two interferometers, we find that $\Delta \varphi_{BI} \ll \Delta \varphi_{MZ}$ for the same change in surface flatness and parallelism. Therefore, change in fringe pattern of a birefringence interferometer will be very minor as compared to corresponding change in the fringe pattern of a Mach-Zehnder interferometer. In other words, we can say that birefringence interferometer is more tolerant to variations in the surface flatness and parallelism as compared to a Mach-Zehnder interferometer.

3.4. Michelson Interferometry

A Michelson interferometer is a potential instrument for surface topography studies. In this interferometer, a laser beam is split into two beams at right angles to each in 50:50 intensity ratio. One of the beams traverses towards a mirror and gets reflected back to the beam splitter (reference beam), while the second beam is reflected from the surface of the crystal sample under test (test beam). These two reflected beams recombine at the plane of the beam splitter to produce an interference pattern. The intensity of the test beam reflected from the surface of the crystal is very feeble as compared to the reference beam. Therefore, in order to obtain a good interference pattern, the intensity of the reference beam is reduced by using a neutral density filter. Figure 4(a, b) show the schematic of the optical set up and the photograph of the Michelson interferometer. Depending upon the alignment of the interferometer, two types of fringes are obtained [Verma, 2009]:

i. *Fizeau fringes:* These fringes are obtained when a slight wedge is introduced between the two interfering beams by tilting one of the mirrors of the interferometer. These fringes are parallel and equidistant straight lines, and also called fringes of equal thickness (Figure 5a).

ii. *Haidinger fringes:* When the input laser beam is expanded to get a source of large angular size and the interfering wavefronts obtained from it are exactly parallel to each other, we observe rings in the interference pattern due to an air gap of constant thickness between them (Figure 5b). These fringes are also referred to as fringes of equal inclination. If the alignment of the beam splitter is exactly 45° to the incident beam, we get perfectly circular fringes. If the inclination shifts from 45°, the fringes first become elliptical and with further misalignment turn into hyperbolas.

The Michelson interferograms provide a geographical description of the surface of the crystal sample under test. Initially the interferometer is aligned to get fringes of equal thickness (*Fizeau fringes*), which are then separated to yield only one fringes in the field of view (i.e. infinite fringe interferogram). After this stage, the interferograms that appear in the field of view are due to the microscopic features on the crystal surface. Although this interferometer has been extensively used in optical metrology [Malacara, 1992], however it has not been exploited in crystal surface metrology.

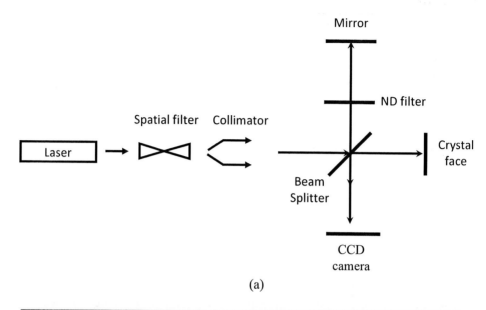

(a)

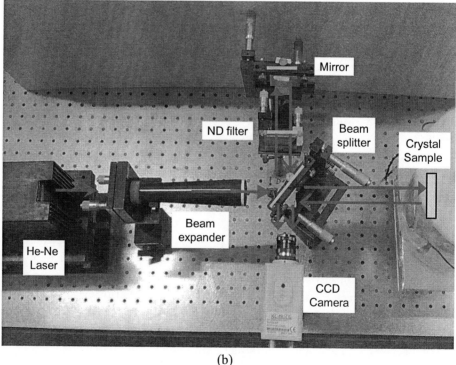

(b)

Figure 4. Optical schematic of the Michelson interferometer (a), and the photograph of the interferometer (b) [Verma, 2010].

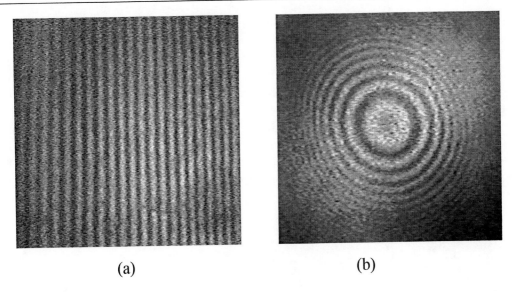

Figure 5. Michelson interferograms showing the Fizeau fringes (a), and the Haidinger fringes (b) [Verma, 2009]

4. APPLICATIONS

The application of the interferometric techniques discussed in Section 3 are presented in this section. Several inorganic, semiorganic and organic crystals grown by different growth techniques have been investigated.

4.1. Conoscopic Studies

Conoscopy is commonly used for detection of the orientation of optic axes and to get a qualitative assessment of the optical homogeneity of the crystals. The crystals of pure and doped $LiNbO_3$ were studied for the optical homogeneity and optic axis position. It was found that the crystals retained their good optical quality when doped with nominal dopant concentration. Figure 6a shows the conoscopic interference pattern of the as-grown pure $LiNbO_3$ crystal. The shape of the interference pattern reveals information about the homogeneity and the strain inside the crystal. The symmetrical fringe pattern is a signature of good optical homogeneity of the crystal [Kar, 2004]. Conoscopy pattern of the vapor transport equilibrated (VTE) sample is shown in Figure 6b. The symmetrical fringe patterns for VTE modified near stoichiometric lithium niobate (SLN) crystals confirm that the crystals are strain free and homogenous in composition [Kar, 2006]. Conoscopy was done to determine the position and sign of the optic axis of the grown Fe and Mn doped $LiNbO_3$ crystal. Figure 7a shows the conoscopic pattern of Fe:Mn:$LiNbO_3$ crystal along the optic axis. It consists of circular fringes of interference colors (isochromes) and a black cross (isogyres) in the centre that is superimposed on the circular fringes. Isochromes are formed due to varying retardation of the converging beam of light while passing through the crystal. The number of isochromes depends on the thickness of the sample. The point where isogyres cross is called melatope which represents the point of optic axis. In order to determine the sign of the optic axis a quartz retardation plate (530 nm) was introduced between the crystal plate and the analyzer. Figure 7b shows the conoscopic pattern obtained when the retardation plate is introduced. The interference colors in the 1st and 3rd quadrant decreases (i.e. move to the left on the interference color chart), whereas they increase in the 2nd and 4th quadrant. This specific pattern of color change in the above four quadrants is characteristic of optically negative crystals. The symmetrical shape of this pattern reveals that the grown crystal has good refractive index homogeneity [Kar, 2009A]. Figure 8 shows the conoscopic pattern of the lithium tetra borate (Li$_2$B$_4$O$_7$) crystal along its c- axis. The symmetrical fringe pattern shows good optical homogeneity of the crystal [Kar, 2008B].

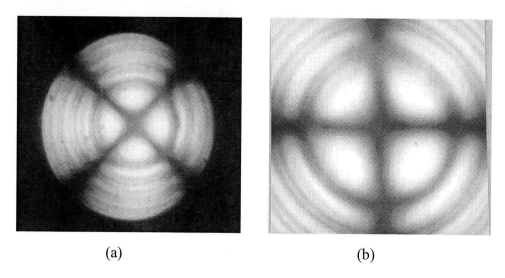

Figure 6. (a) Conoscopy pattern of undoped LiNbO$_3$ crystal. (b) Conoscopy pattern of VTE treated sample. Sharpness of the fringes is indicative of the strain-free nature of the crystal [Kar, 2004 & 2006].

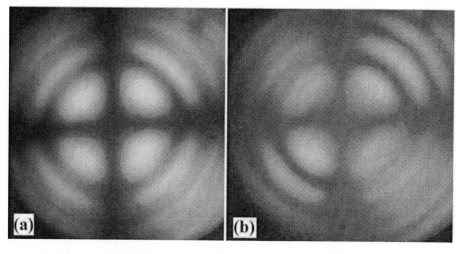

Figure 7. Conoscopic interference pattern for Fe:Mn:LiNbO$_3$ crystal. (a) Along optic axis, and (b) when a retardation plate (530 nm) was introduced to check the sign of optic axis [Kar, 2009A].

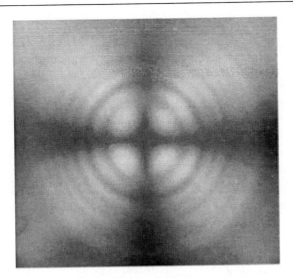

Figure 8. Conoscopy pattern of c-cut lithium tetra borate (Li$_2$B$_4$O$_7$) crystal plate [Kar, 2008B].

Recently, Dinakaran et al. reported growth of KDP crystal directly in phase matching direction [Dinakaran, 2010A]. The growth was performed in a cylindrical ampoule, and hence was physically constrained. Due to this reason, the authors performed a comparative analysis of the optical homogeneity of the KDP crystal grown in phase matched direction with that grown conventionally without any externally imposed boundaries. A KDP crystal plate normal to the [100] direction was cut and polished for conoscopy. Figure 9a shows the conoscopic pattern of the KDP plate. A retardation plate of 530 nm was introduced between the crystal plate and the analyzer oriented at 45° with respect to the polarizer axis to confirm the optical sign of the crystal. Figure 9b shows the conoscopy pattern taken after inserting retardation plate [Dinakaran, 2010B]. Both the conoscopy patterns show that the crystal quality is very good and is not appreciably affected by the constrained growth.

Conoscopy was also used for determining the position of optic axis in zinc tris (thiourea) sulphate (ZTS) crystal. ZTS crystal belongs to the orthorhombic system and is optically biaxial. Therefore, its optical indicatrix is an ellipsoid with three different principal refractive index axes. The two optic axes are symmetrical about the Z axis (or n_γ -axis) and the angle between the two is called optic angle or $2Vz$ angle. The interference pattern obtained when the acute bisectrix is oriented perpendicular to the microscope stage is shown in Figure 10a. The melatopes lie outside the field of view of the conoscopic image, which happens when the $2Vx$ angle is large. The isochromes surround the melatope on the left and right edges of the image. The isogyres coincide with the crosshairs of the microscope. On rotating the microscope stage by 45°, the melatope cross splits and the isogyres form two hyperboles that lie in the opposite quadrants of the conoscopic image (Figure 10b). When the crystal plate is oriented such that one of the optic axes is perpendicular to the microscope stage, circular isochromes are visible along with a single isogyres in the conoscopic image (Figure 10c). When a wave plate was inserted (530 nm), the interference colors were found to increase due to addition of path difference (Figure 10d). The characteristic changes in the interference color of the conoscopic image confirm that the crystal is optically negative [Dinakaran, 2010C].

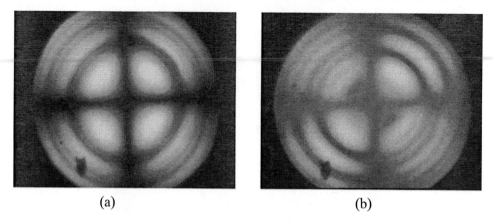

Figure 9. (a) Conoscopic pattern of KDP crystal plate, and (b) after inserting retardation plate (530 nm) [Dinakaran, 2010B].

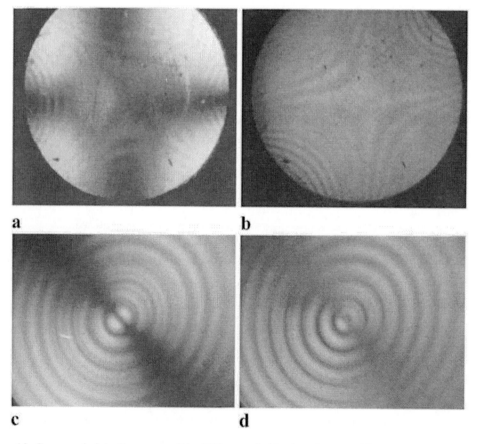

Figure 10. Conoscopic interferograms of the ZTS crystal, (a) when the acute bisectrix is perpendicular to the microscope stage, (b) when the acute bisectrix plate is rotated by 45°, (c) when the optic axis is perpendicular to the microscope stage and (d) optic axis interference pattern when a wave plate (530 nm) is inserted between the sample and the analyzer [Dinakaran, 2010C].

4.2. Mach-Zehnder Interferometric Studies

A Mach-Zehnder interferometer aligned in an infinite fringe mode was used to test the crystal element for refractive index inhomogeneities in the bulk of the sample. The samples were made in the form of parallel plates and both the surfaces were polished to make them optically flat and parallel to each other. It may be pointed out that in addition to the local inhomogeneities we can also get an estimate of the deviation of the crystal faces from the parallelism, i.e. an estimate of the wedge angle. This is a crucial parameter when the element being tested has end application where the wedge angle desired is almost zero.

Figure 11(a, b) shows the Mach-Zehnder interferogram for lithium tetra borate crystal when the respective samples (labelled 1 and 2) are placed in the test arm of the interferometer. Since this interferometer functions in transmission mode, therefore, it gives information of the refractive index homogeneity of the bulk of the sample. Ideally, under the conditions of no inhomogeneities and no wedge angle, the interferogram should have only one fringes (either dark or bright) showing uniform phase distribution along the sample size. However, as shown in Figure 11a, there are a large number of parallel and equispaced fringes representing that the sample has a good overall homogeneity but has a wedge formed by the opposite surfaces. This is attributed to the low precision in preparing the parallelism. The prime motive of the study was to check the optical quality of the grown crystal rather than to control of the wedge angle of the element. However, zooming down to individual fringes shows fringe irregularities at microscopic level, which could be attributed to local refractive index inhomogeneities. Figure 11b shows the fringe pattern for the second sample. In this case the fringes are irregularly spaced and are curved at the edges, indicating inhomogeneities in the sample as well as that the sample surfaces are curved at the edges. This sample is having some residual stress. However less number of fringes per unit length indicates a relatively smaller wedge in the second sample [Kar, 2008B].

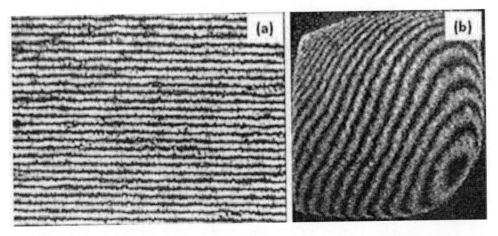

Figure 11. Mach-Zehnder interferogram in infinite setting mode when the LTB samples are placed in one of the arms of the interferometer. (a) Sample-1, dimensions $16 \times 14 \times 5$ mm^3, the fringes correspond to only 2×1.5 mm^2 area. (b) Sample-2, dimension $9 \times 8 \times 2$ mm^3, the interferogram correspond to full sample dimension [Kar, 2008B].

Mach–Zehnder interferometry was also used to assess the refractive-index homogeneity of KDP crystal at the microscopic level. This technique provides information about residual stresses inside the crystal arising due to growth conditions. The interferogram for as grown KDP crystals grown in forced convection and free convection conditions are shown in Figure 12(a, b). The interferogram shows parallel and equispaced fringes, which indicates that the sample has good refractive-index homogeneity and is free from stress [Dinakaran, 2010D].

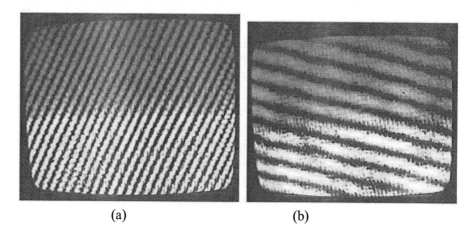

(a) (b)

Figure 12. Mach–Zehnder interferogram of KDP crystal, (a) grown under forced convection and (b) grown under free convection [Dinakaran, 2010D].

ZTS crystals were also examined for refractive-index homogeneity using Mach-Zehnder interferometry. Figure 13 shows the Mach-Zehnder interferogram for ZTS crystal plate parallel to (100) face [Dinakaran, 2010C]. The fringes are straight on a global level but are not so when monitored more closely. This shows that although there are no major continuous change in the refractive index of the crystal, however there are minor variations locally. These could be attributed to local variations in the dislocation density, strain in the crystal due to lattice mismatch and compositional inhomogeneities. These lead to variations in the refractive index of the sample locally.

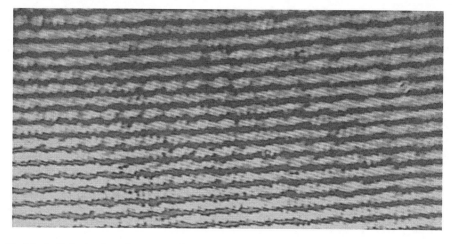

Figure 13. Mach–Zehnder interferogram of ZTS crystal plate parallel to (100) face [Dinakaran, 2010C].

Figure 14 shows the Mach-Zehnder interferogram for a plate cut from a crystal of lithium paranitrophenolate trihydrate [Dinakaran, 2011]. The parallel and equispaced fringes imply that the sample has a good overall refractive index homogeneity and is free from any strain. The fringes appear due to a wedge angle between the opposite faces of the crystal plate. Less number of fringes per unit length indicates a relatively smaller wedge in the second sample.

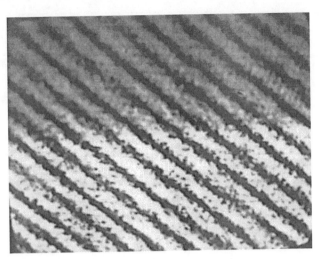

Figure 14. Mach-Zehnder interferogram of lithium paranitrophenolate trihydrate crystal [Dinakaran, 2011].

4.3. Birefringence Interferometric Studies

Birefringence interferometry is a robust and sensitive technique to investigate the optical quality of crystals. Henningsen & Singh [Henningsen, 1989] were the first to design and demonstrate a birefringence interferometer in the infrared range for assessing the optical quality of crystals required for optical and acoustic devices. They used it for evaluating the optical quality of thallium arsenic selenide (Tl_3AsSe_3), lead bromide ($PbBr_2$), and mercurous halide (Hg_2Cl_2 and Hg_2Br_2) crystals [Singh, 1989, 1990, 1992 & 1993]. We have adopted a similar interferometer in the visible range for assessing the quality of crystals required for optical and laser applications [Kar, 2008A, 2009A & 2009B; Dinakaran, 2010B, 2010C, 2010D & 2011; Senthil, 2011]. A few examples of the use of this interferometer for the above application are cited below.

The transition metal doping in NLO materials like $LiNbO_3$ is known to produce crystalline defects in the lattice. These defects affect the optical homogeneity of the crystals at microscopic level. Fe doped $LiNbO_3$ crystal element was studied for optical homogeneity using birefringence interferometry. The birefringence interferogram is shown in Figure 15(a). The variation in birefringence for a z-cut Fe doped $LiNbO_3$ crystal sample of cross-section 10 × 10 mm^2 and 2 mm thickness along the beam propagation was found to be 0.001898 [Kar, 2008A]. Similarly, the birefringence homogeneity for doubly doped (Fe and Mn) $LiNbO_3$ crystal was also measured by using a sample of approximately 10 x 8 mm^2 and 1.1 mm thickness along the beam propogation direction. Figure 15(b) shows the representative birefringence interferogram for doubly doped Fe:Mn:LiNbO₃ crystal. The variation in birefringence calculated rom the interferogram was 0.004602 [Kar, 2009A].

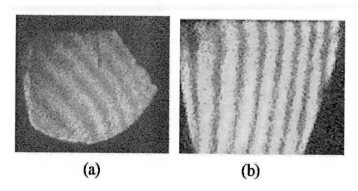

Figure 15. Birefringence interferogram for Fe:LiNbO₃ (a), and Fe:Mn:LiNbO₃ (b) crystal plates [Kar, 2008A and 2009A].

Figure 16(a, b) shows the interferograms taken for Mn:Li₂B₄O₇ crystal plate of dimension 10 x 7 mm at different incidence angles [Kar, 2009B]. When the angle of incidence of the polarized beam falling on the sample is varied, the intensity of the two components getting transmitted through the analyzer changes resulting in low contrast interference pattern (Figure 16b). Simultaneously, this results in increase of the optical path length, and hence the number of fringes in the interferogram. The number of fringes appearing in an interferogram is used to quantify the variation in birefringence along the direction of propagation of the beam. The variation in birefringence calculated from the first interferogram for a crystal plate of cross-section 10 × 6 mm² and 0.97 mm thickness along the beam propagation was 0.003894. This precise measurement of $\Delta(n_e-n_o)$, proves the usefulness of birefringence interferometry for quantifying the variation of birefringence of the crystal along any desired direction.

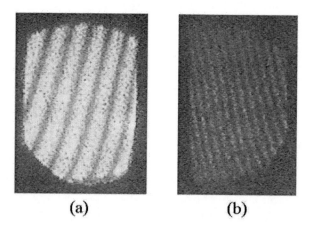

Figure 16(a, b). Birefringence interferograms for Mn:Li₂B₄O₇ crystal at different incident angles [Kar, 2009B].

Using this technique, the exact direction of the optic axis of the grown crystal could also be determined. This further extends the use of birefringence interferometry in obtaining seeds for directional growth along the optic axis. From the shape of the interference fringes we infer that the dislocation density, which affects refractive index homogeneity at microscopic level is uniform throughout the crystal. This is because an abrupt increase or decrease in the dislocation density in the sample manifests itself in the form of discontinuity in the birefringence interference fringes [Rao, 1991].

Figure 17a shows the birefringence interferogram for a KDP crystal grown directly in phase matched direction. The variation in birefringence for a crystal plate having 10 mm diameter and 1.34 mm thickness along the beam propagation was calculated as 0.00753 [Dinakaran, 2010B]. Similarly, KDP crystal grown under free and forced convection conditions by unidirectional growth technique were compared for their birefringence homogeneity. The interferograms for the crystal plates obtained from the KDP crystal grown along [001] direction under forced and free convection conditions respectively are shown in Figure 17(b, c) [Dinakaran, 2010D]. The variation in birefringence for a crystal plate of cross-section 10 × 8 mm^2 and 1.87 mm thickness obtained from the KDP crystal grown under forced convection condition was computed as 0.0081 and the similar variation for a plate of cross-section 10 × 9 mm^2 and 1.56 mm thickness obtained from the KDP crystal grown under free convection was computed as 0.0089 [Dinakaran, 2010D].

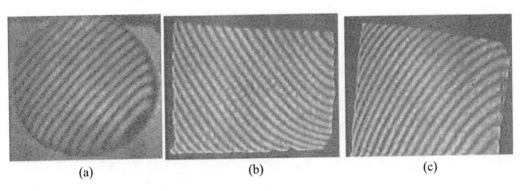

(a) (b) (c)

Figure 17. (a)Birefringence interferogram of KDP crystal. (a) grown in phase matched direction [Dinakaran, 2010B], (b) for a plate taken from a crystal grown under forced convection, and (c) for a plate taken from a crystal grown under free convection [Dinakaran, 2010 D].

Three plates from the ZTS crystal perpendicular to its principal crystallographic axes were cut and polished for birefringence homogeneity measurements. Figure 18(a–c) show the birefringence interferogram of ZTS crystal plates along (100), (010) and (001) respectively. The periodicity of the fringes suggests that the crystal is free from any abrupt local variation in the defects such as dislocation density, inclusions, or stress. The crystal plate cut along a-, b-, and c- direction had cross-section of roughly 10 × 10 mm^2 and thicknesses of 1.52 mm, 1.1 mm and 1.81 mm respectively. The variation in birefringence for the three crystal plates were computed from the interferograms as 0.001665, 0.012616 and 0.009761 respectively [Dinakaran, 2010C].

A plate cleaved perpendicular to [001] direction of sodium acid phthalate (NaAP) crystal was used for birefringence studies. The birefringence interferogram obtained for NaAP crystal plate of 10 mm diamter and 0.52 mm thickness along [001] direction is shown in

Fig.19a [Senthil, 2011]. The variation in birefringence calculated for the above crystal plate was $\Delta(n_e-n_o)$= 0.0146030. Similarly the optical homogeneity of another semiorganic crystal, namely lithium paranitrophenolate trihydrate (LPNPT) was also investigated. A plate of cross-section 15 × 10 mm^2 and 1 mm thickness along b-axis was cut from the crystal. Figure 19b shows the birefringence interferogram of the crystal plate [Dinakaran, 2011]. The change in birefringence for the plate computed from the interferogram was 0.033538.

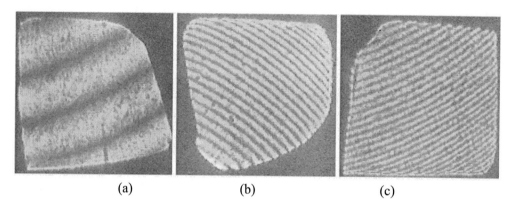

Figure 18. Birefringence interferograms of plates cut from the ZTS crystal parallel to (100) (a), (010) (b) and (001) face (c) [Dinakaran, 2010C].

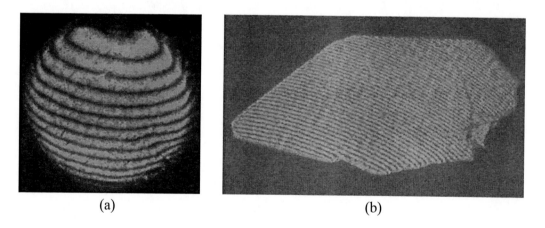

Figure 19. Birefringence interferograms. (a) for NaAP crystal [Senthil, 2011] and (b) for LPNPT crystal [Dinakaran, 2011].

5. CONCLUSION

In this chapter we have described various optical interferometric techniques used for characterizing the optical homogeneity, birefringence, optical indicatrix and the surface quality of crystals. Our primary focus has been on crystals having optical or laser applications. The reason being the stringent requirement on the quality of the crystal for these applications which makes it increasingly important to deploy non-destructive techniques for the purpose. With this motivation we set up different optical interferometric techniques. In particular, the techniques of conoscopy, the Mach-Zehnder interferometry, the birefringence interferometry and the Michelson interferometry has been set up. The principle of these techniques along with the details of optical instrumentation has been reported in this chapter. To demonstrate their applicability and potential, several examples have been discussed in the last section.

REFERENCES

[Bloss, 1961] F.D. Bloss, "*An Introduction to the Methods of Optical Crystallography*", (Holt, Rinehart & Winston, New York, 1961).

[Born, 1980] M. Born and E. Wolf, *"Principles of Optics"*, (Pergamon Press, Oxford, U.K., 1980) 808 pp.

[Byrappa, 2003] K. Byrappa and T. Ohachi (eds.), *"Crystal Growth Technology"*, (William Andrew Inc. and Norwich, New York, USA, 2003).

[Davidson, 2005] M.W. Davidson and M. Abramowitz, Introduction to Optical Microscopy, Digital Imaging, and Photomicrography, see *http://www.micro.magnet.fsu.edu/primer/*

[Dinakaran, 2010A] S. Dinakaran, Sunil Verma, S. Jerome Das, S. Kar, K.S. Bartwal and P.K. Gupta, Investigations for obtaining enhanced SHG element of KH_2PO_4 crystal, *Physica-B* 405 (2010) 1809-1812

[Dinakaran, 2010B] S. Dinakaran, Sunil Verma, S. Jerome Das, G. Bhagavannarayana, S. Kar and K.S. Bartwal, Investigations on crystalline and optical perfection of SHG oriented KDP crystals, *Appl. Phys. A* 99 (2) (2010) 445-450.

[Dinakaran, 2010C] S. Dinakaran, Sunil Verma, S. Jerome Das, G. Bhagavannarayana, S. Kar and K.S. Bartwal, Determination of crystalline perfection, optical indicatrix, birefringence and refractive index homogeneity of ZTS crystals, *Appl. Phys. B* 103 (2011) 345-349.

[Dinakaran, 2010D] S. Dinakaran, Sunil Verma, S. Jerome Das, S. Kar and K.S. Bartwal, Influence of forced convection on unidirectional growth of crystals, *Physica B* 405 (2010) 3919-3923.

[Dinakaran, 2011] S. Dinakaran, Sunil Verma and S. Jerome Das, Solubility, crystal growth, morphology, crystalline perfection and optical homogeneity of lithium paranitrophenolate trihydrate, a semiorganic NLO crystal, *CrystEngComm* 13 (2011) 2375-2380.

[Hariharan, 1985] P. Hariharan, *"Optical Interferometry"*, (Academic Press, Sydney, 1985) 303 pp.

[Henningsen, 1989] T. Henningsen and N.B. Singh, Crystal characterization by use of birefringence interferometry, *J. Cryst. Growth* 96 (1989) 114-118.

[Hurle, 1994] D.T.J. Hurle (ed.), *"Handbook of Crystal Growth"*, Vols. 1-3 (North-Holland, Amsterdam, 1993 & 1994).

[Kar, 2004] S. Kar, R. Bhatt, K.S. Bartwal and V.K. Wadhawan, Optimisation of chromium doping in $LiNbO_3$ single crystals, *Cryst. Res. & Technol.* 39 (2004) 230-234.

[Kar, 2006] S. Kar, R. Bhatt, V. Shukla, R. K. Choubey, P. Sen and K.S. Bartwal, Optical behaviour of VTE treated near stoichiometric $LiNbO_3$ crystals, *Solid State Commun.* 137 (2006) 283-287.

[Kar, 2008A] S. Kar, Sunil Verma and K.S. Bartwal, Growth optimization and optical characteristics of Fe doped LiNbO$_3$ crystals, *Cryst. Growth Des.*, 8 (12) (2008) 4424-4427

[Kar, 2008B] S. Kar, Sunil Verma and K.S. Bartwal, Optical inhomogeneities in Li$_2$B$_4$O$_7$ single crystals, *Cryst. Res. Technol.*, 43 (4) (2008) 438-442.

[Kar, 2009A] S. Kar, Sunil Verma, M.S. Khan and K.S. Bartwal, Growth and optical homogeneity investigations on Fe and Fe:Mn codoped LiNbO$_3$ crystals, *Cryst. Res. Technol.* 44 (12) (2009) 1303-1307.

[Kar, 2009B] S. Kar, Sunil Verma and K.S. Bartwal, Structural and optical investigations on Mn doped Li$_2$B$_4$O$_7$ crystals, *Cryst. Res. Technol.*, 44 (3) (2009) 305-308.

[Malacara, 1992] Daniel Malacara (Ed.), *"Optical Shop Testing"*, 2nd ed., (John Wiley and Sons, New York, 1992) 773 pp.

[Mayinger, 1994] F. Mayinger (ed.), *"Optical Measurements: Techniques and Applications"*, (Springer-Verlag, Berlin, 1994) 463 pp.

[Mullin, 2001] J.W. Mullin, *"Crystallization"*, 4th ed., (Butterworth-Heinemann, Oxford, 2001).

[Rao, 1991] S.M. Rao, C. Cao, A.K. Batra and R.B. Lal, Ground based experiments on the growth and characterization of L-Arginine Phosphate (LAP) crystals, *SPIE* 1557 (1991) 283 - 292.

[Senthil, 2011] A. Senthil, P. Ramasamy and Sunil Verma, Investigations on the SR method growth, etching, birefringence, laser damage threshold and dielectric characterization of sodium acid phthalate single crystals, *J. Cryst. Growth* 318 (2011) 757-761.

[Singh, 1989] N.B. Singh, D.K. Davies, M. Gottlieb, T. Henningsen, R. Mazelsky and M.E. Glickman, Effect of temperature gradient on the optical quality of mercurous chloride crystals, *J. Cryst. Growth* 96 (1989) 969-972.

[Singh, 1990] N.B. Singh, T. Henningsen, Z.K. Kun, K.C. Yoo, R.H. Hopkins and R. Mazelsky, Growth and characterization of thallium arsenic selenide crystals for nonlinear optical applications, *Prog. Crystal Growth and Charact.* 20 (1990) 175-188.

[Singh, 1992] N.B. Singh, M. Gottlieb, T. Henningsen, R.H. Hopkins, R. Mazelsky, M.E. Glickman, S.R. Coriell, W.M.B. Duval and G.J. Santoro, Effect of growth conditions on the quality of lead bromide crystals, *J. Cryst. Growth* 123 (1992) 227-235.

[Singh, 1993] N.B. Singh, M. Gottlieb, R.H. Hopkins, R. Mazelsky, W.M.B. Duval and M.E. Glickman, Physical vapor transport growth of mercurous chloride crystals, Prog. *Crystal Growth and Charact.* 27 (1993) 201-231.

[Verma, 2008] Sunil Verma and Paul J. Shlichta, Imaging techniques for mapping of solution parameters, growth rate, and surface features during the growth of crystals from solution, *Prog. Cryst. Growth & Charact. Materials* 54 (2008) 1-120.

[Verma, 2009] Sunil Verma and K. Muralidhar, Convection, concentration, and surface features analysis during crystal growth from solution using optical diagnostics, in "*Recent Research Developments in Crystal Growth*", 5 (2009) 141-314 (Transworld Research Network, India).

[Verma, 2010] Sunil Verma, In-situ and real time monitoring of process parameters during growth of KDP crystal, an important ferroic material, *Phase Transitions* 83 (2010) 714-727.

INDEX

A

access, 2, 264, 532
accounting, 338, 343
acid, 42, 45, 47, 48, 383, 538, 557, 560
acquisitions, 198
actuators, 2, 32, 56, 60
adaptation, 88, 182, 188
adaptations, 184
adjustment, 18, 105, 113, 157, 538
ADP, 79, 385, 411
advancement, 220
aerospace, 1, 18, 472
aesthetic, 532
AFM, 96
age, 528
agencies, 335
air temperature, 499
Alaska, 203, 216
algorithm, x, xii, 6, 141, 163, 181, 186, 199, 200, 211, 243, 244, 245, 246, 247, 248, 249, 250, 251, 252, 253, 254, 256, 258, 260, 262, 263, 264, 265, 278, 289, 293, 339, 415, 416, 417, 418, 1, 420, 421, 424, 425, 426, 427, 428, 429, 431, 432, 433, 435, 436, 437, 438, 439, 443, 444, 445, 447, 448, 449, 450, 451, 452, 453, 460, 461, 465, 466, 467, 468, 469, 504
alters, 172, 261
aluminium, 11, 12, 13, 14, 15, 42
amine, 474
amines, 474
amplitude, vii, x, xi, 55, 56, 61, 62, 67, 73, 75, 78, 86, 90, 106, 134, 157, 158, 161, 163, 171, 177, 178, 181, 182, 184, 185, 186, 193, 195, 196, 201, 217, 223, 224, 226, 227, 228, 229, 230, 231, 232, 237, 239, 243, 244, 246, 247, 248, 249, 250, 251, 253, 258, 261, 263, 264, 265, 266, 269, 272, 278, 284, 285, 296, 304, 305, 306, 340, 350, 355, 378, 417, 427, 433, 436, 437, 449, 480, 481, 482, 483, 496, 498, 505, 516, 517, 521, 522, 523, 524
anisotropy, 18, 30, 58, 190
APC, 62, 89, 90
aqueous solutions, 42, 391
Argentina, 471, 488
arsenic, 554, 560
ASI, 204, 212
Asia, 205
assessment, x, xv, 193, 444, 468, 514, 537, 548
astigmatism, 127, 237, 253, 260
asymmetry, 103
atmosphere, 115, 117, 118, 340, 405
atoms, 32, 105, 114, 516
Austria, 209
automation, 515
automobiles, 32

B

background radiation, 343, 345
bandwidth, 134, 321
banks, 202
barium, 385, 538, 539
base, 116, 117, 129, 197, 199, 338, 392, 393, 394, 397, 398, 474, 492
behaviors, 143, 295
Beijing, 412
Belgium, 407, 415, 491
bending, 7, 8, 9, 11, 13, 14, 51, 68, 277, 360, 365, 366, 520
benefits, 183, 221, 234
bias, 116, 445, 468
biological systems, 492
birefringence, xv, 537, 540, 544, 545, 546, 554, 555, 556, 557, 558, 559, 560
birefringence measurement, 546, 557

bismuth, 538
blackbody radiation, 516
blends, 473, 474, 486
bonding, 68
bone, 40
boreholes, 27
bounds, 344, 346, 350
brass, 69
bulk materials, 56, 59

C

cables, 105
calculus, 6, 26, 185, 256
calibration, 100, 101, 112, 141, 274, 275, 279, 291, 373, 448
campaigns, 532
candidates, 200, 201
capillary, 485
carbon, 41, 308
carbon dioxide, 41
case studies, xiv, 514
case study, 208, 209, 212, 214, 410
category a, 41
cation, 492
cell assembly, 494
ceramic, 62, 90
cerium, 538
cesium, 104, 105, 111, 114, 539
challenges, 108
charge coupled device, 296
chemical, vii, xiii, 1, 37, 41, 45, 96, 236, 237, 354, 357, 471, 534
chemical characteristics, 41
chemical properties, 96
chemicals, 506
China, 295, 316, 317, 412
chromium, 559
circularly polarized light, 165
circulation, 116, 494
cladding, 484
cladding layer, 484
clarity, xiv, 110, 228, 514
classes, xv, 383, 417, 537, 538
classification, 1
cleaning, 40, 515, 532, 534
climate, 195, 535
CO2, 115
coal, 212, 213
coatings, 76
coding, 320, 426, 530
coherence, xiii, 73, 98, 104, 105, 106, 125, 132, 134, 141, 142, 143, 151, 176, 200, 201, 204, 205, 216, 217, 245, 297, 341, 352, 355, 428, 513, 515, 518, 519, 522
collaboration, 405
collisions, 516
coma, 237, 253
combustion, 220, 239, 492
commercial, 156, 200, 201, 204, 277, 472, 479, 515
communication, 183, 340, 344, 350, 351, 352
communication systems, 351
community, 439
compaction, 207
comparative analysis, 550
compensation, viii, xi, 7, 56, 57, 65, 66, 85, 86, 87, 90, 192, 319, 325, 335, 361, 392, 393, 405, 443, 448
complementarity, 235, 517
complexity, xiv, xv, 465, 514, 532
compliance, 58, 66
composites, 472, 473, 474
composition, 37, 79, 82, 133, 354, 472, 492, 493, 523, 548
compounds, vii, 1
compression, 7, 40, 195
computation, 465, 467
computer, xii, 9, 10, 60, 76, 99, 100, 109, 132, 133, 137, 142, 147, 253, 275, 337, 370, 405, 409, 415, 461, 478, 494, 499, 525
computer simulations, xii, 415, 461
computer technology, 337
computing, 249, 370
conception, 515, 516
conditioning, 531
conduction, xii, 40, 354, 357, 383, 392, 393, 402
conductivity, 397, 493, 503
configuration, vii, viii, 82, 123, 124, 160, 161, 163, 164, 172, 174, 175, 177, 178, 179, 180, 181, 182, 183, 186, 187, 188, 221, 222, 223, 229, 231, 234, 246, 254, 261, 262, 301, 302, 303, 308, 309, 311, 322, 324, 331, 333, 361, 383, 388, 395, 398, 492, 532
conservation, xiv, 514, 515, 517, 528, 530, 531, 532, 533, 534
construction, 18, 29, 526, 530
contamination, 103, 385
contour, 77
contradiction, 276
convention, 99, 285
convergence, 246, 248, 249, 251, 253, 262
conversion rate, 472
cooling, 68, 69, 70, 103, 384, 385, 495, 539
copper, 27, 28, 90, 492, 493, 494, 499, 502, 503, 504

correlation, vii, 1, 41, 42, 43, 44, 54, 108, 109, 110, 111, 113, 116, 200, 218, 296, 305, 306, 346, 350, 534, 535
correlation analysis, 116
correlation coefficient, 350
corrosion, vii, 1, 40, 41, 45, 46, 47, 49, 96
cost, 2, 29, 101, 221, 222, 235, 239, 244, 382, 438, 473, 539
counterfeiting, 239
covering, 496, 528
CPT, 200, 206
cracks, 31, 42, 296, 309, 312, 529, 530, 546
creep, 118, 213, 217
critical value, 402
crop, 505
crown, 76
crystal growth, xii, 354, 368, 381, 383, 384, 385, 388, 389, 390, 391, 392, 405, 408, 409, 410, 411, 412, 413, 414, 539, 559, 560
crystal quality, 538, 550
crystal structure, 384
crystalline, 539, 555, 559
crystallization, 383, 384, 388, 538, 540
crystals, viii, xv, 31, 53, 55, 57, 58, 65, 70, 79, 80, 85, 86, 87, 383, 384, 385, 386, 387, 388, 389, 407, 409, 410, 411, 412, 414, 537, 538, 539, 540, 542, 548, 553, 554, 555, 556, 558, 559, 560
cure, xiii, 471, 472, 473, 474
curing process, xiii, 471, 475, 482, 486
cycles, 109, 320, 329, 461
Czech Republic, 55, 447, 467

D

data analysis, 110, 111, 360
data collection, 473
data processing, 108, 409
decay, 505
decoding, 426
deduction, 476
defect formation, 407
defects, xiv, 31, 40, 42, 296, 306, 308, 309, 384, 385, 415, 514, 528, 530, 531, 539, 540, 546, 555
deficit, 403
deflation, 204
deformation, xiv, 1, 5, 9, 10, 11, 13, 19, 20, 21, 22, 31, 32, 34, 57, 59, 66, 68, 70, 118, 196, 198, 199, 200, 201, 202, 204, 207, 208, 210, 211, 212, 213, 214, 216, 217, 218, 296, 299, 300, 301, 311, 513, 514
degradation, 134

Denmark, 103
dentin, 5
depth, viii, 42, 95, 97, 99, 100, 109, 110, 119, 211, 240, 382, 383, 426, 428, 508, 509
derivatives, 176, 192, 345, 347, 401
detachment, 529
detectable, 70, 73, 375, 398
detection, x, xii, xiii, 43, 59, 60, 70, 90, 176, 177, 179, 185, 189, 193, 198, 202, 210, 211, 213, 239, 254, 256, 274, 275, 278, 279, 287, 296, 309, 337, 340, 342, 357, 445, 447, 459, 460, 461, 465, 468, 528, 548
detection system, 254, 340
detonation, 220
deviation, 69, 76, 77, 102, 118, 126, 128, 141, 221, 358, 372, 397, 430, 432, 500, 552
DFT, 422
diaphragm, x, 60, 74, 219
dielectric constant, 70, 88
diffusion, xiii, 118, 176, 239, 384, 385, 390, 391, 392, 406, 409, 410, 486, 491, 492, 500, 502, 504, 506, 508, 509
diffusion process, 506
diffusivities, 388, 413
diffusivity, 500
dimethacrylate, 473
diode laser, 468
direct measure, 535
discharges, 90, 217
discontinuity, 239, 375, 555
dislocation, 539, 546, 553, 555
dispersion, 341
disposition, 7, 35
distortions, x, 96, 193, 195, 196, 204, 220
distribution, xiv, 16, 43, 63, 75, 77, 162, 166, 169, 181, 186, 187, 188, 222, 230, 231, 247, 259, 270, 278, 281, 284, 285, 286, 287, 289, 296, 339, 341, 342, 343, 370, 378, 381, 384, 388, 389, 395, 396, 397, 398, 401, 405, 407, 408, 417, 427, 429, 449, 450, 451, 460, 461, 473, 478, 496, 497, 498, 499, 500, 502, 503, 504, 514, 523, 524, 529, 532, 552
diversity, 244, 263, 416
DOI, 207, 535, 559
doping, 555, 559
drawing, 60, 70, 71, 72, 89, 97, 117, 152, 358, 359, 398, 494
duality, 517
dyes, 357

E

earthquakes, x, 193
egg, 414
elaboration, 531
elastic deformation, vii, 1
election, 19
electric field, viii, 3, 55, 56, 58, 59, 60, 61, 66, 67, 76, 78, 79, 80, 81, 82, 83, 84, 86, 88, 166, 354, 355
electrodes, 42, 57, 63, 65, 66, 68, 90
electromagnetic, 194, 338, 354, 356, 473, 481, 515, 516
electromagnetic waves, 356
electron, 47, 239, 411, 517
electrons, 304, 341, 516, 517, 521
electro-optical properties, 79
emission, xii, 104, 337, 342, 345, 474, 515, 516
employment, 478
encoding, xiv, 426, 514
encouragement, 405
energy, 41, 114, 255, 343, 383, 397, 398, 516, 517, 521, 528, 538
engineering, xii, 1, 3, 7, 40, 295, 296, 353, 354, 357, 381, 383, 385, 447, 448
England, 215
entropy, 408
environment, 41, 355, 388, 389, 391, 410, 411, 479, 525, 528, 530
environmental contamination, 312
environmental effects, 221, 239, 240
epoxy resins, 473
equality, 18
equilibrium, 5, 11, 117, 483, 503, 528, 539
erosion, 49
error estimation, 199
ESO, 51, 443
etching, 101, 539, 560
ethics, xiv, 514
Europe, 205
evaporation, 384, 393
evidence, xiv, 30, 265, 514
evolution, x, xiii, 47, 173, 182, 193, 199, 201, 210, 262, 264, 387, 471, 478, 481, 483, 486, 487, 488, 492, 506, 509
excess stress, 493
excitation, 105, 296, 481, 483, 527, 528, 530
execution, 2
exercise, 539
experimental condition, 172, 261, 273, 402, 480, 501
experimental design, 254
exploitation, 196
exposure, 19, 23, 28, 30, 35, 42, 44, 380, 388, 404, 527, 530
extinction, 342, 528
extraction, 176, 188, 215, 254, 283, 287, 289, 293, 444
extracts, 265

F

fabrication, 32, 235, 240, 472
families, 416, 417, 1
fear, 223
FFT, 156, 248
fiber, xi, xiii, 6, 106, 232, 233, 234, 306, 308, 313, 323, 324, 337, 338, 340, 344, 350, 389, 412, 471, 472, 473, 476, 477, 478, 479, 481, 484, 485
fiber optics, xiii, 412, 471
fidelity, 322, 531
fillers, 472
films, 33, 62, 65, 66, 72, 97
filters, 29, 163, 165, 170, 171, 172, 173, 175, 199, 201, 292, 297
financial, 265
financial support, 265
finite element method, 296, 306
fitness, 249
flame, 190, 220, 232, 234, 238, 362
flatness, 2, 118, 443, 494, 542, 546
flexibility, 360, 499, 532
flights, 388
flooding, 208
flow field, 381, 398, 399
fluctuations, 42, 60, 73, 98, 107, 110, 116, 220, 221, 240, 340, 341, 384, 402, 427, 472, 495, 504, 505, 506, 542
fluid, xii, 40, 213, 217, 220, 353, 354, 356, 357, 358, 362, 364, 369, 373, 376, 381, 384, 385, 389, 391, 392, 394, 395, 396, 397, 398, 400, 402, 403, 404, 408, 410, 411, 485, 492, 493
foils, 281
force, 5, 7, 8, 9, 96, 305, 307, 338, 384, 385, 405, 517, 525
formation, x, xii, xiii, xv, 190, 192, 193, 200, 221, 236, 237, 270, 292, 353, 354, 355, 360, 379, 385, 392, 395, 443, 472, 513, 514, 515, 523, 525, 529, 531, 532, 540, 544
formula, ix, 6, 124, 149, 247, 248, 287, 365, 373, 418, 432, 451, 452, 454
Fourier analysis, 271, 423
Fourier transform technique, 495
fractures, 18
France, 208, 211, 212, 213, 214, 336, 532

free energy, 4
free radicals, 472
freedom, 65, 323
friction, 96
fusion, 239, 508

G

gamma rays, 515
gel, xiii, 471, 483
gelation, 473
geometrical optics, 223
geometry, 20, 100, 102, 116, 125, 142, 147, 195, 275, 278, 329, 385, 387, 390, 391, 392, 398, 402, 481, 493, 502, 503, 523, 524, 525
Germany, 51, 52, 54, 202, 211, 215
glass transition, xiii, 471
glass transition temperature, xiii, 471
glasses, 32
glue, 69, 101, 102, 103, 111, 116, 117, 118, 119
glycol, 473
Gori, 534
GPS, 205, 206, 210
grades, 196
graduate students, 405
grain size, 239
graph, 109, 160, 475
graphite, 27, 30
gratings, ix, xi, 155, 156, 158, 161, 162, 163, 169, 177, 178, 181, 182, 183, 184, 185, 186, 190, 269, 270, 273, 274, 283, 481, 485, 520
gravitational field, 384
gravity, 173, 208, 365, 398, 406, 409, 410, 492, 503
grazing, 231
Greece, 203, 215, 216, 513
grids, ix, 155, 156, 161, 162, 182, 190
growth, xv, 358, 383, 384, 385, 386, 387, 388, 389, 391, 405, 406, 407, 409, 410, 411, 412, 537, 538, 539, 548, 550, 553, 556, 559, 560
growth rate, 384, 385, 386, 387, 389, 412, 538, 539, 560
guidelines, 509

H

halogen, 105, 107
halos, 221
hardness, 472
Hawaii, 191, 203, 217, 218
hazards, 217

heat transfer, xii, 353, 357, 372, 382, 383, 394, 401, 402, 409, 410, 411, 480, 538
height, viii, 7, 37, 46, 95, 97, 98, 99, 100, 101, 103, 104, 108, 109, 110, 111, 112, 113, 114, 115, 116, 117, 118, 119, 202, 376, 398, 426, 429, 434, 483, 493, 500, 502, 504, 507, 531
helium, 69, 236, 354
high strength, 306
history, xiv, 304, 340, 384, 514, 524, 530
hologram, 236, 237, 239, 270, 296, 378, 382, 389
homogeneity, xv, 13, 83, 90, 134, 278, 282, 291, 383, 537, 539, 540, 542, 548, 550, 552, 553, 554, 555, 556, 557, 558, 559, 560
host, 473, 485, 538
House, 352, 489
human, xiv, 220, 278, 514
human body, 278
humidity, 115, 116, 117, 118
hybrid, 306
hydrogen, 45, 516
hypothesis, 516
hysteresis, 67, 68
hysteresis loop, 67, 68

I

Iceland, 203, 217, 218
icon, 527
ideal, 76, 164, 165, 167, 172, 175, 252, 288, 320, 321, 382, 416, 429, 502
identification, 90, 200
identity, 6, 35, 66, 450, 451
illumination, xiv, 10, 42, 105, 157, 176, 221, 237, 253, 269, 274, 275, 276, 278, 282, 283, 286, 287, 291, 340, 378, 514, 533
imagery, 520
imaging systems, 245, 248, 259
immersion, 182, 278, 279
immunity, 351
improvements, 493
incidence, viii, 41, 43, 105, 123, 124, 125, 126, 127, 130, 133, 142, 150, 231, 278, 406, 556
incompatibility, 382
independence, x, 193, 343
Independence, 195
India, 54, 123, 219, 241, 353, 390, 391, 392, 407, 409, 411, 412, 413, 537, 560
induction, 538
industrial sectors, 2, 40
industries, 472
industry, 1, 18, 29, 40, 41, 43, 337
inequality, 285
inertia, 9, 18, 342, 357, 402, 539

inflation, 204
information processing, 19
inhomogeneity, 221, 532
initial state, 46, 499
initiation, 42, 385
INS, 202
insertion, 494
insulation, 29, 495
insulators, 29
integrated optics, 539
integration, 61, 141, 208, 288, 356, 357, 359, 373, 544
intellectual property, 265
intensity values, 134, 449
interface, 278, 372, 384, 388, 389, 390, 391, 397, 398, 410, 476, 516, 538
International Space Station, 389
intervention, xiv, 514
inversion, 285
ion-exchange, 539
ions, 56
Iran, 210
iridium, 538
iron, 36, 44, 45, 46, 48
irradiation, 474, 480, 481, 483, 485, 488
Islam, 241
isolation, 132, 254, 382, 532, 542
isotherms, 363, 397, 405
isotope, 492
Israel, 240
issues, 31, 212, 472
Italy, 120, 203, 204, 207, 208, 209, 210, 211, 213, 214, 215, 217, 218
iteration, 200, 247, 248, 249

J

Japan, 50, 93, 203, 215, 216, 308

K

kinetics, 387, 407, 411, 414

L

laminar, 220, 385, 386
L-arginine, 389
laser ablation, 534
lasers, 73, 341, 342, 354, 428, 515, 523
laws, 223, 516, 517, 518, 522
lead, 18, 29, 31, 33, 107, 112, 118, 274, 323, 338, 354, 472, 538, 553, 554, 560

leakage, xi, 319, 320, 321, 322, 325, 326, 328, 331, 335, 472
LED, 474, 475, 479, 480, 486
legs, 27
leisure, 382
lens, ix, 9, 10, 18, 22, 23, 27, 35, 36, 37, 39, 40, 43, 44, 45, 60, 66, 76, 97, 98, 99, 100, 111, 123, 129, 132, 133, 135, 137, 142, 144, 145, 146, 147, 148, 149, 150, 151, 156, 177, 178, 184, 186, 187, 190, 220, 221, 227, 236, 240, 245, 246, 250, 254, 255, 256, 257, 258, 259, 260, 261, 270, 272, 275, 285, 358, 359, 360, 492, 494, 501, 502, 525
lifetime, 41
light beam, xiii, xiv, 19, 46, 105, 106, 114, 115, 133, 142, 220, 240, 354, 356, 357, 362, 501, 513, 518, 520, 544
light scattering, 30, 220, 240
linear function, 4, 135
Lion, 316
liquids, 79, 176, 182, 239, 372, 377, 402, 404, 408, 409, 410, 412, 492, 500, 507
lithium, 538, 539, 548, 550, 552, 554, 557, 559
lithography, 101
loci, 363, 371, 544
locus, 363
logging, 494
longevity, 472
Luo, 92, 93
lying, 112, 116, 372
lysozyme, 389, 414

M

machinery, 7, 18, 40, 41
macromolecules, 383, 492
magnesium, 171, 289
magnet, 559
magnetic field, vii, 1, 2, 32, 33, 34, 35, 36, 37, 354, 389, 414, 518
magnetic materials, 2, 53
magnetic properties, 32
magnetization, 32, 239
magnetostriction, 2, 19, 32, 33, 34, 35, 37
magnitude, 7, 37, 61, 76, 220, 248, 249, 252, 322, 325, 327, 328, 330, 365, 375, 382, 385, 493, 508, 538
majority, 90, 338
management, 200, 210, 352, 493
manipulation, 5, 182
manufacturing, xiii, 306, 338, 471
mapping, 214, 216, 356, 358, 372, 373, 374, 376, 382, 385, 386, 390, 412, 560

masking, 103
mass, xii, 7, 10, 11, 13, 42, 45, 47, 48, 209, 210, 220, 353, 354, 357, 361, 372, 373, 376, 383, 384, 388, 391, 394, 404, 408, 473, 491, 502, 505, 507, 517, 521
mass loss, 42, 45, 47, 48
material surface, 46, 56
materials, vii, viii, xiii, 1, 2, 3, 6, 7, 18, 32, 40, 41, 42, 55, 56, 57, 58, 66, 68, 73, 78, 79, 88, 90, 101, 239, 296, 306, 312, 383, 413, 471, 472, 473, 503, 514, 528, 529, 532, 535, 538, 555
matrix, xiii, 26, 35, 99, 100, 166, 340, 346, 347, 348, 349, 350, 370, 381, 471, 472, 504
matter, iv, 373, 502, 516, 517, 520
mechanical properties, xiii, 471
mechanical stress, 32, 58, 88
media, x, 96, 116, 219, 220, 221, 354, 355, 373, 445, 475, 496, 506, 516
medical, xiv, 513
medicine, 40
melt, xv, 101, 537, 538
melting, 29, 538
melting temperature, 29
melts, 538
membranes, 296
memory, xi, 50, 337, 350, 499
meridian, 259
metals, 97
meter, 105
methodology, 28, 32, 37, 362, 368, 372, 373, 377, 382, 389, 531, 532
Mexico, 211
microelectronics, 337
microgravity, 388, 389, 391, 408, 409, 412
micrometer, 96, 97, 103, 357, 479, 480, 539
microscope, ix, 46, 47, 97, 98, 100, 101, 103, 106, 111, 116, 117, 123, 135, 137, 173, 182, 191, 275, 280, 281, 479, 540, 541, 550, 551
microscopy, 33, 47, 56, 96, 97, 188, 338, 530
microwaves, 194, 515
miniature, 96
Ministry of Education, 265, 467
mission, 197
missions, 202, 389
mixing, xi, 190, 319, 320, 323, 331
model system, 408
modelling, 209
models, 218, 338, 342, 516
modifications, xi, 243, 244, 245, 246, 249
modulus, 5, 8, 9, 11, 14, 17, 25, 28, 178, 185, 250, 285, 307, 308, 342
modus operandi, 44

moisture, 117
moisture content, 117
mole, 406
molecular structure, 474
molecules, 473, 492, 528, 538
momentum, 357, 397, 398, 403, 502, 517, 520
monomers, 474
morphology, 385, 387, 389, 407, 531, 559
Moscow, 351, 352
motivation, 100, 558
motor control, 106
multiples, 518, 524, 525
multiplication, 158, 197

N

nanoelectronics, 2
nanometer, 320
nanometers, 484
nanotechnology, 120
neglect, 79, 86, 363, 477
neon, 236, 354
Netherlands, 51, 212, 316, 407
neutral, 9, 105, 546
neutrons, 521
next generation, xiv, 514
nickel, 37
Niels Bohr, 516, 517
nodes, 158
nucleation, 384
nucleus, 477
null, 147, 304, 306
numerical aperture, 234, 240
nutrient, 539

O

obstacles, 223
oil, 40, 173, 174, 181, 182, 275, 278, 279, 289, 290, 383, 396, 397, 398, 401, 402
one dimension, 12, 352
operations, xii, 353, 368
optical fiber, 57, 105, 232, 233, 234, 296, 351, 407, 477, 484, 485
optical parameters, xv, 11, 537
optical properties, 376, 506
optical systems, 127, 243, 245
optimization, 191, 265, 381, 472, 502, 505, 560
optimization method, 381
optoelectronics, xi, xv, 79, 239, 337, 537
originality, 530
orthogonality, 320

oscillation, 62, 90, 109, 422, 480, 481, 482, 483
overlap, 129, 131, 179, 235, 246, 250, 420
oxygen, 41

P

paints, 43
PAN, 91
parallel, 9, 14, 23, 33, 34, 41, 58, 65, 75, 82, 88, 89, 96, 97, 100, 101, 102, 103, 116, 117, 119, 124, 125, 127, 128, 129, 130, 138, 139, 140, 141, 143, 144, 145, 146, 147, 148, 189, 220, 232, 274, 278, 279, 280, 313, 354, 360, 362, 365, 370, 371, 472, 477, 478, 479, 546, 552, 553, 554, 556, 558
parallel processing, 220
parallelism, 101, 103, 134, 542, 546, 552
parameter estimates, 342
parasite, 73, 76
parents, 50
parity, 159, 160, 169, 170
partition, 529
pellicle, 147
periodicity, 557
peri-urban, 210
permit, 382
personal communication, 443
pharmaceutical, 40
phase objects, 188, 220, 222, 223
phase shifts, ix, 6, 75, 125, 126, 129, 141, 155, 156, 157, 166, 169, 175, 179, 180, 188, 256, 277, 278, 279, 283, 291, 444, 448, 461, 466, 468
phase transformation, 3
phase transitions, viii, 55, 57, 416
Philadelphia, 410
phosphate, 383, 389, 414, 538, 539
photocells, 106
photodetectors, 340, 460
photoemission, 341
photographs, 22, 35, 202
photonics, 538
photons, 516, 517, 520, 521, 526
photopolymerization, 478, 480
photosensitivity, 472
physical aging, 515
physical properties, 29, 295, 472
physics, xiii, 52, 352, 360, 389, 408, 503, 513, 515
piano, 212
piezoelectric crystal, 42
plane waves, 518
platform, 323, 324, 386, 390, 392

platinum, 65, 538
playing, 354
point defects, 539
point spread function, 159, 169
Poisson equation, 359
polar, 313, 385, 387, 397, 407
polarization, ix, 6, 58, 63, 70, 72, 74, 76, 81, 82, 87, 88, 89, 123, 124, 125, 126, 127, 129, 134, 136, 137, 149, 150, 151, 155, 156, 157, 159, 164, 165, 166, 169, 174, 175, 176, 177, 178, 179, 183, 185, 188, 190, 192, 254, 255, 256, 270, 283, 284, 289, 291, 320, 321, 323, 341, 360, 411, 472
polymer, xiii, 80, 309, 471, 472, 473, 485, 534
polymer films, 80
polymerization, 472, 473, 474, 479, 480, 481, 482, 483, 486
polymerization process, 472, 473, 481, 482
polymers, xiii, 471, 492, 534
potassium, 383, 414, 538, 539
preparation, iv, 19, 475, 542
preservation, xiv, xv, 514, 515
pressure gauge, 29
primary products, 522
principles, iv, vii, xii, xiv, 125, 132, 144, 220, 296, 315, 353, 378, 405, 413, 513, 515, 516, 517, 520, 533, 540
probability, 108, 111, 338, 339, 341, 343, 384, 517
probe, 65, 96, 106, 356
process control, 534
profilometer, 2, 41, 445
project, 202, 204, 210, 528, 532
proliferation, 204
propagation, 144, 146, 237, 244, 246, 362, 365, 459, 477, 515, 517, 529, 530, 546, 556, 557
propane, 473
proportionality, 520
protection, 41, 210
protein crystallization, 409, 411
publishing, 448
purification, 505
purity, 538

Q

quanta, 516
quantification, 475
quantization, 252, 343, 344
quantum electrodynamics, 517
quantum mechanics, 516, 517
quantum optics, 104, 341
quartz, 2, 27, 29, 30, 31, 62, 69, 406, 503, 549

R

radar, xi, 199, 207, 208, 210, 211, 212, 213, 214, 215, 216, 217, 218, 337, 338, 351
radial distance, 313
radiation, x, 5, 29, 41, 44, 133, 193, 194, 195, 220, 271, 341, 354, 357, 375, 383, 515, 516, 518, 522, 523, 538, 539
Radiation, 343, 344, 351, 517
radio, xi, 337, 338, 351, 515
radiography, 530
radius, 77, 149, 150, 477
Raman spectroscopy, 70
random errors, 448, 459, 460
RB1, 128, 129
reactions, 42
reading, 152, 404
real time, 6, 265, 336, 375, 410, 448
reality, 501
reasoning, 525
reception, 338, 340
recognition, 270
recombination, 222, 269
recommendations, iv, 335
reconstruction, 183, 192, 237, 238, 244, 246, 249, 250, 251, 253, 260, 261, 262, 263, 265, 266, 270, 278, 281, 282, 291, 293, 378, 379, 381, 382, 407, 411, 413, 449, 524, 525, 533
recovery, 243
redundancy, 204
reference frame, 45
reference system, 3
reflectivity, 76, 97, 229, 315, 530
refraction index, 475, 476
refractive index variation, 225, 391, 404, 484, 493, 501, 505, 542
refractive indices, 79, 83, 115, 377, 389, 540
regression, 26, 30, 35, 37, 53, 200
regression method, 37
regression model, 30, 35
relaxation, 492, 504, 506, 507
relevance, 156
reliability, xii, 5, 26, 278, 282, 291, 337, 461
relief, 195
remote sensing, 195
repair, 306, 308
replication, xi, 169, 269, 286
reproduction, 109, 523, 531
requirements, xv, 2, 40, 74, 246, 254, 255, 259, 291, 354, 514, 515, 532
researchers, 358
residual error, 201
resins, 473, 474, 478, 481, 486, 487

resistance, 40, 42, 439
resolution, x, xi, xii, xiii, 2, 46, 96, 97, 110, 156, 195, 201, 202, 221, 239, 240, 244, 282, 319, 324, 351, 353, 357, 368, 375, 378, 382, 389, 402, 405, 426, 471, 481, 494, 539, 546
response, xii, 5, 9, 19, 46, 57, 58, 66, 67, 73, 76, 272, 285, 320, 353, 357, 460, 465, 466, 481, 485, 531, 539
restoration, 183, 191, 351, 472, 531, 532, 534
restorative material, 472
restorative materials, 472
retardation, 163, 165, 166, 167, 170, 171, 177, 184, 549, 550, 551
rings, 516, 546
risk, 18, 238, 530, 531
robotics, 32
room temperature, 69, 70, 80, 82, 83, 85, 100, 392
root, 2, 101, 119, 423, 427, 429, 451, 461
root-mean-square, 101, 119, 429, 461
rotations, 26, 37, 103, 256, 538
roughness, vii, viii, 1, 2, 19, 40, 41, 42, 43, 44, 45, 46, 49, 54, 95, 96, 101, 117, 119
routines, xv, 514, 527
Royal Society, 220
rubber, 103, 479, 480, 485, 493, 503
rules, 339
Russia, 337

S

safety, 18, 40, 306, 539
saturation, 32, 33, 34, 37, 368, 385
savings, 41
scaling, 111, 163, 181, 183, 186, 247, 416
scatter, 111
scattering, 240, 363, 404, 405, 506, 516, 539
schema, ix, 155, 279, 529, 530
science, xii, 2, 32, 40, 96, 220, 338, 381, 447, 448, 515, 518
scope, xi, 60, 76, 105, 114, 119, 136, 269
security, 239
seed, 385, 387, 538
semiconductor, 32
sensing, 183, 212, 213, 243, 264, 265, 448, 478
sensitivity, x, 57, 59, 62, 72, 73, 192, 219, 221, 227, 240, 244, 296, 304, 315, 382, 384, 385, 389, 405, 407, 416, 1, 434, 452, 461, 462, 473, 492, 546
sensors, x, xiii, 32, 56, 191, 193, 195, 202, 471, 473, 486, 499, 500, 501, 535
shade, 98

shape, xiv, 5, 7, 11, 21, 34, 40, 41, 47, 48, 62, 68, 69, 76, 77, 82, 96, 106, 118, 144, 220, 221, 260, 261, 277, 279, 296, 309, 310, 312, 313, 344, 374, 402, 425, 513, 514, 529, 548
shear, viii, ix, 58, 63, 118, 123, 124, 125, 126, 127, 129, 130, 131, 133, 137, 142, 145, 149, 150, 151, 152, 156, 176, 177, 178, 179, 181, 182, 183, 185, 186, 187, 188, 189, 190, 191, 192, 237, 269, 308, 413
shear deformation, 58
shock, 220
showing, 98, 180, 181, 186, 224, 232, 252, 431, 548, 552
signals, xi, 61, 73, 106, 107, 110, 119, 252, 253, 264, 265, 300, 320, 322, 325, 337, 338, 340, 341, 342, 344, 351
signal-to-noise ratio, 427
signs, 47, 141, 157, 159, 160, 161, 162, 163, 416, 438, 439
silica, 29, 76, 124, 125, 130, 135, 493
silicon, 101, 102
silver, 229, 238
simulation, xii, 14, 251, 263, 330, 331, 334, 337, 430, 432, 433, 434, 437, 459, 461, 464, 503
simulations, 248, 249, 250, 251, 252, 253, 264, 331, 416, 426, 427, 428, 431, 432, 433, 436, 438, 469
sine wave, 98, 99, 428
Singapore, 241
single crystals, 387, 407, 559, 560
SiO2, 229
skeleton, 368, 398
skewness, 41
smoothing, 108, 278
smoothness, 76
sodium, 385, 407, 557, 560
software, 90, 111, 116, 199, 200, 201, 204, 264, 416, 532
solid phase, 405
solid state, 104
solidification, 376, 408, 472, 480, 481, 482, 483
solubility, 384, 389
solution, xii, xv, 9, 25, 26, 42, 46, 57, 75, 90, 101, 204, 250, 252, 265, 288, 339, 340, 354, 358, 360, 372, 376, 383, 384, 385, 386, 387, 388, 389, 390, 391, 392, 395, 406, 407, 410, 411, 412, 413, 414, 450, 451, 452, 504, 537, 538, 560
solvents, 493
space environment, 411
space shuttle, 388
space station, 18

Spain, 1, 53, 193, 195, 197, 208, 210, 212, 214, 243
spatial frequency, 157, 161, 222, 255, 259, 270, 274, 285, 423, 496, 529
special relativity, 517, 520
species, 354, 357, 492, 505, 508
specific surface, 201
spectroscopy, 156
speed of light, 356, 516
spindle, 148
Spring, 409
stability, viii, ix, 5, 62, 73, 74, 83, 90, 95, 100, 102, 128, 155, 156, 182, 188, 200, 201, 270, 278, 283, 291, 298, 319, 341, 383, 385, 427, 460, 461, 493, 494, 500, 501, 508, 542
stabilization, 69, 83, 475
standard deviation, 100, 101, 112, 113, 115, 118, 200, 252, 427, 460
state, vii, 1, 11, 28, 33, 35, 40, 41, 47, 48, 49, 128, 129, 137, 211, 270, 303, 305, 341, 383, 384, 392, 400, 402, 472, 483, 485, 492, 499, 504, 505, 506, 514, 517, 527, 528, 530, 538, 539, 556
states, 11, 42, 166, 284, 321, 341, 412
statistical processing, 41
statistics, 338, 339, 341, 352, 517
steel, 33, 41, 42, 54, 101, 102, 103, 104, 105, 106, 108, 113, 115, 116, 117, 120, 485
STM, 96
storage, 118, 183, 200, 338, 344, 492, 538
storage media, 538
stratification, 388
stress, 1, 4, 5, 14, 18, 42, 263, 296, 307, 312, 472, 546, 552, 553, 557
stress fields, 14
stress intensity factor, 307, 312
stretching, 6, 7, 51
structural modifications, 534
structure, xiii, 18, 42, 114, 246, 247, 312, 338, 384, 389, 471, 474, 481, 494, 515, 528
structuring, 528
substitutions, 539
substrate, vii, 55, 56, 57, 62, 63, 65, 66, 68, 171, 234, 539
substrates, 62, 65, 90, 289, 534
subtraction, 62, 305
succession, 476
sulfate, 391
Sun, 152, 293, 414
superimposition, 165, 244
supplier, 11
suppression, 287, 288
surface layer, 385

surface structure, 389
susceptibility, 136, 448, 459, 466
suspensions, 409
swelling, 102, 116, 117, 118
Switzerland, 218
symmetry, 29, 58, 78, 79, 281, 313, 401, 402

T

Taiwan, 207, 211, 295, 316, 317
target, xi, 214, 319, 320, 322, 325, 335, 478
technologies, xiii, 471
technology, 1, 2, 32, 40, 213, 220, 228, 239, 337, 338, 350, 492, 514, 534, 538
teeth, 5
telephone, 29
temperature dependence, 79, 83, 90
tension, 338, 342
tensions, 10
testing, 79, 125, 135, 136, 156, 176, 183, 189, 190, 191, 192, 220, 221, 239, 243, 270, 290, 292, 310, 407, 413, 443, 444, 448, 515, 528
texture, 532
thallium, 554, 560
thermal analysis, 2
thermal deformation, 18, 19, 22
thermal energy, 357
thermal expansion, vii, viii, 2, 18, 19, 20, 22, 24, 25, 26, 27, 28, 29, 30, 31, 55, 56, 57, 58, 88, 89, 90, 112, 113, 115, 116, 117, 120, 319, 481, 485
thermal properties, 5, 29, 503
thermal stability, 494
thermal treatment, 37
thermalization, 499
thermoregulation, 501
thin films, 5, 33, 56, 57, 62, 65, 67, 72, 90, 97, 141, 143
thinning, 398
third dimension, 357
three-dimensional representation, 429
Tibet, 203, 217
time series, 211
time use, 497
tin, 445
tissue, 40
titanate, 538
tokamak, 240
tones, 278
total energy, 521
total internal reflection, 484
tracks, 498
trade, 227, 379, 382
trade-off, 227, 379, 382
trajectory, 194
transducer, 2, 5, 9, 10, 11, 135, 374, 449
transference, 477
transformation, 21, 287
transformations, 338, 423
transition metal, 555
transition temperature, 70
translation, 3, 4, 22, 25, 26, 28, 34, 116, 117, 118, 147, 179, 185, 230, 274, 283, 286, 325
transmission, 165, 166, 171, 172, 175, 177, 239, 284, 285, 323, 388, 475, 476, 552
transparency, 181, 539
transparent medium, 354, 357
transport, xii, xiii, 244, 354, 357, 381, 383, 388, 403, 413, 491, 501, 502, 506, 508, 548, 560
transport processes, 357, 381, 388
transportation, 532, 539
treatment, xiv, 37, 46, 503, 514
trial, 111
triangulation, 201
turbulence, 190, 335, 425, 428
Turkey, 203, 217
twist, 313

U

UK, 50, 52, 53, 121, 534
ultrasound, 2
UNESCO, 532
unification, 517
uniform, 42, 231, 234, 239, 253, 286, 287, 354, 358, 390, 402, 426, 473, 480, 493, 499, 531, 542, 552, 555, 557
urban, 205, 207, 208, 210, 212
urban arcas, 205
USA, 50, 51, 52, 53, 90, 202, 203, 207, 213, 215, 236, 407, 410, 411, 413, 473, 479, 533, 559
USSR, 54

V

vacuum, 40, 110, 111, 114, 115, 309, 310, 311, 354, 356, 406
validation, 213, 216, 336
vapor, xv, 105, 114, 115, 385, 409, 537, 538, 539, 548, 560
variables, 18, 38, 40, 49, 354, 357, 376, 417, 435
variations, xii, 5, 42, 141, 171, 220, 225, 232, 233, 235, 238, 239, 305, 353, 373, 376, 377, 389, 433, 448, 449, 451, 484, 485, 491, 492, 501, 542, 546, 553

vector, ix, 5, 15, 19, 20, 22, 25, 26, 33, 155, 168, 197, 224, 274, 303, 322, 326, 340, 354, 365, 498
vegetation, 202
vehicles, 295, 308
vein, 340
velocity, xi, 106, 107, 109, 201, 239, 319, 321, 322, 324, 325, 326, 330, 335, 336, 398, 402
versatility, 2, 382
vertical dimensions, viii, 95
vibration, vii, xi, 27, 32, 55, 57, 61, 62, 63, 65, 66, 68, 69, 72, 78, 82, 129, 132, 134, 136, 295, 296, 299, 303, 304, 305, 306, 312, 313, 315, 335, 382, 425, 427, 443, 452, 469, 473, 479, 481, 488, 532, 542
viscosity, 398, 480, 481
vision, 195, 196
visualization, xii, 220, 231, 238, 353, 357, 358, 378, 402, 408, 412, 527

W

walking, 116
war, 497
waste, 209
water, 27, 28, 36, 102, 103, 113, 115, 117, 118, 205, 206, 207, 208, 211, 212, 368, 373, 383, 392, 393, 394, 397, 398, 406, 494, 503
wave number, 43, 247, 320
wave vector, 522

wavelengths, 41, 43, 97, 105, 107, 109, 114, 354, 376, 377, 515, 523, 525
wear, 40
welding, 41
wind turbines, 308
windows, ix, 125, 130, 151, 155, 156, 157, 158, 159, 160, 161, 162, 163, 164, 165, 168, 169, 170, 171, 173, 174, 175, 176, 179, 183, 270, 271, 272, 273, 274, 279, 282, 283, 284, 285, 289, 291
wires, 233
workers, 377, 385, 389, 391, 532
worldwide, 515
worms, 530

X

X-ray diffraction, 539
x-rays, xiv, 514

Y

yield, 31, 41, 321, 401, 547
yttrium, 538

Z

zinc, 383, 385, 387, 407, 538, 550
ZnO, 67